T0178168

Broad Learning Through Fusions

Jiawei Zhang • Philip S. Yu

Broad Learning Through Fusions

An Application on Social Networks

 Springer

Jiawei Zhang
Department of Computer Science
Florida State University
Tallahassee, FL, USA

Philip S. Yu
Department of Computer Science
University of Illinois
Chicago, IL, USA

ISBN 978-3-030-12530-1   ISBN 978-3-030-12528-8   (eBook)
https://doi.org/10.1007/978-3-030-12528-8

This Springer imprint is published by the registered company Springer Nature Switzerland AG.
The registered company address is: Gewerbestrasse 11, 6330 Cham, Switzerland

To my dear parents
Yinhe Zhang and Zhulan Liu.

Jiawei Zhang

To my family.

Philip S. Yu

Preface

This textbook is written for the readers who are interested in broad learning, especially in information fusion and knowledge discovery across multiple fused information sources. Broad learning is a general learning problem, which can be studied in various disciplines. Meanwhile, to illustrate the problem settings and the learning algorithms more clearly, this book uses the online social network as an example. To make the textbook self-contained, the book also provides an overview of necessary background knowledge for the readers. If it is the first time for the readers to read a textbook related to broad learning, machine learning, data mining, and social network mining, the readers will find this book to be very easy to follow.

Overview of the Book

There are 12 chapters in this textbook, which are divided into four main parts: Part I covers Chaps. 1–3, which introduce the overview of broad learning, machine learning, and social networks for the readers; Part II covers Chaps. 4–6, which include the existing social network alignment problems and algorithms; Part III covers Chaps. 7–11, which provide a comprehensive description about the recent social media mining problems across multiple information sources; Part IV covers Chap. 12, which indicates some potential future development of broad learning for the readers. Readers and instructors can use this textbook according to the guidance provided in Chap. 1.

Except Chap. 12, each chapter also has ten exercises for the readers. The exercise questions are divided into three levels: easy, medium, and hard, which are on the basic concepts, theorem proofs, algorithm details, as well as exercises on algorithm implementations. Some of the exercises can be used in the after-class homework, and some can be used as the course projects instead. The instructors can determine how to use the exercises according to their difficulty levels, as well as the needs of the courses.

Broad learning is a very large topic, and we cannot cover all the materials in this textbook. For the readers who want to further explore other related areas, at the end of each chapter, we provide a section about the bibliography notes. The readers can refer to the cited articles for more detailed information about the materials to your interests.

Acknowledgments

This book would not have been possible without many contributors whose names did not make it to the cover. We will mention them here according to the appearance order of their contributed works in this book. The active network alignment algorithm introduced in Sect. 6.4 is based on the collaborative work with Junxing Zhu (National University of Defense Technology) and the work with Yuxiang

Ren (Florida State University). The large-scale network synergistic community detection algorithm introduced in Sect. 8.5 is based on the collaborative work with Songchang Jin (National University of Defense Technology). The information diffusion algorithm introduced in Sect. 9.4 and viral marketing algorithm introduced in Sect. 10.4 are based on the collaborative works with Qianyi Zhan (Jiangnan University).

This book has benefitted from the significant feedbacks from our students, friends, and colleagues. We would like to thank many people who help to read and review the book which greatly improve both the organization of the book and the detailed contents covered in the book. We would like to thank Lin Meng (Florida State University) for helping review Chaps. 3, 6, 7, and 11; Yuxiang Ren (Florida State University) for reviewing Chaps. 2, 4, 5, and 6; and Qianyi Zhan (Jiangnan University) for reviewing Chaps. 9 and 10.

Jiawei would like to thank his long-term collaborators, including (sorted according to their last names) Charu C. Aggarwal, Yi Chang, Jianhui Chen, Bowen Dong, Yanjie Fu, Lifang He, Qingbo Hu, Songchang Jin, Xiangnan Kong, Moyin Li, Taisong Li, Kunpeng Liu, Ye Liu, Yuanhua Lv, Guixiang Ma, Xiao Pan, Weixiang Shao, Chuan Shi, Weiwei Shi, Lichao Sun, Pengyang Wang, Sen-Zhang Wang, Yang Yang, Chenwei Zhang, Qianyi Zhan, Lei Zheng, Shi Zhi, Junxing Zhu, and Zhu-Hua Zhou. Jiawei also wants to thank his PhD advisor Philip S. Yu for his guidance during the early years as a researcher and the members of his IFM Lab (Information Fusion and Mining Laboratory). Finally, Jiawei would like to thank his respected parents, *Yinhe Zhang* and *Zhulan Liu*, for their selfless love and support. The book grabs so much time that should be spent in accompanying them.

Philip would like to thank his collaborators, including past and current members of his BDSC (Big Data and Social Computing) Lab at UIC.

Special thanks are due to Melissa Fearon and Caroline Flanagan at Springer Publishing Company who convinced us to go ahead with this project and were constantly supportive and patient in the face of recurring delays and missed deadlines. We thank Paul Drougas from Springer, our editor, for the help in improving the book considerably.

This book is partially supported by NSF through grants IIS-1763365 and IIS-1763325.

Some Other Words

Broad learning is a fast-growing research area, and few people can have a very deep understanding about all the detailed materials covered in this book. Although the authors have several years of exploration experiences about the frontier of this area, there may still exist some mistakes and typos in this book inevitably in writing. We are grateful if the readers can inform the authors about such mistakes they find when reading the book, which will help improve the book in the coming editions.

Tallahassee, FL, USA   Jiawei Zhang
Chicago, IL, USA   Philip S. Yu
November 11, 2018

Contents

Part I Background Introduction

1 Broad Learning Introduction 3
  1.1 What Is Broad Learning 3
  1.2 Problems and Challenges of Broad Learning 4
    1.2.1 Cross-Source Information Fusion 4
    1.2.2 Cross-Source Knowledge Discovery 6
    1.2.3 Challenges of Broad Learning 6
  1.3 Comparison with Other Learning Tasks 7
    1.3.1 Broad Learning vs. Deep Learning 7
    1.3.2 Broad Learning vs. Ensemble Learning 8
    1.3.3 Broad Learning vs. Transfer Learning vs. Multi-Task Learning 8
    1.3.4 Broad Learning vs. Multi-View, Multi-Source, Multi-Modal, Multi-Domain Learning 9
  1.4 Book Organization 10
    1.4.1 Part I 10
    1.4.2 Part II 11
    1.4.3 Part III 11
    1.4.4 Part IV 12
  1.5 Who Should Read This Book 12
  1.6 How to Read This Book 13
    1.6.1 To Readers 13
    1.6.2 To Instructors 13
    1.6.3 Supporting Materials 14
  1.7 Summary 14
  1.8 Bibliography Notes 14
  1.9 Exercises 15
  References 15

2 Machine Learning Overview 19
  2.1 Overview 19
  2.2 Data Overview 20
    2.2.1 Data Types 20
    2.2.2 Data Characteristics 26
    2.2.3 Data Pre-processing and Transformation 26

	2.3	Supervised Learning: Classification	32
		2.3.1 Classification Learning Task and Settings	33
		2.3.2 Decision Tree	34
		2.3.3 Support Vector Machine	40
	2.4	Supervised Learning: Regression	46
		2.4.1 Regression Learning Task	46
		2.4.2 Linear Regression	46
		2.4.3 Lasso	47
		2.4.4 Ridge	48
	2.5	Unsupervised Learning: Clustering	48
		2.5.1 Clustering Task	49
		2.5.2 K-Means	50
		2.5.3 DBSCAN	52
		2.5.4 Mixture-of-Gaussian Soft Clustering	54
	2.6	Artificial Neural Network and Deep Learning	56
		2.6.1 Artificial Neural Network Overview	57
		2.6.2 Deep Learning	62
	2.7	Evaluation Metrics	66
		2.7.1 Classification Evaluation Metrics	66
		2.7.2 Regression Evaluation Metrics	68
		2.7.3 Clustering Evaluation Metrics	69
	2.8	Summary	71
	2.9	Bibliography Notes	72
	2.10	Exercises	72
	References	73
3	**Social Network Overview**	**77**
	3.1	Overview	77
	3.2	Graph Essentials	78
		3.2.1 Graph Representations	78
		3.2.2 Connectivity in Graphs	80
	3.3	Network Measures	82
		3.3.1 Degree	82
		3.3.2 Centrality	85
		3.3.3 Closeness	91
		3.3.4 Transitivity and Social Balance	96
	3.4	Network Categories	98
		3.4.1 Homogeneous Network	99
		3.4.2 Heterogeneous Network	102
		3.4.3 Aligned Heterogeneous Networks	106
	3.5	Meta Path	111
		3.5.1 Network Schema	111
		3.5.2 Meta Path in Heterogeneous Social Networks	111
		3.5.3 Meta Path Across Aligned Heterogeneous Social Networks	114
		3.5.4 Meta Path-Based Network Measures	116
	3.6	Network Models	117
		3.6.1 Random Graph Model	118
		3.6.2 Preferential Attachment Model	120

3.7	Summary		121
3.8	Bibliography Notes		122
3.9	Exercises		123
References			124

Part II   Information Fusion: Social Network Alignment

4   Supervised Network Alignment 129
  4.1   Overview 129
  4.2   Supervised Network Alignment Problem Definition 130
  4.3   Supervised Full Network Alignment... 131
      4.3.1   Feature Extraction for Anchor Links 132
      4.3.2   Supervised Anchor Link Prediction Model 138
      4.3.3   Stable Matching 139
  4.4   Supervised Partial Network Alignment 142
      4.4.1   Partial Network Alignment Description............................... 142
      4.4.2   Inter-Network Meta Path Based Feature Extraction 143
      4.4.3   Class-Imbalance Classification Model 146
      4.4.4   Generic Stable Matching 149
  4.5   Anchor Link Inference with Cardinality Constraint 151
      4.5.1   Loss Function for Anchor Link Prediction 151
      4.5.2   Cardinality Constraint Description................................... 153
      4.5.3   Joint Optimization Function 154
      4.5.4   Problem and Algorithm Analysis 156
      4.5.5   Distributed Algorithm 158
  4.6   Summary 160
  4.7   Bibliography Notes 160
  4.8   Exercises 161
  References 163

5   Unsupervised Network Alignment 165
  5.1   Overview 165
  5.2   Heuristics Based Unsupervised Network Alignment 166
      5.2.1   User Names Based Network Alignment Heuristics 166
      5.2.2   Profile Based Network Alignment Heuristics 169
  5.3   Pairwise Homogeneous Network Alignment 170
      5.3.1   Heuristics Based Network Alignment Model 171
      5.3.2   IsoRank 173
      5.3.3   IsoRankN 175
      5.3.4   Matrix Inference Based Network Alignment......................... 176
  5.4   Multiple Homogeneous Network Alignment with Transitivity Penalty 177
      5.4.1   Multiple Network Alignment Problem Description 178
      5.4.2   Unsupervised Multiple Network Alignment 179
      5.4.3   Transitive Network Matching 184
  5.5   Heterogeneous Network Co-alignment.. 186
      5.5.1   Network Co-alignment Problem Description 187
      5.5.2   Anchor Link Co-inference... 188
      5.5.3   Network Co-matching 195

5.6   Summary 198
5.7   Bibliography Notes 198
5.8   Exercises 199
References 200

6   **Semi-supervised Network Alignment** 203
    6.1   Overview 203
    6.2   Semi-supervised Learning: Overview 204
            6.2.1   Semi-supervised Learning Problem Setting..................... 204
            6.2.2   Semi-supervised Learning Models 205
            6.2.3   Active Learning 209
            6.2.4   Positive and Unlabeled (PU) Learning 211
    6.3   Semi-supervised Network Alignment 212
            6.3.1   Loss Function for Labeled and Unlabeled Instances 212
            6.3.2   Cardinality Constraint on Anchor Links 214
            6.3.3   Joint Objective Function for Semi-supervised Network Alignment 215
    6.4   Active Network Alignment 215
            6.4.1   Anchor Link Label Query Strategy 216
            6.4.2   Active Network Alignment Objective Function 219
    6.5   Positive and Unlabeled (PU) Network Alignment 221
            6.5.1   PU Network Alignment Problem Formulation and Preliminary 221
            6.5.2   PU Network Alignment Model 222
    6.6   Summary 224
    6.7   Bibliography Notes 224
    6.8   Exercises 225
    References 225

Part III   Broad Learning: Knowledge Discovery Across Aligned Networks

7   **Link Prediction** 229
    7.1   Overview 229
    7.2   Traditional Single Homogeneous Network Link Prediction 230
            7.2.1   Unsupervised Link Prediction 230
            7.2.2   Supervised Link Prediction 232
            7.2.3   Matrix Factorization Based Link Prediction 233
    7.3   Heterogeneous Network Collective Link Prediction 235
            7.3.1   Introduction to LBSNs 235
            7.3.2   Collective Link Prediction 236
            7.3.3   Information Accumulation and Feature Extraction.............. 236
            7.3.4   Collective Link Prediction Model............................. 239
    7.4   Cold Start Link Prediction for New Users 242
            7.4.1   New User Link Prediction Problem Description 242
            7.4.2   Cold Start Link Prediction Problem Formulation 244
            7.4.3   Link Prediction Within Target Network 245
            7.4.4   Cold-Start Link Prediction 250

7.5   Spy Technique Based Inter-Network PU Link Prediction   251
        7.5.1   Cross-Network Concurrent Link Prediction Problem   251
        7.5.2   Concurrent Link Prediction Problem Formulation   252
        7.5.3   Social Meta Path Definition and Selection   253
        7.5.4   Spy Technique Based PU Link Prediction   256
        7.5.5   Multi-Network Concurrent PU Link Prediction Framework   258
7.6   Sparse and Low Rank Matrix Estimation Based PU Link Prediction   259
        7.6.1   Problem Description   259
        7.6.2   Intra-Network Link Prediction   261
        7.6.3   Inter-Network Link Prediction   262
        7.6.4   Proximal Operator Based CCCP Algorithm   267
7.7   Summary   269
7.8   Bibliography Notes   270
7.9   Exercises   271
References   271

8   Community Detection   275
8.1   Overview   275
8.2   Traditional Homogeneous Network Community Detection   276
        8.2.1   Node Proximity Based Community Detection   277
        8.2.2   Modularity Maximization Based Community Detection   278
        8.2.3   Spectral Clustering Based Community Detection   280
8.3   Emerging Network Community Detection   283
        8.3.1   Background Knowledge   284
        8.3.2   Problem Formulation   285
        8.3.3   Intimacy Matrix of Homogeneous Network   286
        8.3.4   Intimacy Matrix of Attributed Heterogeneous Network   287
        8.3.5   Intimacy Matrix Across Aligned Heterogeneous Networks   290
        8.3.6   Approximated Intimacy to Reduce Dimension   292
        8.3.7   Clustering and Weight Self-adjustment   293
8.4   Mutual Community Detection   295
        8.4.1   Background Knowledge   295
        8.4.2   Problem Formulation   296
        8.4.3   Meta Path Based Social Proximity Measure   297
        8.4.4   Network Characteristic Preservation Clustering   298
        8.4.5   Discrepancy Based Clustering of Multiple Networks   299
        8.4.6   Joint Mutual Clustering of Multiple Networks   301
8.5   Large-Scale Network Synergistic Community Detection   304
        8.5.1   Problem Formulation   304
        8.5.2   Distributed Multilevel k-Way Partitioning   304
        8.5.3   Distributed Synergistic Partitioning Process   306
8.6   Summary   308
8.7   Bibliography Notes   310
8.8   Exercises   311
References   312

9   Information Diffusion 315
    9.1   Overview 315
    9.2   Traditional Information Diffusion Models 316
            9.2.1   Threshold Based Diffusion Model 317
            9.2.2   Cascade Based Diffusion Model 319
            9.2.3   Epidemic Diffusion Model 321
            9.2.4   Heat Diffusion Models 323
    9.3   Intertwined Diffusion Models 324
            9.3.1   Intertwined Diffusion Models for Multiple Topics 325
            9.3.2   Diffusion Models for Signed Networks 328
    9.4   Inter-Network Information Diffusion 331
            9.4.1   Network Coupling Based Cross-Network Information Diffusion 332
            9.4.2   Random Walk Based Cross-Network Information Diffusion 334
    9.5   Information Diffusion Across Online and Offline World 337
            9.5.1   Background Knowledge 337
            9.5.2   Preliminary 339
            9.5.3   Online Diffusion Channel 340
            9.5.4   Offline Diffusion Channel 341
            9.5.5   Hybrid Diffusion Channel 342
            9.5.6   Channel Aggregation 343
            9.5.7   Channel Weighting and Selection 344
    9.6   Summary 345
    9.7   Bibliography Notes 346
    9.8   Exercises 347
    References 347

10   Viral Marketing 351
    10.1   Overview 351
    10.2   Traditional Influence Maximization 352
            10.2.1   Influence Maximization Problem 352
            10.2.2   Approximated Seed User Selection 353
            10.2.3   Heuristics-Based Seed User Selection 356
    10.3   Intertwined Influence Maximization 357
            10.3.1   Conditional TIM 358
            10.3.2   Joint TIM 360
    10.4   Cross-Network Influence Maximization 365
            10.4.1   Greedy Seed User Selection Across Networks 366
            10.4.2   Dynamic Programming-Based Seed User Selection 369
    10.5   Rumor Initiator Detection 373
            10.5.1   The ISOMIT Problem 375
            10.5.2   NP-Hardness of Exact ISOMIT Problem 375
            10.5.3   A Special Case: k-ISOMIT-BT Problem 376
            10.5.4   RID Method for General Networks 377
    10.6   Summary 380
    10.7   Bibliography Notes 381
    10.8   Exercises 382
    References 382

11 Network Embedding 385
  11.1   Overview 385
  11.2   Relation Translation Based Graph Entity Embedding 386
        11.2.1   TransE 386
        11.2.2   TransH 387
        11.2.3   TransR 389
  11.3   Homogeneous Network Embedding 390
        11.3.1   DeepWalk 390
        11.3.2   LINE 393
        11.3.3   node2vec 395
  11.4   Heterogeneous Network Embedding 397
        11.4.1   HNE: Heterogeneous Information Network Embedding 398
        11.4.2   Path-Augmented Heterogeneous Network Embedding 400
        11.4.3   HEBE: HyperEdge Based Embedding 400
  11.5   Emerging Network Embedding Across Networks 403
        11.5.1   Concept Definition and Problem Formulation....................... 404
        11.5.2   Deep DIME for Emerging Network Embedding...................... 405
  11.6   Summary 410
  11.7   Bibliography Notes 411
  11.8   Exercises 412
  References 412

Part IV   Future Directions

12 Frontier and Future Directions 417
  12.1   Overview 417
  12.2   Large-Scale Broad Learning 417
  12.3   Multi-Source Broad Learning 418
  12.4   Broad Learning Applications 418
  12.5   Summary 418
  12.6   Exercises 418
  References 419

Part I
Background Introduction

Broad Learning Introduction

1

1.1   What Is Broad Learning

It was six men of Indostan
To learning much inclined,
"Who went to see the Elephant"
(Though all of them were blind),
That each by observation
Might satisfy his mind.

The First approached the Elephant,
And happening to fall
Against his broad and sturdy side,
At once began to bawl:
"God bless me! but the Elephant
Is very like a WALL!"

The Second, feeling of the tusk,
Cried, "Ho, what have we here,
So very round and smooth and sharp?
To me 'tis mighty clear
This wonder of an Elephant
Is very like a SPEAR!"

The Third approached the animal,
And happening to take
The squirming trunk within his hands,
Thus boldly up and spake:
"I see," quoth he, "the Elephant
Is very like a SNAKE!"

The Fourth reached out an eager hand,
And felt about the knee
"What most this wondrous beast is like
Is mighty plain," quoth he:
"Tis clear enough the Elephant
Is very like a TREE!"

The Fifth chanced to touch the ear,
Said: "E'en the blindest man
Can tell what this resembles most;
Deny the fact who can,
This marvel of an Elephant
Is very like a FAN!"

The Sixth no sooner had begun
About the beast to grope,
Than seizing on the swinging tail
That fell within his scope,
"I see," quoth he, "the Elephant
Is very like a ROPE!"

And so these men of Indostan
Disputed loud and long,
Each in his own opinion
Exceeding stiff and strong,
Though each was partly in the right,
And all were in the wrong!

— John Godfrey Saxe (1816–1887)

We would like to start this book with an ancient story about *"The Blind Men and the Elephant"* from *John Godfrey Saxe*. This story is a famous Indian fable about six blind sojourners who come across different parts of an elephant in their life journeys. In turn, each blind man creates his own version of reality from that limited experiences and perspectives. Instead of explaining its philosophical meanings, we indent to use this story to illustrate the current situations that both the academia and industry are facing about *artificial intelligence*, *machine learning*, and *data mining*.

© Springer Nature Switzerland AG 2019
J. Zhang, P. S. Yu, *Broad Learning Through Fusions*,
https://doi.org/10.1007/978-3-030-12528-8_1

In the real world, on the same information entities, e.g., products, movies, POIs (points-of-interest), and even human beings, a large amount of information can actually be collected from various sources. These information sources are usually of different varieties, like Walmart vs. Amazon for commercial products, IMDB vs. Rotten Tomatoes for movies, Yelp vs. Foursquare for POIs, and various online social media websites for human beings. Each information source provides a specific signature of the same entity from a unique underlying aspect. Meanwhile, in most of the cases, these information sources are normally separated from each other without any correlations. An effective fusion of these different information sources will provide an opportunity for researchers and practitioners to understand the entities more comprehensively, which renders *broad learning* to be introduced in this book an extremely important problem.

Broad learning initially proposed in [52, 54, 56] is a new type of learning task, which focuses on fusing multiple large-scale information sources of diverse varieties together and carrying out synergistic data mining tasks across these fused sources in one unified analytic. Fusing and mining multiple information sources are both the fundamental problems in broad learning studies. Broad learning investigates the principles, methodologies, and algorithms for synergistic knowledge discovery across multiple fused information sources, and evaluates the corresponding benefits. Great challenges exist in broad learning for the effective fusion of relevant knowledge across different aligned information sources, which depends upon not only the relatedness of these information sources but also the target application problems. Broad learning aims at developing general methodologies, which will be shown to work for a diverse set of applications, while the specific parameter settings can be effectively learned from the training data. A recent survey article about broad learning is available at [39].

1.2   Problems and Challenges of Broad Learning

Broad learning is a novel yet challenging learning task. The main problems covered in broad learning include *information fusion* and *knowledge discovery* of multiple data sources. Meanwhile, there exist great challenges to address these two tasks due to both the *diverse data inputs* and *various application scenarios* in the real-world problem settings.

1.2.1   Cross-Source Information Fusion

One of the key tasks in broad learning is the fusion of information from different sources, which can be done at different levels, e.g., *raw data level*, *feature space level*, *model level*, and *output level*. The specific fusion techniques used at different levels may have significant differences.

- *Raw data level*: In the case where the data from different sources are of the same modality, e.g., textual data, and have no significant information distribution differences, such kinds of data can be effectively fused at the raw data level.
- *Feature space level*: Based on the data from different sources, a set of features can be effectively extracted, which can be fused together via simple feature vector concatenation (if there exist no significant distribution differences) or feature vector embedding and transformation (to accommodate the information distribution differences).
- *Model level*: In some cases, with the information from different sources, a set of learning models can be effectively trained for each of the sources, so as to fuse the multi-source information at the model level.
- *Output level*: Based on the learned results from each of the sources, they can be effectively fused to output a joint result, which will define the information fusion task at the output level.

In broad learning, the "multiple sources" term is actually a general concept, which may refer to the different information *views*, *categories*, *modalities*, *concrete sources*, and *domains* depending on the specific application settings. We will illustrate these concepts with examples as follows, respectively.

- *Multi-view*: In the traditional webpage ranking problem, information about webpages can be represented in two different views: *textual contents* and *hyperlinks*, which provide complimentary information for learning the ranking scores of the webpages.
- *Multi-categories*: In online social networks, users can perform various categories of social activities, e.g., *making friends, writing posts, checking-in at locations, etc.* Each of these social activities will generate a category of information providing crucial information to help discover the social patterns of users.
- *Multi-modality*: In recent multimedia studies, the information entities (e.g., news articles) usually have information represented in different modalities, including the *textual title, textual content, images*, and *live interview videos*. Integration of such diverse information modalities together will improve the learning performance of models designed for the news articles greatly.
- *Multi-source*: Nowadays, to enjoy more social network services, users are usually involved in multiple online social networks simultaneously, e.g., Facebook, LinkedIn, Twitter, and Foursquare. Information in each of these social network sources can reveal the users' social behaviors from different perspectives.
- *Multi-domain*: Compared with the concepts aforementioned, *domain* is usually used in mathematics, which refers to a set of possible values of independent variables that a function/relation is defined on. Multi-domain learning is a special case of broad learning tasks which fuses the information from different groups of relevant or irrelevant information entities.

Meanwhile, as illustrated in Fig. 1.1, depending on the information flow directions, information fusion in broad learning can be done in different manners, e.g., *information immigration*, *information exchange*, and *information integration*.

- *Information immigration*: In many applications of broad learning, there will exist a specific target source, which can be very sparse and short of useful information. Immigration of information from external mature sources to the target source can hopefully resolve such a problem.
- *Information mutual exchange*: Mutual information exchange is another common application scenario in broad learning, where information will be immigrated among all these sources mutually. With the information from all these sources, application problems studied in all these data sources can benefit from the abundant data simultaneously.
- *Information integration*: Another common application setting of broad learning is the integration of information from all the data sources together, where there exists no target source at all. Such an application setting is normally used in the profiling problem of the information entities shared across different data sources, the fused information from which will lead to a more comprehensive knowledge about these entities.

In this book, we will take online social networks as an example to illustrate the broad learning problems and algorithms. Formally, the information fusion across multiple online social networks is also called the *network alignment* problem [18, 42, 43, 47, 50, 51], which aims at inferring the set of *anchor links* [18, 46] mapping users across networks.

| Information Immigration | Information Exchange | Information Integration |

Fig. 1.1 Information immigration vs. information exchange vs. information integration

1.2.2 Cross-Source Knowledge Discovery

Based on the fused data sources, various concrete application problems can be studied, which will also benefit from the fused data greatly. For instance, with the multiple fused online social network data sets, research problems, like *link prediction* [21, 44–46, 51, 53], *community detection* [7, 15, 40, 41], *information diffusion* [10, 37, 38, 48, 49], *viral marketing* [16, 37, 38, 48, 49], and *network embedding* [2, 27, 52] can all be improved significantly.

- *Link prediction*: With the fused information from multiple aligned network data sources, we will have a better understanding about the connection patterns among the social network users. The link prediction results will be greatly improved with the fused network data sets.
- *Community detection*: According to the connections accumulated from the multiple online social networks, we can also obtain a clearer picture about the network community structures formed by the users. For the social networks with extremely sparse social connections, the data transferred from external sources can hopefully recover the true communities of users in the network.
- *Information diffusion*: Via the social interactions among users, information of various topics can propagate among users through various diffusion channels. Besides the intra-network information diffusion channels, due to the cross-network information sharing, information can also propagate across social networks as well, which will lead to broader impacts actually.
- *Viral marketing*: To carry out commercial advertising and product promotion activities, a necessary marketing strategy is required (e.g., the initial seed users selected to propagate the product information), which can guide the commercial actions during the promotion process. Due to the cross-source information diffusion, the marketing strategy design should consider the information from multiple networks simultaneously.
- *Network embedding*: In recent years, due to the surges of deep learning, representation learning has become a more and more important research problem, which aims at learning the feature vectors characterizing the properties of information entities. Based on the fused social networks, more information about the users can be collected, which can be used for more effective representation feature vector learning for the users.

1.2.3 Challenges of Broad Learning

In the two tasks covered in the broad learning problem mentioned above, great challenges may exist in both fusing and mining the multiple information sources. We categorize the challenges into two main groups as follows:

- *How to fuse*: The data fusion strategy is highly dependent on the data types, and different data categories may require different fusion methods. For instance, to fuse image sources about the same entities, a necessary entity recognition step is required; to combine multiple online social networks, inference of the potential anchor links mapping the shared users across networks will be a key task; meanwhile, to integrate diverse textual data, concept entity extraction or topic modeling can both be the potential options. In many cases, the fusion strategy is also correlated with the specific applications to be studied, which may pose extra constraints or requirements on the fusion results. More information about related data fusion strategies will be introduced later in Part II of this book.

- *How to learn*: To learn useful knowledge from the fused data sources, there also exist many great challenges. In many of the cases, not all the data sources will be helpful for certain application tasks. For instance, in the social community detection task, the fused information about the users' credit card transaction will have less correlation with the social communities formed by the users. On the other hand, the information diffusion among users is regarded as irrelevant with the information sources depicting the daily commute routes of people in the real world. Among all these fused data sources, picking the useful ones is not an easy task. Several strategies, like meta path weighting/selection, information source importance scaling, and information domain embedding, will be described in the specific application tasks to be introduced in Part III of this book.

1.3   Comparison with Other Learning Tasks

There exist great differences of broad learning with other existing learning tasks, e.g., *deep learning* [9], *ensemble learning* [55], *transfer learning* [26], *multi-task learning* [4], and *multi-view* [36], *multi-source* [5], *multi-modal* [25], *multi-domain* [24, 35] learning tasks. In this part, we will provide a brief comparison of these learning tasks to illustrate their correlations and differences.

1.3.1   Broad Learning vs. Deep Learning

As illustrated in Fig. 1.2, deep learning [9] is a rebranding of multi-layer neural network research works. Deep learning is "deep" in terms of the model hidden layers connecting the input to the output space. Generally, the learning process in most machine learning tasks is to achieve a good projection between the input and the output space. In addition, for lots of the application tasks, such a mapping can be very complicated and is usually non-linear. With more hidden layers, the deep learning models will be capable to capture such a complicated projection, which has been demonstrated with the successful application of deep learning in various areas, e.g., speech and audio processing [6, 13], language modeling and processing [1, 23], information retrieval [11, 29], objective recognition and computer vision [20], as well as multi-modal and multi-task learning [32, 33].

However, broad learning actually focuses on a very different perspective instead. Broad learning is "broad" in terms of both its input data variety and the learning model components. As introduced before, the input data sources of broad learning are of very "broad" varieties, including text, image, video, speech, and graph. Meanwhile, to handle such diverse input sources, the broad learning model should have broad input components to process them simultaneously. In these input raw data representation learning components, deep learning and broad learning can also work hand in hand. Deep learning can convert these different modality representations of text, image, video, speech, and graph all into feature vectors, which can make them easier to fuse in broad learning via the

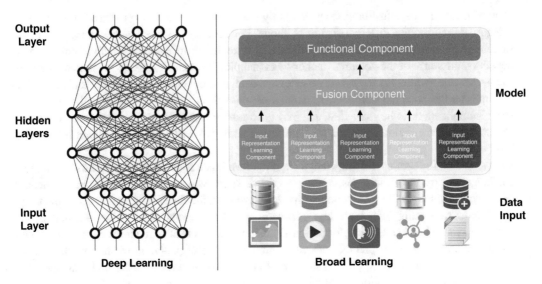

Fig. 1.2 An comparison of deep learning vs. broad learning

fusion component as shown in Fig. 1.2. Viewed in such a perspective, broad learning is broad in both its model input components and information fusion component, where deep learning can serve an important role for the broad input representation learning.

1.3.2   Broad Learning vs. Ensemble Learning

As shown in Fig. 1.3, ensemble learning [55] is a machine learning paradigm where multiple unit models will be trained and combined to solve the same problem. Traditional machine learning models usually employ one hypothesis about the training data, while ensemble learning methods propose to construct a set of hypotheses on the training data and combine them together for the learning purposes. In most of the cases, ensemble learning is carried out based on one single training set, with which a set of unit learning models will be trained and combined to generate the consensus output.

Generally speaking, broad learning works in a very different paradigm compared with ensemble learning. Broad learning aims at integrating diverse data sources for learning and mining problems, but ensemble learning is usually based on one single data input. Meanwhile, broad learning also has a very close correlation with ensemble learning, as ensemble learning also involves the "information integration" step in the model building. Depending on the specific layers for information fusion, many techniques proposed in ensemble learning can also be used for broad learning tasks. For instance, when the information fusion is actually done at the output level in broad learning, the existing techniques, e.g., *boosting* and *bagging*, used in ensemble learning can be adopted as the fusion method for generating the output in broad learning.

1.3.3   Broad Learning vs. Transfer Learning vs. Multi-Task Learning

Traditional transfer learning [26] focuses on the immigration of knowledge from one source to the other source, which may share the same information entities or can be totally irrelevant. Transfer learning problems and models are mostly proposed based on the assumption that the data in different

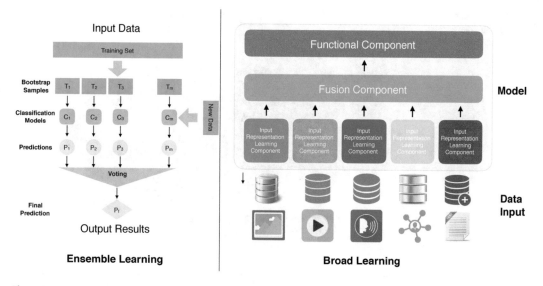

Fig. 1.3 A comparison of ensemble learning vs. broad learning

domains may follow a similar distribution (either in the original feature space or in a certain latent feature space), where the models trained based on external data sources can also be used in the target source. Transfer learning has demonstrated its advantages in overcoming the information shortage problems in many applications.

Meanwhile, for the multi-task learning task [4], it focuses on studying the simultaneous learning of multiple tasks at the same time, where each task can optimize its learning performance through some shared knowledge with other tasks. Furthermore, in the case where there exists a sequential relationship among these learning tasks, the problem will be mapped to the lifelong learning [30] instead.

In a certain sense, both transfer learning and multi-task learning can be treated as a special case of broad learning. Transfer learning aims at immigrating knowledge across sources, which is actually one of the learning paradigms of broad learning as introduced before. Meanwhile, according to the descriptions in Sect. 1.2.1, broad learning allows more flexible information flow directions in the learning process. For the multi-task learning, it studies the learning problem of multiple tasks simultaneously, which can be reduced to the broad learning problem with each task studied based on a separate data set, respectively.

1.3.4 Broad Learning vs. Multi-View, Multi-Source, Multi-Modal, Multi-Domain Learning

Considering that the multiple data "sources" studied in broad learning can have different physical meanings depending on the specific learning settings as introduced in Sect. 1.2.1, which provides a great flexibility for applying broad learning in various learning scenarios. Generally, as illustrated in Fig. 1.4, the multi-view [36], multi-source [5], multi-modal [25], and multi-domain [24, 35] learning tasks can all be effectively covered in the broad learning concept, and the research works in these learning tasks will also greatly enrich the development of broad learning.

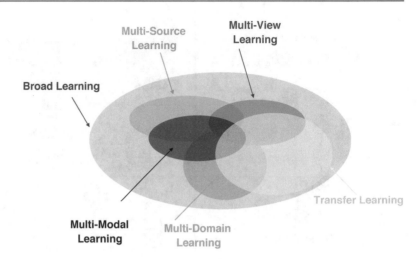

Fig. 1.4 An comparison of ensemble learning vs. broad learning

1.4   Book Organization

This book has four main parts. The first part, covering Chaps. 1–3, will provide the basic essential background knowledge about *broad learning*, *machine learning*, and *social networks*. In the following parts of this book, we will use online social networks as an application setting to introduce the problems and models of broad learning. In the second part, i.e., Chaps. 4–6, this book mainly focuses on introducing the existing online social network alignment concepts, problems, and algorithms, which is also the prerequisite step for broad learning across multiple online social networks. After that, Chaps. 7–11 will make up the third part of this book, which will talk about the application problems that can be studied across multiple online aligned social networks, including *link prediction*, *community detection*, *information diffusion*, *viral marketing*, and *network embedding*. Finally, in the fourth part of this book, i.e., Chap. 12, some potential future research directions and opportunities about broad learning will be provided as a conclusion for this book.

1.4.1   Part I

The first part of the book covers three chapters, and will provide some basic background knowledge of *broad learning*, *machine learning*, and *social network* to make this book self-contained.

- **Chapter 1: Broad Learning Introduction.** Broad learning is a new type of learning and knowledge discovery task that emerge in recent years. The first chapter of this book illustrates the definitions of broad learning, the motivations to study broad learning problems, and also provides the main problems and challenges covered in broad learning in various concrete applications. To help the readers understand this book better, Chap. 1 also includes two sections of reading instructions, including the potential readers of this book as well as how to read this book.
- **Chapter 2: Machine Learning Overview.** Broad learning introduced in this book is based on the existing machine learning works. Chapter 2 introduces some basic background knowledge about machine learning, including data representation, supervised learning, unsupervised learning, deep learning, and some frequently used evaluation metrics.

- **Chapter 3: Social Network Overview.** Online social networks can be formally represented as graphs involving both various kinds of nodes and complex links. Before talking about the problems and models, some essential knowledge about social networks will be provided in Chap. 3, including graph essentials, network categories and measures, network models, and the meta path concept.

1.4.2 Part II

The second part of the book includes Chaps. 4–6. Depending on the availability of training data, different categories of network alignment models have been proposed to solve the social network alignment problem based on the supervised, unsupervised, and semi-supervised learning settings, respectively.

- **Chapter 4: Supervised Network Alignment.** In the case where a set of existing and non-existing anchor links are labeled and available, supervised network alignment models can be built based on the labeled data, where the existing and non-existing anchor links can be used as the positive and negative instances, respectively. Chapter 4 will introduce several online social network alignment models based on the supervised learning setting.
- **Chapter 5: Unsupervised Network Alignment.** In the real scenarios, the anchor links are very expensive to label, which will introduce very large costs in terms of time, money, and labor resources. In such a case, unsupervised network alignment models requiring no training data at all can be a great choice, which will be introduced in Chap. 5 in great detail.
- **Chapter 6: Semi-supervised Network Alignment.** In many other cases, a small proportion of the anchor link labels can be retrieved subject to certain cost constraints, while majority of the remaining anchor links are still unlabeled. Inferring the other potential anchor links aligning the networks with both labeled and unlabeled anchor links is called the semi-supervised network alignment problem, which will be introduced in Chap. 6.

1.4.3 Part III

By aligning different online social networks together, various application tasks can benefit from the information fused from these different information sources. This part will introduce five different knowledge discovery application problems across aligned online social networks.

- **Chapter 7: Link Prediction.** Given a screenshot of an online social network, inferring the connections to be formed among users in the future is named as the link prediction problem. Various real-world services can be cast as the link prediction problem. For instance, friend recommendation can be modeled as a friendship link prediction problem, while location check-in inference can be treated as a location link prediction problem. Chapter 7 covers a comprehensive introduction of the link prediction problem within and across aligned online social networks.
- **Chapter 8: Community Detection.** Social communities denote groups of users who are strongly connected with each other inside each groups but have limited connections to users in other groups. Discovering the social groups formed by users in online social networks is named as the community detection problem, and correctly detected social communities can be important for many social network services. Chapter 8 will introduce the community detection problem within and across aligned online social networks in detail.
- **Chapter 9: Information Diffusion.** Via the social interactions among users, information will propagate from one user to another, which is modeled as the information diffusion process in online

social networks. Chapter 9 focuses on introducing the information diffusion problem specifically, where different categories of diffusion models will be talked about. Across aligned social networks, several state-of-the-art information diffusion models will also be covered in Chap. 9.

- **Chapter 10: Viral Marketing.** Based on the information diffusion model, influence can be effectively propagated among users. By selecting a good group of initial seed users who will spread out the information, real-world product promotions and election campaigns usually aim at achieving the maximum influence in online social networks. Chapter 10 covers the viral marketing problems across online social networks.
- **Chapter 11: Network Embedding.** Most existing machine learning algorithms usually take the feature representation data as the input, which can hardly be applied to handle the network structured data directly. One way to resolve such a problem is to apply network embedding to extract the latent feature representations of the nodes, where those extracted features should also preserve the network structure information. Chapter 11 will focus on the network embedding problem in detail.

1.4.4   Part IV

Broad learning is a novel yet important area, and there exist adequate opportunities and new research problems to be studied. The current broad learning research works also suffer from several big problems, which will be the future research directions as well. In the last part of the book, a big picture of broad learning will be provided and some potential future development directions in this area will also be illustrated, which altogether will form the conclusion of this book.

- **Chapter 12: Frontier and Future Work Directions.** Chapter 12 will provide some other frontier broad learning applications in various areas. Besides the works introduced in the previous chapters of this book, several promising future development directions of broad learning will be illustrated in Chap. 12. The data sets available in the real world are usually of a very large scale, which can involve millions even billions of data instances and attributes. Broad learning across multiple aligned data sources renders the scalability problem more severe. What's more, existing broad learning algorithms are mostly based on pairwise aligned data sources. The more general broad learning problems across multiple (more than two) aligned data sources simultaneously can be another potential development direction. In addition, this book mainly focuses on broad learning in online social networks, but broad learning is a general multiple aligned source learning problem, can actually be applied in many other different domains as well. Finally, at the end of this book, Chap. 12 will draw a conclusion about broad learning.

1.5   Who Should Read This Book

The readers of this book are oriented at the senior undergraduate, graduate students, academic researchers, industrial practitioners, and project managers interested in broad learning and social network mining, data mining, and machine learning.

Readers with a computer science background and have some basic knowledge about data structure, programming, graph theory, and algorithms will find this book to be easily accessible. Some basic knowledge about linear algebra, matrix computation, statistics, and probability theory will be helpful for readers to understand the technique details of the models and algorithms covered in the book. Having prior knowledge about data mining, machine learning, computational social science is a plus, but not necessary.

This book can be used as the textbook in the computational social science, social media mining, data science, data mining, and machine learning application courses for both undergraduate and graduate students. This book is organized in a way that can be taught in one semester to students with background about computer science, statistics, and probability. This book can also be used as the seminar course reading materials for graduate students interested in doing research about data mining, machine learning, and social media mining. The learning materials (including course slides, tutorials, data sets, and toolkit package) will be provided at the IFM Lab broad learning webpage.[1]

Moreover, this book can also be used as the reference book for researchers, practitioners, and project managers of related fields who are interested in learning the basics and tangible examples of this emerging field. This book can help these readers to understand the potentials and opportunities of broad learning in academic research, system/model development, and applications in the industry.

1.6 How to Read This Book

As introduced in the previous sections, this book has four parts. Part I (Chaps. 1–3) covers the introductory materials of broad learning, machine learning, and social networks. Part II (Chaps. 4–6) covers the network alignment problems and models based on different learning settings. Part III (Chaps. 7–11) covers the application problems across multiple aligned social networks. Part IV (Chap. 12) covers the discussions about the future development directions and opportunities of broad learning.

1.6.1 To Readers

We recommend the readers to read this book in the order as how this book is organized from Part I–IV, but the chapters covered in each part can be read in any order (Exception: Chap. 9 had better be read ahead of or together with Chap. 10). Depending on the expertise of the students, some chapters can be skipped or skimmed through quickly. For the readers with background about machine learning, Chap. 2 can be skipped. For readers having background about social media and graph theory, Chap. 3 can be skipped. For readers doing research in social media mining projects on link prediction, community detection, information diffusion, marketing strategy, and network embedding, the introductory materials of traditional single-network based models provided at the beginning of Chaps. 7–11 can be skipped.

1.6.2 To Instructors

This book can be taught in one semester (28 courses, 1.25 h per course) in the order as the how the book is organized from Parts I to IV. The chapters covered in each part can be taught in any sequence (Exception: Chap. 9 had better be delivered ahead of or together with Chap. 10). Depending on the prior knowledge that the students have, certain chapters can be skipped or leave for the students to read after class. The first three introductory chapters of Part I can be delivered in six courses, and Chaps. 2 and 3 together can take five courses. The three chapters about network alignment in Part II can be taught in six courses, where each chapter takes two courses. These five chapters covered in Part III can be delivered in 15 courses, and each chapter takes 3 courses. The last chapter can leave free for the students to read if they are interested in and also plan to further explore this area.

---

[1]http://www.ifmlab.org/BroadLearning.html.

1.6.3 Supporting Materials

Updates to chapters, teaching materials (including source slides, tutorials, slides, courses, data sets, learning toolkit packages, and other resources) are available at the IFM Lab broad learning webpage (see footnote 1) as mentioned in Sect. 1.5.

1.7 Summary

In this chapter, we have provided the introduction to *broad learning*, which is a new type of learning tasks focusing on fusing multiple information sources together for synergistic knowledge discovery. The motivation to study *broad learning* is due to the scattered distributions of data in the real world. An effective fusion of these different information sources will provide an opportunity for researchers and practitioners to understand the target information entities more comprehensively.

We presented the main problems covered in *broad learning*, which include *information fusion* and *cross-source knowledge discovery*. Great challenges exist in addressing these two tasks due to both the diverse data inputs and various application scenarios in the real-world problem settings. We briefly introduced various information fusion methods and manners. We also took online social networks as an example to illustrate five different application problems that we can study across the fused information sources.

We showed the comparison of *broad learning* with several other learning tasks, including *deep learning*, *ensemble learning*, *transfer learning*, *multi-task learning*, as well as *multi-view*, *multi-source*, *multi-modal*, and *multi-domain* learning tasks. We introduced the differences of these learning tasks, and also illustrated the correlations among these different learning problems.

We provided a brief introduction about the organization of this book, which involves four main parts: (1) Part I (three chapters): background knowledge introduction; (2) Part II (three chapters): social network alignment problem and algorithms; (3) Part III (five chapters): application problems studied across the aligned social networks; and (4) Part IV (one chapter): potential future development directions.

We also indicated the potential readers of this book, including senior undergraduate, graduate students, academic researchers, industrial practitioners, and project managers interested in broad learning. For both the readers and the instructors, we provided the reading and teaching guidance as well. Useful supporting materials would also be provided at the book webpage, which could help the readers, instructors, and students to follow the contents covered in this book more easily.

1.8 Bibliography Notes

The *broad learning* concept initially proposed by the authors of this textbook in their research papers [52, 54, 56] is an important problem in both academia and industry. Lots of prior works have been done by the book authors to study both the *fusion* of multiple information sources, e.g., the fusion of online social networks via alignment [18, 42, 43, 47, 50, 51], and the application problems that can be analyzed based on the fused information sources, e.g., *link prediction* [44–46, 51, 53], *community detection* [15, 40, 41], *information diffusion* [37, 38, 48, 49], *viral marketing* [37, 38, 48, 49], and *network embedding* [52]. In the following chapters of this book, we will introduce these research works in great detail.

The essence of *deep learning* is to compute hierarchical feature representations of the observational data [9, 20]. With the surge of deep learning research in recent years, lots of research works have appeared to apply the deep learning methods, like deep belief network [12], deep Boltzmann machine [29], deep neural network [14, 17, 19], and deep autoencoder model [31], in various applications. *Ensemble learning* [22, 28] aims at using multiple learning algorithms to obtain better predictive performance than could be obtained from any of the constituent learning algorithms alone. There exist a number of common ensemble types, which include *bootstrap aggregating (bagging)* [3], *boosting* [8], and *stacking* [34]. A more detailed introduction to the *ensemble learning* is available in the book [55].

Transfer learning focuses on storing knowledge gained while solving one problem and applying it to a different but related problem. A comprehensive survey about *transfer learning* problems and algorithms is available in [26]. Meanwhile, by studying multiple tasks together, the *multi-task learning* problem was introduced. For the formal introduction to *multi-task learning*, please refer to [4]. Several other learning problems mentioned in this chapter include *multi-view* [36], *multi-source* [5], *multi-modal* [25], and *multi-domain* [24, 35] learning tasks, which are all strongly correlated with the *broad learning* task covered in this book.

1.9   Exercises

1. (Easy) Please describe what is *broad learning*, and why *broad learning* is an important learning problem.
2. (Easy) What are the two main problems covered in *broad learning* tasks?
3. (Easy) Please enumerate the information fusion methods at different levels, and try to list some application examples in which these information fusion methods can be used.
4. (Easy) Based on the information flow directions, how many different information fusion manners exist? Please enumerate these approaches, and provide some application settings that these approaches can be applied.
5. (Easy) Besides the *link prediction, community detection, information diffusion, viral marketing,* and *network embedding* problems, can you list some other learning problems that can be studied based on the *broad learning* setting? Please also mention what are the input information sources.
6. (Medium) What are the differences and correlations between *deep learning* and *broad learning*?
7. (Medium) What are the differences and correlations between *ensemble learning* and *broad learning*?
8. (Medium) Can *ensemble learning* models be used to handle *broad learning* problems? Please briefly explain how to do that.
9. (Medium) As introduced in this chapter, *transfer learning* and *multi-task learning* can both be viewed as a special case of *broad learning*. Can you briefly introduce why?
10. (Medium) Please briefly describe the relationship between the *multi-view, multi-modal, multi-source, multi-domain learning* with *broad learning*.

References

1. E. Arisoy, T. Sainath, B. Kingsbury, B. Ramabhadran, Deep neural network language models, in *Proceedings of the NAACL-HLT 2012 Workshop: Will We Ever Really Replace the N-gram Model? On the Future of Language Modeling for HLT (WLM '12)* (Association for Computational Linguistics, Stroudsburg, 2012), pp. 20–28
2. A. Bordes, N. Usunier, A. Garcia-Durán, J. Weston, O. Yakhnenko, Translating embeddings for modeling multi-relational data, in *Advances in Neural Information Processing Systems* (2013)
3. L. Breiman, Bagging predictors. Mach. Learn. **24**(2), 123–40 (1996)
4. R. Caruana, Multitask learning. Mach. Learn. **28**(1), 41–75 (1997)

5. W. Dai, Q. Yang, G. Xue, Y. Yu, Boosting for transfer learning, in *Proceedings of the 24th International Conference on Machine Learning* (ACM, New York, 2007), pp. 193–200
6. L. Deng, G. Hinton, B. Kingsbury, New types of deep neural network learning for speech recognition and related applications: an overview, in *2013 IEEE International Conference on Acoustics, Speech and Signal Processing* (IEEE, Piscataway, 2013)
7. S. Fortunato, Community detection in graphs. Phys. Rep. **486**(3–5), 75–174 (2010)
8. Y. Freund, R. Schapire, A short introduction to boosting. J. Jpn. Soc. Artif. Intell. **14**(5), 771–780 (1999)
9. I. Goodfellow, Y. Bengio, A. Courville, *Deep Learning* (MIT Press, Cambridge, 2016). http://www.deeplearningbook.org
10. D. Gruhl, R. Guha, D. Liben-Nowell, A. Tomkins, Information diffusion through blogspace, in *Proceedings of the 13th International Conference on World Wide Web* (ACM, New York, 2004), pp. 491–501
11. S. Hill, Elite and upper-class families, in *Families: A Social Class Perspective* (2012)
12. G. Hinton, S. Osindero, Y. Teh, A fast learning algorithm for deep belief nets. Neural Comput. **18**(7), 1527–1554 (2006)
13. G. Hinton, L. Deng, D. Yu, G. Dahl, A. Mohamed, N. Jaitly, A. Senior, V. Vanhoucke, P. Nguyen, T. Sainath, B. Kingsbury, Deep neural networks for acoustic modeling in speech recognition. IEEE Signal Process. Mag. **29**(6), 82–97 (2012)
14. H. Jaeger, Tutorial on training recurrent neural networks, covering BPPT, RTRL, EKF and the "echo state network" approach, Technical report (2002)
15. S. Jin, J. Zhang, P. Yu, S. Yang, A. Li, Synergistic partitioning in multiple large scale social networks, in *2014 IEEE International Conference on Big Data (Big Data)* (IEEE, Piscataway, 2014)
16. D. Kempe, J. Kleinberg, É Tardos, Maximizing the spread of influence through a social network, in *Proceedings of the Ninth ACM SIGKDD International Conference on Knowledge Discovery and Data Mining* (ACM, New York, 2003), pp. 137–146
17. Y. Kim, Convolutional neural networks for sentence classification, in *Proceedings of the 2014 Conference on Empirical Methods in Natural Language Processing (EMNLP)* (Association for Computational Linguistics, 2014), pp. 1746–1751
18. X. Kong, J. Zhang, P. Yu, Inferring anchor links across multiple heterogeneous social networks, in *Proceedings of the 22nd ACM international conference on Information and Knowledge Management* (ACM, New York, 2013), pp. 179–188
19. A. Krizhevsky, I. Sutskever, G. Hinton, Imagenet classification with deep convolutional neural networks, in *Advances in Neural Information Processing Systems* (2012)
20. Y. LeCun, Y. Bengio, G. Hinton, Deep learning. Nature **521**, 436–444 (2015). http://dx.doi.org/10.1038/nature14539
21. D. Liben-Nowell, J. Kleinberg, The link prediction problem for social networks, in *Proceedings of the Twelfth International Conference on Information and Knowledge Management* (ACM, New York, 2003), pp. 556–559
22. R. Maclin, D. Opitz, Popular ensemble methods: an empirical study. J. Artif. Intell. Res. (2011). arXiv:1106.0257
23. A. Mnih, G. Hinton, A scalable hierarchical distributed language model, in *Advances in Neural Information Processing Systems 21 (NIPS 2008)* (2009)
24. H. Nam, B. Han, Learning multi-domain convolutional neural networks for visual tracking. Comput. Vis. Pattern Recognit. (2015). arXiv:1510.07945
25. J. Ngiam, A. Khosla, M. Kim, J. Nam, H. Lee, A. Ng, Multimodal deep learning, in *Proceedings of the 28th International Conference on Machine Learning (ICML-11)* (2011), pp. 689–696
26. S. Pan, Q. Yang, A survey on transfer learning. IEEE Trans. Knowl. Data Eng. **22**(10), 1345–1359 (2010)
27. B. Perozzi, R. Al-Rfou, S. Skiena, Deepwalk: online learning of social representations, in *Proceedings of the 20th ACM SIGKDD International Conference on Knowledge Discovery and Data Mining* (ACM, New York, 2014), pp. 701–710
28. R. Polikar, Ensemble based systems in decision making. IEEE Circuits Syst. Mag. **6**(3), 21–45 (2006)
29. R. Salakhutdinov, G. Hinton, Semantic hashing. Int. J. Approx. Reason. **50**(7), 969–978 (2009)
30. S. Thrun, *Lifelong Learning Algorithms* (1998)
31. P. Vincent, H. Larochelle, I. Lajoie, Y. Bengio, P. Manzagol, Stacked denoising autoencoders: learning useful representations in a deep network with a local denoising criterion. J. Mach. Learn. Res. **11**, 3371–3408 (2010)
32. J. Weston, S. Bengio, N. Usunier, Large scale image annotation: learning to rank with joint word-image embeddings. J. Mach. Learn. Res. **81**(1), 21–35 (2010)
33. J. Weston, S. Bengio, N. Usunier, Wsabie: scaling up to large vocabulary image annotation, in *Proceedings of the International Joint Conference on Artificial Intelligence (IJCAI)* (2011)
34. D. Wolpert, Stacked generalization. Neural Netw. **5**, 241–259 (1992)
35. T. Xiao, H. Li, W. Ouyang, X. Wang, Learning deep feature representations with domain guided dropout for person re-identification. Comput. Vis. Pattern Recognit. (2016). arXiv:1604.07528
36. C. Xu, D. Tao, C. Xu, A survey on multi-view learning. Mach. Learn. (2013). arXiv:1304.5634

37. Q. Zhan, J. Zhang, S. Wang, P. Yu, J. Xie, Influence maximization across partially aligned heterogeneous social networks, in *Pacific-Asia Conference on Knowledge Discovery and Data Mining* (Springer, Berlin, 2015), pp. 58–69

38. Q. Zhan, J. Zhang, X. Pan, M. Li, P. Yu, Discover tipping users for cross network influencing, in *2016 IEEE 17th International Conference on Information Reuse and Integration (IRI)* (IEEE, Piscataway, 2016)

39. J. Zhang, Social network fusion and mining: a survey. Soc. Inf. Netw. (2018). arXiv:1804.09874

40. J. Zhang, P. Yu, Community detection for emerging networks, in *Proceedings of the 2015 SIAM International Conference on Data Mining* (Society for Industrial and Applied Mathematics, Philadelphia, 2015)

41. J. Zhang, P. Yu, MCD: mutual clustering across multiple social networks, in *2015 IEEE International Congress on Big Data* (IEEE, Piscataway, 2015)

42. J. Zhang, P. Yu, Multiple anonymized social networks alignment, in *2015 IEEE International Conference on Data Mining* (IEEE, Piscataway, 2015)

43. J. Zhang, P. Yu, PCT: partial co-alignment of social networks, in *Proceedings of the 25th International Conference on World Wide Web* (ACM, New York, 2016), pp. 749–759

44. J. Zhang, X. Kong, P. Yu, Predicting social links for new users across aligned heterogeneous social networks, in *2013 IEEE 13th International Conference on Data Mining* (IEEE, Piscataway, 2013)

45. J. Zhang, X. Kong, P. Yu, Transferring heterogeneous links across location-based social networks, in *Proceedings of the 7th ACM International Conference on Web Search and Data Mining* (ACM, New York, 2014), pp. 303–312

46. J. Zhang, P. Yu, Z. Zhou, Meta-path based multi-network collective link prediction, in *Proceedings of the 20th ACM SIGKDD International Conference on Knowledge Discovery and Data Mining* (ACM, New York, 2014), pp. 1286–1295

47. J. Zhang, W. Shao, S. Wang, X. Kong, P. Yu, PNA: partial network alignment with generic stable matching, in *2015 IEEE International Conference on Information Reuse and Integration* (IEEE, Piscataway, 2015)

48. J. Zhang, S. Wang, Q. Zhan, P. Yu, Intertwined viral marketing in social networks, in *2016 IEEE/ACM International Conference on Advances in Social Networks Analysis and Mining (ASONAM)* (IEEE, Piscataway, 2016)

49. J. Zhang, P. Yu, Y. Lv, Q. Zhan, Information diffusion at workplace, in *Proceedings of the 25th ACM International on Conference on Information and Knowledge Management* (ACM, New York, 2016), pp. 1673–1682

50. J. Zhang, Q. Zhan, P. Yu, Concurrent alignment of multiple anonymized social networks with generic stable matching, in *Theoretical Information Reuse and Integration* (Springer, Cham, 2016), pp. 173–196

51. J. Zhang, J. Chen, J. Zhu, Y. Chang, P. Yu, Link prediction with cardinality constraints, in *Proceedings of the Tenth ACM International Conference on Web Search and Data Mining* (ACM, New York, 2017), pp. 121–130

52. J. Zhang, C. Xia, C. Zhang, L. Cui, Y. Fu, P. Yu, BL-MNE: emerging heterogeneous social network embedding through broad learning with aligned autoencoder, in *Proceedings of the 2017 IEEE International Conference on Data Mining* (IEEE, Piscataway, 2017)

53. J. Zhang, J. Chen, S. Zhi, Y. Chang, P. Yu, J. Han, Link prediction across aligned networks with sparse and low rank matrix estimation, in *2017 IEEE 33rd International Conference on Data Engineering (ICDE)* (IEEE, Piscataway, 2017)

54. J. Zhang, L. Cui, P. Yu, Y. Lv, BL-ECD: broad learning based enterprise community detection via hierarchical structure fusion, in *Proceedings of the 2017 ACM on Conference on Information and Knowledge Management* (ACM, New York, 2017), pp. 859–868

55. Z. Zhou, *Ensemble Methods: Foundations and Algorithms*, 1st edn. (Chapman & Hall/CRC, London, 2012)

56. J. Zhu, J. Zhang, L. He, Q. Wu, B. Zhou, C. Zhang, P. Yu, Broad learning based multi-source collaborative recommendation, in *Proceedings of the 2017 ACM on Conference on Information and Knowledge Management* (ACM, New York, 2017), pp. 1409–1418

Machine Learning Overview

2

2.1   Overview

Learning denotes the process of acquiring new declarative knowledge, the organization of new knowledge into general yet effective representations, and the discovery of new facts and theories through observation and experimentation. Learning is one of the most important skills that mankind can master, which also renders us different from the other animals on this planet. To provide an example, according to our past experiences, we know the sun rises from the east and falls to the west; the moon rotates around the earth; 1 year has 365 days, which are all knowledge we derive from our past life experiences.

As computers become available, mankind has been striving very hard to implant such skills into computers. For the knowledge which are clear for mankind, they can be explicitly represented in program as a set of simple reasoning rules. Meanwhile, in the past couple of decades, an extremely large amount of data is being generated in various areas, including the World Wide Web (WWW), telecommunication, climate, medical science, transportation, etc. For these applications, the knowledge to be detected from such massive data can be very complex that can hardly be represented with any explicit fine-detailed specification about what these patterns are like. Solving such a problem has been, and still remains, one of the most challenging and fascinating long-range goals of machine learning.

Machine learning is one of the disciplines, which aims at endowing programs with the ability to learn and adapt. In machine learning, experiences are usually represented as data, and the main objective of machine learning is to derive models from data that can capture the complicated hidden patterns. With these learned models, when we feed them with new data, the models will provide us with the inference results matching the captured patterns. Generally, to test the effectiveness of these learned models, different evaluation metrics can be applied to measure the performance of the inference results.

Existing machine learning tasks have become very diverse, but based on the supervision/label information used in the model building, they can be generally categorized into two types: "supervised learning" and "unsupervised learning." In supervised learning, the data used to train models are pre-labeled in advance, where the labels indicate the categories of different data instances. The representative examples of supervised learning task include "classification" and "regression." Meanwhile, in unsupervised learning, no label information is needed when building the models, and the representative example of unsupervised learning task is "clustering." Between unsupervised

© Springer Nature Switzerland AG 2019
J. Zhang, P. S. Yu, *Broad Learning Through Fusions*,
https://doi.org/10.1007/978-3-030-12528-8_2

learning and supervised learning, there also exists another type of learning tasks actually, which is named as the "semi-supervised learning." Semi-supervised learning is a class of learning tasks and techniques that make use of both labeled and unlabeled data for training, typically a small amount of labeled data with a large amount of unlabeled data. Meanwhile, besides these aforementioned learning tasks, there also exist many other categorizations of the learning tasks, like "transfer learning," "sparse learning," "reinforcement learning," and "ensemble learning."

To make this book self-contained, the goal of this chapter is to provide a rigorous, yet easy to follow, introduction to the main concepts underlying machine learning, including the detailed data representations, supervised learning and unsupervised learning tasks and models, and several classic evaluation metrics. Considering the popularity of deep learning [14] in recent years, we will also provide a brief introduction to deep learning models in this chapter. Many other learn tasks, like semi-supervised learning [7, 57], transfer learning [31], sparse learning [28], etc., will be introduced in the following chapters when discussing the specific research problems in detail.

2.2 Data Overview

Data is a physical representation of information, from which useful knowledge can be effectively discovered by machine learning and data mining models. A good data input can be the key to the discovery of useful knowledge. In this section, we will introduce some background knowledge about data, including data types, data quality, data transformation and processing, and data proximity measures, respectively.

2.2.1 Data Types

A data set refers to a collection of data instances, where each data instance denotes the description of a concrete information entity. In the real scenarios, the data instances are usually represented by a number of attributes capturing the basic characteristics of the corresponding information entity. For instance, let's assume we see a group of Asian and African elephants (i.e., elephant will be the information entity). As shown in Table 2.1, each elephant in the group is of certain weight, height, skin smoothness, and body shape (i.e., different attributes), which can be represented by these attributes as an individual data instance. Generally, as shown in Fig. 2.1, the mature Asian elephant is in a smaller size compared with the mature African elephant. African elephants are discernibly larger in size, about 8.2–13 ft (2.5–4 m) tall at the shoulder, and they weigh between 5000 and 14,000 lbs (2.5–7 t). Meanwhile, Asian elephants are more diminutive, measuring about 6.6–9.8 ft (2–3 m) tall at the shoulder and weighing between 4500 and 12,000 lbs (2.25–6 t). In addition, from their ear size,

Table 2.1 An example of the elephant data set (ear size: 1 denotes large; 0 denotes small)

Elephant ID	Weight (t)	Height (m)	Skin	Ear size	Trunk "finger"	$\cdots$	Category
1	6.8	4.0	Wrinkled	1	2	$\cdots$	African
2	5.8	3.5	Wrinkled	1	2	$\cdots$	African
3	4.5	2.1	Smooth	0	1	$\cdots$	Asian
4	5.8	3.1	Wrinkled	0	1	$\cdots$	Asian
5	4.8	2.7	Wrinkled	1	2	$\cdots$	African
6	5.6	2.8	Smooth	1	1	$\cdots$	Asian
$\cdots$	$\cdots$	$\cdots$	$\cdots$	$\cdots$	$\cdots$	$\cdots$	$\cdots$

Asian Elephant   **African Elephant**

Fig. 2.1 A comparison of Asian elephant vs African elephant

skin smoothness, and trunk "finger" number, etc., we can also effectively differentiate them from each other. The group of the elephant data instances will form an elephant data set, as illustrated in Table 2.1.

Instance-attribute style data format is a general way for data representation. For different types of data sets, the attributes used to describe the information entities will be different. In the following parts of this subsection, we will talk about the categories of attributes first, and then introduce different data types briefly.

2.2.1.1 Attribute Types

Attribute is the basic element in describing information entities, and we provide its formal definition as follows.

Definition 2.1 (Attribute) Formally, an attribute denotes a basic property or characteristic of an information entity. Generally, attribute values can vary from one information entities to another in a provided data set.

In the example provided in Table 2.1, the elephant body weight, height, skin smoothness, ear size, and trunk "finger" number are all the attributes of the elephant information entities. Among these attributes, both body weight and height are the attributes with continuous values, which can take values in a reasonable range (e.g., [1 t, 7 t] for weight and [1 m, 5 m] for height, respectively); trunk "finger" number is a discrete attribute instead, which takes values from set {1, 2}; ear size is a transformed attribute, which maps the ear size into 2 levels (1: large; 0: small); and skin smoothness is an attribute with categorical values from set {Wrinkled, Smooth}. These attributes listed above are all the facts about elephants.

In Table 2.1, we use integers to denote the elephant ID, ear size, and trunk "finger" number. However, for the same number appearing in these three kinds of attributes, they will have different physical meanings. For instance, the elephant ID 1 of the 1st row in the table denotes the unique identifier of the elephant in the table; meanwhile, the ear size 1 of rows 1, 2, 5, and 6 in the table denotes the elephants have a large ear size; and the trunk "finger" number 1 of rows 3, 4, and 6 denotes the count of trunk fingers about the elephant instances. Therefore, to interpret these data attributes, we need to know their specific attribute types and the corresponding physical meanings in advance.

Fig. 2.2 A systematic categorization of attribute types

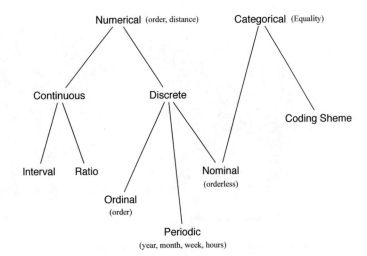

As shown in Fig. 2.2, the attribute types used in information entity description can usually be divided into two main categories, *categorical attributes* and *numeric attributes*. For instance, in Table 2.1, skin smoothness is a *categorical attribute* and weight is a *numeric attribute*. As to the ear size, originally this attribute is a *categorical attribute*, where the values have inherent orders in terms of the elephant ear size. Meanwhile, the transformed attribute (i.e., the integer numbers) becomes a numerical attribute instead, where the numbers display such order relationships specifically. We will talk about the attribute transformation later.

Generally, the *categorical attributes* are the qualitative attributes, while the *numeric attributes* are the quantitative attributes, both of which can be further divided into several sub-categories as well. For instance, depending on whether the attributes have order relationships or not, we can divide the *categorical attributes* into *nominal attributes* and *ordinal attributes*. On the other hand, depending on the continuity of attribute values, the *numeric attributes* can be further divided into *discrete attributes* and *continuous attributes*. Detailed descriptions about these above attribute categories are provided as follows.

- **Nominal Categorical Attribute**: Nominal categorical attributes provide enough information to distinguish one data instance from another, which don't have any internal relationships. For instance, for the elephant instances described in Table 2.1, the involved *nominal categorical attributes* include *elephant category* ({Asian, African}) and *elephant skin smoothness* ({Wrinkled, Smooth}). Besides this example, some other representative *nominal categorical attributes* include *colors* ({Red, Green, ..., Purple}), *names* (people name: {Alice, Bob, ..., Zack}, country name: {Afghanistan, Belgium, China, ..., Zambia}), *ID numbers* ({1, 2, ..., 99} or {id001, id002, ..., id099}).
- **Ordinal Categorical Attributes**: Similar to the nominal categorical attributes, the ordinal categorical attributes also provide enough information to distinguish one object from another. Furthermore, the ordinal attribute values also bear order information. Representative examples of *ordinal categorical attributes* include *elephant ear size* ({Small, Large}), *goodness* ({Good, Better, Best}), *sweetness* ({Sour, Slightly Sour, ..., Slightly Sweet, Sweet}), grades ({A, B, C, D, F}).
- **Discrete Numeric Attributes**: Discrete numeric attribute is one type of numeric attribute from a finite or countably infinite set of values. Representative examples of *discrete numeric attributes* include the attributes about *counts* (e.g., elephant trunk "finger" number, population of countries, number of employees in companies, number of days).

- **Continuous Numeric Attributes**: Continuous numeric attributes have real numbers as the attribute values. Representative examples include temperature, mass, size (like length, width, and height). Continuous numeric attributes are usually denoted as float-point real numbers in the concrete representations.

2.2.1.2 Types of Data Sets

Data can be collected from different disciplines, and the original storage representations of the data can be very diverse. The data storage types depend on various factors in data gathering, like the equipment applied to gather the data, the preprocessing and cleaning operations, the storage device format requirements, and specific application domains. To help provide a big picture about data set types used in this book, we propose to categorize data into *record data*, *graph data*, and *ordered data*. In this part, we will provide a brief introduction of these data types together with some illustrative examples.

- **Record Data**: Many data sets can be represented as a set of independent data records described by certain pre-specified attributes, like the elephant data set shown in Table 2.1. In the record style data sets, the records are mostly independent of each other with no dependence or correlation relationships among them. Besides the fixed attributes, for many other cases, the record attributes can be dynamic and may change differently from one record to another. A representative example of such a kind of data set is the "Market Transaction Data" (as illustrated in plot a of Fig. 2.3), which is a set of market shopping transaction records. Each record denotes the purchased items in the transaction, which may vary significantly for different transactions. In addition, in Fig. 2.3,

A

TID	ITEMS
1	Bread, Butter, Milk
2	Beer, Diaper
3	Beer, Diaper, Bread, Milk
4	Soda, Diaper, Milk

B

AID	Age	Has_Job	Own_House	Credit	Approval
1	Young	TRUE	FALSE	Good	Yes
2	Young	FALSE	FALSE	Fair	No
3	Middle	TRUE	TRUE	Excellent	Yes
4	Senior	FALSE	TRUE	Good	Yes

C

	computer	software	linux	knuth	love	mac	program	windows
Document 1	3	2	0	5	1	0	3	5
Document 2	0	0	1	3	2	0	2	0
Document 3	0	5	4	6	8	0	6	0
Document 4	7	2	7	0	0	5	0	4

Fig. 2.3 Examples of record data ((**a**) shopping transaction record; (**b**) loan application record; (**c**) document-word representation)

we also show two other examples of the record data sets, where one is about the loan application record and the other one is about the document-word representation table. In the loan application record, besides the application IDs, the involved attributes include "*Age*," "*Has Job*," "*Own House*," "*Credit*," and "*Approval*," where the first four attributes are about applicant profile and the last attribute is about the application decision. For the document-word representation table, the words are treated as the attributes and the numbers in the table denote the number of appearance of the words in certain documents.

- **Graph Data**: For the information entities with close correlations, graph will be a powerful and convenient representation for data, where the nodes can represent the information entities while the links indicate the relationships. Generally, in graphs, the links among the information entities may convey very important information. In some cases, the link can also be directed, weighted, or signed, which will indicate the direction, weight and polarity of the relationships among the information entities. As shown in Fig. 2.4, representative examples of graph data include: (1) Online social networks (OSNs), where the nodes denote users and the links represent the friendship/follow links among the users; (2) Chemical molecule graphs, where the nodes denote the atoms and the links represent the bonds among the atoms; and (3) World Wide Web (WWW), where the nodes represent the webpages while the links denote the hyperlinks among the webpages. Graph style data can be represented in the Instance-Attribute format as well, where both the instances and

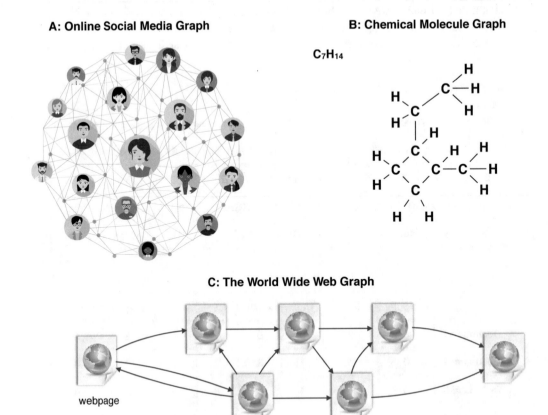

Fig. 2.4 Examples of graph data ((**a**) online social media graph; (**b**) chemical molecule graph; (**c**) the internet webpage graph)

attributes correspond to the information entities in the graphs, respectively. Such a representation will be very similar to the graph adjacency matrix to be introduced in Sect. 3.2.1, whose entries will have value 1 (or a signed weight) if they correspond to the connected information entities and 0 otherwise.

- **Ordered Data**: For the other data types, the instances or the attributes may be ordered in terms of time or space, and such a kind of relationship cannot be captured by either the record data or the graph data types, which will be represented as the ordered data instead. Representative examples of the ordered data include the *sequential transaction data*, *sequence data*, *text data*, *time series data*, and *spatial data*, some of which are shown in Fig. 2.5. In the real world, people tend to shop at the same market for multiple times (like customers C_1 and C_2 in plot a of Fig. 2.5), whose purchase records will have a temporal order relationship with each other, e.g., I_4 and I_5 are always purchased after I_1. For the gene data from biology/bioinformatics, as shown in plot c of Fig. 2.5, it can be represented as a sequence composed by A, T, G, and C (e.g., "GGTTCCTGCTCAAGGCCCGAA"), which defines the code of both human and animals. Data accumulated from the observations about nature or finance like temperature, air pressure, stock prices, and trading volumes can be represented as the ordered time series data (e.g., the Dow Jones Industrial Index as shown in plot b of Fig. 2.5). For the data collected from the offline world, like crime rate, traffic, as well as the weather observations, they will have some relationships in terms of their spatial locations, which can be represented as the ordered data as well.

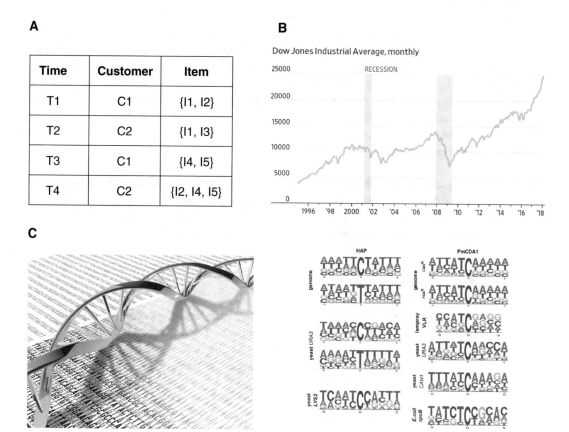

Fig. 2.5 Examples of ordered data ((**a**) sequential transaction record; (**b**) monthly Dow Jones industrial average; (**c**) DNA double helix and interpreted sequence data)

2.2.2 Data Characteristics

For machine learning tasks, several characteristics of the data sets can be very important and may affect the learning performance greatly, which include *quality*, *dimensionality*, and *sparsity*.

- **Quality**: Few data sets are perfect and real-world data sets will contain errors in the collection, processing, and storage process due to various reasons, like human errors, flaws in data processing, and limitations of devices. The data errors can also be categorized into various types, which include noise, artifacts, outlier data instances, missing value in data representation, inconsistent attribute values, and duplicate data instances. "*Garbage in, Garbage out*" is a common rule in machine learning. All these data errors will degrade the data set quality a lot, and may inevitably affect the learning performance.
- **Dimensionality**: Formally, the dimensionality of a data set denotes the number of attributes involved in describing each data instance. Data set dimensionality depends on both the data set and the data processing methods. Generally, data sets of a larger dimensionality will be much more challenging to handle, which is also referred to as the "*Curse of Dimensionality*" [2, 3, 52]. To overcome such a challenge, data processing techniques like dimension reduction [51] and feature selection [15] have been proposed to reduce the data dimensionality effectively by either projecting the data instances to a low-dimensional space or selecting a small number of attributes to describe the data instances.
- **Sparsity**: In many cases, in the Instance-Attribute data representations, a large number of the entries will have zero values. Such an observation is very common in application scenarios with a large attribute pool but only a small number of the attributes are effective in describing each data instance. Information sparsity is a common problem for many data types, like record data (e.g., transaction data), graph data (e.g., social network data and WWW data), and ordered data (e.g., text data and sequential transaction data). Great challenges exist in handling the data set with a large sparsity, which has very little information available for learning and model building. A category of learning task named "*sparse learning*" [28] has been proposed to handle such a problem, and we will introduce it in Sect. 7.6.

Before carrying out the machine learning tasks, these aforementioned characteristics need to be analyzed on the data set in advance. For the data sets which cannot meet the requirements or assumptions of certain machine learning algorithms, necessary data transformation and pre-processing can be indispensable, which will be introduced in the following subsection in detail.

2.2.3 Data Pre-processing and Transformation

To make the data sets more suitable for certain learning tasks and algorithms, several different common data pre-processing and transformation operations will be introduced in this part. To be more specific, the data operations to be introduced in this subsection include *data cleaning and pre-processing* [49], *data aggregation and sampling* [49], *data dimensionality reduction and feature selection* [15, 51], and *data transformation* [49].

2.2.3.1 Data Cleaning and Pre-processing

Data cleaning and pre-processing [49] focus on improving the quality of the data sets to make them more suitable for certain machine learning tasks. As introduced in the previous subsection, regular errors that will degrade the data quality include *noise, outliers, missing values, inconsistent values,* and *duplicate data*. Many data cleaning and pre-processing techniques have been proposed to address these problems to improve the data quality.

Data noise denotes the random factor that will distort the data instance attribute values or the addition of spurious instances. Noise is actually very hard to be distinguished from non-noise data, which are normally mixed together. In the real-world learning tasks, noise is extremely challenging to detect, measure, and eliminate. Existing works on handling noise mainly focus on improving the learning algorithms to make them robust enough to handle the noise. Noise is very common in ordered data, like time series data and geo-spatial data, where redundant noisy signals can be gathered in data collection due to the problems with the device bandwidth or data collection techniques.

Outliers denote the data instances that have unique characteristics, which are different from the majority of normal data instances in terms of the instance itself or only certain attributes. Outlier detection has been a key research problem in many areas, like fraud detection, spam detection, and network intrusion detection, which all aim at discovering certain unusual data instances which are different from the regular ones. Depending on the concrete learning tasks and settings, different outlier detection techniques have been proposed already, e.g., supervised outlier detection methods (based on the extracted features about outliers) and unsupervised outlier detection methods (based on clustering algorithms to group instances into clusters, and the isolated instances will be outliers).

In data analysis, missing value is another serious problem, causes of which are very diverse, like unexpected missing data due to device fault in data collection, and intentional missing data due to privacy issues in questionnaire filling. Different techniques can be applied to handle the missing values, e.g., simple elimination of the data instance or attribute containing missing values, missing value estimation, and ignoring missing values in data analysis and model building. Elimination of the data instances or attributes is the simplest way to deal with the missing value problem, but it will also lead to problems in removing important data instances/attributes from the data set. As to the missing value estimation, methods like random missing value guess, mean value refilling, and majority value refilling can be effectively applied. Ignoring the missing values in data analysis requires the learning algorithms to be robust enough in data analysis, which requires necessary calibrations of the models and is out of the scope of data pre-processing.

Data inconsistency is also a common problem in the real-world data analysis, which can be caused by problems in data collection and storage, e.g., mis-reading of certain storage areas or failure in writing protection of some variables in the system. For any two data instances with inconsistent attribute values, a simple way will be to discard one, but extra information may be required in determining which instance to remove. Another common problem in data analysis is data duplication, which refers to the multiple-time occurrence of data instances corresponding to the same information entities in the data set. Data duplication is hard to measure, as it is very challenging to distinguish real duplicated data instances from legitimated data instances corresponding to different information entities. One way to address such a problem will be information entity resolution to identify the data instances actually corresponding to the same information entities.

2.2.3.2 Data Aggregation and Sampling

For many learning tasks, the data set available can be very big involving a large number of data instances. Learning from such a large-scale data set will be very challenging for many learning algorithms, especially those with a high time complexity. To accommodate the data sets for these existing learning algorithms, two data processing operations can be applied: data aggregation and data sampling [49].

Data aggregation denotes the operation of combining multiple data instances into one. As aforementioned, one motivation to apply data aggregation is to reduce the data size as well as the data analysis time cost. Based on the aggregated data sets, many expensive (in terms of space and time costs) learning algorithms can be applied to analyze the data. Another motivation to apply data aggregation is to analyze the data set from a hierarchical perspective. Such an operation is especially useful for data instances with hierarchical attributes, e.g., sale transaction records in different markets, counties, cities, states in the USA. The sales of certain target product can be quite limited in a specific market, and the market level transaction record data will be very sparse. However, by aggregating (i.e., summing up) the sale transactions from multiple markets in the same counties, cities, and even states, the aggregated statistical information will be more dense and meaningful. Data aggregation can also have disadvantages, as it can lead to information loss inevitably, where the low-level data patterns will be no longer available in the aggregated high-level data sets.

Another operation that can be applied to handle the large-scale data sets is data sampling. Formally, sampling refers to the operation of selecting a subset of data instances from the complete data set with certain sampling methods. Data sampling has been used in data investigation for a long time, as processing the complete data set can be extremely time consuming, especially for those with a large size (e.g., billions of records). Selecting a subset of the records allows an efficient data analysis in a short period of time. Meanwhile, to preserve the original data distribution patterns, the sampled data instances should be representative enough to cover the properties about the original data set. Existing data sampling approaches can be divided into two main categories:

- **Random Sampling**: Regardless of the data instances, random sampling selects the data instances from data sets randomly, which is the simplest type of sampling approach. For such a kind of sampling approach, depending on whether the selected instances will be replaced or not, there exist two different variants: (1) random data sampling without replacement, and (2) random data sampling with replacement. In the approach with instance replacement, all the selected instances will be replaced back in the original data set and can be selected again in the rounds afterwards. Random sampling approaches will work well for most regular data sets with a similar size of instances belonging to different types (i.e., class balanced data sets).
- **Stratified Sampling**: However, when handling the data sets with imbalanced class distributions, the random data sampling approach will suffer from many problems. For instance, given a data set with 90% positive instances and 10% negative instances (positive and negative here denote two different classes of data instances, which will be introduced in Sect. 2.3 in detail), random sampling approach is applied to sample 10 of the instances from the original data set. In the sampled data instances, it is highly likely that very few negative instances will be selected due to their scarcity in the original data set. To overcome such a problem, the *stratified sampling* can be applied instead. Stratified sampling will select instances from both positive and negative instance sets separately, i.e., 9 positive instances and 1 negative data instance will be selected finally.

2.2.3.3 Data Dimensionality Reduction and Feature Selection

Besides the number of data instances, the number of attributes used to describe the data instances can be very large as well, which renders many learning algorithms fail to work. Due to the *"curse of dimensionality"* [2, 3, 52], the increase of data dimensionality will make the data much harder to handle. Both the classification and clustering (to be introduced later) tasks will suffer from such high-dimensional data sets due to the large number of variables to be learned in the models and the lack of meaningful evaluation metrics. There exist two classic methods to reduce the data dimensionality, i.e., *dimensionality reduction* and *feature selection* [15, 51], which will be introduced as follows.

Conventional data dimensionality reduction approaches include *principal components analysis (PCA)*, *independent component analysis (ICA)*, and *linear discriminant analysis (LDA)*, etc., which apply linear algebra techniques to project data instances into a lower-dimensional space. PCA is a statistical procedure that uses an orthogonal transformation to convert the observations of possibly correlated variables into a set of linearly uncorrelated variables, which are called the principal components. The objective continuous attributes to be discovered in PCA should be (1) a linear combination of original attributes, (2) orthogonal to each other, and (3) able to capture the maximum amount of variation in the observation data. Different from PCA, ICA aims at projecting the data instances into several independent components, where the directions of these projections should have the maximum statistical independence. Several metrics can be used to measure the independence of these projection directions, like mutual information and non-Gaussianity. LDA can be used to perform supervised dimensionality reduction. LDA projects the input data instances to a linear subspace consisting of the directions which maximize the separation between classes. In LDA, the dimensions of the output are necessarily less than the number of classes. Besides these approaches introduced, there also exist so many other dimensionality reduction approaches, like *canonical correlation analysis (CCA)* and *singular value decomposition (SVD)*, which will not be introduced here since they are out of the scope of this book. A comprehensive review of these dimensionality reduction methods is available in [51].

Dimensionality reduction approaches can effectively project the data instances into a lower-dimensional space. However, the physical meanings of the objective space can be very hard to interpret. Besides dimensionality reduction, another way to reduce the data dimension will be to select a subset of representative attributes from the original attribute set to represent the data instances, i.e., *feature selection* [15]. Among the original attributes, many of them can be either *redundant* or *irrelevant* with each other. Here, the *redundant attributes* denote the attributes sharing duplicated information with the other attributes in the data, while the *irrelevant attributes* represent the attributes which are not useful for the machine learning tasks actually. The physical meanings of these selected attributes will be still the same as the original ones. As illustrated in Fig. 2.6, existing feature selection approaches can be categorized into three groups:

- **Filter Approaches**: Before starting the learning and mining tasks, filter approaches will select the features in advance, which are independent of the learning and mining tasks.

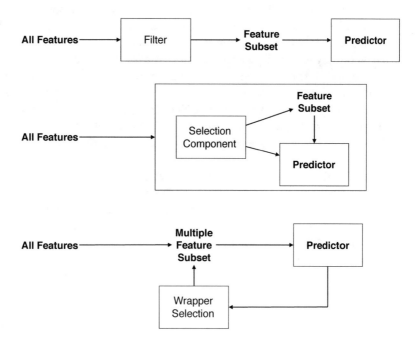

Fig. 2.6 Feature selection approaches (filter approach, embedded approach, and wrapper approach)

- **Embedded Approaches**: Some learning models have the ability to do feature selection as one component in the models themselves. Such feature selection approaches are named as the embedded approaches.
- **Wrapper Approaches**: Many other feature selection approaches use the learning model as a black box to find the best subset of attributes which are useful for the objective learning tasks. Such a feature selection approach involves model learning for many times and will not enumerate all the potential attribute subsets, which is named as the *wrapper approach*.

As to the feature subset search algorithms, there exist two classic methods, i.e., the *forward feature selection* and the *backward feature selection*. Forward feature selection approaches begin with an empty set and keep adding feature candidates into the subset, while backward feature selection approaches start with a full set and keep deleting features from the set. To determine which feature to add or delete, different strategies can be applied, like sequential selection and greedy selection. If the readers are interested in feature selection methods, you are suggested to refer to [15] for a more complete literature.

2.2.3.4 Data Transformation

In many cases, the input data set cannot meet the requirements of certain learning algorithms, and some basic data transformation operations will be needed. Traditional data transformation operations include *binarization*, *discretization*, and *normalization* [49].

- **Binarization**: For categorical attributes, we usually need to quantify them into binary representations before feeding them to many learning algorithms. Normally, there are many different categorical attribute binarization methods. For instance, given a categorical attribute with m potential values, we can quantify them into binary codes of length $\log_2^m$. For instance, for an apple sweetness attribute with sweetness degrees {Sour, Slightly Sour, Tasteless, Slight Sweet, Sweet}, it can be quantified into a code of length 3 (like Sour: 001; Slightly Sour: 010; Tasteless: 011; Slight Sweet: 100; Sweet: 101). However, such a quantification method will introduce many problems, as it will create some correlations among the attributes (like "Sour" and "Sweet' will be very similar in their code representations, i.e., "001" vs "101," which share two common digits). A more common way to quantify categorical attributes (of m different values) is to represent them with a code of length m instead. For instance, for the five different sweetness degrees, we use "00001" to represent "Sour," "00010" to represent "Slightly Sour," "00100" to represent "Tasteless," "01000" to represent "Slightly Sweet," and "10000" to represent "Sweet." Such a quantification approach will transform the categorical attributes into independent binary codes. The shortcoming of such a binarization lies in its code length: for the categorical attribute with lots of values, the code will be very long.
- **Discretization**: In many cases, we need to transform continuous attributes into discrete ones instead for easier classification or association analysis, and such a process is called the attribute discretization. For instance, given an attribute with continuous values in range [min, max], we want to discretize the attribute into n discrete values. An easy way to achieve such an objective will be to select $n - 1$ splitting points (e.g., $x_1, x_2, \ldots, x_{n-1}$) to divide the value range into n bins (i.e., [min, x_1], (x_1, x_2], $\ldots$, (x_{n-1}, max]), where the attribute values in each bin will be denoted as a specific discrete value. Depending on whether the supervision information is used in the splitting points selection or not, existing attribute discretization approaches can be divided into two categories: *supervised attribute discretization* and *unsupervised attribute discretization*. Conventional unsupervised attribute discretization approaches include *equal width* and *equal depth* based splitting points selection, which aims at dividing the data points into intervals of the same

Fig. 2.7 An example of attribute normalization

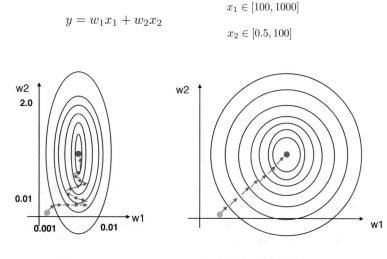

$$y = w_1 x_1 + w_2 x_2$$

$$x_1 \in [100, 1000]$$
$$x_2 \in [0.5, 100]$$

Without Attribute Normalization   **With Attribute Normalization**

interval length and the same data point numbers in each interval, respectively. Meanwhile, the supervised attribute discretization approaches will use the supervision information in splitting points selection, and a representative approach is called the *entropy-based* attribute discretization.

- **Normalization**: Attribute normalization is an operation used to standardize the range of independent variables or attributes of data. Since the range of values of raw data varies widely, in some machine learning algorithms, the objective functions can be extremely challenging to solve properly without normalization. The simplest attribute normalization method is Min-Max normalization. The general formula of the Min-Max rescaling approach is given as: $x_{new} = \frac{x - x_{min}}{x_{max} - x_{min}}$, where x denotes the value of an attribute, x_{min} and x_{max} denote the minimum and maximum values of the correspond attribute, respectively. Besides the Min-Max normalization approach, some other normalization approach includes the Mean-Std normalization approach, whose general formula can be represented as $x_{new} = \frac{x - \bar{x}}{\sigma}$, where $\bar{x}$ and σ represent the mean and standard deviation about an objective attribute.

To illustrate the motivations of attribute normalization, in Fig. 2.7, we show an example about solving the objective optimization function based on datasets with and without attribute normalization, respectively. Here, we aim at building a linear regression model $y = w_1 x_1 + w_2 x_2$, where $y \in [0, 1]$ and $x_1 \in [100, 1000]$, $x_2 \in [0.5, 100]$. To learn the model, the objective function will aim at identifying the optimal weight variables w_1 and w_2, which can minimize the model learning loss, i.e., the red dots at the center of the curve. As shown in the left plot of Fig. 2.7, without attribute normalization, the objective function curve is in an oval shape, where w_2 has a relatively wider feasible range compared with w_1. Searching for the optimal point based on such a curve with conventional optimization algorithms, e.g., gradient descent, will be extremely slow. However, with attribute normalization, the model variables w_1 and w_2 will have a similar feasible range and the objective function curve will be very close to a circular shape. The optimal point can be efficiently identified along the function gradient with a very small number of search rounds instead.

This section has covered a brief introductory description about data, data characteristics, and data transformations, which will be used for data analysis and processing in machine learning. Based on the processed data set, we will introduce different kinds of learning tasks in detail in the following sections, respectively.

2.3   Supervised Learning: Classification

Supervised learning tasks aim at discovering the relationships between the input attributes (also called features) and a target attribute (also called labels) of data instances, and the discovered relationship is represented as either the structures or the parameters of the supervised learning models. Supervised learning tasks are mainly based on the assumption that the input features and objective labels of both historical and future data instances are independent and identically distributed (i.i.d.). Based on such an assumption, the supervised learning model trained with the historical data can also be applied to the objective label prediction on the future data. Meanwhile, in many real-world supervised learning tasks, such an assumption can hardly be met and may be violated to a certain degree.

In supervised learning tasks, as illustrated in Fig. 2.8, the data set will be divided into three subsets: training set, validation set (optional), and testing set. For the data instances in the training set and the validation set, their labels are known, which can be used as the supervision information for learning the models. After a supervised learning model has been trained, it can be applied to the data instances in the testing set to infer their potential labels. Depending on the objective label types, the existing supervised learning tasks can be divided into *classification tasks* and *regression tasks*, respectively, where the label set of classification tasks is usually a pre-defined class set $\mathcal{Y}$, while that of regression tasks will be the real number domain $\mathbb{R}$ instead.

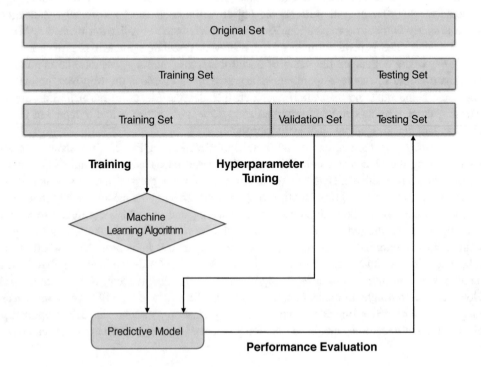

Fig. 2.8 Training set, validation set, and testing set split

For both classification and regression tasks, different machine learning models[1] have been proposed already, like classic classification models: *decision tree* [37] and *support vector machine* [9], and classic regression models: *linear regression* [55], *lasso* [50], and *ridge* [20]. In this section, we will mainly focus on introducing the classification learning task, including its learning settings and two classic classification models: *decision tree* and *support vector machine*. A brief introduction to the classic regression models will be provided in the next section.

2.3.1 Classification Learning Task and Settings

In classification problem settings, the data sets are divided into three disjoint subsets, which include a training set, a validation set, and a testing set. Generally, the data instances in the training set are used to train the models, i.e., to learn the model structure or parameters. The validation set is used for some hyperparameter selection and tuning. And the testing set is mainly used for evaluating the learning performance of the built models. Many different methods have been proposed to split the data set into these subsets, like *multiple random sampling* and *cross validation* [25].

- **Multiple Random Sampling**: Given a data set containing n data instances, *multiple random sampling* strategy will generate two separate subsets of instance for model training and validation purposes, where the remaining instances will be used for model testing only. In some cases, such a data set splitting method will encounter the unreliability problem, as the testing set can be too small to be representative for the overall data set. To overcome this problem, such a process will be performed n times, and in each time, different training and testing sets will be produced.
- **Cross Validation**: When the data set is very small, *cross validation* will be a common strategy for splitting the data set. There are different variants of cross validation, like *k-fold cross validation* and *leave-one-out cross validation*. As shown in Fig. 2.9, in the *k-fold cross validation*, the data set is divided into k equal sized subsets, where each subset can be picked as a testing set while the remaining $k - 1$ subsets are used as the training set (as well as the validation set). Such a process runs for k times, and in each time a different subset will be used as the testing set. The *leave-one-out cross validation* works in a similar way, which picks one single instance as the testing set and the remaining instances will be used as the training set in each round. For the data sets of a large scale, the *leave-one-out cross validation* approach will suffer from the high time cost problem. In the real-world model learning, the *n-fold cross validation* is used more frequently for data set splitting compared with the other strategies.

Formally, in the classification task, we can denote the feature domain and label domain as $\mathcal{X}$ and $\mathcal{Y}$, respectively. The objective of classification tasks is to build a model with the training set and the validation set (optional), i.e., $f : \mathcal{X} \to \mathcal{Y}$, to project the data instances from their feature representations to their labels. Formally, the data instances in the training set can be represented as a set of n feature-label pairs $\mathcal{T} = \{(\mathbf{x}_1, y_1), (\mathbf{x}_2, y_2), \ldots, (\mathbf{x}_n, y_n)\}$, where $\mathbf{x}_i \in \mathcal{X}$ denotes the feature vector of the ith data instance and $y_i \in \mathcal{Y}$ represents its corresponding label. The validation set $\mathcal{V}$ can be represented in a similar way with both features and known labels for the data instances. Meanwhile, the data instances in the testing set are different from those in the training and validation sets, which only have the feature representation only without any known labels. Formally, the testing set involving

---

[1]Machine learning models usually denote the well trained learning algorithms by some training data. In the sequel of this book, we will not differentiate the differences between machine learning models and machine learning algorithms by default.

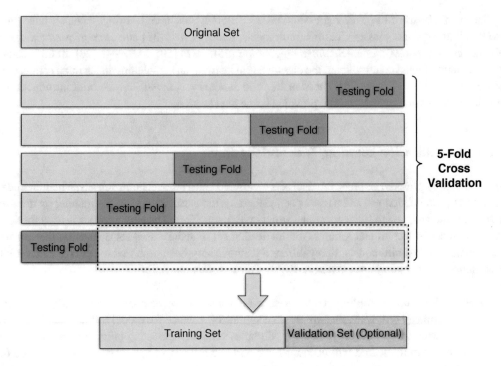

Fig. 2.9 An example of five-fold cross validation

m data instances can be represented as a set $\mathcal{S} = \{\mathbf{x}_{n+1}, \mathbf{x}_{n+2}, \ldots, \mathbf{x}_{n+m}\}$. Based on the well-trained models, we will be able to identify the labels of the data instances in the testing set.

Next, we will take the binary classification task as an example (i.e., $\mathcal{Y} = \{+1, -1\}$ or $\mathcal{Y} = \{+1, 0\}$) and introduce some classic classification algorithms, including *decision tree* [37] and *support vector machine* [9].

2.3.2 Decision Tree

In this section, we will introduce the classic decision tree classification algorithm [37], and will take the binary classification problem as an example. For binary classification tasks, the objective class domain involves two different pre-defined class labels $\{+1, -1\}$ (i.e., the positive class label and negative class label). The procedure of classifying data instances into positive and negative classes actually involves a series of decision-making about questions, like "whether this elephant has two trunk fingers?" or "is the elephant weight greater than 6 tons?" As indicated by the name, decision tree is such a kind of classification model, which is based on a series of decision-making procedure. Before achieving the final decision (i.e., data instances belonging to positive/negative classes), a group of pre-decisions will be made in advance. The final learned model can be represented as a tree structured diagram, where each internal node represents a "question" and each branch denotes a decision option. Decision tree is one of the most widely used machine learning algorithms for classification tasks due to its effectiveness, efficiency, and simplicity.

Fig. 2.10 An example of decision tree model built for the elephant data set

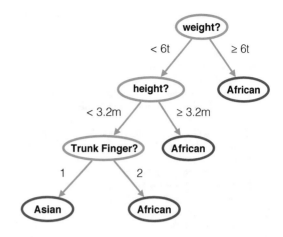

2.3.2.1 Algorithm Overview

To help illustrate the decision tree model more clearly, based on the elephant dataset shown in Table 2.1, we provide a trained decision tree model in Fig. 2.10.

Example 2.1 For the elephant classification example shown in Fig. 2.10, the attribute of interest is the "*Elephant Category*" attribute, which serves as the classification objective label and takes value from {Asian, African}. Based on the "*Weight*" attribute, the data instances can be divided into two subsets: (1) elephant (data instances) with "*Weight*" no less than 6 t (i.e., instance {1}), and (2) elephant (data instances) with "*Weight*" less than 6 t (i.e., {2, 3, 4, 5, 6}). For the data instances in the first group (i.e., weight ≥6 t), they all belong to the African elephant category; while the second group involves both Asian and African elephants. Meanwhile, based on the elephant height (i.e., height ≥3.2 m or height <3.2 m), we can further divide the second group into two sub-groups: {2} and {3, 4, 5, 6}, where instance 2 belongs to the African elephant category. The remaining elephants in {3, 4, 5, 6} can be precisely divided into {5} and {3, 4, 6} based on the trunk "finger" number attribute (i.e., 1 trunk finger or 2 trunk fingers), where the instance in the first group has the Asian elephant label and those in the second group all have the African elephant label.

From the built decision tree model shown in Fig. 2.10, we observe that there exist two types of nodes in the tree: (1) decision node (i.e., the non-leaf node in blue color), and (2) result node (i.e., the leaf node in red color). Each decision node represents an attribute test, which divides the data instances into two groups (i.e., two branches). In the ideal case, each result node represents a classified label, which is usually pure in terms of the label values. In other words, the training data instances classified into each result node should all have the same label. For instance, in the tree shown in Fig. 2.10, we can exactly determine the "*Elephant Category*" attribute of all the data instances in Table 2.1. In other words, the decision tree model in Fig. 2.10 is well trained and can be applied to precisely classify the elephant data instances in Table 2.1.

Decision tree applies the *divide-and-conquer* strategy in model learning, which partitions the data instances into different groups based on their attributes. The pseudo-code of decision tree learning is provided in Algorithm 1. From the pseudo-code, we observe that the training process of the decision tree model involves a recursive process. In each recursion, the returning conditions of the algorithm include:

1. All the data instances belong to the same class and no need for further division.
2. The attribute set is empty and the data instances cannot be further divided.

Algorithm 1 DecisionTree

Require: Training Set $\mathcal{T} = \{(\mathbf{x}_1, y_1), (\mathbf{x}_2, y_2), \cdots, (\mathbf{x}_n, y_n)\}$
     Attribute Set $\mathcal{A} = \{a_1, a_2, \cdots, a_d\}$.
Ensure: A Trained Decision Tree Model
1: generate a node N
2: **if** instances in $\mathcal{T}$ belong to the same class **then**
3:     mark node N as the result node; **Return**
4: **end if**
5: **if** $\mathcal{A} == \emptyset$ **OR** instances in $\mathcal{T}$ take the same values in attribute set $\mathcal{A}$ **then**
6:     mark node N as the result node, whose corresponding label is the majority class of instances in $\mathcal{T}$; **Return**
7: **end if**
8: select the optimal partition attribute a^* from $\mathcal{A}$
9: **for** all values a_v^* for attribute a^* **do**
10:     generate a branch node $N_{a_v^*}$ for node N
11:     select an instance subset $\mathcal{T}_{a_v^*} \subset \mathcal{T}$ taking value a_v^* for attribute a^*
12:     **if** $\mathcal{T}_{a_v^*} == \emptyset$ **then**
13:         mark branch node $N_{a_v^*}$ as the result node, whose corresponding label will be the majority class of instances in $\mathcal{T}$; **Continue**
14:     **else**
15:         $SubTree = $ DecisionTree$(\mathcal{T}_{a_v^*}, \mathcal{A} \setminus \{a^*\})$
16:         assign root node of $SubTree$ to the branch node $N_{a_v^*}$
17:     **end if**
18: **end for**
19: **Return** a decision tree model with N as the root

---

3. All the data instances take the same values in the provided attribute set, which cannot be further divided.

4. The provided data instance set is empty and cannot be further divided.

In addition, for these different cases, the operations to be performed in Algorithm 1 will be different.

- In case 1, the current node N will be marked as a result node, and its corresponding label will be the common class label of data instances in $\mathcal{T}$. The current node N will also be returned as a single-node decision tree to the upper level function call.
- In cases 2 and 3, the current node N is marked as the result node, and its corresponding label will be the majority class of instances in $\mathcal{T}$. The current node N will be returned as a single-node decision tree to the upper level function call.
- In case 4, the generated node $N_{a_v^*}$ is marked as a result node, and its corresponding label will be the majority class of instances in $\mathcal{T}$. The algorithm will continue to the next possible value of the selected optimal division attribute a^*.
- Otherwise, the algorithm will make a recursive call of the DecisionTree function, whose returned root node will be assigned to the branch node $N_{a_v^*}$.

2.3.2.2 Attribute Selection Metric: Information Gain

According to Algorithm 1, the structure of the built decision tree is heavily dependent on the selection of the optimal division node a^* in each recursive step, which will affect the performance of the model greatly. Generally, via a sequential attribute test from the decision tree root node to the result node, the data instances will be divided into several different groups and each decision tree result node corresponds to one data instance group. In general, we may want the data instances in each divided group to be as pure as possible. Here, the concept "pure" denotes that most/all the data instances in each group should have the same class label, and attributes that can introduce the maximum "*purity* *increase*" will be selected first by the model. Viewed in such a perspective, the quantification of "*purity*" and "*purity increase*" will be very important for the decision tree model building.

Given the current training set $\mathcal{T}$, depending on the data instance labels, the ratio of the data instances belong to the lth class can be denoted as p_l ($l \in \{1, 2, \ldots, |\mathcal{Y}|\}$). Meanwhile, based on a certain attribute $a_i \in \mathcal{A}$ in the attribute set, the data instances in $\mathcal{T}$ can be partitioned into k groups $\mathcal{T}_{a_{i,1}}, \mathcal{T}_{a_{i,2}}, \ldots, \mathcal{T}_{a_{i,k}}$, where $\{a_{i,1}, a_{i,2}, \ldots, a_{i,k}\}$ denotes the value set of attribute a_i. Therefore, the ratio of data instances taking value $a_{i,j}$ for attribute a_i in the current data instance can be represented as $\frac{|\mathcal{T}_{a_{i,j}}|}{|\mathcal{T}|}$.

Based on these notations, different "*purity*" and "*purity increase*" concept quantification metrics have been proposed for the optimal attribute selection already, like *Entropy* [47] and *Information Gain* [39, 47]. Formally, given a data set $\mathcal{T}$ together with the data instance class distribution ratios $\{p_l\}_{l \in \{1,2,\ldots,|\mathcal{Y}|\}}$, the "*purity*" of all the data instances can be represented with the *information entropy* concept as follows:

$$Entropy(\mathcal{T}) = -\sum_{l=1}^{|\mathcal{Y}|} p_l \log_2 p_l. \tag{2.1}$$

Meanwhile, after picking an attribute a_i, the data instances in $\mathcal{T}$ will be further divided into k subsets $\mathcal{T}_{a_{i,1}}, \mathcal{T}_{a_{i,2}}, \ldots, \mathcal{T}_{a_{i,k}}$. If the class label distribution in each of these k subsets is pure, i.e., the overall entropy is small, the selected attribute a_i will be a good choice. Formally, the *information gain* [39, 47] introduced by attribute a_i on data instance set $\mathcal{T}$ can be represented as

$$Gain(\mathcal{T}, a_i) = Entropy(\mathcal{T})$$

$$-\sum_{a_{i,j}} \frac{|\mathcal{T}_{a_{i,j}}|}{|\mathcal{T}|} Entropy(\mathcal{T}_{a_{i,j}}). \tag{2.2}$$

The optimal attribute in the current attribute set $\mathcal{A}$ that can lead to maximum *information gain* can be denoted as

$$a^* = \arg_{a_i \in \mathcal{A}} \max Gain(\mathcal{T}, a_i). \tag{2.3}$$

The famous ID3 Decision Tree algorithm [37] applies *information gain* as the division attribute selection metric. Besides *information gain*, many other measures can be used as the best attribute selection metrics as well, e.g., *information gain ratio* and *Gini index*, which will be introduced in the following part.

2.3.2.3 Other Attribute Selection Metrics

Actually, *information gain* is a biased attribute selection metric, which favors the attributes with more potential values to take and may have negative effects on the attribute selection. To overcome such a disadvantage, a new metric named *information gain ratio* [39] is proposed, which normalizes the *information gain* by the corresponding attributes' *intrinsic values*. Formally, based on the current data instance set $\mathcal{T}$, the *intrinsic value* of an attribute a_i can be represented as

$$IV(\mathcal{T}, a_i) = -\sum_{a_{i,j}} \frac{|\mathcal{T}_{a_{i,j}}|}{|\mathcal{T}|} \log \frac{|\mathcal{T}_{a_{i,j}}|}{|\mathcal{T}|}. \tag{2.4}$$

Based on the above notation, the *information gain ratio* introduced by attribute a_i can be formally represented as follows:

$$GainRatio(\mathcal{T}, a_i) = \frac{Gain(\mathcal{T}, a_i)}{IV(\mathcal{T}, a_i)}. \tag{2.5}$$

The optimal attribute to be selected in each round will be those which can introduce the maximum *information gain ratio* instead, and the selection criterion can be formally represented as

$$a^* = \arg_{a_i \in \mathcal{A}} \max GainRatio(\mathcal{T}, a_i). \tag{2.6}$$

However, *information gain ratio* is also shown to be a biased metric. Different from the *information gain*, the *information gain ratio* metric favors the attributes with less potential values instead. Therefore, the famous C4.5 Decision Tree algorithm [38] applies heuristics to pre-select the attribute candidates with *information gain* that is larger than the average, and then applies *information gain ratio* to select the optimal attribute.

Another regularly used division node selection metric is *Gini index* [39]. Formally, the *Gini* metric of a given data instance set $\mathcal{T}$ can be represented as

$$Gini(\mathcal{T}) = \sum_{i=1}^{|\mathcal{Y}|} \sum_{i'=1, i' \neq i}^{|\mathcal{Y}|} p_i p_{i'} = 1 - \sum_{i=1}^{|\mathcal{Y}|} p_i^2. \tag{2.7}$$

Here, p_i denotes the data instance ratio belonging to the ith class. In general, mixed data instance set with class labels evenly distributed will have larger *Gini* scores. Based on the *Gini* metric, we can define the *Gini index* [39] of the data set about attribute $a_i \in \mathcal{A}$ as

$$GiniIndex(\mathcal{T}, a_i) = \sum_{a_{i,j}} \frac{|\mathcal{T}_{a_{i,j}}|}{|\mathcal{T}|} Gini(\mathcal{T}_{a_{i,j}}). \tag{2.8}$$

The optimal selection of the division attribute will be that introducing the minimum *Gini index*, i.e.,

$$a^* = \arg_{a_i \in \mathcal{A}} \min GiniIndex(\mathcal{T}, a_i). \tag{2.9}$$

2.3.2.4 Other Issues
Besides the optimal attribute selection, there also exist many other issues that should be studied in the decision tree algorithm.

1. Overfitting and Branch Pruning *Overfitting* [17] is a common problem encountered in the learning tasks with supervision information. Model overfitting denotes the phenomenon that the model fits the training data "too good," which treats and captures some specific pattern in the training set as a common pattern in the whole data set, but will achieve a very bad performance when being applied to some unseen data instances. Formally, given the training data set, testing set, and two trained decision tree models f_1 and f_2, model f_1 is said to overfit the data set $\mathcal{T}$, if the other f_1 achieves higher accuracy on the training set than f_2, but performs much worse than f_2 on the unseen testing set. In the overfitting scenario, the built decision tree can be very deep with so many branches, which will classify all the instances in the training set perfectly without making any mistake, but can hardly be generalized to the unseen data instances.

To overcome such a problem, different techniques can be applied in the decision tree model building process, like *branch pruning* [38]. The *branch pruning* strategy applied in decision tree training process includes both *pre-pruning* and *post-pruning*, as introduced in [38]. For the *pre-pruning* strategy, in building the decision tree model, before generating a child node for the decision nodes, certain tests will be performed. If dividing the current decision node into several child nodes will not generalize the model to improve its performance, the current decision node will be marked as the result node. On the other hand, for the *post-pruning* strategy, after the original decision tree model has been built, the strategy will check all the decision nodes. If replacing the decision node as a result node will generalize the mode to improve the performance, the sub-tree rooted at the current decision node will be replaced by a result node instead.

Generally, *pre-pruning* strategy can be dangerous, as it is actually not clear what will happen if the tree is extended further without *pre-pruning*. Meanwhile, the *post-pruning* strategy is more useful, as it is based on the complete built decision tree model, and it is clear which branch of the tree is useful and which one is not. *Post-pruning* has been applied in many existing decision tree learning algorithms.

2. Missing Value In many cases, there can exist some missing values for the data instances in the data set, which is very common for data obtained from areas, like social media and medical care. Due to the personal privacy protection concerns, people may hide some personal important information or sensitive information in the questionnaire. The classic way to handle the missing values in data mining will be to fill in the entries with some special values, like "*Unknown.*" In addition, if the attribute takes discrete values, we can also fill in the missing entries with the most frequent value of that attribute; if the attribute is continuous, we can fill in the missing entries with mean value of the attribute.

Meanwhile, for the decision tree model, besides these common value filling techniques, some other methods can also be applied to handle the missing value problem. In the classic decision tree algorithm C4.5 [38], at a tree decision node regarding a certain attribute, it can distribute the training data instances with missing values for that attribute to each branch of the tree proportionally according to the distribution of the training instances. For example, let's take a as an attribute to be dealt with at the current decision tree, and assume $\mathcal{T}$ to be the current data instance set and $\mathcal{T}_{a_i} \subset \mathcal{T}$ to be a subset of data instances with value a_i for attribute a. For each data instances $\mathbf{x} \in \mathcal{T}$, a weight $w_{\mathbf{x}}$ will be assigned to $\mathbf{x}$. We propose to define the ratio of data instances with value a_i to be

$$r_{a_i} = \frac{\sum_{\mathbf{x} \in \mathcal{T}_{a_i}} w_{\mathbf{x}}}{\sum_{\mathbf{x} \in \mathcal{T}} w_{\mathbf{x}}}. \tag{2.10}$$

If data instance $\mathbf{x}$ has no value for attribute a, $\mathbf{x}$ will be assigned to all the child nodes of a, i.e., the branch corresponding to values $\{a_1, a_2, \ldots, a_k\}$. What's more, the weight of $\mathbf{x}$ for the branch corresponding to value a_i will be $r_{a_i} \cdot w_{\mathbf{x}}$.

3. Multi-Variable Decision Tree Traditional decision tree model tests one single attribute at each of the decision node once. If we take each attribute as a coordinate of a data instance, the decision boundaries outlined by the decision tree model will be parallel to the axes, which renders the classification results interpretable but the decision boundary can involve too many small segments to fit real decision boundaries.

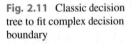 **Fig. 2.11** Classic decision tree to fit complex decision boundary

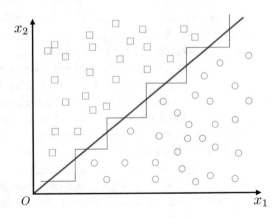

Example 2.2 For instance, as shown in Fig. 2.11, given the data instances with two features $\mathbf{x} = [x_1, x_2]^\top$ and one label y with decision boundary denoted by the red line, where instances lying at the top left are positive (i.e., the blue squares) while those at the bottom right are negative (i.e., the red circles) instead. To fit such a decision boundary, the decision tree model will involve a series of zig-zag segments (i.e., the black lines) as the decision boundary learned from the data set.

One way to resolve such a problem is to involve multiple variables in the tests of decision nodes. Formally, the decision test at each decision node can be represented as $\sum_i w_i a_i = t$, where w_i denotes the weight of the ith attribute a_i. Formally, all the involved variables, i.e., weight w_i for $\forall a_i$ together with the threshold value t, can be learned from the data. Formally, the decision tree algorithm involving multiple variables at each decision node test is named as the *multi-variable decision tree* [6]. Different from the classic *single-variable decision tree* algorithm, at each decision node, instead of selecting the optimal attribute for division, *multi-variable decision tree* aims at finding the optimal linear classifier, i.e., the optimal weights and threshold. Based on the *multi-variable decision tree*, the data instances shown in Fig. 2.11 can be classified by test function $x_1 - x_2 = 0$ perfectly, where instances with attributes $x_1 - x_2 < 0$ will be partitioned into one branch and those with attributes $x_1 - x_2 > 0$ will be partitioned into another branch instead.

2.3.3 Support Vector Machine

In this part, we will introduce another well-known classification model, which is named as the support vector machine (*SVM*) [9]. *SVM* is a supervised learning algorithm that can analyze data used for classification tasks. Given a set of training instances belonging to different classes, the *SVM* learning algorithm aims at building a model that assigns the data instances to one class or the other, making it a non-probabilistic binary linear classifier. In *SVM*, the data instances are represented as the points in a feature space, which can be divided into two classes by a clear hyperplane learned by the model, where the gap between the division boundary should be as wide as possible. New data instances will be mapped into the same space and classified into a class depending on which side of the decision boundary they fall in. In addition to performing linear classification, *SVM* can also efficiently perform a non-linear classification using the kernel trick [9], implicitly mapping their inputs into a high-dimensional feature space. In this part, we will introduce the *SVM* algorithm in detail, including its objective function, dual problem, and the kernel trick.

2.3.3.1 Algorithm Overview

Given a training set involving n labeled instances $\mathcal{T} = \{(\mathbf{x}_1, y_1), (\mathbf{x}_2, y_2), \ldots, (\mathbf{x}_n, y_n)\}$ belonging to binary classes $\{+1, -1\}$, the *SVM* model aims at identifying a hyperplane to separate the data instances. Formally, in the feature space, a division hyperplane can be represented as a linear function

$$\mathbf{w}^\top \mathbf{x} + b = 0, \tag{2.11}$$

where $\mathbf{w} = [w_1, w_2, \ldots, w_d]^\top$ denotes the scalars of each feature dimension. Scalar vector $\mathbf{w}$ also denotes the direction of the hyperplane, and b indicates the shift of the hyperplane from the original point.

For any data instance in the training set, e.g., $(\mathbf{x}_i, y_i) \in \mathcal{T}$, the distance from point $\mathbf{x}_i$ to the division hyperplane can be mathematically represented as

$$d_{\mathbf{x}_i} = \frac{|\mathbf{w}^\top \mathbf{x}_i + b|}{\|\mathbf{w}\|}. \tag{2.12}$$

For a good division hyperplane, it should be able to divide the data instances into different classes correctly. In other words, for the data instance $\mathbf{x}_i$ above the hyperplane, i.e., $\mathbf{w}^\top \mathbf{x}_i + b > 0$, it should belong to one class, e.g., the *positive class* with $y_i = +1$. Meanwhile, for the data instance below the hyperplane, it should belong to the other class, e.g., the *negative class* with $y_i = -1$. To select the optimal division hyperplane, *SVM* rescales the variables $\mathbf{w}$ and b to define two other hyperplanes (namely, the *positive* and *negative* hyperplanes) with the following equations:

$$H_+ : \mathbf{w}^\top \mathbf{x} + b = +1, \tag{2.13}$$

$$H_- : \mathbf{w}^\top \mathbf{x} + b = -1, \tag{2.14}$$

such that the following equations can hold:

$$\begin{cases} \mathbf{w}^\top \mathbf{x}_i + b \geq +1, & \forall (\mathbf{x}_i, y_i) \in \mathcal{T}, \text{if } y_i = +1, \\ \mathbf{w}^\top \mathbf{x}_i + b \leq +1, & \forall (\mathbf{x}_i, y_i) \in \mathcal{T}, \text{if } y_i = -1. \end{cases} \tag{2.15}$$

We know that the division hyperplane is parallel to the *positive* and *negative* hyperplanes defined above with equal distance between them. Meanwhile, for the data instances which actually lie in the *positive* and *negative* hyperplanes, they are called the *support vectors*. The distance between the positive and negative hyperplanes can be formally represented as

$$d_{+,-} = \frac{2}{\|\mathbf{w}\|}. \tag{2.16}$$

The *SVM* model aims at finding a classification hyperplane which can maximize the distance between the positive and negative hyperplanes $\left(\text{or minimize } \frac{\|\mathbf{w}\|^2}{2} \text{ equivalently}\right)$, while ensuring all the data instances are correctly classified. Formally, the objective function of the *SVM* model can be represented as

$$\min_{\mathbf{w}, b} \frac{\|\mathbf{w}\|^2}{2},$$

$$s.t. \ y_i(\mathbf{w}^\top \mathbf{x}_i + b) \geq 1, \forall (\mathbf{x}_i, y_i) \in \mathcal{T}. \tag{2.17}$$

The objective function is actually a convex quadratic programming problem involving $d + 1$ variables and n constraints, which can be solved with the existing convex programming toolkits. However, solving the problem can be very time consuming. A more efficient way to address the problem is to transform the problem to its dual problem and solve the dual problem instead, which will be talked about in the following part.

2.3.3.2 Dual Problem

The primal objective function of *SVM* is convex. By applying the Lagrangian multiplier method, the corresponding Lagrange function of the objective function can be represented as

$$L(\mathbf{w}, b, \boldsymbol{\alpha}) = \frac{1}{2} \|\mathbf{w}\|^2 + \sum_{i=1}^{n} \alpha_i (1 - y_i(\mathbf{w}^\top \mathbf{x}_i + b)), \qquad (2.18)$$

where $\boldsymbol{\alpha} = [\alpha_1, \alpha_2, \ldots, \alpha_n]^\top$ $(\alpha_i \geq 0)$ denotes the vector of multipliers.

By taking the partial derivatives of $L(\mathbf{w}, b, \boldsymbol{\alpha})$ with regard to $\mathbf{w}, b$ and making them equal to 0, we will have

$$\frac{\partial L(\mathbf{w}, b, \boldsymbol{\alpha})}{\partial \mathbf{w}} = \mathbf{w} - \sum_{i=1}^{n} \alpha_i y_i \mathbf{x}_i = 0 \quad \Rightarrow \quad \mathbf{w} = \sum_{i=1}^{n} \alpha_i y_i \mathbf{x}_i, \qquad (2.19)$$

$$\frac{\partial L(\mathbf{w}, b, \boldsymbol{\alpha})}{\partial b} = -\sum_{i=1}^{n} \alpha_i y_i = 0 \quad \Rightarrow \quad \sum_{i=1}^{n} \alpha_i y_i = 0. \qquad (2.20)$$

According to the representation, by replacing $\mathbf{w}$ and $\sum_{i=1}^{n} \alpha_i y_i$ with $\sum_{i=1}^{n} \alpha_i y_i \mathbf{x}_i$ and 0 in Eq. (2.18), respectively, we will have

$$L(\mathbf{w}, b, \boldsymbol{\alpha}) = \sum_{i=1}^{n} \alpha_i - \frac{1}{2} \sum_{i,j=1}^{n} \alpha_i \alpha_j y_i y_j \mathbf{x}_i^\top \mathbf{x}_j. \qquad (2.21)$$

With the above derivatives, we can achieve a new representation of $L(\mathbf{w}, b, \boldsymbol{\alpha})$ together with the constraints $\alpha_i \geq 0, \forall i \in \{1, 2, \ldots, n\}$ and $\sum_{i=1}^{n} \alpha_i y_i = 0$, which actually defines the dual problem of the objective function:

$$\max_{\boldsymbol{\alpha}} \sum_{i=1}^{n} \alpha_i - \frac{1}{2} \sum_{i,j=1}^{n} \alpha_i \alpha_j y_i y_j \mathbf{x}_i^\top \mathbf{x}_j$$

$$s.t. \sum_{i=1}^{n} \alpha_i y_i = 0,$$

$$\alpha_i \geq 0, \forall i \in \{1, 2, \ldots, n\}. \qquad (2.22)$$

Why should we introduce the dual problem? To answer this question, we need to introduce an important function property as follows. For function $L(\mathbf{w}, b, \boldsymbol{\alpha})$, we have

$$\max_{\boldsymbol{\alpha}} \min_{\mathbf{w}, b} L(\mathbf{w}, b, \boldsymbol{\alpha}) \leq \min_{\mathbf{w}, b} \max_{\boldsymbol{\alpha}} L(\mathbf{w}, b, \boldsymbol{\alpha}). \qquad (2.23)$$

The proof to the above property will be left as an exercise at the end of this chapter.

In other words, the optimal dual problem actually defines an upper bound of the optimal solution to the primal problem. Here, we know $\frac{1}{2}\|\mathbf{w}\|^2$ is convex and $\mathbf{w}^\top\mathbf{x} + b$ is affine. In Eq. (2.23), the equal sign $=$ can be achieved iff $\mathbf{w}$, b, and $\boldsymbol{\alpha}$ can meet the following KKT (Karush-Kuhn-Tucker) [9] conditions:

$$
\begin{cases}
\alpha_i \geq 0, \\
y_i(\mathbf{w}^\top\mathbf{x}_i + b) - 1 \geq 0, \\
\alpha_i(y_i(\mathbf{w}^\top\mathbf{x}_i + b) - 1) = 0.
\end{cases}
\tag{2.24}
$$

In other words, in such a case, by solving the dual problem, we will be able to achieve the primal problem as well.

According to the third KKT conditions, we observe that $\forall(\mathbf{x}_i, y_i) \in \mathcal{T}$, at least one of the following two equations must hold:

$$
\alpha_i = 0,
\tag{2.25}
$$

$$
y_i(\mathbf{w}^\top\mathbf{x}_i + b) = 1.
\tag{2.26}
$$

For the data instances with the corresponding $\alpha_i = 0$, they will not actually appear in the objective function of neither the primal nor the dual problem. In other words, these data instances are *"useless"* in determining the model variables (or the decision boundary). Meanwhile, for the data instances with $y_i(\mathbf{w}^\top\mathbf{x}_i + b) = 1$, i.e., those data points lying in the *positive* and *negative* hyperplanes H_+ and H_-, we need to learn their corresponding optimal multiplier scalar α_i, which will affect the final learned models. Therefore, the *SVM* model variables will be mainly determined by these *support vectors*, which is also the reason why the model is named as the *support vector machine*.

As we can observe, the dual problem is also a quadratic programming problem involving n variables and n constraints. However, in many cases, solving the dual problem is still much more efficiently than solving the primal, especially when $d \gg n$. Solving the dual objective function doesn't depend on the dimension of the feature vectors, which is very important for feature vectors of a large dimension or the application of *kernel tricks* when the data instances are not linearly separable.

Therefore, by deriving and addressing the dual problem, we will be able to understand that *support vectors* play an important role in determining the classification boundary of the *SVM* model. In addition, the *dual problem* also provides the opportunity for the efficient model learning especially with the *kernel tricks* to be introduced in the following part.

Some efficient learning algorithms have been proposed to solve the dual objective function, like SMO (sequential minimal optimization) algorithm [36], which will further reduce the learning cost. We will not introduce SMO here, since it is not the main focus of this textbook. Based on the learned optimal $\boldsymbol{\alpha}^*$, we can obtain the optimal $\mathbf{w}^*$ and b^* variables of the *SVM* model, which will be used to classify the future data instances (e.g., featured by vector $\mathbf{x}$) based on the sign of function $(\mathbf{w}^*)^\top\mathbf{x}+b^*$.

2.3.3.3 Kernel Trick

In the case when the data instances are not linearly separable, one effective way to handle the problem is to project the data instances to a high-dimensional feature space, in which the data instances will be linearly separable by a hyperplane, and such a technique is called the *kernel trick* (or *kernel method*) [9]. The *kernel trick* has been shown to be very effective when applied in *SVM*, which allows *SVM* to classify the data instances following very complicated distributions.

Fig. 2.12 An example of
kernel function in SVM

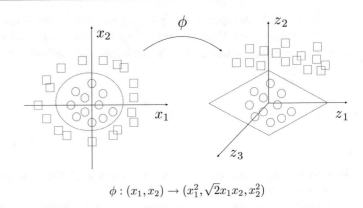

$$\phi : (x_1, x_2) \rightarrow (x_1^2, \sqrt{2}x_1 x_2, x_2^2)$$

Example 2.3 In Fig. 2.12, we show an example to illustrate the *kernel trick* with *SVM*. Given a group
of data instances in two different classes, where the *red circle* denotes the *positive class* and the *blue
square* denotes the *negative class*. According to the data instance distribution in the original feature
space (represented by two features x_1 and x_2), we observe that they cannot be linearly separated
by drawing a line actually. To divide these data instances, a non-linear division boundary will be
needed, i.e., the dashed line in black. Meanwhile, if we project the data instances from the two-
dimensional feature space to a three-dimensional feature space with the indicated kernel function
$\phi : (x_1, x_2) \rightarrow (x_1^2, \sqrt{2}x_1 x_2, x_2^2)$, we can observe that those data instances will become linearly
separable with a hyperplane in the new feature space.

Formally, let $\phi : \mathcal{R}^d \rightarrow \mathcal{R}^D$ be a mapping that projects the data instances from a d-dimensional
feature space to another D-dimensional feature space. In the new feature space, let's assume the data
instances can be linearly separated by a hyperplane in the *SVM* model. Formally, the hyperplane that
can separate the data instances can be represented as

$$\mathbf{w}^\top \phi(\mathbf{x}) + b = 0, \tag{2.27}$$

where $\mathbf{w} = [w_1, w_2, \ldots, w_D]^\top$ and b are the variables to be learned in the model.

According to Sect. 2.3.3.2, the primal and dual optimization objective functions of the *SVM* model
in the new feature space can be formally represented as

$$\min_{\mathbf{w}} \frac{1}{2} \|\mathbf{w}\|^2$$
$$s.t. \; y_i(\mathbf{w}^\top \phi(\mathbf{x}_i) + b) \geq 1, i = 1, 2, \ldots, n; \tag{2.28}$$

and

$$\max_{\boldsymbol{\alpha}} \sum_{i=1}^{n} \alpha_i - \frac{1}{2} \sum_{i,j=1}^{n} \alpha_i \alpha_j y_i y_j \phi(\mathbf{x}_i)^\top \phi(\mathbf{x}_j)$$
$$s.t. \; \sum_{i=1}^{n} \alpha_i y_i = 0,$$
$$\alpha_i \geq 0, \forall i \in \{1, 2, \ldots, n\}. \tag{2.29}$$

Here, $\phi(\mathbf{x}_i)^\top \phi(\mathbf{x}_j)$ denotes the inner projection of two projected feature vectors, calculation cost of which will be very expensive if the new feature space dimension D is very large. For simplicity, we introduce a new notation $\kappa(\mathbf{x}_i, \mathbf{x}_j)$ to represent $\phi(\mathbf{x}_i)^\top \phi(\mathbf{x}_j)$, and rewrite the above dual objective function as follows:

$$\max_{\boldsymbol{\alpha}} \sum_{i=1}^{n} \alpha_i - \frac{1}{2} \sum_{i,j=1}^{n} \alpha_i \alpha_j y_i y_j \kappa(\mathbf{x}_i, \mathbf{x}_j)$$

$$s.t. \sum_{i=1}^{n} \alpha_i y_i = 0,$$

$$\alpha_i \geq 0, \forall i \in \{1, 2, \ldots, n\}. \tag{2.30}$$

By solving the above function, we can obtain the optimal $\boldsymbol{\alpha}^*$, based on which the classifier function can be represented as

$$f(\mathbf{x}) = (\mathbf{w}^*)\phi(\mathbf{x}) + b^*$$

$$= \sum_{i=1}^{n} \alpha_i^* y_i \phi(\mathbf{x}_i)^\top \phi(\mathbf{x}) + b$$

$$= \sum_{i=1}^{n} \alpha_i^* y_i \kappa(\mathbf{x}_i, \mathbf{x}) + b, \tag{2.31}$$

where $\mathbf{w}^* = \sum_{i=1}^{n} \alpha_i^* y_i \phi(\mathbf{x}_i)^\top$ according to the derivatives in Eq. (2.19).

We can observe that in both the training and testing processes, we don't really need the concrete representations of the projected feature vectors $\{\phi(\mathbf{x}_i)\}_{\mathbf{x}_i}$ but a frequent calculation of $\kappa(\cdot, \cdot)$ will be needed instead. The representation of $\kappa(\cdot, \cdot)$ is determined by the definition of the projection function $\phi(\cdot)$. Formally, the function $\kappa(\cdot, \cdot)$ is defined as the *kernel function* in *SVM* (it has also been widely applied in many other learning algorithms). If the calculation cost of $\kappa(\cdot, \cdot)$ is lower than that of $\phi(\mathbf{x}_i)^\top \phi(\mathbf{x}_j)$, based on the *kernel function*, the overall learning cost of non-linear *SVM* will be reduced greatly.

Example 2.4 Let $\phi([x_1, x_2]^\top) = [x_1^2, \sqrt{2}x_1x_2, x_2^2]^\top$ be a function which projects the data instances from a two-dimensional feature space to a three-dimensional feature space. Let $\mathbf{x} = [x_1, x_2]^\top$ and $\mathbf{z} = [z_1, z_2]^\top$ be two feature vectors, we can represent the inner product of $\phi(\mathbf{x})$ and $\phi(\mathbf{z})$ as

$$\phi(\mathbf{x})^\top \phi(\mathbf{z}) = \left[x_1^2, \sqrt{2}x_1x_2, x_2^2\right]\left[z_1^2, \sqrt{2}z_1z_2, z_2^2\right]^\top$$

$$= x_1^2 z_1^2 + 2x_1x_2z_1z_2 + x_2^2z_2^2$$

$$= (x_1z_1 + x_2z_2)^2$$

$$= (\mathbf{x}^\top \mathbf{z})^2. \tag{2.32}$$

Computing the inner product with the kernel function $\kappa(\mathbf{x}, \mathbf{z}) = (\mathbf{x}^\top \mathbf{z})^2$ involves an inner product operation in a two-dimensional feature space (i.e., $\mathbf{x}^\top \mathbf{z} = [x_1, x_2][z_1, z_2]^\top$) and a real-value square operation (i.e., $(\mathbf{x}^\top \mathbf{z})^2$), whose cost is lower than that introduced in feature vector projection and the inner product operation in the 3-dimension space with equation, i.e., $\phi(\mathbf{x})$, $\phi(\mathbf{z})$ and $\phi(\mathbf{x})^\top \phi(\mathbf{z}) = [x_1^2, x_2^2, \sqrt{2}x_1 x_2][z_1^2, z_2^2, \sqrt{2}z_1 z_2]^\top$.

The advantages of applying the kernel function in training non-linear SVM will be more significant in the case where $d \ll D$. Normally, instead of defining the projection function $\phi(\cdot)$, we can define the kernel function $\kappa(\cdot, \cdot)$ directly. Some frequently used kernel functions in *SVM* are listed as follows:

- **Polynomial Kernel**: $\kappa(\mathbf{x}, \mathbf{z}) = (\mathbf{x}^\top \mathbf{z} + \theta)^d$ $(d \geq 1)$.
- **Gaussian RBF Kernel**: $\kappa(\mathbf{x}, \mathbf{z}) = \exp\left(-\frac{\|\mathbf{x} - \mathbf{z}\|^2}{2\sigma^2}\right)$ $(\sigma > 0)$.
- **Laplacian Kernel**: $\kappa(\mathbf{x}, \mathbf{z}) = \exp\left(-\frac{\|\mathbf{x} - \mathbf{z}\|}{2\sigma}\right)$ $(\sigma > 0)$.
- **Sigmoid Kernel**: $\kappa(\mathbf{x}, \mathbf{z}) = \tanh(\beta \mathbf{x}^\top \mathbf{z} + \theta)$ $(\beta > 0, \theta < 0)$.

Besides these aforementioned functions, there also exist many other *kernel functions* used in either *SVM* or the other learning models, which will not be introduced here.

2.4   Supervised Learning: Regression

Besides the classification problem, another important category of supervised learning tasks is *regression*. In this section, we will introduce the *regression* learning task, as well as three well-known regression models, i.e., *linear regression* [55], *Lasso*, [50] and *Ridge* [20], respectively.

2.4.1   Regression Learning Task

Regression differs from *classification* tasks in the domain of labels. Instead of inferring the pre-defined classes that the data instances belong to, *regression* tasks aim at predicting some real-value attributes for the data instances, like the box office of movies, price of stocks, and population of countries. Formally, given the training data $\mathcal{T} = \{(\mathbf{x}_1, y_1), (\mathbf{x}_2, y_2), \ldots, (\mathbf{x}_n, y_n)\}$, where $\mathbf{x}_i = [x_{i,1}, x_{i,1}, \ldots, x_{i,d}]^\top \in \mathbb{R}^d$ and $y_i \in \mathbb{R}$, *regression* tasks aim at building a model that can predict the real-value labels y_i based on the feature vector $\mathbf{x}_i$ representation of the data instances. In this section, we will take the regression models which combine the features linearly as an example, whose predicted label can be represented as

$$\hat{y}_i = w_0 + w_1 x_{i,1} + w_2 x_{i,2} + \cdots + w_d x_{i,d}, \tag{2.33}$$

where $\mathbf{w} = [w_0, w_1, w_2, \ldots, w_d]^\top$ denotes the weight and bias variables of the regression model.

Generally, by minimizing the difference between the predicted labels and the ground truth labels, we can learn the parameter $\mathbf{w}$ in the model. Depending on the loss objective function representations, three different regression models, i.e., *linear regression* [55], *Lasso* [50], and *Ridge* [20], will be introduced as follows.

2.4.2   Linear Regression

Given the training set $\mathcal{T} = \{(\mathbf{x}_1, y_1), (\mathbf{x}_2, y_2), \ldots, (\mathbf{x}_n, y_n)\}$, the linear regression model adopts the mean square error as the loss function, which computes the average square error between the prediction labels and the ground-truth labels of the training data instances. Formally, the optimal parameter $\mathbf{w}^*$ can be represented as

$$\mathbf{w}^* = \arg_{\mathbf{w}} \min E(\hat{\mathbf{y}}, \mathbf{y})$$

$$= \arg_{\mathbf{w}} \min \frac{1}{n} \sum_{i=1}^{n} (\hat{y}_i - y_i)^2$$

$$= \arg_{\mathbf{w}} \min \frac{1}{n} \|\mathbf{X}\mathbf{w} - \mathbf{y}\|_2^2, \tag{2.34}$$

where $\mathbf{X} = \left[\bar{\mathbf{x}}_1^\top, \bar{\mathbf{x}}_2^\top, \ldots, \bar{\mathbf{x}}_n^\top\right] \in \mathbb{R}^{n \times (d+1)}$ and $\mathbf{y} = [y_1, y_2, \ldots, y_n]^\top$ denote the feature matrix and label vectors of the training instances. In the representation, vector $\bar{\mathbf{x}}_i = [x_{i,1}, x_{i,1}, \ldots, x_{i,d}, 1]^\top \in \mathbb{R}^{(d+1)}$, where a dummy feature $+1$ is appended to the feature vectors so as to incorporate the bias term w_0 also as a feature weight in model learning.

The mean square error used in the *linear regression* model has very a good mathematical property. Meanwhile, the method of computing the minimum loss based on the mean square error is also called the *least square method*. In linear regression, the *least square method* aims at finding a hyperplane, the sum of distance between which and the training instances can be minimized. The above objective function can be resolved by making derivative of the error term regarding the parameter $\mathbf{w}$ equal to 0, and we can have

$$\frac{\partial E(\hat{\mathbf{y}}, \mathbf{y})}{\partial \mathbf{w}} = 2\mathbf{X}^\top (\mathbf{X}\mathbf{w} - \mathbf{y}) = 0$$

$$\Rightarrow \mathbf{X}^\top \mathbf{X} \mathbf{w} = \mathbf{X}^\top \mathbf{y}, \tag{2.35}$$

Here, to obtain the closed-form optimal solution of $\mathbf{w}^*$, it needs to involve matrix inverse operation of $\mathbf{X}^\top \mathbf{X}$. Depending on whether $\mathbf{X}^\top \mathbf{X}$ is invertible or not, there will be different solutions to the above objective function:

- If matrix $\mathbf{X}^\top \mathbf{X}$ is of full rank or $\mathbf{X}^\top \mathbf{X}$ is positive definite, we have

$$\mathbf{w}^* = (\mathbf{X}^\top \mathbf{X})^{-1} \mathbf{X}^\top \mathbf{y}. \tag{2.36}$$

- However, in the case that $\mathbf{X}^\top \mathbf{X}$ is not full rank, there will be multiple solutions to the above objective function. For the linear regression model, all these solutions will lead to the minimum loss. Meanwhile, in some cases, there will be some preference about certain types of parameter $\mathbf{w}$, which can be represented as the regularization term added to the objective function, like the *Lasso* and *Ridge* regression models to be introduced as follows.

2.4.3 Lasso

Lasso [50] is also a linear model that estimates sparse coefficients. It is useful in some contexts due to its tendency to prefer solutions with fewer parameter values, effectively reducing the number of variables upon which the given solution is dependent. For this reason, *Lasso* and its variants are fundamental to the field of compressed sensing. Under certain conditions, it can recover the exact set of non-zero weights. Mathematically, the objective function of *Lasso* can be represented as

$$\arg_{\mathbf{w}} \min \frac{1}{2n} \|\mathbf{Xw} - \mathbf{y}\|_2^2 + \alpha \cdot \|\mathbf{w}\|_1 , \tag{2.37}$$

where the coefficient vector $\mathbf{w}$ is regularized by its L_1-norm, i.e., $\|\mathbf{w}\|_1$.

Considering that the L_1-norm regularizer $\|\mathbf{w}\|_1$ is not differentiable, and no closed-form solution exists for the above objective function. Meanwhile, a wide variety of techniques from convex analysis and optimization theory have been developed to extremize such functions, which include the subgradient methods [10], least-angle regression (LARS) [11], and proximal gradient methods [32]. As to the choice of scalar α, it can be selected based on a validation set of the data instances.

In addition to the regression models in the linear form, there also exist a large number of regression models in the high-order polynomial representation, as well as more complicated representations. For more information about the other regression models, their learning approaches and application scenarios, please refer to [41] for more detailed information.

2.4.4 Ridge

Ridge [20] addresses some of the problems of ordinary least squares (OLS) by imposing a penalty on the size of coefficient $\mathbf{w}$. The ridge coefficients minimize a penalized residual sum of squares, i.e.,

$$\arg_{\mathbf{w}} \min \|\mathbf{Xw} - \mathbf{y}\|_2^2 + \alpha \cdot \|\mathbf{w}\|_2^2 , \tag{2.38}$$

where $\alpha > 0$ denotes the complexity parameter used to control the size shrinkage of coefficient vector $\mathbf{w}$.

Generally, a larger α will lead to a greater shrinkage of coefficient $\mathbf{w}$ and thus the coefficient will be more robust to collinearity. Learning of the optimal coefficient $\mathbf{w}^*$ of the *ridge regression* model is the same as the learning process of the *linear regression* model, and the optimal $\mathbf{w}^*$ can be represented as

$$\mathbf{w}^* = (\mathbf{X}^\top \mathbf{X} + \alpha \mathbf{I})^{-1} \mathbf{X}^\top \mathbf{y} . \tag{2.39}$$

Here, we observe that the optimal solution depends on the choice of the scalar α a lot. As $\alpha \to 0$, the optimal solution $\mathbf{w}^*$ will be equal to the solution to the *linear regression* model, while as $\alpha \to \infty$, $\mathbf{w}^* \to \mathbf{0}$. In the real-world practice, the optimal scalar α can usually be selected based on the validation set partitioned from cross validation.

2.5   Unsupervised Learning: Clustering

Different from the supervised learning tasks, in many cases, no supervision information is available to guide the model learning and the data instances available have no specific object labels at all. The tasks of learning some inner rules and patterns from such unlabeled data instances are formally called the unsupervised learning tasks, which actually provide some basic information for further data analysis. Unsupervised learning involves very diverse learning tasks, among which *clustering* can be the main research focus and has very broad applications. In this section, we will introduce the *clustering learning tasks* and cover three well-used clustering algorithms, including *K-Means* [16], *DBSCAN* [12], and *Mixture-of-Gaussian* [40].

2.5.1 Clustering Task

Clustering tasks aim at partitioning the data instances into different groups, where instances in each cluster are more similar to each other compared with those from other clusters. For instance, movies can be divided into different genres (e.g., comedy vs tragedy vs Sci-fi), countries can be partitioned into different categories (e.g., developing countries vs developed countries), a basket of fruits can be grouped into different types (e.g., apple vs orange vs banana). Generally, in real-world applications, clustering can be used as a single data mining procedure or a data pre-processing step to divide the data instances into different categories.

Formally, in the clustering tasks, the unlabeled data instances can be represented as set $\mathcal{D} = \{\mathbf{x}_1, \mathbf{x}_2, \ldots, \mathbf{x}_n\}$, where each data instance can be represented as a d-dimensional feature vector $\mathbf{x}_i = [x_{i,1}, x_{i,2}, \ldots, x_{i,d}]^\top$. Clustering tasks aim at partitioning the data instances into k disjoint groups $\mathcal{C} = \{\mathcal{C}_1, \mathcal{C}_2, \ldots, \mathcal{C}_k\}$, where $\mathcal{C}_i \subset \mathcal{D}, \forall i \in \{1, 2, \ldots, k\}$, while $\mathcal{C}_i \cap \mathcal{C}_j = \emptyset, \forall i, j \in \{1, 2, \ldots, k\}, i \neq j$, and $\bigcup_i \mathcal{C}_i = \mathcal{D}$. Generally, we can use the cluster index $y_i \in \{1, 2, \ldots, k\}$ to indicate the cluster ID that instance $\mathbf{x}_i$ belongs to, and the cluster labels of all the data instances in $\mathcal{D}$ can be represented as a vector $\mathbf{y} = [y_1, y_2, \ldots, y_n]^\top$.

In clustering tasks, a formal definition of the distance among the data instances is required, and many existing clustering algorithms heavily rely on the data instance distance definition. Given a function $dist(\cdot, \cdot) : \mathcal{D} \times \mathcal{D} \to \mathbb{R}$, it can be defined as a distance measure iff it can meet the following properties:

- **Non-negative**: $dist(\mathbf{x}_i, \mathbf{x}_j) \geq 0$,
- **Identity**: $dist(\mathbf{x}_i, \mathbf{x}_j) = 0$ iff. $\mathbf{x}_i = \mathbf{x}_j$,
- **Symmetric**: $dist(\mathbf{x}_i, \mathbf{x}_j) = dist(\mathbf{x}_j, \mathbf{x}_i)$,
- **Triangular Inequality**: $dist(\mathbf{x}_i, \mathbf{x}_j) \leq dist(\mathbf{x}_i, \mathbf{x}_k) + dist(\mathbf{x}_k, \mathbf{x}_j)$.

Given two data instances featured by vectors $\mathbf{x}_i = [x_{i,1}, x_{i,2}, \ldots, x_{i,d}]^\top$ and $\mathbf{x}_j = [x_{j,1}, x_{j,2}, \ldots, x_{j,d}]^\top$, a frequently used distance measure is the *Minkowski distance*:

$$dist(\mathbf{x}_i, \mathbf{x}_j) = \left(\sum_{k=1}^{d} |x_{i,k} - x_{j,k}|^p \right)^{\frac{1}{p}}. \tag{2.40}$$

The *Minkowski distance* is a general distance representation, and it covers various well-known distance measures depending on the selection of value p:

- **Manhattan Distance**: in the case that $p = 1$, we have $dist_m(\mathbf{x}_i, \mathbf{x}_j) = \sum_{k=1}^{d} |x_{i,k} - x_{j,k}|$.

- **Euclidean Distance**: in the case that $p = 2$, we have $dist_e(\mathbf{x}_i, \mathbf{x}_j) = \left(\sum_{k=1}^{d} |x_{i,k} - x_{j,k}|^2 \right)^{\frac{1}{2}}$.

- **Chebyshev Distance**: in the case that $p \to \infty$, we have $dist_c(\mathbf{x}_i, \mathbf{x}_j) = \max(\{|x_{i,k} - x_{j,k}|\}_{k=1}^{d})$.

Besides the *Minkowski distance*, there also exist many other distance measures, like VDM (value difference metric), which works well for the unordered attributes. In many cases, different attributes actually play a different role in the distance calculation and should be weighted differently. Some distance metrics assign different weights to the attributes to differentiate them. For instance, based on the *Minkowski distance*, we can define the weighted *Minkowski distance* to be

$$dist_w = \left(\sum_{k=1}^{d} w_k \cdot |x_{i,k} - x_{j,k}|^p \right)^{\frac{1}{p}}, \tag{2.41}$$

where $w_i \geq 0, \forall i \in \{1, 2, \ldots, d\}$ and $\sum_{k=1}^{d} w_i = 1$.

Generally, clustering tasks aim at grouping similar data instances (i.e., with a smaller distance) into the same cluster, while different data instances (i.e., with a larger distance) into different clusters. Meanwhile, depending on the data instance-cluster belonging relationships, the clustering tasks can be categorized into two types: (1) *hard clustering*: each data instance belongs to exact one cluster, and (2) *soft clustering*: data instances can belong to multiple clusters with certain confidence scores. In the following part of this section, we will introduce two *hard clustering* algorithms: *K-Means* [16] and *DBSCAN* [12], and one *soft clustering* algorithm: *Mixture-of-Gaussian* [40].

2.5.2 K-Means

In the unlabeled data set, the data instances usually belong to different groups. In each of the groups, there can exist a representative data instance which can outline the characteristics of the group internal data instances. Therefore, after identifying the prototype data instances, the clusters can be discovered easily. Meanwhile, how to define the cluster prototypes is an open problem, and many different algorithms have been proposed already, among which *K-Means* [16] is one of the most well-known clustering algorithms.

Let $\mathcal{D} = \{\mathbf{x}_1, \mathbf{x}_2, \ldots, \mathbf{x}_n\}$ be the set of unlabeled data instances, and *K-Means* algorithm aims at partitioning $\mathcal{D}$ into k clusters $\mathcal{C} = \{\mathcal{C}_1, \mathcal{C}_2, \ldots, \mathcal{C}_k\}$. For the data instances belonging to each cluster, the cluster prototype data instance is defined as the center of the cluster in *K-Means*, i.e., the mean of the data instance vectors. For instance, for cluster $\mathcal{C}_i$, its center can be represented as follows formally:

$$\boldsymbol{\mu}_i = \frac{1}{|\mathcal{C}_i|} \sum_{\mathbf{x} \in \mathcal{C}_i} \mathbf{x}, \forall \mathcal{C}_i \in \mathcal{C}. \tag{2.42}$$

There can exist different ways to partition the data instances, to select the optimal one from which a clear definition of clustering quality will be required. In *K-Means*, the quality of the clustering result $\mathcal{C} = \{\mathcal{C}_1, \mathcal{C}_2, \ldots, \mathcal{C}_k\}$ can be measured by the introduced square of the Euclidean distance between the data instances and the prototypes, i.e.,

$$E(\mathcal{C}) = \sum_{i=1}^{k} \sum_{\mathbf{x} \in \mathcal{C}_i} \|\mathbf{x} - \boldsymbol{\mu}_i\|_2^2. \tag{2.43}$$

Literally, the clustering result which can minimize the above loss term will bring about a very good partition about the data instances. However, identifying the optimal prototypes and the cluster partition by minimizing the above square loss is a very challenging problem, which is actually NP-hard. To address such a challenge, the *K-Means* algorithm adopts a greedy strategy to find the prototypes and cluster division iteratively. The pseudo-code of the *K-Means* algorithm is available in Algorithm 2. As shown in the algorithm, in the first step, *K-Means* algorithm first randomly picks k data instances

---

Algorithm 2 KMeans

Require: Data Set $\mathcal{D} = \{\mathbf{x}_1, \mathbf{x}_2, \cdots, \mathbf{x}_n\}$;
    Cluster number k.
Ensure: Clusters $\mathcal{C} = \{\mathcal{C}_1, \mathcal{C}_2, \cdots, \mathcal{C}_k\}$
 1: Randomly pick k data instances from $\mathcal{D}$ as the prototype data instances $\{\boldsymbol{\mu}_1, \boldsymbol{\mu}_2, \cdots, \boldsymbol{\mu}_k\}$
 2: Stable-tag = $False$
 3: **while** Stable-tag == $False$ **do**
 4:    Let $\mathcal{C}_i = \emptyset, \forall i = 1, 2, \cdots, k$
 5:    **for** data instance $\mathbf{x}_i \in \mathcal{D}$ **do**
 6:        $\lambda = \arg_{j \in \{1,2,\cdots,k\}} \min dist(\mathbf{x}_i - \boldsymbol{\mu}_j)$
 7:        $\mathcal{C}_\lambda = \mathcal{C}_\lambda \cup \{\mathbf{x}_i\}$
 8:    **end for**
 9: **end while**
10: **for** $j \in \{1, 2, \cdots, k\}$ **do**
11:    compute the new prototype data instance $\boldsymbol{\mu}'_j = \frac{1}{|\mathcal{C}_j|} \sum_{\mathbf{x} \in \mathcal{C}_j} \mathbf{x}$
12:    **if** $\{\boldsymbol{\mu}_1, \boldsymbol{\mu}_2, \cdots, \boldsymbol{\mu}_k\} == \{\boldsymbol{\mu}'_1, \boldsymbol{\mu}'_2, \cdots, \boldsymbol{\mu}'_k\}$ **then**
13:        Stable-tag = $True$
14:    **else**
15:        $\{\boldsymbol{\mu}_1, \boldsymbol{\mu}_2, \cdots, \boldsymbol{\mu}_k\} = \{\boldsymbol{\mu}'_1, \boldsymbol{\mu}'_2, \cdots, \boldsymbol{\mu}'_k\}$
16:    **end if**
17: **end for**
18: **Return** $\mathcal{C} = \{\mathcal{C}_1, \mathcal{C}_2, \cdots, \mathcal{C}_k\}$

---

from $\mathcal{D}$ as the prototypes, and the data instances will be assigned to the clusters centered by these prototypes. For each data instance $\mathbf{x}_i \in \mathcal{D}$, *K-Means* selects the prototype $\boldsymbol{\mu}_j$ closest to $\mathbf{x}_i$, and adds $\mathbf{x}_i$ to cluster $\mathcal{C}_j$, where the distance measure $dist(\mathbf{x}_i - \boldsymbol{\mu}_j)$ can be the Euclidean distance between $\mathbf{x}_i$ and $\boldsymbol{\mu}_j$ (or the other distance measures we define before). Based on the obtained clustering result, *K-Means* re-computes the prototype centers of each newly generated clusters, which will be treated as the new prototypes. If in the current round, none of the prototypes changes, *K-Means* will return the current data instance partition as the final clustering result.

Example 2.5 In Fig. 2.13, we use an example to further illustrate the *K-Means* algorithm. For instance, given a group of data instance as shown in Fig. 2.13a, *K-Means* first randomly selects three points as the centers from the space as shown in Fig. 2.13b. All the data instances will be assigned to their nearest centers, and they will form three initial clusters (in three different colors) as illustrated in Fig. 2.13c. Based on the initial clustering results, *K-Means* recomputes the centers as the average of the cluster internal data instances in Fig. 2.13d, and further partitions the data set with the new centers in Fig. 2.13e. As the results converge and there exist no changes for partition, the *K-Means* algorithm will stop and output the final data partition results.

The *K-Means* algorithm is very powerful in handling data sets from various areas, but may also suffer from many problems. According to the algorithm descriptions and the example, we observe that the performance of *K-Means* is quite sensitive to the initial selection of prototypes, especially in the case when the number of data instance is not so large. Furthermore, in *K-Means*, the hyperparameter k needs to be selected beforehand, which is actually very challenging to infer from the data without knowing the distributions of the data. It will also introduce lots of parameter tuning works for *K-Means*. Next, we will introduce a density-based clustering algorithm, named *DBSCAN*, which doesn't need the cluster number parameter k as the input.

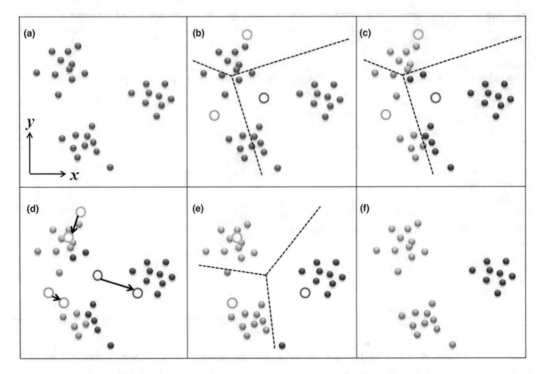

Fig. 2.13 An example of the K-Means algorithm in partitioning data instances ((**a**) input data instances; (**b**)–(**f**) steps of K-Means in clustering the data input)

2.5.3   DBSCAN

DBSCAN [12] is short for *density-based spatial clustering of applications with noise*, and it is a density-based clustering algorithm, which assumes that the cluster number and cluster structure can be revealed by the data instance distribution density. Generally, density-based clustering algorithms are based on the connectivity relationships among data instances, via which data instances can keep expanding until achieving the final clustering results. To characterize the connectivity relationships among data instances, *DBSCAN* introduces a concept called *neighborhood* parameterized by (ϵ, η). Given a data set $\mathcal{D} = \{\mathbf{x}_1, \mathbf{x}_2, \ldots, \mathbf{x}_n\}$ involving n data instances, several important concepts used in *DBSCAN* are defined as follows:

- ϵ-**Neighborhood**: For data instance $\mathbf{x}_i \in \mathcal{D}$, its ϵ-neighborhood represents a subset of data instances from $\mathcal{D}$, with a distance shorter than ϵ from $\mathbf{x}_i$, i.e., $N_\epsilon(\mathbf{x}_i) = \{\mathbf{x}_j \in \mathcal{D} | dist(\mathbf{x}_i, \mathbf{x}_j) \leq \epsilon\}$ (function $dist(\cdot, \cdot)$ denotes the Euclidean distance by default).
- **Core Objects**: Based on the data instances and their corresponding ϵ-neighborhoods, the core objects denote the data instances whose ϵ-neighborhood contains at least η data instances. In other words, data instance $\mathbf{x}_i$ is a core object iff $|N_\epsilon(\mathbf{x}_i)| \geq \eta$.
- **Directly Density-Reachable**: If data instance $\mathbf{x}_j$ lies in the ϵ-neighborhood of $\mathbf{x}_i$, $\mathbf{x}_j$ is said to be *directly density-reachable* from $\mathbf{x}_i$, i.e., $\mathbf{x}_i \to \mathbf{x}_j$.
- **Density-Reachable**: Given two data instances $\mathbf{x}_i$ and $\mathbf{x}_j$, $\mathbf{x}_j$ is said to be *density-reachable* from $\mathbf{x}_i$ iff there exists a path $\mathbf{v}_1 \to \mathbf{v}_2 \to \cdots \to \mathbf{v}_{k-1} \to \mathbf{v}_k$ connecting $\mathbf{x}_i$ with $\mathbf{x}_j$, where $\mathbf{v}_l \in \mathcal{D}$ and $\mathbf{v}_1 = \mathbf{x}_i$, $\mathbf{v}_k = \mathbf{x}_j$. Sequence $\mathbf{v}_l \to \mathbf{v}_{l+1}$ represents that $\mathbf{v}_{l+1}$ is *directly density-reachable* from $\mathbf{v}_l$.

Algorithm 3 DBSCAN

Require: Data Set $\mathcal{D} = \{\mathbf{x}_1, \mathbf{x}_2, \cdots, \mathbf{x}_n\}$;
  Neighborhood parameters (ϵ, η).
Ensure: Clusters $\mathcal{C} = \{\mathcal{C}_1, \mathcal{C}_2, \cdots, \mathcal{C}_k\}$
 1: Initialize core object set $\Omega = \emptyset$
 2: **for** $\mathbf{x}_i \in \mathcal{D}$ **do**
 3:    Obtain the ϵ-neighborhood of $\mathbf{x}_i$: $N_\epsilon(\mathbf{x}_i)$
 4:    **if** $|N_\epsilon(\mathbf{x}_i)| \geq \eta$ **then**
 5:      $\Omega = \Omega \cup \{\mathbf{x}_i\}$
 6:    **end if**
 7: **end for**
 8: Initialize cluster number $k = 0$ and unvisited data instance set $\Gamma = \mathcal{D}$
 9: **while** $\Omega \neq \emptyset$ **do**
10:    Keep a record of current unvisited data instances $\Gamma' = \Gamma$
11:    Randomly select a data instance $\mathbf{o} \in \Omega$ to initialize a queue $Q = (\mathbf{o})$
12:    $\Gamma = \Gamma \setminus \{\mathbf{o}\}$
13:    **while** $Q \neq \emptyset$ **do**
14:      Get the head data instance $\mathbf{q} \in Q$
15:      **if** $|N_\epsilon(\mathbf{q})| \geq \eta$ **then**
16:        Add $N_\epsilon(\mathbf{q}) \cap \Gamma$ into queue Q
17:        $\Gamma = \Gamma \setminus N_\epsilon(\mathbf{q})$
18:      **end if**
19:    **end while**
20:    $k = k + 1$, and generate cluster $\mathcal{C}_k = \Gamma' \setminus \Gamma$
21:    $\Omega = \Omega \setminus \mathcal{C}_k$
22: **end while**
23: **Return** $\mathcal{C} = \{\mathcal{C}_1, \mathcal{C}_2, \cdots, \mathcal{C}_k\}$

- **Density-Connected**: Given two data instance $\mathbf{x}_i$ and $\mathbf{x}_j$, $\mathbf{x}_i$ is said to be *density-connected* to $\mathbf{x}_j$, iff $\exists \mathbf{x}_k \in \mathcal{D}$ that $\mathbf{x}_i$ and $\mathbf{x}_j$ are both *density-reachable* from $\mathbf{x}_k$.

DBSCAN aims at partitioning the data instances into several densely distributed regions. In *DBSCAN*, a cluster denotes a maximal subset of data instances, in which the data instances are *density-connected*. Formally, a subset $\mathcal{C}_k \subset \mathcal{D}$ is a cluster detected by *DBSCAN* iff both the following two properties hold:

- **Connectivity**: $\forall \mathbf{x}_i, \mathbf{x}_j \in \mathcal{C}_k$, $\mathbf{x}_i$ and $\mathbf{x}_j$ are *density-connected*.
- **Maximality**: $\forall \mathbf{x}_i \in \mathcal{C}_k$, $\mathbf{x}_j \in \mathcal{D}$, $\mathbf{x}_j$ is *density-reachable* from $\mathbf{x}_i$ implies $\mathbf{x}_j \in \mathcal{C}_k$.

To illustrate how *DBSCAN* works, we provide its pseudo-code in Algorithm 3. According to the algorithm, at the beginning, *DBSCAN* selects a set of core objects as the seeds, from which *DBSCAN* expands the clusters based on the ϵ-neighborhood concept introduced before iteratively to find the *density-reachable* clusters. Such a process continues until all the core objects have been visited. The performance of *DBSCAN* depends on the selection order of the seed data instances a lot, and different selection orders will lead to totally different clustering results. *DBSCAN* doesn't need the cluster number as the input parameter, but requires (ϵ, η) to define the ϵ-*neighborhood* and *core objects* of the data instances.

Both the *K-Means* and *DBSCAN* algorithms are actually *hard clustering* algorithms, where instances are partitioned into different groups and each data instance only belongs to one single cluster. Besides such a type of clustering algorithms, there also exist some other clustering algorithms that allow data instances to belong to multiple clusters at the same time, like the *Mixture-of-Gaussian* algorithm to be introduced in the next subsection.

2.5.4 Mixture-of-Gaussian Soft Clustering

Different from the previous two clustering algorithms, the *Mixture-of-Gaussian* clustering algorithm [40] uses probability to model the cluster prototypes. Formally, in the d-dimensional feature space, given a feature vector $\mathbf{x}$ that follows a certain distribution, e.g., the Gaussian distribution, its probability density function can be represented as

$$p(\mathbf{x}|\boldsymbol{\mu}, \boldsymbol{\Sigma}) = \frac{1}{(2\pi)^{\frac{d}{2}}|\boldsymbol{\Sigma}|^{\frac{1}{2}}}$$
$$\times \exp\left(-\frac{1}{2}(\mathbf{x}-\boldsymbol{\mu})^{\top}\boldsymbol{\Sigma}(\mathbf{x}-\boldsymbol{\mu})\right), \qquad (2.44)$$

where $\boldsymbol{\mu}$ denotes the d-dimensional mean vector and $\boldsymbol{\Sigma}$ is a $d \times d$ covariance matrix.

In the *Mixture-of-Gaussian* clustering algorithm, each cluster is represented by a Gaussian distribution, where $\boldsymbol{\mu}_j$ and $\boldsymbol{\Sigma}_j$ can denote the parameters of the jth Gaussian distribution. Formally, the probability density function of the *Mixture-of-Gaussian* distribution can be represented as

$$p_M(\mathbf{x}) = \sum_{j=1}^{k} \alpha_j \cdot p(\mathbf{x}|\boldsymbol{\mu}_j, \boldsymbol{\Sigma}_j), \qquad (2.45)$$

where α_j denotes the mixture coefficient and $\sum_{j=1}^{k}\alpha_j = 1$.

The data instances in the training set $\mathcal{D} = \{\mathbf{x}_1, \mathbf{x}_2, \ldots, \mathbf{x}_n\}$ are assumed to be generated from the *Mixture-of-Gaussian* distribution. Given a data instance $\mathbf{x}_i \in \mathcal{D}$, the probability that the data instance is generated by the jth Gaussian distribution (i.e., its cluster label $y_i = j$) can be represented as

$$p_M(y_i = j|\mathbf{x}_i) = \frac{p(y_i = j) \cdot p(\mathbf{x}_i|y_i = j)}{p_M(\mathbf{x}_i)}$$
$$= \frac{\alpha_j \cdot p(\mathbf{x}_i|\boldsymbol{\mu}_j, \boldsymbol{\Sigma}_j)}{\sum_{l=1}^{k}\alpha_l \cdot p(\mathbf{x}_i|\boldsymbol{\mu}_l, \boldsymbol{\Sigma}_l)}. \qquad (2.46)$$

Meanwhile, in the *Mixture-of-Gaussian* clustering algorithm, a set of parameters $\{\boldsymbol{\mu}_j\}_{j=1}^{k}, \{\boldsymbol{\Sigma}_j\}_{j=1}^{k}$ and $\{\alpha_j\}_{j=1}^{k}$ are involved, which can be inferred from the data. Formally, based on the data set $\mathcal{D}$, we can represent its log-likelihood for the data instances to be distributed by following the *Mixture-of-Gaussian* to be

$$\mathcal{L}(\mathcal{D}) = \ln\left(\prod_{i=1}^{n} p_M(\mathbf{x}_i)\right)$$
$$= \sum_{i=1}^{n} \ln\left(\sum_{j=1}^{k}\alpha_j \cdot p(\mathbf{x}_i|\boldsymbol{\mu}_j, \boldsymbol{\Sigma}_j)\right). \qquad (2.47)$$

For the parameters which can maximize the above log-likelihood function will be the optimal solution to the *Mixture-of-Gaussian* model.

The EM (Expectation Maximization) algorithm can be applied to learn the optimal parameters for the *Mixture-of-Gaussian* clustering algorithm. By taking the derivatives of the objective function with regard to $\boldsymbol{\mu}_j$ and $\boldsymbol{\Sigma}_j$ and making them equal to 0, we can have

$$
\begin{cases}
\boldsymbol{\mu}_j &= \dfrac{\sum_{i=1}^{n} p_M(y_i=j|\mathbf{x}_i)\mathbf{x}_i}{\sum_{i=1}^{n} p_M(y_i=j|\mathbf{x}_i)}, \\[2mm]
\boldsymbol{\Sigma}_j &= \dfrac{\sum_{i=1}^{n} p_M(y_i=j|\mathbf{x}_i)(\mathbf{x}_i-\boldsymbol{\mu}_j)(\mathbf{x}_i-\boldsymbol{\mu}_j)^\top}{\sum_{i=1}^{n} p_M(y_i=j|\mathbf{x}_i)}.
\end{cases}
\tag{2.48}
$$

As to the weights $\{\alpha_j\}_{j=1}^{k}$, besides the objective function, there also exist some constraints $\alpha_j \geq 0$ and $\sum_{j=1}^{k} \alpha_j = 1$ on them. We can represent the Lagrange function of the objective function to be

$$
L(\{\alpha_j\}_{j=1}^{k}, \lambda) = \sum_{i=1}^{n} \ln \left(\sum_{j=1}^{k} \alpha_j \cdot p(\mathbf{x}_i|\boldsymbol{\mu}_j, \boldsymbol{\Sigma}_j) \right)
$$
$$
+ \lambda \left(\sum_{j=1}^{k} \alpha_j - 1 \right).
\tag{2.49}
$$

By making the derivative of the above function to α_j equal to 0, we can have

$$
\sum_{i=1}^{n} \frac{p(\mathbf{x}_i|\boldsymbol{\mu}_j, \boldsymbol{\Sigma}_j)}{\sum_{l=1}^{k} \alpha_l \cdot p(\mathbf{x}_i|\boldsymbol{\mu}_l)} + \lambda = 0.
\tag{2.50}
$$

Meanwhile, by multiplying the above equation by α_j and summing up the equations by enumerating all Gaussian distribution prototypes, we have

$$
\sum_{j=1}^{k} \sum_{i=1}^{n} \frac{\alpha_j \cdot p(\mathbf{x}_i|\boldsymbol{\mu}_j, \boldsymbol{\Sigma}_j)}{\sum_{l=1}^{k} \alpha_l \cdot p(\mathbf{x}_i|\boldsymbol{\mu}_l, \boldsymbol{\Sigma}_l)} + \sum_{j=1}^{k} \alpha_j \cdot \lambda = 0
$$
$$
\Rightarrow \sum_{j=1}^{k} \sum_{i=1}^{n} p_M(y_i = j|\mathbf{x}_i) + \lambda = 0
$$
$$
\Rightarrow \lambda = -n.
\tag{2.51}
$$

Furthermore, we have

$$
\sum_{i=1}^{n} \frac{\alpha_j \cdot p(\mathbf{x}_i|\boldsymbol{\mu}_j, \boldsymbol{\Sigma}_j)}{\sum_{l=1}^{k} \alpha_l \cdot p(\mathbf{x}_i|\boldsymbol{\mu}_l, \boldsymbol{\Sigma}_l)} + \lambda \cdot \alpha_j = 0
$$
$$
\Rightarrow \sum_{i=1}^{n} p_M(y_i = j|\mathbf{x}_i) + \lambda \cdot \alpha_j = 0
$$
$$
\Rightarrow \alpha_j = \frac{1}{-\lambda} \sum_{i=1}^{n} p_M(y_i = j|\mathbf{x}_i)
$$
$$
= \frac{1}{n} \sum_{i=1}^{n} p_M(y_i = j|\mathbf{x}_i).
\tag{2.52}
$$

Algorithm 4 Mixture-of-Gaussian

Require: Data Set $\mathcal{D} = \{\mathbf{x}_1, \mathbf{x}_2, \cdots, \mathbf{x}_n\}$;
  Gaussian distribution prototype number k.
Ensure: Clusters $\mathcal{C} = \{\mathcal{C}_1, \mathcal{C}_2, \cdots, \mathcal{C}_k\}$
 1: Initialize the parameters $\{\alpha_j, \boldsymbol{\mu}_j, \boldsymbol{\Sigma}_j)\}_{j=1}^k$
 2: $Converge - tag = False$
 3: **while** $Converge - tag == False$ **do**
 4: **for** $i \in \{1, 2, \cdots, n\}$ **do**
 5:   Calculate the posterior probability $p_M(y_i = j|\mathbf{x}_i)$
 6: **end for**
 7: **for** $j \in \{1, 2, \cdots, k\}$ **do**
 8:   Calculate new mean vector $\boldsymbol{\mu}'_j = \frac{\sum_{i=1}^n p_M(y_i=j|\mathbf{x}_i)\mathbf{x}_i}{\sum_{i=1}^n p_M(y_i=j|\mathbf{x}_i)}$
 9:   Calculate new covariance matrix $\boldsymbol{\Sigma}'_j = \frac{\sum_{i=1}^n p_M(y_i=j|\mathbf{x}_i)(\mathbf{x}_i-\boldsymbol{\mu}_j)(\mathbf{x}_i-\boldsymbol{\mu}_j)^\top}{\sum_{i=1}^n p_M(y_i=j|\mathbf{x}_i)}$
 10:   Calculate new weight $\alpha'_j = \frac{1}{n}\sum_{i=1}^n p_M(z_i = j|\mathbf{x}_i)$
 11: **end for**
 12: **if** $\{\alpha_j, \boldsymbol{\mu}_j, \boldsymbol{\Sigma}_j)\}_{j=1}^k == \{\alpha'_j, \boldsymbol{\mu}'_j, \boldsymbol{\Sigma}'_j)\}_{j=1}^k$ **then**
 13:   $Converge - tag = True$
 14: **else**
 15:   $\{\alpha_j, \boldsymbol{\mu}_j, \boldsymbol{\Sigma}_j)\}_{j=1}^k = \{\alpha'_j, \boldsymbol{\mu}'_j, \boldsymbol{\Sigma}'_j)\}_{j=1}^k$
 16: **end if**
 17: **end while**
 18: Initialize $\mathcal{C} = \{\mathcal{C}_j\}_{j=1}^k$, where $\mathcal{C}_j = \emptyset$
 19: **for** $\mathbf{x}_i \in \mathcal{D}$ **do**
 20: Determine the cluster label $y_i^* = \arg_{y\in\{1,2,\cdots,k\}} p_M(y_i = y|\mathbf{x}_i)$
 21: $\mathcal{C}_{y_i^*} = \mathcal{C}_{y_i^*} \cup \{\mathbf{x}_i\}$
 22: **end for**
 23: **Return** $\mathcal{C} = \{\mathcal{C}_1, \mathcal{C}_2, \cdots, \mathcal{C}_k\}$

According to the analysis, the expected values of $\boldsymbol{\mu}_j$ and $\boldsymbol{\Sigma}_j$ are the weighted sum of the mean and covariance of the data instances in the provided data set. Meanwhile, weight α_j is the average posterior probability of data instance belonging to the jth Gaussian distribution. The pseudo-code of parameter learning and cluster inference parts of the *Mixture-of-Gaussian* clustering algorithm is provided in Algorithm 4.

2.6   Artificial Neural Network and Deep Learning

Artificial neural network (ANN) is a computational algorithm aiming at modeling the way that a biological brain solves problems with large clusters of connected biological neurons. Artificial neural network models include both supervised and unsupervised learning models, which can work on both classification and regression tasks. Artificial neural networks as well as the recent deep learning [14] models have achieved a remarkable success in addressing various difficult learning tasks. Therefore, we lift them up as a separate section to talk about. In this part, we will provide a brief introduction to the artificial neural networks, including the basic background knowledge, *perceptron* [30, 42], *multi-layer feed-forward neural network* [48], *error back propagation algorithm* [43–45, 54], and several well-known deep learning models, i.e., *deep autoencoder* [14,53], *deep recurrent neural network* [14], and *deep convolutional neural network* [14, 26, 27].

Fig. 2.14 A picture of
neuron

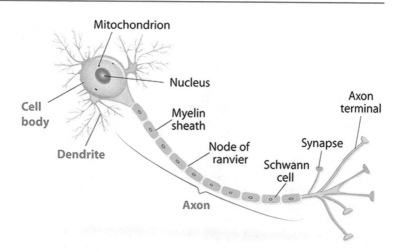

2.6.1 Artificial Neural Network Overview

As shown in Fig. 2.14, human brains are composed by a group of *neurons*, which are connected
by the *axons* and *dendrites*. In human brains, signal can transmit from a neuron to another one via
the *axon* and *dendrite*. The connecting point between *axon* and *dendrite* is called the *synapse*, where
human brains can learn new knowledge by changing the connection strength of these *synapses*. Similar
to the biological brain network, artificial neural networks are composed of a group of connected
artificial neurons, which receive inputs from the other neurons. Depending on the inputs and the
activation thresholds, artificial neurons can be either activated to transmit information to the other
artificial neuron or stay inactive.

In Fig. 2.15a, we show an example of the classic McCulloch-Pitts (M-P) neuron model [29]. As
shown in the model, the neuron takes the inputs $\{x_1, x_2, \ldots, x_n\}$ from n other neurons and its output
can be represented as y. The connection weights between other neurons and the current neuron can
be denoted as $\{w_1, w_2, \ldots, w_n\}$, and the inherent activating threshold of the current neuron can be
represented as θ. The output y of the current neuron depends on both the inputs $\{x_1, x_2, \ldots, x_n\}$,
connection weights $\{w_1, w_2, \ldots, w_n\}$, and the activating threshold θ. Formally, the output of the
current neuron can be represented as

$$y = f\left(\sum_{i=1}^{n} w_i x_i - \theta\right),\tag{2.53}$$

where $f(\cdot)$ is usually called the *activation function*. *Activation function* can project the input signals
to the objective output, and many different *activation functions* can be used in the real-world
applications, like the *sign function* and *sigmoid function*.

Formally, the *sign function* can be represented as

$$f(x) = \begin{cases} 1, & \text{if } x > 0; \\ 0, & \text{otherwise.} \end{cases}\tag{2.54}$$

In the case where the *sign function* is used as the *activation function*, if the weighted input sum
received from the other neurons is greater than the current neuron's threshold, the current neuron will

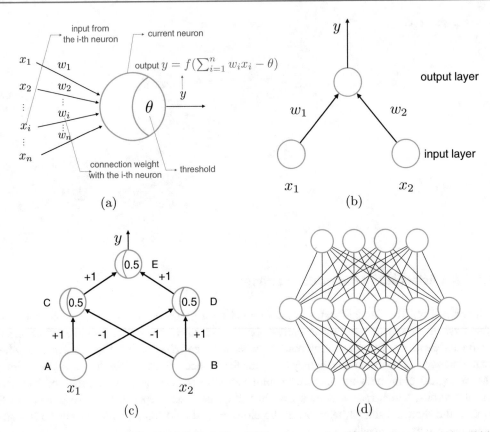

Fig. 2.15 Neural network models: **(a)** McCulloch-Pitts (M-P) neuron model, **(b)** perceptron mode, **(c)** multi-layer perceptron model, and **(d)** multi-layer feed-forward model

be activated and output 1; otherwise, the output will be 0. The *sign function* is very simple, and works well in modeling the active/inactive states of the neurons. However, the mathematical properties, like *discontinuous* and *nonsmooth*, of the *sign function* are not good, which render the *sign function* rarely used in real-world artificial neural network models.

Different from the *signed function*, the *sigmoid function* is a continuous, smooth, and differentiable function. Formally, the *sigmoid function* can be represented as

$$f(x) = \frac{1}{1 + e^{-x}}, \tag{2.55}$$

which outputs a value in the (0, 1) range for all inputs $x \in \mathbb{R}$. When the *sigmoid function* is used as the *activation function*, if the input is greater than the neuron's *activation threshold*, the output will be greater than 0.5 and will approach 1 as the weighted input sum further increases; otherwise, the output will be smaller than 0.5 and approaches 0 when the weighted input sum further decreases.

2.6.1.1 Perceptron and Multi-Layer Feed-Forward Neural Network

Based on the M-P neuron, several neural network algorithms have already been proposed. In this part, we will introduce two classic artificial neural network models: (1) *perceptron* [30, 42] and (2) *multi-layer feed-forward neural network* [48].

The architecture of *perceptron* [30, 42] is shown in Fig. 2.15c. *Perceptron* consists of two layers of neurons: (1) the input layer, and (2) the output layer. The input layer neurons receive external inputs and transmit them to the output layer, while output layer is an M-P neuron which receives the input and uses the *sign function* as the *activation function*. Given the training data, the parameters involved in *perceptron*, like weights $\mathbf{w} = [w_1, w_2, \ldots, w_n]^\top$ and threshold θ, can all be effectively learned. To simplify the learning process, here, we can add one extra dummy input feature "-1" for the M-P neuron whose connection weight can be denoted as θ. In this way, we can unify the activating threshold with the connection weights as $\mathbf{w} = [w_1, w_2, \ldots, w_n, \theta]^\top$. In the learning process, given a training instance $(\mathbf{x}, y)$, the weights of the *perceptron* model can be updated by the following equations until convergence:

$$w_i = w_i + \partial w_i, \tag{2.56}$$

$$\partial w_i = \eta(y - \hat{y})x_i, \tag{2.57}$$

where $\eta \in (0, 1)$ represents the learning rate and $\hat{y}$ denotes output value of *perceptron*.

Perceptron is one of the simplest artificial neural network models, and can only be used to implement some simple linear logical operations, like *and*, *or* and *not*, where the above learning process will converge to the optimal variables.

- **AND** $(x_1 \wedge x_2)$: Let $w_1 = w_2 = 1$ and $\theta = 1.9$, we have $y = f(1 \cdot x_1 + 1 \cdot x_2 - 1.9)$, which achieves value $y = 1$ iff $x_1 = x_2 = 1$.
- **OR** $(x_1 \vee x_2)$: Let $w_1 = w_2 = 1$ and $\theta = 0.5$, we have $y = f(1 \cdot x_1 + 1 \cdot x_2 - 0.5)$, which achieves value $y = 1$ if $x_1 = 1$ or $x_2 = 1$.
- **NOT** $(\neg x_1)$: Let $w_1 = -0.6$, $w_2 = 0$ and $\theta = -0.5$, we have $y = f(-0.6 \cdot x_1 + 0 \cdot x_2 + 0.5)$, which achieves value $y = 1$ if $x_1 = 0$; and value $y = 0$ if $x_1 = 1$.

However, as pointed out by Minsky in [30], *perceptron* cannot handle non-linear operations, like *xor* (i.e., $x_1 \oplus x_2$), as no convergence can be achieved with the above weight updating equations.

To overcome such a problem, *multi-layer perceptron* [48] has been introduced, which can classify the instances that are not linearly separable. Besides the input and output layers, the *multi-layer perceptron* also involves a hidden layer. For instance, to fit the *XOR* function, the *multi-layer perceptron* architecture is shown in Fig. 2.15c, where the connection weights and neuron thresholds are also clearly indicated.

- **XOR** $(x_1 \oplus x_2)$: Between the input layer and hidden layer, let the weights $w_{A,C} = w_{B,D} = 1$, $w_{A,D} = w_{B,C} = -1$, thresholds $\theta_C = \theta_D = 0.5$. We can have $y_C = f(1 \cdot x_1 - 1 \cdot x_2 - 0.5)$ and $y_D = f(-1 \cdot x_1 + 1 \cdot x_2 - 0.5)$. Between the hidden layer and output layer, let $w_{C,E} = w_{D,E} = 1$ and threshold $\theta_E = 0.5$, we will have the model output $y = f(1 \cdot y_C + 1 \cdot y_D - 0.5) = f(f(x_1 - x_2 - 0.5) + f(-x_1 + x_2 - 0.5) - 0.5)$. If $x_1 = x_2 = 0$ or $x_1 = x_2 = 1$, we have $y = 0$, while if $x_1 = 1, x_2 = 0$ or $x_1 = 0, x_2 = 1$, we have $y = 1$.

In addition to the *multi-layer perceptron*, a more general multi-layer neural network architecture is shown in Fig. 2.15d, where multiple neuron layers are involved. Between different adjacent layers, the neurons are connected, while those in the non-adjacent layers (i.e., the same layer or skipped layers) are not connected. The artificial neural networks in such an architecture are named as the *feed-forward neural networks* [48], which receive input from the input layer, process the input via the hidden layer, and output the result via the output layer. For the *feed-forward neural networks* shown in Fig. 2.15d, it involves one single hidden layer, which is called the *single hidden-layer feed-forward neural network* in this book. Meanwhile, for the neural networks involving multiple hidden layers, they will be called the *multi-layer neural networks*.

Besides *perceptron, feed-forward neural network* models, there also exist a large number of other neural network models with very diverse architectures, like the *cascade-correlation neural network* [13], *Elman neural network* [14], and *Boltzmann neural network* [46]. We will not introduce them here, since they are out of the scope of this textbook. Interested readers may refer to the cited literatures for more information about these models. Generally, for the neural networks involving hidden layers, the model variable learning algorithm will be more challenging and different from *perceptron*. In the following part, we will introduce the well-known *error back propagation algorithm* [43–45, 54] for neural network model learning.

2.6.1.2 Error Back Propagation Algorithm

To this context so far, the most successful learning algorithm for *multi-layer neural networks* is the *Error Back Propagation (BP)* algorithm [43–45, 54]. The *BP* algorithm has been shown to work well for many different types of *multi-layer neural network* as well as the recent diverse *deep neural network* models. In the following part, we will use the *single hidden-layer feed-forward neural network* as an example to illustrate the *BP* algorithm.

As shown in Fig. 2.16, in the *single-hidden layer feed-forward neural network*, there exist d different input neurons, q hidden neurons, and l output neurons. Let the data set used for training the model be $\mathcal{T} = \{(\mathbf{x}_1, \mathbf{y}_1), (\mathbf{x}_2, \mathbf{y}_2), \ldots, (\mathbf{x}_n, \mathbf{y}_n)\}$. Given a data instance featured by vector $\mathbf{x}_i$ as the input, the neural network model will generate a vector $\hat{\mathbf{y}}_i = [\hat{y}_i^1, \hat{y}_i^2, \ldots, \hat{y}_i^l]^\top \in \mathbb{R}^l$ of length l as the output. We will use θ_j to denote the activating threshold of the jth output neuron, and γ_h to denote the activating threshold of the hth hidden neuron. Between the input layer and hidden layer, the connection weight between the ith input neuron and the hth hidden neuron can be represented as $v_{i,h}$. Meanwhile, between the hidden layer and the output layer, the connection weight between the hth hidden neuron and the jth output neuron can be represented as $w_{h,j}$. For the hth hidden neuron, its received input can be represented as $\alpha_h = \sum_{i=1}^{d} v_{i,h} x_i$, and for the jth output neuron, its received input can be denoted as $\beta_j = \sum_{h=1}^{q} w_{h,j} b_h$. Here, we assume the *sigmoid function* is used as the *activation function*, which will project the input to the output in both hidden layer and output layer.

Fig. 2.16 An example of error back propagation algorithm

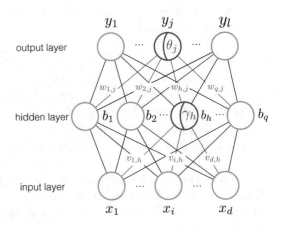

Given a training data instance $(\mathbf{x}_i, \mathbf{y}_i)$, let's assume the output of the neural network to be $\hat{\mathbf{y}}_i = [\hat{y}_i^1, \hat{y}_i^2, \ldots, \hat{y}_i^l]^\top$, and square error is used as the evaluation metric. We can represent the output $\hat{y}_i^j$ and training error as

$$\hat{y}_i^j = f(\beta_j - \theta_j), \tag{2.58}$$

$$E_i = \frac{1}{2} \sum_{j=1}^{l} (\hat{y}_i^j - y_i^j)^2, \tag{2.59}$$

where $\frac{1}{2}$ is added in the error function to remove the constant factors when computing the partial derivatives.

BP algorithm proposes to update the model variables with the gradient descent approach. Formally, in gradient descent approach, the model variables will be updated with multiple rounds, and the updating equation in each round can be represented as follows:

$$w \leftarrow w + \eta \cdot \partial w, \tag{2.60}$$

where η denotes the learning rate and it is usually a small constant value. Term ∂w represents the negative gradient of the error function regarding variable w.

We can take $w_{h,j}$ as an example to illustrate the learning process of the BP algorithm. Formally, based on the given training instance $(\mathbf{x}_i, \mathbf{y}_i)$ and the chain rule of derivatives, we can represent the partial derivative of the introduced error regarding variable $w_{h,j}$ as follows:

$$\partial w_{h,j} = -\frac{\partial E_i}{\partial w_{h,j}} = -\frac{\partial E_i}{\partial \hat{y}_i^j} \cdot \frac{\partial \hat{y}_i^j}{\partial \beta_j} \cdot \frac{\partial \beta_j}{\partial w_{h,j}}. \tag{2.61}$$

Furthermore, according to the definition of E_i, β_j and the property of sigmoid function (i.e., $f'(x) = f(x)(1 - f(x))$), we have

$$\frac{\partial E_i}{\partial \hat{y}_i^j} = \hat{y}_i^j - y_i^j, \tag{2.62}$$

$$\frac{\partial \hat{y}_i^j}{\partial \beta_j} = \hat{y}_i^j (1 - \hat{y}_i^j), \tag{2.63}$$

$$\frac{\partial \beta_j}{\partial w_{h,j}} = b_h. \tag{2.64}$$

Specifically, to simplify the representations, we will introduce a new notation g_j, which denotes

$$g_j = -\frac{\partial E_i}{\partial \hat{y}_i^j} \cdot \frac{\partial \hat{y}_i^j}{\partial \beta_j} = (y_i^j - \hat{y}_i^j) \cdot \hat{y}_i^j (1 - \hat{y}_i^j). \tag{2.65}$$

By replacing the above terms into Eq. (2.61), we have

$$\partial w_{h,j} = g_j \cdot b_h. \tag{2.66}$$

In a similar way, we can achieve the updating equation for variable θ_j as follows:

$$\theta_j \leftarrow \theta_j + \eta \cdot \partial\theta_j, \text{ where } \partial\theta_j = -g_j. \tag{2.67}$$

Furthermore, by propagating the error back to the hidden layer, we will also be able to update the connection weights between the input layer and hidden layer, as well as the activating threshold of the hidden neurons. Formally, we can represent the updating equation of connection weight $v_{i,h}$ and γ_h as follows:

$$v_{i,h} = v_{i,h} + \eta \cdot \partial v_{i,h}, \text{ where } \partial v_{i,h} = e_h \cdot x_i, \tag{2.68}$$

$$\gamma_h = \gamma_h + \eta \cdot \partial\gamma_h, \text{ where } \partial\gamma_h = -e_h. \tag{2.69}$$

In the above equations, term e_h is an introduced new representation, which can be represented as follows:

$$
\begin{aligned}
e_h &= -\frac{\partial E_i}{\partial b_h} \cdot \frac{\partial b_h}{\partial \alpha_h} \\
&= -\sum_{j=1}^{l} \frac{\partial E_i}{\partial \beta_j} \cdot \frac{\partial \beta_j}{\partial b_h} f'(\alpha_h - \gamma_h) \\
&= b_h(1 - b_h) \sum_{j=1}^{l} w_{h,j} \cdot g_j.
\end{aligned}
\tag{2.70}
$$

In the updating equation, the learning rate $\eta \in (0, 1)$ can actually be different in updating different variables. Furthermore, according to the updating equations, the term g_j calculated in updating the weights between the hidden layer and output layer will also be used to update the weights between input layer and hidden layer. It is also the reason why the method is called the *Error Back Propagation* algorithm [43–45, 54].

2.6.2   Deep Learning

In recent years, deep learning [14], as a rebranding of artificial neural networks, has become very popular and many different types of deep learning models have been proposed already. These *deep learning* models generally have a large *capacity* (in terms of containing information), as they are usually in a very complex and deep structure involving a large number of variables. For this reason, *deep learning* models are capable to capture very complex projections from the input data to the objective outputs. Meanwhile, due to the availability of massive training data, powerful computational facilities and diverse application domains, training *deep learning* models is becoming a feasible task, which has dominated both academia and industry in these years.

In this part, we will introduce several popular *deep learning* models briefly, which include *deep autoencoder* [14, 53], *deep recurrent neural network (RNN)* [14], and *deep convolutional neural network (CNN)* [14, 26, 27]. Here, we need to indicate that the *deep autoencoder* model is actually an unsupervised *deep learning* models, and the *deep RNN* and *deep CNN* are both supervised *deep learning* models. The training of *deep learning* models is mostly based on the *error back propagation* algorithm, and we will not cover the *deep learning* model learning materials in this part any more.

2.6.2.1 Deep Autoencoder

Deep autoencoder is an unsupervised neural network model, which can learn a low-dimensional representation of the input data via a series of non-linear mappings. *Deep autoencoder* model involves two main steps: encoder and decoder. The encoder step projects the original input to the objective low-dimensional feature space, while the decoder step recovers the latent feature representation to a reconstruction space. In the *deep autoencoder* model, we generally need to ensure that the original feature representation of data instances should be as similar to the reconstructed feature representation as possible.

The architecture of the *deep autoencoder* model is shown in Fig. 2.17a. Formally, let $\mathbf{x}_i$ represent the original input feature representation of a data instance, $\mathbf{y}_i^1, \mathbf{y}_i^2, \ldots, \mathbf{y}_i^o$ be the latent feature representation of the instance at hidden layers $1, 2, \ldots, o$ in the encoder step, and $\mathbf{z}_i \in \mathbb{R}^d$ be the representation in the low-dimensional object space. Formally, the relationship between these vector variables can be represented with the following equations:

$$\begin{cases} \mathbf{y}_i^1 &= \sigma(\mathbf{W}^1 \mathbf{x}_i + \mathbf{b}^1), \\ \mathbf{y}_i^k &= \sigma(\mathbf{W}^k \mathbf{y}_i^{k-1} + \mathbf{b}^k), \forall k \in \{2, 3, \ldots, o\}, \\ \mathbf{z}_i &= \sigma(\mathbf{W}^{o+1} \mathbf{y}_i^o + \mathbf{b}^{o+1}). \end{cases} \tag{2.71}$$

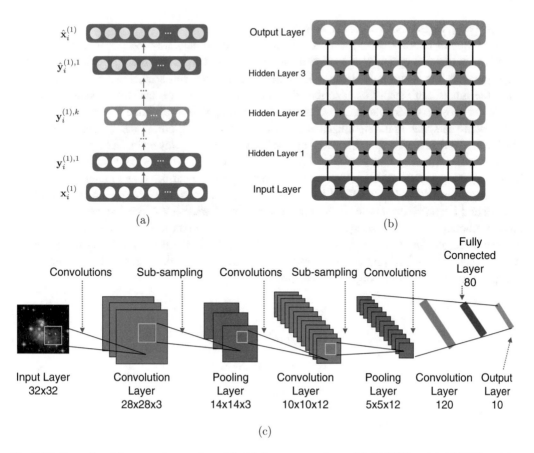

Fig. 2.17 Examples of deep neural network models: (**a**) deep autoencoder model, (**b**) RNN model, (**c**) CNN model

Meanwhile, in the decoder step, the input will be the latent feature vector $\mathbf{z}_i$ (i.e., the output of the encoder step) instead, and the final output will be the reconstructed vector $\hat{\mathbf{x}}_i$. The latent feature vectors at each hidden layers in the decoder step can be represented as $\hat{\mathbf{y}}_i^o, \ldots, \hat{\mathbf{y}}_i^2, \hat{\mathbf{y}}_i^1$. The relationship among these vector variables can be denoted as follows:

$$
\begin{cases}
\hat{\mathbf{y}}_i^o & = \sigma(\hat{\mathbf{W}}^{o+1}\mathbf{z}_i + \hat{\mathbf{b}}^{o+1}), \\
\hat{\mathbf{y}}_i^{k-1} & = \sigma(\hat{\mathbf{W}}^k\hat{\mathbf{y}}_i^k + \hat{\mathbf{b}}^k), \forall k \in \{2, 3, \ldots, o\}, \\
\hat{\mathbf{x}}_i & = \sigma(\hat{\mathbf{W}}^1\hat{\mathbf{y}}_i^1 + \hat{\mathbf{b}}^1).
\end{cases}
\tag{2.72}
$$

The objective of the *deep autoencoder* model is to minimize the loss between the original feature vector $\mathbf{x}_i$ and the reconstructed feature vector $\hat{\mathbf{x}}_i$ of all the instances in the network. Formally, the objective function of the *deep autoencoder* model can be represented as

$$
\arg_{\mathbf{W}, \hat{\mathbf{W}}, \mathbf{b}, \hat{\mathbf{b}}} \min \sum_i \left\| \mathbf{x}_i - \hat{\mathbf{x}}_i \right\|_2^2,
\tag{2.73}
$$

where L_2 norm is used to define the loss function. In the objective function, terms $\mathbf{W}$, $\hat{\mathbf{W}}$ and $\mathbf{b}$, $\hat{\mathbf{b}}$ represent the variables involved in the encoder and decoder steps in *deep autoencoder*, respectively.

2.6.2.2 Deep Recurrent Neural Network

Recurrent neural network (RNN) is a class of artificial neural network where connections between units form a directed cycle. *RNN* has been successfully used for some special learning tasks, like language modeling, word embedding, handwritten recognition, and speech recognition. For these tasks, the inputs can usually be represented as an ordered sequence, and there exists a temporal correlation between the inputs. *RNN* can capture such a temporal correlation effectively.

In recent years, due to the significant boost of GPUs' computing power, the representation ability of *RNN* models has been greatly improved by involving more hidden layers into a deeper architecture, which is called the *deep recurrent neural network* [33]. In Fig. 2.17b, we show an example of a 3-hidden layer *deep RNN* model, which receives input from the input layer, and outputs the result to the output layer. In the *deep RNN* model, the states of the neurons depends on both the lower-layer neurons and the previous neuron (in the same layer). For the *deep RNN* model shown in Fig. 2.17b, given a training instance $((\mathbf{x}_1, \mathbf{y}_1), (\mathbf{x}_2, \mathbf{y}_2), \ldots, (\mathbf{x}_T, \mathbf{y}_T))$, we can represent the hidden states of neurons corresponding to input $\mathbf{x}_t$ (where $t \in \{1, 2, \ldots, T\}$) in the three hidden layers and output layer as vectors $\mathbf{h}_t^1, \mathbf{h}_t^2, \mathbf{h}_t^3$ and $\mathbf{y}_t$, respectively. The dynamic correlations among these variables can be represented with the following equations formally:

$$
\mathbf{h}_t^1 = f_h(\mathbf{x}_t, \mathbf{h}_{t-1}^1; \boldsymbol{\theta}_h),
\tag{2.74}
$$

$$
\mathbf{h}_t^2 = f_h(\mathbf{h}_t^1, \mathbf{h}_{t-1}^2; \boldsymbol{\theta}_h),
\tag{2.75}
$$

$$
\mathbf{h}_t^3 = f_h(\mathbf{h}_t^2, \mathbf{h}_{t-1}^3; \boldsymbol{\theta}_h),
\tag{2.76}
$$

$$
\hat{\mathbf{y}}_t = f_o(\mathbf{h}_t^3; \boldsymbol{\theta}_o),
\tag{2.77}
$$

where $f_h(\cdot; \boldsymbol{\theta}_h)$ and $f_o(\cdot; \boldsymbol{\theta}_o)$ denote the hidden state transition function and output function parameterized by variables $\boldsymbol{\theta}_h$ and $\boldsymbol{\theta}_o$, respectively. These functions can be defined in different ways, depending on the unit models used in depicting the neuron states, e.g., traditional M-P neuron or the

LSTM (i.e., long short-term memory) unit [19], GRU (i.e., gated recurrent unit) [8], and the recent GDU (gated diffusive unit) [56].

For the provided training instance $((\mathbf{x}_1, \mathbf{y}_1), (\mathbf{x}_2, \mathbf{y}_2), \ldots, (\mathbf{x}_T, \mathbf{y}_T))$, the loss introduced by the prediction result compared against the ground truth can be denoted as

$$J(\boldsymbol{\theta}_h, \boldsymbol{\theta}_o) = \sum_{t=1}^{T} d(\hat{\mathbf{y}}_t, \mathbf{y}_t), \tag{2.78}$$

where $d(\cdot, \cdot)$ denotes the difference between the provided variables, e.g., Euclidean distance or cross-entropy.

Given a set of n training instances $\mathcal{T} = \{((\mathbf{x}_1^i, \mathbf{y}_1^i), (\mathbf{x}_2^i, \mathbf{y}_2^i), \ldots, (\mathbf{x}_T^i, \mathbf{y}_T^i))\}_{i=1}^{n}$, by minimizing the training loss, we will be able to learn the variables $\boldsymbol{\theta}_h$ and $\boldsymbol{\theta}_o$ of the *deep RNN* model. The learning process of the *deep RNN* models with the classic M-P neuron may usually suffer from the gradient vanishing/exploding problem a lot. To overcome such a problem, some new unit neuron models have been proposed, including LSTM, GRU, and GDU. More detailed descriptions about these unit models are available in [8, 19, 56].

2.6.2.3 Deep Convolutional Neural Network

The deep *convolutional neural network* (*CNN*) model [27] is a type of feed-forward artificial neural network, in which the connectivity pattern between the neurons is inspired by the organization of the animal visual cortex. *CNN* has been shown to be effective in a lot of applications, especially in image and computer vision related tasks. The concrete applications of *CNN* include image classification, image semantic segmentation, and object detection in images. In recent years, some research works also propose to apply *CNN* for the textual representation learning and classification tasks [24]. In this part, we will use image classification problem as an example to illustrate the *CNN* model. As shown in Fig. 2.17c, given an image input, the image classification task aims at determining the labels for the image.

According to the architecture, the *CNN* model is formed by a stack of distinct layers that transform the input image into the output labels. Depending on their functions, these layers can be categorized into the *input layer, convolutional layer, pooling layer, ReLU layer, fully connected layer*, and *output layer*, which will be introduced as follows, respectively:

- **Input Layer**: In the input layer, the neurons receive the image data input, and represent it as a stack of matrices (e.g., a high-order tensor). For instance, as shown in Fig. 2.17c, for an input image of size 32×32 by pixels, we can represent it as a 3-way tensor of dimensions $32 \times 32 \times 3$ if the image is represented in the RGB format.
- **Convolutional Layer**: In the convolutional layer, convolution kernels will applied to extract patterns from the current image representations. For instance, if matrix $\mathbf{K} = \begin{bmatrix} 1 & 2 & 1 \\ 0 & 0 & 0 \\ 1 & 2 & 1 \end{bmatrix}$ is used as the convolution kernel, by applying it to extract features from the images, we will be able to identify the horizontal edges from the image, where the pixels above and below the edge differ a lot. In Fig. 2.17c, a 5×5 kernel is applied to the input images, which brings about a representation of dimensions $28 \times 28 \times 3$.
- **Pooling Layer**: Another important layer in the *CNN* model is the pooling layer, which performs non-linear down-sampling of the feature representation from the convolution layers. There are several different non-linear functions to implement pooling, among which *max pooling* and *mean*

pooling are the most common pooling techniques. Pooling partitions the feature representation into a set of non-overlapping rectangles, and for each such sub-region, it outputs the maximum number (if *max-pooling* is applied). The intuition of pooling is that once a feature has been found, its exact location isn't as important as its rough location relative to other features. Pooling greatly reduces the size of the representation, which can reduce the amount of parameters and computation in the model and hence also control overfitting. Pooling is usually periodically inserted in between successive convolution layers in the *CNN* architecture. For instance, in Fig. 2.17c, we perform the pooling operation on a 2×2 sub-region in the feature representation layer of dimensions $28 \times 28 \times 3$, which will lead to a pooling layer of dimensions $14 \times 14 \times 3$.

- **ReLU Layer**: In *CNN*, when applying the *sigmoid function* as the *activation function*, it will suffer from the *gradient vanish* problem [1, 34] a lot (just like deep *RNN*), which may make the gradient based learning method (e.g., SGD) fail to work. In real-world practice, a non-linear function

$$
f(x) = \begin{cases} 0, & \text{if } x < 0, \\ x, & \text{otherwise.} \end{cases} \tag{2.79}
$$

is usually used as the *activation function* instead. Neurons using such a kind of *activation function* is called the *Rectified Linear Unit* (*ReLU*), and the layers involving the *ReLU* as the units are called the *ReLU Layer*. The introduction of *ReLU* to replace sigmoid is an important change in *CNN*, which significantly reduces the difficulty in learning *CNN* variables and also greatly improves its performance.

- **Fully Connected Layer**: Via a series of *convolution layers* and *pooling layers*, the input image will be transformed into a vector representation. The classification task will be performed by a *fully connected layer* and an *output layer*, which together will compose a *single-hidden layer feed-forward neural network* actually. For instance, in Fig. 2.17c, with convolution and sampling operations, we can obtain a feature vector of dimension 120, which together with a *fully connected layer* (of dimension 84) and the output layer (of dimension 10) will perform the classification task finally.

- **Output Layer**: The classification result will be achieved from the *output layer*, which involves 10 neurons in the example in Fig. 2.17c.

2.7   Evaluation Metrics

We have introduced the classification, regression, clustering, and deep learning tasks together with their several well-known algorithms already in the previous sections, and also talked about how to build the modes with the available data set. By this context so far, we may have several questions in mind: (1) how is quality of the built models, (2) how to compare the performance of different models. To answer these two questions, we will introduce some frequently used evaluation metrics in this part for the classification, regression, and clustering task, respectively.

2.7.1   Classification Evaluation Metrics

Here, we take the binary classification task as an example. Let $\mathbf{y} = [y_1, y_2, \ldots, y_n]^\top$ ($y_i \in \{+1, -1\}$) be the true labels of n data instances, and $\hat{\mathbf{y}} = [\hat{y}_1, \hat{y}_2, \ldots, \hat{y}_n]^\top$ ($\hat{y}_i \in \{+1, -1\}$) be the labels predicted by a classification model. Based on vectors $\mathbf{y}$ and $\hat{\mathbf{y}}$, we can introduce a new concept named

Table 2.2 Confusion matrix

	Predicted positive	Predicted negative
Actual positive	TP	FN
Actual negative	FP	TN

confusion matrix, as shown in Table 2.2. In the table, depending on the true and predicted labels, the data instances can be divided into 4 main categories as follows:

- **TP (True Positive)**: the number of correctly classified positive instances;
- **FN (False Negative)**: the number of incorrectly classified positive instances;
- **FP (False Positive)**: the number of incorrectly classified negative instances;
- **TN (True Negative)**: the number of correctly classified negative instances.

Based on the confusion matrix, we can define four different classification evaluation metrics as follows:

Accuracy:

$$Accuracy = \frac{TP + TN}{TP + FN + FP + TN}, \tag{2.80}$$

Precision:

$$Precision = \frac{TP}{TP + FP}, \tag{2.81}$$

Recall:

$$Recall = \frac{TP}{TP + FN}, \tag{2.82}$$

F_β**-Score**:

$$F_\beta = (1 + \beta^2) \cdot \frac{Precision \cdot Recall}{\beta^2 \cdot Precision + Recall}, \tag{2.83}$$

where F_1-Score (with $\beta = 1$) is normally used in practice.

Among these 4 metrics, *Accuracy* considers both TP and TN simultaneously in the computation, which works well for the *class balanced* scenarios (i.e., the amount of positive and negative data instances is close) but may suffer from a serious problem in evaluating the learning performance of classification models in the *class imbalanced* scenario.

Example 2.6 Let's assume we have one million patient data records. Among these patient records, only 100 records indicate that the patient have the Alzheimer disease (AD), which is also the objective label that we want to predict. If we treat the patient records with AD as the positive instance, and those without AD as the negative instance, then the dataset will form a *class imbalanced scenario*.

Given two models with the following performance, where *Model 1* is well-trained but *Model 2* simply predicts all the data instance to be *negative*.

- **Mode 1**: with TP = 90, FP = 10, FN = 900, TN = 999,000;
- **Model 2**: with TP = 0, FP = 100, FN = 0, TN = 999,900.

By comparing the performance, *Model 1* is definitely much more useful than *Model 2* in practice. However, due to the *class imbalance* problem, we may have different evaluation for these two models with *accuracy*. According to the results, we observe that *Model 1* can identify most (90%) of the AD patients from the data but also mis-identify 900 normal people as AD patients. *Model 1* can achieve an *accuracy* about $\frac{90+999,000}{1,000,000} = 99.909\%$. Meanwhile, for *Model 2*, by predicting all the patient records to be negative, it cannot identify the AD patients at all. However, it can still achieve an *accuracy* at $\frac{999,900}{1,000,000} = 99.99\%$, which is even larger than the *accuracy* of *Model 1*.

According to the above example, the *accuracy* metric will fail to work when handling the *class imbalanced* data set. However, *Precision*, *Recall*, and F_1 metrics can still work well in such a scenario. We will leave the computation of *Precision*, *Recall*, and F_1 for this example as an exercise for the readers.

Besides these four metrics, two other curve based evaluation metrics can also be defined based on the confusion matrix, which are called the *Precision-Recall curve* and *ROC curve* (receiver operating characteristic curve). Given the n data instances together with their true and predicted labels, we can obtain a $(precision, recall)$ pair at each of the data instances $\mathbf{x}_i$ based on the cumulative prediction results from the beginning to the current position. The *Precision-Recall curve* is a plot obtained based on such $(precision, recall)$ pairs. Meanwhile, as to the *ROC curve*, it is introduced based on two new concepts named *true positive rate* (TPR) and *false positive rate* (FPR):

$$TPR = \frac{TP}{TP + FN}, \tag{2.84}$$

$$FPR = \frac{FP}{TN + FP}. \tag{2.85}$$

Based on the predicted and true labels of these n data instances, a series of (TPR, FPR) pairs can be calculated, which will plot the *ROC curve*. In practice, a larger area under the curves generally indicates a better performance of the models. Formally, the area under the *ROC curve* is defined as the *ROC AUC* metric (AUC is short for *area under curve*), and the area under the *Precision-Recall curve* is called the *PR AUC*.

2.7.2   Regression Evaluation Metrics

For the regression tasks, the predicted and true labels are actually real values, instead of pre-defined class labels in classification tasks. For the regression tasks, given the predicted and true label vectors $\hat{\mathbf{y}}$ and $\mathbf{y}$ of data instances, some frequently used evaluation metrics include

Explained Variance Regression Score

$$EV(\mathbf{y}, \hat{\mathbf{y}}) = 1 - \frac{Var(\mathbf{y} - \hat{\mathbf{y}})}{Var(\mathbf{y})}, \tag{2.86}$$

where $Var(\cdot)$ denotes the variance of the vector elements.

Mean Absolute Error

$$MAE(\mathbf{y}, \hat{\mathbf{y}}) = \frac{1}{n} \sum_{i=1}^{n} |y_i - \hat{y}_i|. \tag{2.87}$$

Mean Square Error

$$MSE(\mathbf{y}, \hat{\mathbf{y}}) = \frac{1}{n} \sum_{i=1}^{n} (y_i - \hat{y}_i)^2. \tag{2.88}$$

Median Absolute Error

$$MedAE(\mathbf{y}, \hat{\mathbf{y}}) = median(|y_1 - \hat{y}_1|, |y_2 - \hat{y}_2|, \dots, |y_n - \hat{y}_n|). \tag{2.89}$$

R^2 Score

$$R^2(\mathbf{y}, \hat{\mathbf{y}}) = 1 - \frac{\sum_{i=1}^{n} (y_i - \hat{y}_i)^2}{\sum_{i=1}^{n} (y_i - \bar{y})^2}, \tag{2.90}$$

where $\bar{y} = \frac{1}{n} \sum_{i=1}^{n} y_i$.

2.7.3 Clustering Evaluation Metrics

For the clustering tasks, the output results are actually not real classes but just identifier of clusters. Given a data set $\mathcal{D} = \{\mathbf{x}_i\}_{i=1}^{n}$ involving n data instances, we can denote the ground truth clusters and inferred clustering result as $\mathcal{C} = \{\mathcal{C}_1, \dots, \mathcal{C}_k\}$ and $\hat{\mathcal{C}} = \{\hat{\mathcal{C}}_1, \dots, \hat{\mathcal{C}}_k\}$, respectively. Furthermore, for all the data instances in $\mathcal{D}$, we can represent their inferred cluster labels and the real cluster labels as $\hat{\mathbf{y}}$ and $\mathbf{y}$, respectively. For any pair of data instances $\mathbf{x}_i, \mathbf{x}_j \in \mathcal{D}$, based on the predicted and true cluster labels, we can divide the instance pairs into four categories (S: Same; D: Different):

SS: Set $SS = \{(\mathbf{x}_i, \mathbf{x}_j) | y_i = y_j, \hat{y}_i = \hat{y}_j, i \neq j\}$, and we denote $|SS| = a$.
SD: Set $SD = \{(\mathbf{x}_i, \mathbf{x}_j) | y_i = y_j, \hat{y}_i \neq \hat{y}_j, i \neq j\}$, and we denote $|SD| = b$.
DS: Set $DS = \{(\mathbf{x}_i, \mathbf{x}_j) | y_i \neq y_j, \hat{y}_i = \hat{y}_j, i \neq j\}$, and we denote $|DS| = c$.
DD: Set $DD = \{(\mathbf{x}_i, \mathbf{x}_j) | y_i \neq y_j, \hat{y}_i \neq \hat{y}_j, i \neq j\}$, and we denote $|DD| = d$.

Here, set SS contains the data instance pairs that are in the same cluster in both the prediction results and the ground truth; set SD contains the data instance pairs that are in the same cluster in the ground truth but in different clusters in the prediction result; set DS contains the data instance pairs that are in different clusters in the ground truth but in the same cluster in the prediction result; and set DD contains the data instance pairs that are in different clusters in both the prediction results and the ground truth. We have these set sizes sum up to be $a + b + c + d = n(n-1)$.

Based on these notations, metrics like *Jaccard's Coefficient*, *FM Index* (Fowlkes and Mallows Index), and *Rand Index* can be defined

Jaccard's Coefficient

$$JC(\mathbf{y}, \hat{\mathbf{y}}) = \frac{a}{a + b + c}. \tag{2.91}$$

FM Index

$$FMI(\mathbf{y}, \hat{\mathbf{y}}) = \sqrt{\frac{a}{a + b} \cdot \frac{a}{a + c}}. \tag{2.92}$$

Rand Index

$$RI(\mathbf{y}, \hat{\mathbf{y}}) = \frac{2(a + d)}{a + b + c + d} = \frac{2(a + d)}{n(n - 1)}. \tag{2.93}$$

These above evaluation metrics can be used in the case where the clustering ground truth is available, i.e., $\mathcal{C}$ or $\mathbf{y}$ is known. In the case that no cluster ground truth can be obtained for the data instances, we will introduce another set of evaluation metrics based on the distance measures among the data instances. Formally, given a clustering result $\mathcal{C} = \{\mathcal{C}_1, \ldots, \mathcal{C}_k\}$, we introduce the following 4 concepts based on the instance distance measures

- **Average Distance**:

$$dist(\mathcal{C}_i) = \frac{2}{|\mathcal{C}_i|(|\mathcal{C}_i| - 1)} \sum_{j,l \in \{1,2,\ldots,|\mathcal{C}_i|\}, j < l} dist(\mathbf{x}_j, \mathbf{x}_l), \tag{2.94}$$

- **Diameter**:

$$diam(\mathcal{C}_i) = \max_{j,l \in \{1,2,\ldots,|\mathcal{C}_i|\}, j < l} dist(\mathbf{x}_j, \mathbf{x}_l), \tag{2.95}$$

- **Inter-Cluster Distance**:

$$d_{\min}(\mathcal{C}_i, \mathcal{C}_j) = \min_{\mathbf{x}_i \in \mathcal{C}_i, \mathbf{x}_j \in \mathcal{C}_j} dist(\mathbf{x}_i, \mathbf{x}_j), \tag{2.96}$$

- **Inter-Cluster Center Distance**:

$$d_{cen}(\mathcal{C}_i, \mathcal{C}_j) = dist(\boldsymbol{\mu}_i, \boldsymbol{\mu}_j). \tag{2.97}$$

Here $\boldsymbol{\mu}_i$ and $\boldsymbol{\mu}_j$ denote the centers of clusters $\mathcal{C}_i$ and $\mathcal{C}_j$, respectively. Based on these concepts, we will introduce two distance based clustering evaluation metrics: *DB Index* (Davies-Bouldin Index), and *Dunn Index*, which are frequently used in evaluating the clustering results quality.

DB Index

$$DBI = \frac{1}{k} \sum_{i=1}^{k} \max_{j \neq i} \left(\frac{dist(\mathcal{C}_i) + dist(\mathcal{C}_j)}{d_{cen}(\mathcal{C}_i, \mathcal{C}_j)} \right), \tag{2.98}$$

Dunn Index

$$DI = \min_{i \in \{1,2,\dots,k\}} \left\{ \min_{j \neq i} \left(\frac{d_{\min}(\mathcal{C}_i, \mathcal{C}_j)}{\max_{l \in \{1,2,\dots,k\}} diam(\mathcal{C}_l)} \right) \right\}. \tag{2.99}$$

Besides these metrics introduced in this section, there also exist a large number of evaluation metrics proposed for different application scenarios, which will not be introduced here since they are out of the scope of this book. We may introduce some of them in the following chapters when talking about the specific social network fusion or knowledge discovery problems.

2.8 Summary

In this chapter, we provided an overview of data operations and machine learning tasks, which aims at endowing computers with the ability to "learn" and "adapt." In machine learning, experiences are usually represented as data, and the main objective of machine learning is to derive "models" from data that can capture the complicated hidden patterns. These new models can be fed to the new data, which can provide us with the inference results matching the captured patterns.

We have also introduced the basic knowledge about data, including the data attributes and data types. The data attributes can be categorized into various types, including *numerical* and *categorical*. *Record data*, *graph data*, and *ordered data* together will form the three main data types, that will be studied in this book. We have also talked about the data characteristics, including *quality*, *dimensionality*, and *sparsity*. Several basic data processing operations, e.g., *data cleaning and pre-processing*, *data aggregating and sampling*, *data dimensionality reduction and feature selection* as well as *data transformation*, have been introduced in this chapter as well.

We divided the supervised learning tasks into *classification tasks* and *regression tasks*, respectively. For the classification part, we have introduced its problem setting, training/testing set splitting approach, and two well-known models, i.e., *decision tree* and *SVM*. Meanwhile, for the regression part, we provide the description for the problem setting and three regression models in the linear form, i.e., *linear regression*, *Lasso*, and *Ridge*. We mainly focused on *clustering* when introducing the unsupervised learning tasks in this chapter. Three clustering algorithms were introduced in this chapter, including two hard clustering approaches: *K-Means* and *DBSCAN*, and one soft clustering approach: *Mixture-of-Gaussian*.

We also provided an introduction to the neural network research works and the recent deep learning models in this chapter. Starting from the *classic M-P neuron model* to *perceptron*, *multi-layer perceptron model*, and the *multi-layer feed-forward neural network model*, we provided an overview about the neural network development history. To train the models, we covered the well-known *error back-propagation* algorithm by taking the *single hidden-layer feed-forward neural network model* as an example. The latest development of deep learning models is also introduced in this chapter, including *deep autoencoder*, *deep RNN*, and *deep CNN* models, respectively.

This chapter was concluded with the descriptions of several evaluation metrics for measuring the performance of classification, regression, and clustering models, respectively.

2.9   Bibliography Notes

The data mining textbook [49] provides a comprehensive introduction about data types, data characteristics, and data processing operations. The readers may refer to the textbook, especially the chapters regarding the data introduction part, for more information regarding certain topics that you are interested in.

For the decision tree, the most famous representative models include ID3 [37], C4.5 [38], and CART [5], and the readers may check these three papers as a guidance when reading the decision tree section. In selecting the optimal attributes to construct the decision tree internal nodes, *information gain*, *information gain ratio*, and *Gini index* [39,47] are usually the most frequently used metrics. The tree branch pruning strategies mentioned in this chapter have a detailed introduction in [38], and the interested readers may check that article for more information.

SVM initially published in [9] has dominated the machine learning research area for a long time, which also serves as the foundation of *statistical learning* later on. SVM can achieve an outstanding performance on textual classification task [22]. Assisted with the kernel trick, SVM can be applied to handle the non-linearly separable data instances. Meanwhile, the selection of the kernels is still an open problem by this context so far.

The K-Means algorithm introduced in this chapter actually has so many different variants, like K-Medoids [23] which uses data instances as the prototypes and K-Modes [21] that can handle discrete attributes. To detect soft clustering results, K-Means can be extended to the Fuzzy C-Means [4], where each data instance can belong to multiple clusters. Some methods have been proposed for selecting the optimal cluster number K [35]. However, in the real practice, parameter K is normally selected by trying multiple different values and selecting the best one from them.

Neural network is a *black-box* model, whose learning performance is extremely challenging to explain. If the readers are interested in neural networks, you are suggested to read the textbook [18], which provides a systematic introduction about neural network models. Meanwhile, in recent years, due to the surge of deep learning, the latest deep learning book [14] has also become very popular among researchers. BP was initially proposed by Werbos in [54] and later re-introduced by Rumelhart in [43–45]. By this context so far, the BP algorithm is still used as the main learning algorithm for training neural network models.

2.10   Exercises

1. (Easy) According to the attribute type categorization provided in Fig. 2.2, please indicate the types of attributes used in Table 2.1.
2. (Easy) What's the *"curse of dimensionality"*? Please briefly explain the concept and indicate the potential problems caused by large data dimensionality, as well as the existing methods introduced to resolve such a problem.
3. (Easy) Please compute the *Precision*, *Recall*, and F_1 for both *Model 1* and *Model 2* in Example 2.6, whose performance statistics are provided as follows:
   - **Mode 1**: with TP = 90, FP = 10, FN = 900, TN = 999,000;
   - **Model 2**: with TP = 0, FP = 100, FN = 0, TN = 999,900.
4. (Easy) Compare the *pre-pruning* and *post-pruning* strategies used in *decision tree* mode training, and indicate their advantages and disadvantages.
5. (Medium) Please briefly summarize the *kernel trick* used in *SVM*, and explain why *kernel trick* is helpful for training *SVM*.

6. (Medium) Please compare the L_1-norm and L_2-norm used in *Lasso* and *Ridge*, respectively. Since they both can regularize the model variables, please try to provide their advantages and disadvantages.

7. (Medium) Please explain why the *single hidden-layer feed-forward neural network* model provided in Fig. 2.15c can address the *XOR* problem.

8. (Hard) Based on the technique introduced in Sect. 2.3.2.2, please try to construct a *decision tree* for the data records provided in Table 2.1.

9. (Hard) Please prove that there exists the closed form solution to the *Ridge* regression model, i.e., matrix $(\mathbf{X}^\top \mathbf{X} + \alpha \mathbf{I})$ in Eq. (2.39) is invertible.

10. (Hard) Please try to prove that Eq. (2.23) introduced in Sect. 2.3.3.2 holds, i.e., "*Maxmin no greater than Minmax.*"

$$\max_{\boldsymbol{\alpha}} \min_{\mathbf{w}, b} L(\mathbf{w}, b, \boldsymbol{\alpha}) \leq \min_{\mathbf{w}, b} \max_{\boldsymbol{\alpha}} L(\mathbf{w}, b, \boldsymbol{\alpha}). \tag{2.100}$$

References

1. Y. Bengio, P. Simard, P. Frasconi, Learning long-term dependencies with gradient descent is difficult. IEEE Trans. Neural Netw. **5**(2), 157–166 (1994)
2. T. Bengtsson, P. Bickel, B. Li, Curse-of-dimensionality revisited: collapse of the particle filter in very large scale systems, in *Probability and Statistics: Essays in Honor of David A. Freedman*, vol. 2 (2008), pp. 316–334
3. S. Berchtold, C. Bohm, H. Kriegel, The pyramid-technique: towards breaking the curse of dimensionality, in *Proceedings of the 2002 ACM SIGMOD International Conference on Management of Data (SIGMOD '02)*, vol. 27, pp. 142–153 (1998)
4. J. Bezdek, *Pattern Recognition with Fuzzy Objective Function Algorithms* (Kluwer Academic Publishers, Norwell, 1981)
5. L. Breiman, J. Friedman, R. Olshen, C. Stone, *Classification and Regression Trees* (Wadsworth and Brooks, Monterey, 1984)
6. C. Brodley, P. Utgoff, Multivariate decision trees. Mach. Learn. **19**(1), 45–77 (1995)
7. O. Chapelle, B. Schlkopf, A. Zien, *Semi-supervised Learning*, 1st edn. (MIT Press, Cambridge, 2010)
8. J. Chung, Ç. Gülçehre, K. Cho, Y. Bengio, Empirical evaluation of gated recurrent neural networks on sequence modeling. CoRR, abs/1412.3555 (2014)
9. C. Cortes, V. Vapnik, Support-vector networks. Mach. Learn. **20**(3), 273–297 (1995)
10. J. Duchi, E. Hazan, Y. Singer, Adaptive subgradient methods for online learning and stochastic optimization. J. Mach. Learn. Res. **12**, 2121–2159 (2011)
11. B. Efron, T. Hastie, I. Johnstone, R. Tibshirani, Least angle regression. Ann. Stat. **32**, 407–499 (2004)
12. M. Ester, H. Kriegel, J. Sander, X. Xu, A density-based algorithm for discovering clusters a density-based algorithm for discovering clusters in large spatial databases with noise, in *Proceedings of the Second International Conference on Knowledge Discovery and Data Mining* (AAAI Press, Menlo Park, 1996)
13. S. Fahlman, C. Lebiere, The cascade-correlation learning architecture, in *Advances in Neural Information Processing Systems 2* (Morgan-Kaufmann, Burlington, 1990)
14. I. Goodfellow, Y. Bengio, A. Courville, *Deep Learning* (MIT Press, Cambridge, 2016). http://www.deeplearningbook.org
15. I. Guyon, A. Elisseeff, An introduction to variable and feature selection. J. Mach. Learn. Res. **3**, 1157–11182 (2003)
16. J. Hartigan, M. Wong, A k-means clustering algorithm. JSTOR Appl. Stat. **28**(1), 100–108 (1979)
17. D. Hawkins, The problem of overfitting. J. Chem. Inf. Comput. Sci. **44**(1), 1–12 (2004)
18. S. Haykin, *Neural Networks: A Comprehensive Foundation*, 2nd edn. (Prentice Hall PTR, Upper Saddle River, 1998)
19. S. Hochreiter, J. Schmidhuber, Long short-term memory. Neural Comput. **9**(8), 1735–1780 (1997)
20. A. Hoerl, R. Kennard, Ridge regression: biased estimation for nonorthogonal problems. Technometrics **42**(1), 80–86 (2000)
21. Z. Huang, Extensions to the k-means algorithm for clustering large data sets with categorical values. Data Min. Knowl. Discov. **2**(3), 283–304 (1998)

22. T. Joachims, Text categorization with support vector machines: learning with many relevant features, in *European Conference on Machine Learning* (Springer, Berlin, 1998)
23. L. Kaufmann, P. Rousseeuw, *Clustering by Means of Medoids* (North Holland/Elsevier, Amsterdam, 1987)
24. Y. Kim, Convolutional neural networks for sentence classification, in *Proceedings of the 2014 Conference on Empirical Methods in Natural Language Processing (EMNLP)* (Association for Computational Linguistics, Doha, 2014)
25. R. Kohavi, A study of cross-validation and bootstrap for accuracy estimation and model selection, in *International Joint Conference on Artificial Intelligence (IJCA)* (Morgan Kaufmann Publishers Inc., San Francisco, 1995)
26. A. Krizhevsky, I. Sutskever, G. Hinton, Imagenet classification with deep convolutional neural networks, in *Proceedings of the 25th International Conference on Neural Information Processing Systems (NIPS'12)* (Curran Associates Inc., Red Hook, 2012)
27. Y. Lecun, L. Bottou, Y. Bengio, P. Haffner, Gradient-based learning applied to document recognition, in *Proceedings of the IEEE* (IEEE, Piscataway, 1998)
28. J. Liu, S. Ji, J. Ye, SLEP: sparse learning with efficient projections. Technical report (2010)
29. W. McCulloch, W. Pitts, A logical calculus of the ideas immanent in nervous activity. Bull. Math. Biophys. **5**(4), 115–133 (1943)
30. M. Minsky, S. Papert, *Perceptrons: Expanded Edition* (MIT Press, Cambridge, 1988)
31. S. Pan, Q. Yang, A survey on transfer learning. IEEE Trans. Knowl. Data Eng. **22**(10), 1345–1359 (2010)
32. N. Parikh, S. Boyd, Proximal algorithms. Found. Trends Optim. **1**(3), 123–231 (2014)
33. R. Pascanu, C. Gulcehre, K. Cho, Y. Bengio, How to construct deep recurrent neural networks. CoRR, abs/1312.6026 (2013)
34. R. Pascanu, T. Mikolov, Y. Bengio, On the difficulty of training recurrent neural networks, in *Proceedings of the 30th International Conference on International Conference on Machine Learning (ICML'13)* (2013)
35. D. Pelleg, A. Moore, X-means: extending k-means with efficient estimation of the number of clusters, in *Proceedings of the 17th International Conference on Machine Learning, Stanford* (2000)
36. J. Platt, Sequential minimal optimization: a fast algorithm for training support vector machines. Technical report. Adv. Kernel Methods Support Vector Learning 208 (1998)
37. J. Quinlan, Induction of decision trees. Mach. Learn. **1**(1), 81–106 (1986)
38. J. Quinlan, *C4.5: Programs for Machine Learning* (Morgan Kaufmann Publishers Inc., San Francisco, 1993)
39. L. Raileanu, K. Stoffel. Theoretical comparison between the Gini index and information gain criteria. Ann. Math. Artif. Intell. **41**(1), 77–93 (2004)
40. C. Rasmussen, The infinite Gaussian mixture model, in *Advances in Neural Information Processing Systems 12* (MIT Press, Cambridge, 2000)
41. J. Rawlings, S. Pantula, D. Dickey, *Applied Regression Analysis*, 2nd edn. (Springer, Berlin, 1998)
42. F. Rosenblatt, The perceptron: a probabilistic model for information storage and organization in the brain. Psychol. Rev. **65**, 386 (1958)
43. D. Rumelhart, G. Hinton, R. Williams, Learning internal representations by error propagation, in *Parallel Distributed Processing: Explorations in the Microstructure of Cognition* (MIT Press, Cambridge, 1986)
44. D. Rumelhart, G. Hinton, R. Williams, Learning representations by back-propagating errors, in *Neurocomputing: Foundations of Research* (MIT Press, Cambridge, 1988)
45. D. Rumelhart, R. Durbin, R. Golden, Y. Chauvin, Backpropagation: the basic theory, in *Developments in Connectionist Theory. Backpropagation: Theory, Architectures, and Applications* (Lawrence Erlbaum Associates, Inc., Hillsdale, 1995)
46. R. Salakhutdinov, G. Hinton, Deep Boltzmann machines, in *Proceedings of the Twelfth International Conference on Artificial Intelligence and Statistics* (2009)
47. C. Shannon, A mathematical theory of communication. SIGMOBILE Mob. Comput. Commun. Rev. **5**(1), 3–55 (2001)
48. D. Svozil, V. Kvasnicka, J. Pospichal, Introduction to multi-layer feed-forward neural networks. Chemom. Intell. Lab. Syst. **39**(1), 43–62 (1997)
49. P. Tan, M. Steinbach, V. Kumar, *Introduction to Data Mining (First Edition)* (Addison-Wesley Longman Publishing Co., Inc., Boston, 2005)
50. R. Tibshirani, The lasso method for variable selection in the cox model. Stat. Med. **16**, 385–395 (1997)
51. L. Van Der Maaten, E. Postma, J. Van den Herik, Dimensionality reduction: a comparative review. J. Mach. Learn. Res. **10**, 66–71 (2009)
52. M. Verleysen, D. François, The curse of dimensionality in data mining and time series prediction, in *Computational Intelligence and Bioinspired Systems. International Work-Conference on Artificial Neural Networks* (Springer, Berlin, 2005)
53. P. Vincent, H. Larochelle, I. Lajoie, Y. Bengio, P. Manzagol, Stacked denoising autoencoders: learning useful representations in a deep network with a local denoising criterion. J. Mach. Learn. Res. **11**, 3371–3408 (2010)

54. P. J. Werbos, *Beyond Regression: New Tools for Prediction and Analysis in the Behavioral Sciences*. PhD thesis, Harvard University, Cambridge, 1974
55. X. Yan, X. Su, *Linear Regression Analysis: Theory and Computing* (World Scientific Publishing Co., Inc., River Edge, 2009)
56. J. Zhang, L. Cui, Y. Fu, F. Gouza, Fake news detection with deep diffusive network model. CoRR, abs/1805.08751 (2018)
57. X. Zhu, Semi-supervised learning literature survey. Comput. Sci. **2**(3), 4 (2006)

Social Network Overview

3

3.1 Overview

Online social networks (OSNs) denote the online platforms that are used by people to build social connections with the other people, who may share similar personal or career interests, backgrounds, or real-life connections. Online social networking sites vary a lot and there exist a large number of online social sites of different categories, including *online sharing sites*, *online publishing sites*, *online networking sites*, *online messaging sites* and *online collaborating sites*. Each category of these online social networks can provide specific featured services for the customers. For instance, Facebook allows users to socialize with each other via making friends, posting text, sharing photos and videos; Twitter focuses on providing micro-blogging services for users to write/read the latest news and messages; Foursquare is a location-based social network offering location-oriented services; and Instagram is a photo and video sharing social site among friends or to the public. To enjoy different kinds of social network services simultaneously, users nowadays are usually involved in multiple online social sites at the same time, in each of which they will all form social connections and generate social information.

Generally, online social networks can be represented as graphs in mathematics. Besides the users, there usually exist many other types of information entities, like posts, photos, videos, and comments, generated by users' online social activities. Information entities in online social networks are extensively connected, and the connections among different types of nodes usually have different physical meanings. The diverse nodes and connections render online social network to be a very complex graph structure. Meanwhile, depending on the categories of information entities and connections involved, the online social networks can be divided into different types, like *homogeneous network* [48], *bipartite network* [65], and *heterogeneous network* [51]. To model the phenomenon that users are involved in multiple networks, a new concept called "*multiple aligned heterogeneous social networks*" [29, 71–73] has been proposed in recent years.

Different online social networks are usually of different characteristics, which can be quantified with some network measures formally. Users in online social networks can have different numbers of connections, which can be quantified as the user *node degree* [2, 8] mathematically. User nodes of a larger degree will be more important (in terms of social connections) generally. A more formal concept indicating the node importance is called the *node centrality* [11], which can be quantified with many different measures. Connections are very important for online social networks, and node connection measures quantifying the linking behaviors of nodes in the networks are of great interests.

J. Zhang, P. S. Yu, *Broad Learning Through Fusions*,
https://doi.org/10.1007/978-3-030-12528-8_3

Based on the connections among nodes, the social closeness measures between pairs of nodes can be calculated, where user nodes who frequently interact with each other will have a larger closeness score. As to the local social connection patters, they may also follow the social balance theory, e.g., "friends of my friend are my friends."

For the networks with simple structures, like the homogeneous networks merely involving users and friendship links, the social patterns are usually easy to study. However, for the networks with complex structures, like the heterogeneous networks, the nodes can be connected by different types of links sequentially, which are of different physical meanings. One general technique for heterogeneous network studies is "*meta path*" [53,73], which specifically depicts certain link sequences connecting the nodes based on the network schema. The meta path concept can also be extended to the *multiple aligned social network* scenario [29,73], which can connect the nodes across different social networks. The machine learning approaches introduced in the previous chapter are very general learning models, which take the feature representation data as the input and output the predicted labels of the data instances. There actually also exist some learning algorithms proposed for the network structured data specifically, like the *random walk* approach [36].

In this chapter, we will provide the definitions of some important concepts that are useful for the social network studies, including the basic graph related concepts, and some advanced social network concepts, like *meta path* [53,73]. A clear categorization of the network types will be provided, and some network measures will be introduced to illustrate the properties of the networks. Finally, an introduction about some network-based models will be provided. These concepts, network categories, network measures, and approaches will be frequently used and mentioned in the following chapters of this book.

3.2   Graph Essentials

In mathematics and computer science, the online social networks are generally represented as graphs [60], where the information entities are denoted as the nodes and the connections among the information entities are represented as the links. In this section, we will provide some basic introductory knowledge about graph, including its representations and the connectivity properties.

3.2.1   Graph Representations

Graphs can be represented in different forms, like a traditional graph definition involving nodes and links, an *adjacency matrix* indicating the connectivity among nodes, *adjacency list* and *link list*.

Definition 3.1 (Graph) Formally, a graph can be represented as $G = (\mathcal{V}, \mathcal{E})$, where $\mathcal{V}$ denotes the set of nodes and $\mathcal{E}$ represents the set of links in the graph G.

Generally, node is the basic entity unit in graphs, which can represent different types of information entities when using the graph definition to represent social networks. For instance, in online social networks, a node can denote a user, a post, a comment, and a photo. The formal representation of the node set $\mathcal{V}$ can be denoted as

$$\mathcal{V} = \{v_1, v_2, \ldots, v_n\}, \tag{3.1}$$

where v_i $(1 \leq i \leq n)$ represents a single node in the graph and the node set size (i.e., the size of the graph) is $|\mathcal{V}| = n$.

Meanwhile, the different kinds of connections among the information entities are represented as the links in the graphs, which bear various physical meanings. For instance, in online social networks, the links among users can denote their friend/follow relationships, the links between users and posts denote the post-writing action, and the links between posts and spatial "(latitude, longitude)" coordinate pairs denote the check-ins attached to the posts. Formally, the set of links in the network can be represented as

$$\mathcal{E} = \{e_1, e_2, \ldots, e_m\}, \tag{3.2}$$

where $e_j = (v_o, v_p)$ $(1 \leq j \leq m)$ denotes a link/node pair in the graph. The size of the link set in the network can be represented as $|\mathcal{E}| = m$.

Besides the aforementioned regular graph definition, a graph can also be represented as an *adjacency matrix*, which indicates the connectivity among the nodes.

Definition 3.2 (Adjacency Matrix) Given a graph $G = (\mathcal{V}, \mathcal{E})$, we can represent its corresponding adjacency matrix as a binary matrix $\mathbf{A} = \{0, 1\}^{n \times n}$, where the rows and columns of the matrix correspond to the nodes in G and entry $A(i, j) = 1$ iff link $(v_i, v_j) \in \mathcal{E}$.

The graph definition and its adjacency matrix representation actually have equivalent representation capacity, and the transformation between which can be achieved very easily. Various properties of the graphs can also be revealed by their adjacency matrices as well. For instance, if a graph has a very small number of connections compared with the number of nodes in it, the corresponding adjacency matrix of the graph will be very sparse [43]. Meanwhile, if the nodes in the graph actually form several communities where the nodes in each community tend to have dense connections compared with those outside the communities, the corresponding graph adjacency matrix will have a lower rank [47].

Besides the *adjacency matrix*, the other graph representations include adjacency list. Let set $\Gamma(u_i) = \{u_j | u_j \in \mathcal{V}, (u_i, u_j) \in \mathcal{E}\} \subset \mathcal{V}$ denote the neighbors that user u_i connects to. The adjacency list representation of graph G can be represented as $\{(u_i, \Gamma(u_i))\}_{u_i \in \mathcal{V}}$.

Example 3.1 For instance, given a graph illustrated in Fig. 3.1a, we can represent the graph as

$$G = (\mathcal{V}, \mathcal{E}), \tag{3.3}$$

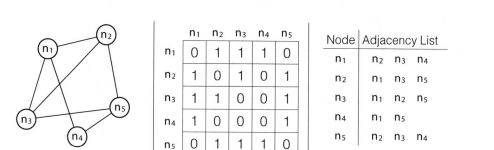

	n_1	n_2	n_3	n_4	n_5
n_1	0	1	1	1	0
n_2	1	0	1	0	1
n_3	1	1	0	0	1
n_4	1	0	0	0	1
n_5	0	1	1	1	0

Node	Adjacency List
n_1	n_2 n_3 n_4
n_2	n_1 n_3 n_5
n_3	n_1 n_2 n_5
n_4	n_1 n_5
n_5	n_2 n_3 n_4

A   B   C

Fig. 3.1 An example of different graph representations. ((**a**) Graph; (**b**) adjacency matrix; (**c**) adjacency list)

where the node set $\mathcal{V} = \{n_1, n_2, n_3, n_4, n_5\}$ contains five nodes and the link set $\mathcal{E} = \{(n_1, n_2), (n_1, n_3), (n_1, n_4), (n_2, n_3), (n_2, n_5), (n_3, n_5), (n_4, n_5)\}$ covers seven links. In the graph, there are five different nodes $\{n_1, n_2, n_3, n_4, n_5\}$, where they are connected by seven links. In the graph, all the nodes are connected with three other nodes, except n_4 which is connected to n_1 and n_5 only. We show its adjacency matrix and adjacency list representations in Fig. 3.1b, c, respectively. For any connected node pairs in the graph, the corresponding entry in the matrix will be filled with value 1; otherwise they have value 0 instead. For instance, link (n_2, n_3) connects nodes n_2 and n_3. In the adjacency matrix, the (2nd, 3rd) entry and the (3rd, 2nd) are both filled with value 1. The graph is also represented as an adjacency list as shown in Fig. 3.1c. For each node in the graph, we provide a list of nodes connected with the nodes. For instance, node n_3 is connected with nodes n_1, n_2, and n_5 simultaneously, which will form the adjacency list of node n_3.

3.2.2 Connectivity in Graphs

Connectivity [8] is an important property of graphs, where nodes are connected with each other via either direct connections or paths consisting of a sequence of links. Formally, given a graph G and a node n in the graph, the set of nodes that are adjacent to n in the graph are called the *adjacent neighbors* of n in the graph G.

Definition 3.3 (Adjacent Neighbor) Given a graph $G = (\mathcal{V}, \mathcal{E})$, the *adjacent neighbors* of node n in G can be represented as $\Gamma(n) = \{n' | n' \in \mathcal{V} \wedge (n, n') \in \mathcal{E}\}$.

Adjacent neighbor set is an important concept in social network studies. For instance, given a social network, the *adjacent neighbor* set of a user denote the online friends that the user is connected to, which is very useful for analyzing the socialization patterns and preference of users in the social network.

Meanwhile, given a node n in a network G, we can call the set of links incident to n in the graph as the *incident links* of node n.

Definition 3.4 (Incident Link) Given a graph $G = (\mathcal{V}, \mathcal{E})$ and a node $n \in \mathcal{V}$, the set of *incident link* set of n in G can be represented as $\Delta(n) = \{e | e \in \mathcal{E} \wedge \exists n' \in \mathcal{V}, (n, n') = e\}$.

Furthermore, we can also define the *incident relationships* between two links. Formally, given two links (a, b) and (c, d) in graph G, (a, b) is said to be incident to (c, d) iff $a = c \vee a = d \vee b = c \vee b = d$, i.e., they share a common node. Based on this definition, we can define the concepts of *walk*, *path*, *trail*, *tour*, and *cycle* of graph G as follows:

- **Walk**: Formally, a *walk* can be denoted as a sequence of nodes $n_1, n_2, \ldots, n_k$ from set $\mathcal{V}$, where there exists a link between any sequential pairs of nodes in the graph. For any three sequential nodes in the sequence, e.g., n_i, n_{i+1}, n_{i+2}, the links (n_i, n_{i+1}) and (n_{i+1}, n_{i+2}) are *incident* to each other sharing a common node n_{i+1}. Furthermore, if the ending node n_k is the same as the starting node n_1 in the *walk*, then it will be called a *closed walk*; otherwise, it is called an *open walk*. The length of the walk is formally defined as the number of links involved in the walk. For instance, sequence $n_1, n_2, \ldots, n_k$ forms a walk of length $k - 1$.
- **Trail**: A *trail* denotes a *walk* in the graph G, where all the links are distinct. By traveling along a *trail*, each link in the *trail* can be visited once, but the nodes can be visited multiple times. The shortest *trail* in graph G can be just one link in the graph.

- **Tour**: A *closed trail* (i.e., the starting and ending nodes of the *trail* are the same) is called a *tour*.
- **Path**: Given a *walk* in the graph G, if all the nodes and links in the *walk* are distinct, the *walk* will be a *path* in the graph. A *path* is also a *trail* in the graph.
- **Cycle**: A *closed path* is defined as a *cycle* in graphs. A *cycle* is also a special type of *tour*.

To help explain the above concepts, we also provide an example as follows, which lists the *walk*, *trail*, *tour*, *path*, and *cycle* instances from the input graph.

Example 3.2 For instance, based on the graph illustrated in Fig. 3.2, the node sequences

1. "$n_1, n_2, n_3, n_5, n_4, n_1, n_2$" is a *walk* of length 6,
2. "n_1, n_2, n_3, n_1, n_4" is a *trail* of length 4,
3. "n_1, n_2, n_5, n_4, n_1" is a *tour* of length 4,
4. "n_1, n_3, n_5, n_2" is a *path* of length 3,
5. "$n_1, n_2, n_3, n_5, n_4, n_1$" is a *cycle* of length 5

in the graph, respectively.

The above concepts can help correlate the nodes in the graphs which are not directly connected with each other.

Definition 3.5 (Reachable) Formally, given two nodes n_i and n_j in the graph G, n_i is said to be *reachable* from n_j iff there is a *path* from n_j to n_i.

For a subset of nodes, which are *reachable* from each other, they together with the links among them will form a *connected component* in the graph.

Definition 3.6 (Connected Component) Given a graph $G = (\mathcal{V}, \mathcal{E})$, the subgraph $G' = (\mathcal{V}', \mathcal{E}')$ is said to be a connected component of G iff $\mathcal{V}' \subset \mathcal{V}$, $\mathcal{E}' \subset \mathcal{E}$, and for any pair of nodes in $\mathcal{V}'$ they are *reachable* via the links in $\mathcal{E}'$.

Example 3.3 For instance, based on the input graph illustrated in Fig. 3.2, the subgraph $G' = (\{n_1, n_2, n_4, n_5\}, \{(n_1, n_2), (n_2, n_5), (n_4, n_5), (n_1, n_4)\})$ will be a *connected component* of the input graph. Meanwhile, considering that all the nodes in the network are *reachable* to each other, and the original network itself is also a *connected component* actually.

Fig. 3.2 An input graph example

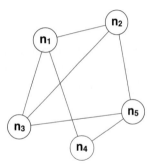

Based on the graph links $\mathcal{E}$, there may exist multiple *paths* of different lengths connecting a certain pair of nodes (e.g., n_i, n_j). Meanwhile, the *path* of the shortest length can be of great importance and has concrete applications in many research problems, like *traffic route planning* [6]. Formally, such a *path* is also named as the *shortest path* in graphs.

Definition 3.7 (Shortest Path) Given a pair of nodes $n_i, n_j \in \mathcal{V}$ in the graph G, the set of *paths* connecting n_i and n_j based on G can be represented as $\mathcal{P}$, in which one of the shortest lengths is called the *shortest path* between n_i and n_j:

$$SP(n_i, n_j) = \min_{p \in \mathcal{P}} |p|, \tag{3.4}$$

where $|p|$ denotes the length of path p.

The *shortest path* between different node pairs in a graph can be of different lengths, where the longest *shortest path* between nodes in graph G is also defined as the *diameter* of the graph.

Definition 3.8 (Graph Diameter) Formally, given a graph G, the *diameter* of graph G can be represented as

$$Diameter(G) = \max_{n_i, n_j \in \mathcal{V}} SP(n_i, n_j). \tag{3.5}$$

Example 3.4 For instance, based on the graph illustrated in Fig. 3.2, the *shortest path* between (1) nodes n_1 and n_2 is "$n_1 \rightarrow n_2$" (of length 1), and (2) nodes n_2 and n_4 is "$n_2 \rightarrow n_5 \rightarrow n_4$" of length 2 (or "$n_2 \rightarrow n_1 \rightarrow n_4$"). For any two nodes selected from the graph, we observe that the *shortest path* length between them are no greater than 2, i.e., the *diameter* of the graph is 2.

3.3   Network Measures

The networks are usually of different structures and will have different properties, which can be indicated by various measures about either the nodes, links, or the overall network structure. In this part, we will introduce a number of measures about the networks, including the *degree* [2, 8] and *centrality* [11] about nodes, *similarity* [67] about node pairs (i.e., the links), and the *transitivity* [19] and *social balance* [20, 57] about the network structures.

3.3.1   Degree

Degree [2, 8] can effectively indicate the number of connections associated with nodes in graphs, which is a very important node measure. In this part, we will introduce the *node degree* concept and the *node degree distribution* [2] in graphs.

3.3.1.1   Node Degree

Given an undirected network $G = (\mathcal{V}, \mathcal{E})$, the node degree denotes the number of edges incident to the nodes, whose formal definition is provided as follows.

Definition 3.9 (Degree) The *degree* of node u in an undirected network $G = (\mathcal{V}, \mathcal{E})$ denotes the number of links incident to it, i.e., $d(u) = |\{(u, v)|v \in \mathcal{V}, (u, v) \in \mathcal{E}\}|$.

In an undirected network, each link will be incident to two nodes, and the total node degree of a network will always be an even number. Furthermore, as to the specific numbers of the degrees, we have the following theorem.

Theorem 3.1 *Given an undirected network $G = (\mathcal{V}, \mathcal{E})$, the total number of node degrees equal to twice the number of links in the network, i.e.,*

$$\sum_{u \in \mathcal{V}} d(u) = 2|\mathcal{E}|. \tag{3.6}$$

Proof In network G, the total node degree can be represented as $\sum_{u \in \mathcal{V}} d(u)$. The removal of link $(u, v) \in \mathcal{E}$, will lower down the degree of nodes u and v by 1, respectively. The total node degree after removing link (u, v) will be equal to $\sum_{u \in \mathcal{V}} d(u) - 2$. After removing all the links (i.e., $|\mathcal{E}|$ links) from the network, the total node degree will be reduced to 0 as all the nodes are isolated without any connections. Therefore, $\sum_{u \in \mathcal{V}} d(u) - 2|\mathcal{E}| = 0$, and we have

$$\sum_{u \in \mathcal{V}} d(u) = 2|\mathcal{E}|. \tag{3.7}$$

In the case that links in the networks are directed, the node degree concept will be further refined into *node in-degree d_{in}* and *node out-degree d_{out}*, which denotes the number of links coming into the nodes and those going out from the nodes, respectively.

Theorem 3.2 *Given a directed network $G = (\mathcal{V}, \mathcal{E})$, the total number of node in-degree and out-degree are both equal to the number of nodes in the network, i.e.,*

$$\sum_{u \in \mathcal{V}} d_{in}(u) = \sum_{u \in \mathcal{V}} d_{out}(u) = |\mathcal{E}|. \tag{3.8}$$

Proof Similarly, we can represent the total node in-degree and out-degree of network G as $\sum_{u \in \mathcal{V}} d_{in}(u)$ and $\sum_{u \in \mathcal{V}} d_{out}(u)$, respectively. From network G, the removal of each link $(u, v) \in \mathcal{E}$ will decrease the out-degree of u and in-degree of v by 1. Therefore, after the removal of link (u, v), the new total node in-degree and out-degree of network G will be $\sum_{u \in \mathcal{V}} d_{in}(u) - 1$ and $\sum_{u \in \mathcal{V}} d_{out}(u) - 1$, respectively. After removing all the links in $\mathcal{E}$, the node in-degree and out-degree will be decreased to 0, and all the nodes will become isolated without any connections. In other words, we have $\sum_{u \in \mathcal{V}} d_{in}(u) - |\mathcal{E}| = \sum_{u \in \mathcal{V}} d_{out}(u) - |\mathcal{E}| = 0$, which implies that

$$\sum_{u \in \mathcal{V}} d_{in}(u) = \sum_{u \in \mathcal{V}} d_{out}(u) = |\mathcal{E}|. \tag{3.9}$$

3.3.1.2 Degree Distribution

Node degree is an important property about the nodes, while the distribution of the node degrees displays an important property of the whole network instead. Given a node degree value d, we can represent the proportion of nodes with degree d as

$$P(d) = \frac{|\{v|v \in \mathcal{V}, d(v) = d\}|}{|\mathcal{V}|}, \tag{3.10}$$

where the numerator denotes the number of nodes with degree d.

All the potential degree values of nodes in the network can be represented as set $\mathcal{D} = \{d(u)|\forall u \in \mathcal{V}\}$. Therefore, the node degrees together with the corresponding proportions will be represented as a tuple set $\{(d, P(d))\}_{d \in \mathcal{D}}$, which can be represented as a distribution plot with degrees as the x axis and the proportions as the y axis.

Example 3.5 For instance, given an undirected network shown in Fig. 3.2, there exist five nodes n_1, n_2, n_3, n_4, n_5 with degrees 3, 3, 3, 2, 3, respectively. Therefore, the node degree and proportion tuples can be represented as $\left\{\left(2, \frac{1}{5}\right), \left(3, \frac{4}{5}\right)\right\}$. We can represent the degree distribution in Fig. 3.3, where majority of the nodes have degree 3 (the largest node degree in the network) and a small proportion of nodes have degree 2 (the smallest node degree in the network).

Such a degree distribution about the toy example shown in Fig. 3.3 is not common in the real-world social networks. In many of the cases, most of the users are regular users with a limited number of friends online (i.e., a small degree), and a small number of celebrities can have a large number of friends (i.e., a large degree).

Example 3.6 In Fig. 3.4, we show the degree distribution plots of two crawled data sets about the Foursquare and Twitter online social networks, where each of them contains about 5000 users. According to the plots, we observe that the user fraction generally drops as the node degree increases in both Foursquare and Twitter. Among all the users, most of the users in both Foursquare and Twitter have a very small degree (less than 10). Compared with Foursquare, users in Twitter have more dense connections and tend to have larger degrees. For instance, the fraction of users with small degrees in Twitter is less than that in Foursquare (i.e., the red dots are below the blue dots for small degrees), while the Twitter user fractions of larger degrees are above those of Foursquare (i.e., the right part of the plot). According to the plot, there also exists one user in the Twitter network with a degree greater

Fig. 3.3 Degree distribution of the example network

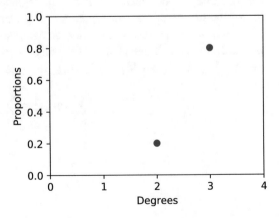

Fig. 3.4 Degree
distribution of the
Foursquare and Twitter
networks

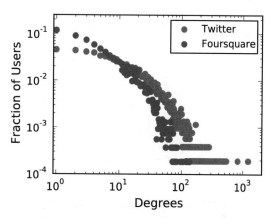

than 1000, i.e., the rightmost red dot, whose is usually a celebrity with a great number of followers in the networks.

3.3.2 Centrality

The concept *centrality* [11] defines how important a node is in the network. To quantify the node importance in the networks, different kinds of metrics can be applied to define the node *centrality*, which will be introduced in this part.

3.3.2.1 Degree Centrality

In the real-world online social networks, the users with lots of connections (i.e., large degrees) tend to be important, as their roles are recognized by other users via the connections with them. Therefore, the node importance can be quantified as the node degrees. Given an undirected network G, the *degree-based centrality* [11, 65] of a node u in the network can be defined as

$$C_d(u) = d(u). \tag{3.11}$$

All the nodes in G can be ordered by their *degree-based centrality*, where the nodes with larger degrees will be more important compared with other nodes with smaller degrees. Meanwhile, given a directed network G, the node centrality can be defined as either their in-degrees, out-degrees, or in-degrees together with out-degrees, which can be formally represented as follows:

$$C_{in}(u) = d_{in}(u), \tag{3.12}$$

$$C_{out}(u) = d_{out}(u), \tag{3.13}$$

$$C_{in/out}(u) = d_{in}(u) + d_{out}(u). \tag{3.14}$$

Example 3.7 For instance in Fig. 3.5, we show a graph with five nodes and five undirected links. Based on the *degree centrality*, among all the nodes in the graph, node n_1 has the largest centrality score, i.e., 3, compared with the remaining nodes. Node n_3 has the smallest centrality score, i.e., 1, and the remaining nodes all have a centrality score of 2.

Fig. 3.5 An input graph
example

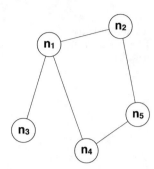

3.3.2.2 Normalized Degree Centrality

Generally, the *degree-based centrality* in different networks is usually of different scale. For example, the Facebook network is of a much larger scale compared with other social networks, like Twitter and Foursquare, and the *degree-based centrality* in Facebook is usually much larger than that in Twitter and Foursquare. To ensure the comparability of the *degree-based centrality* across different networks, one method is to normalize all the centrality measures to a common value interval. Here, different numbers can be used as the denominator for centrality rescaling, e.g., the *maximal degree, sum degree,* and *maximum degree,* which will bring about different *normalized degree centrality* measures.

The maximal number of nodes each node can be connected within a network is $|\mathcal{V}| - 1$, which can be applied to rescale the *degree centrality* to the range [0, 1]. It actually helps define the *maximal degree-based normalized degree centrality*:

$$C_{\max}(u) = \frac{C(u)}{|\mathcal{V}| - 1}. \tag{3.15}$$

Another way to do the normalization will be to define the centrality as the ratio of the degrees with regard to the total degree in the networks, i.e., the *sum degree-based normalized degree centrality*:

$$C_{sum}(u) = \frac{C(u)}{\sum_{v \in \mathcal{V}} d(v)} = \frac{C(u)}{2 \times |\mathcal{E}|}. \tag{3.16}$$

Generally, in the online social networks, few nodes can achieve a degree with values $|\mathcal{V}| - 1$ or $2 \times |\mathcal{E}|$. In other words, these two normalized node degree centrality measure values are highly to be concentrated in a very narrow region $[0, \alpha]$ ($\alpha < 1$ and can be a very small number), where the α will also be different for different online social networks and violate the comparability objective. To resolve such a problem, we propose to normalize the measures with maximum node degree instead, which defines the *maximum degree-based normalized degree centrality*:

$$C_{maximum}(u) = \frac{C(u)}{\max_{v \in \mathcal{V}} d(v)}. \tag{3.17}$$

3.3.2.3 Eigen-Centrality

In the *degree centrality* definition, the users having more friends are assumed to be more important by default. However, in the real world, it can be not the case. Instead of having lots of online friends, *users having more important friends will be more important.* In other words, the users' *centrality* is determined by their online friends' *centrality* [46], i.e.,

$$C(u) = \frac{1}{\lambda} \sum_{v \in \Gamma(u)} C(v), \tag{3.18}$$

where set $\Gamma(u) = \{v | v \in \mathcal{V} \wedge (u, v) \in \mathcal{E}\}$ denotes the set of online neighbors of user u in the network G and λ is a constant scalar.

By organizing the social connections among users in the network as the social adjacency matrix $\mathbf{A} \in \{0, 1\}^{|\mathcal{V}| \times |\mathcal{V}|}$, we can rewrite the above equation as follows:

$$\lambda \mathbf{c} = \mathbf{A}^\top \mathbf{c}, \tag{3.19}$$

where vector $\mathbf{c} = [C(u_1), C(u_2), \ldots, C(u_{|\mathcal{V}|})]^\top$ contains all the centrality values of users in the network.

The above equation indicates that the centrality vector is actually a eigenvector of the social adjacency matrix $\mathbf{A}^\top$, whose corresponding eigenvalue is λ. However, given a matrix $\mathbf{A}^\top$, it will have multiple eigenvectors and eigenvalues. Usually, we prefer to use the positive values to define the centrality measure. According to the Perron–Frobenius theorem [42], given a matrix, there always exists a non-negative eigenvector of the matrix, which corresponds to the largest eigenvalue of $\mathbf{A}$. Therefore, we will use the eigenvector corresponding to the largest eigenvalue of matrix $\mathbf{A}^\top$ to define the *eigen-centrality* [11, 46].

Example 3.8 For example, given an undirected graph shown in Fig. 3.6a, we can represent the adjacency matrix of the undirected input graph as $\mathbf{A} = \begin{bmatrix} 0, 1, 1, 1, 0 \\ 1, 0, 1, 0, 1 \\ 1, 1, 0, 0, 1 \\ 1, 0, 0, 0, 1 \\ 0, 1, 1, 1, 0 \end{bmatrix}$. By decomposing the matrix, we can achieve the eigenvalues of matrix $\mathbf{A}$ to be $[2.856, -2.177, 1.429 \times 10^{-16}, 0.322, -1.0]$. Its largest eigenvalue is 2.856, and the corresponding eigenvector can be represented as

$$\mathbf{c} = \begin{bmatrix} 0.456 \\ 0.491 \\ 0.491 \\ 0.319 \\ 0.456 \end{bmatrix}, \tag{3.20}$$

which denotes the *centrality scores* achieved by the nodes in the graph.

Fig. 3.6 A directed input graph example. (**a**) Undirected graph, (**b**) directed graph

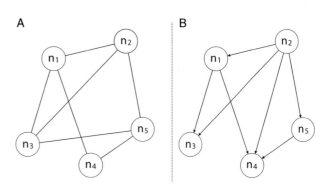

In other words, nodes n_2 and n_3 actually have the largest centrality score among all the nodes in the graph, which is 0.491; the next group will be nodes n_1 and n_5 with a centrality score 0.456; and node n_4 has the lowest centrality score, which is 0.319.

Example 3.9   In Fig. 3.6b, we show an example of a directed input graph with different connections. According to the graph structure, we can represent the graph adjacency matrix as $\mathbf{A} = \begin{bmatrix} 0, 0, 1, 1, 0 \\ 1, 0, 1, 1, 1 \\ 0, 0, 0, 0, 0 \\ 0, 0, 0, 0, 0 \\ 0, 0, 0, 1, 0 \end{bmatrix}$.

By decomposing the matrix, we can achieve its eigenvalues to be 0 for all the nodes in the graph, which may make the *eigen-centrality* fail to work in handling the directed graphs.

3.3.2.4 Katz Centrality

As shown in the previous example, when the networks are directed, the *eigen-centrality* measure may suffer from some serious problems. To overcome such a problem, a new centrality measure, the *Katz centrality* [11], has been proposed, which is defined as follows:

$$\mathbf{c} = \alpha \cdot \mathbf{A}^\top \mathbf{c} + \beta \cdot \mathbf{1}, \tag{3.21}$$

where parameters α and β denote the weights of the *eigen-centrality* and the bias term, respectively. In the case that matrix $\mathbf{I} - \alpha \cdot \mathbf{A}^\top$ is invertible, the *Katz centrality* vector can be formally represented as

$$\mathbf{c} = \beta \cdot (\mathbf{I} - \alpha \cdot \mathbf{A}^\top)^{-1} \cdot \mathbf{1}. \tag{3.22}$$

To ensure the invertibility of matrix $\mathbf{I} - \alpha \cdot \mathbf{A}^\top$, the choice of parameter α can be a little bit tricky. Smaller α tends to unify the *Katz centrality* of all the nodes in the network closer to the value of β, while larger α will reduce the effectiveness of the bias term. In practice, $\alpha < \frac{1}{\lambda_{max}}$ is usually selected, where λ_{max} denotes the maximum eigenvalue of matrix $\mathbf{A}^\top$.

Example 3.10   For instance, given the directed graph shown in Fig. 3.6b, by assigning the parameters $\alpha = \beta = 0.5$, we have the *Katz centrality* vector as follows:

$$\mathbf{c} = \begin{bmatrix} 1.0 \\ 1.875 \\ 0.5 \\ 0.5 \\ 0.75 \end{bmatrix}, \tag{3.23}$$

among which node n_2 has the largest *Katz centrality* (i.e., 1.875) in the input graph.

3.3.2.5 PageRank Centrality

Both *eigen centrality* and *Katz centrality* treat all the neighbor nodes in graphs equally when calculating the centrality scores for the target node. However, in the real world, the impacts of the neighbor nodes are usually different in determining users' centrality score. For example, in online social networks, users like to get connected with celebrities, and these celebrities will be connected with lots of people even though they may not necessary know each other in person. Usually, the celebrities are very important users in online social networks, and they have a large *centrality* score compared against the other users. However, for the users who are connected with these celebrities, we cannot say that they are also important as well. To consider such a phenomenon, a pagerank-based centrality measure has been introduced to provide different neighbors with different weights (determined by their degrees). Formally, the *pagerank centrality* [12] of user u can be defined as

$$C_p(u) = \alpha \cdot \sum_{v \in \Gamma(u)} \frac{C_p(v)}{|\Gamma(v)|} + \beta, \tag{3.24}$$

where the effects from u's neighbors, like $v \in \Gamma(u)$, are weighted by $\frac{1}{|\Gamma(v)|}$. Here, the subscript p denotes the *pagerank*-based *centrality* score.

In other words, for the neighbors with large degrees, their impacts on u will be penalized in the *centrality* score computation, while people with a small degree will have a greater impact on u instead. Formally, the above equation can be rewritten as follows:

$$\mathbf{c} = \alpha \cdot \mathbf{A}^{\top}\mathbf{D}^{-1}\mathbf{c} + \beta \cdot \mathbf{1}, \tag{3.25}$$

where matrix $\mathbf{D} = diag(d_{out}(u_1), d_{out}(u_2), \ldots, d_{out}(u_{|\mathcal{V}|}))$ is a diagonal matrix with the node out-degrees on its diagonal. In the case that matrices $\mathbf{D}$ and $(\mathbf{I} - \alpha\mathbf{A}^{\top}\mathbf{D}^{-1})$ are both invertible, we can have the *pagerank centrality* vector to be

$$\mathbf{c} = \beta \cdot (\mathbf{I} - \alpha \cdot \mathbf{A}^{\top}\mathbf{D}^{-1})^{-1} \cdot \mathbf{1}. \tag{3.26}$$

Parameter α can be selected with similar methods as introduced after Eq. (3.22).

Example 3.11 For example, we can take the directed graph shown in Fig. 3.7 as the input graph, and its *adjacency matrix* together with the *out-degree* diagonal matrix can be represented as

$$\mathbf{A} = \begin{bmatrix} 0,0,1,1,0 \\ 1,0,1,1,1 \\ 0,0,0,0,1 \\ 0,1,1,0,0 \\ 0,0,0,1,0 \end{bmatrix}, \mathbf{D} = \begin{bmatrix} 2,0,0,0,0 \\ 0,4,0,0,0 \\ 0,0,1,0,0 \\ 0,0,0,1,0 \\ 0,0,0,0,1 \end{bmatrix}. \tag{3.27}$$

Fig. 3.7 An input graph for pagerank centrality calculation

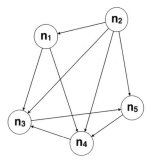

By assigning $\alpha = \beta = 0.5$, we can compute the *pagerank centrality* scores of nodes in the graph to be

$$
\mathbf{c} = \beta \cdot (\mathbf{I} - \alpha \cdot \mathbf{A}^\top \mathbf{D}^{-1})^{-1} \cdot \mathbf{1} = \begin{bmatrix} 0.563 \\ 0.5 \\ 1.406 \\ 1.969 \\ 0.563 \end{bmatrix}. \tag{3.28}
$$

Among all the nodes, n_4 has the largest *pagerank centrality* score compared against the other nodes, and n_2 has the lowest *pagerank centrality* score on the other hand.

3.3.2.6 Betweenness Centrality

The centrality measures aforementioned are mostly defined based on the neighborhood information for the nodes. Another way to define the centrality measure is based on their positions connecting nodes in the networks, which is called the node *betweenness centrality* [11, 16] measure. Generally, if a node u effectively joins the connection paths among nodes in the network, then its position will be more important. Formally, the *betweenness centrality* measure of node u can be defined as

$$
C_b(u) = \sum_{s,t \in \mathcal{V}, s \neq t \neq v} \frac{|\mathcal{P}_{s,t}(u)|}{|\mathcal{P}_{s,t}|}, \tag{3.29}
$$

where $\mathcal{P}_{s,t}(u)$ denotes the set of *shortest paths* between nodes s and t via u in the network, and $\mathcal{P}_{s,t}$ represents the set of all *shortest paths* connecting s and t.

For a node u, it can achieve the maximum *between centrality* if it appears on all the shortest paths $\left(\text{i.e., } \frac{|\mathcal{P}_{s,t}(u)|}{|\mathcal{P}_{s,t}|} = 1\right)$ of all the node pairs in the network, like the central node in the star-structured graph. Formally, in such a case given a network with node set $\mathcal{V}$, the maximum *between centrality* node u achieves can be represented as

$$
\begin{aligned}
C_b^{\max}(u) &= \sum_{s,t \in \mathcal{V}, s \neq t \neq v} \frac{|\mathcal{P}_{s,t}(u)|}{|\mathcal{P}_{s,t}|} \\
&= \sum_{s,t \in \mathcal{V}, s \neq t \neq v} 1 \\
&= 2 \binom{|\mathcal{V}| - 1}{2} \\
&= (|\mathcal{V}| - 1)(|\mathcal{V}| - 2).
\end{aligned} \tag{3.30}
$$

To ensure the *betweenness closeness* measure in different networks are comparable, one effective way will be to rescale the *betweenness centrality* to range [0, 1] with the maximum *between centrality* in the network.

$$
C_{n-b}(u) = \frac{C_{n-b}(u)}{C_b^{\max}(u)} = \frac{\sum_{s,t \in \mathcal{V}, s \neq t \neq v} \frac{|\mathcal{P}_{s,t}(u)|}{|\mathcal{P}_{s,t}|}}{(|\mathcal{V}| - 1)(|\mathcal{V}| - 2)}, \tag{3.31}
$$

To compute the shortest path between all pairs of nodes in a graph $G = (\mathcal{V}, \mathcal{E})$, algorithms, like the Floyd–Warshall algorithm, can be used with an $O(|\mathcal{V}|^3)$ time cost. In the exercises, we will have an example about the *betweenness centrality*, and the readers can try to compute the node centrality scores according to the above definitions.

3.3.3 Closeness

Via the connections, nodes in networks will be closely correlated with each other and have different closeness scores with each other. In this part, we will introduce several frequently used *closeness* [67] measures for the node pairs in networks. To illustrate the measures more clearly, we will use the social networks as an example, where the nodes denote the users and links represent the friendship connections.

3.3.3.1 Local Structure-Based Closeness Measures

Many node closeness measures can calculate the proximity among user nodes with the social network local structure information, like the shared common neighbors. In this part, we will introduce a number of local network structure-based user node closeness metrics as follows, which can effectively measure the social proximity scores among the users.

- **Reciprocity**: For the social networks involving directed links among the nodes (i.e., the link denotes the *follow* relationship), given a pair of nodes u and v in the network, there could exist a link between them inside the networks. For example, if user u follows v in the network, there will exist a directed link $u \rightarrow v$ (i.e., (u, v)) pointing from user u to user v. When measuring the closeness between users u and v, the connected user pairs are generally much closer to each other compared against the disconnected ones. Viewed in this perspective, if u follows v (or v follows u), such a link will indicate the strong closeness between these two users. Meanwhile, in the real-world online social networks, most users tend to follow the celebrities. The follow link between regular users and the celebrities may not necessarily denote they are close in the network, like the celebrities may not even know his/her followers.

  One measure that can denote the closeness between two users, e.g., u and v, in the social networks is the *reciprocal links* [22]. Given that user u follows v in the network (i.e., (u, v) exists in the network), if v also follows u back (i.e., (v, u) also exists), then u and v tend to be very close to each other. Here, link (v, u) will be called the *reciprocal link* of (u, v). The *reciprocal links* can also correctly measure the closeness between regular users and celebrities in social networks. For instance, if regular user u follows a celebrity v in a social network, and v also follows u via a *reciprocal link*, it can indicate that u and v tend to know each other and should be close to each other. Such a measure will also work for two regular users or two celebrity users.

  Formally, the *reciprocity closeness* measure between users u and v can be represented as

  $$C_R(u, v) = \mathbb{I}((u, v) \in \mathcal{E} \wedge (v, u) \in \mathcal{E}), \tag{3.32}$$

  where $\mathcal{E}$ denotes the link set in the social network and $\mathbb{I}(\cdot)$ returns 1 if the condition can hold. Besides measuring the closeness between pairs of user nodes, the *reciprocity* can also be applied to measure the closeness of the whole network G, which can be represented as

  $$C_R(G) = \frac{\sum_{u,v \in \mathcal{V}, u \neq v} C_R(u, v)}{|\mathcal{V}|(|\mathcal{V}| - 1)}$$

  $$= \frac{\sum_{u,v \in \mathcal{V}, u \neq v} \mathbb{I}((u, v) \in \mathcal{E} \wedge (v, u) \in \mathcal{E})}{|\mathcal{V}|(|\mathcal{V}| - 1)}. \tag{3.33}$$

The *reciprocity* of a network denotes among all the potential user pairs in the network, how many percentages of them have the bi-directional follow links. For a network with a larger *reciprocity* score, the connections among users in the network will be stronger, which also indicates closer relationships among the internal nodes.

- **Common Neighbor**: Reciprocity is a closeness measure based on the connections between pairwise user nodes in the network. Actually, besides such pairwise links, via the connections with the other neighbors, many other closeness measures can be defined for user pairs in social networks as well, like the *common neighbor* (CN) [35, 67] closeness measure.

Given two users $u, v \in V$ in an undirected social network, if u and v share lots of common friends, it will indicate that they are highly likely to be close friends and may know each other. Formally, according to the introduction provided in the previous sections, we can formally represent the set of online friends whom users u, v have in the network as sets $\Gamma(u)$ and $\Gamma(v)$, respectively. The *common neighbor* closeness measure between users u and v can be formally represented as

$$C_{CN}(u, v) = |\Gamma(u) \cap \Gamma(v)|. \tag{3.34}$$

For the directed networks, we can define several more refined common neighbor measures, like common in-neighbors (i.e., the common followers), common out-neighbors (i.e., the common followees), common all-neighbors (i.e., the common connected neighbors regardless of the link directions), since the links among users will have a specific direction.

- **Jaccard's Coefficient**: Considering that $CN(u, v)$ can be a very large value merely because the two users both have a lot of neighbors rather than they are strongly related to each other. In other words, the common neighbor measure will have some problems when being used to compute the closeness between certain active users, e.g., the celebrities sharing lots of common fans. Furthermore, the common neighbor measure can neither be used to compare the closeness among the user pairs in different networks, due to the different network scales. One way to overcome these aforementioned problems will be to normalize the common neighbor measures with the users' degrees, which will introduce the following *Jaccard's coefficient* [24, 67] measure.

Given the two users u and v in an undirected network, we can represent the *Jaccard's coefficient* closeness measure between them as

$$C_{JC}(u, v) = \frac{|\Gamma(u) \cap \Gamma(v)|}{|\Gamma(u) \cup \Gamma(v)|}, \tag{3.35}$$

where the denominator denotes the number of users connected to either u or v. Therefore, for the celebrities, users, or networks with a relatively large scales, the user node closeness will be rescaled by assigning them with a larger penalty.

In the case that the networks are directed, different other types of directed versions of Jaccard's coefficient measures can be defined, just like the directed *common neighbor* measures we define before. Jaccard's Coefficient can be treated as a weighted version of common neighbor, where each shared neighbor is assigned with an identical weight $\frac{1}{|\Gamma(u) \cup \Gamma(v)|}$. Many other weights can also be applied actually, like $\frac{1}{|\Gamma(u)| + |\Gamma(v)|}$ used in *Sørensen Index* [50], $\frac{1}{\min\{|\Gamma(u)|, |\Gamma(v)|\}}$ used in *Hub Promoted Index* [44], $\frac{1}{\max\{|\Gamma(u)|, |\Gamma(v)|\}}$ used in the *Hub Depressed Index* [80], and $\frac{1}{|\Gamma(u)| \times |\Gamma(v)|}$ in the *Leicht–Holme–Newman Index* [31].

- **Adamic/Adar**: Meanwhile, in measuring the closeness between users, different common users will play a different role and should have a different weight. To achieve such a goal, a closeness measure *Adamic/Adar* (AA) [1, 67] index is proposed, which penalizes the shared neighbor nodes with larger degrees. Formally, the AA index between users u and v can be defined as

$$C_{AA}(u, v) = \sum_{w \in (\Gamma(u) \cap \Gamma(v))} \frac{1}{\log |\Gamma(w)|}. \tag{3.36}$$

For each of the common neighbor w shared by u and v, the weight assigned to w is $\frac{1}{\log |\Gamma(w)|}$ in AA. The shared common neighbors with smaller degrees will play an important role in indicating the closeness between the user pair. For the directed networks, by considering the link directions, several directed version of AA can be introduced as well. Besides AA, some other similar measures have been proposed, which assign the shared common neighbors with a different weight, like $\frac{1}{|\Gamma(w)|}$ used in the *Resource Allocation Index* (RA) [80].

3.3.3.2 Global Path-Based Closeness Measure

In addition to the local network structure-based closeness measures, many other closeness measures based on paths throughout the network have also been proposed to measure the proximity among the user nodes.

- **Shortest Path**: Generally, the social closeness among users can be measured by the distance among them in the network structure. Given two users who are far away from each other via all the potential paths connecting them (or they are isolated without any paths), they will have a very low closeness score. On the other hand, for the users who are directly connected via a link or a path of a very short length, they should be closer to each other compared with the isolated users. Based on such an intuition, we can define the closeness measure based on the distance of the *shortest path* [67] connecting users in the network:

$$C_{SP}(u, v) = \min\{|p|\}_{p \in \mathcal{P}_{u,v}}, \tag{3.37}$$

  where $\mathcal{P}_{u,v}$ represents the set of paths connecting users u and v inside the network, and $|p|$ denotes the distance of path p.

- **Katz**: Besides the shortest path, all the potential paths connecting user pairs in the networks can indicate their social closeness. Meanwhile, longer paths will show weaker closeness, and shorter paths denote stronger closeness. The *Katz* closeness measure [25, 67] can integrate all these paths together to define the closeness scores among the users in the networks. Formally, the *Katz* closeness between users u and v can be defined as

$$C_{Katz}(u, v) = \sum_{l=1}^{l_{\max}} \beta^l |\mathcal{P}_{u,v}^l|, \tag{3.38}$$

  where $l_{\max}$ denotes the longest path connecting u and v, $\mathcal{P}_{u,v}^l$ denotes the set of paths of length l connecting u and v in the network. Parameter $\beta \in [0, 1]$ is a regularizer term. Normally, smaller β favors shorter paths as β^l can decay very quickly as l increases when β is small, in which case the *Katz* measure will behave like the closeness measures based on the local neighbors introduced before.

3.3.3.3 Random Walk-Based Closeness Measure

In addition to the closeness measures that can be calculated from the network structure directly, there also exist another category of closeness measures that can calculate the closeness scores among users based on *random walk* [36, 67]. In this part, we will introduce the concept of *random walk* first, and provide the introduction to several closeness measures based on it, including *hitting time* [37, 67], *commute time* [33, 67], and *cosine similarity* [23, 67].

Formally, given a network $G = (\mathcal{V}, \mathcal{E})$, let matrix $\mathbf{A} \in \{0, 1\}^{|\mathcal{V}| \times |\mathcal{V}|}$ be the adjacency matrix of network G, where entry $A(i, j) = 1$ iff link $(u_i, u_j) \in \mathcal{E}$. The normalized matrix of $\mathbf{A}$ by rows can be represented as $\mathbf{P} = \mathbf{D}^{-1}\mathbf{A}$, where diagonal matrix $\mathbf{D}$ of $\mathbf{A}$ has value $D(i, i) = \sum_j A(i, j)$ on its diagonal and $P(i, j)$ denotes the probability of stepping on node u_j from node u_i during the walk process. Let vector $\mathbf{x}^{(\tau)}(i)$ denote the probabilities that a random walker is located at user node $u_i \in \mathcal{V}$ at time τ. Then such a probability vector at time $\tau + 1$ will be updated as follows:

$$\mathbf{x}^{(\tau+1)}(i) = \sum_j \mathbf{x}^{(\tau)}(j)P(j, i). \tag{3.39}$$

In other words, the updating equation of vector $\mathbf{x}$ will be as follows, and such an updating process will continue until convergence, i.e.,

$$\text{Updating Equation: } \mathbf{x}^{(\tau+1)} = \mathbf{P}^\top \mathbf{x}^{(\tau)}, \tag{3.40}$$

$$\text{Convergence Equation: } \mathbf{x}^{(\tau+1)} = \mathbf{x}^{(\tau)}, \tag{3.41}$$

which will lead to the final stationary distribution vector $\mathbf{x}$ to be

$$\mathbf{x} = \mathbf{P}^\top \mathbf{x}. \tag{3.42}$$

The above equation denotes that the final stationary probability distribution vector $\mathbf{x}$ of random walk is actually an eigenvector of matrix $\mathbf{P}^\top$ corresponding to eigenvalue 1. Some existing works [15] have pointed out that if a Markov chain is *irreducible* and *aperiodic* then the largest eigenvalue of the transition matrix $\mathbf{P}^\top$ will be equal to 1 and all the other eigenvalues will be strictly less than 1. In addition, in such a condition, there will exist a unique stationary distribution which is vector $\mathbf{x}$ obtained at convergence of the updating equations. Here, we will not cover the proof to the above statement, which will be left as an exercise for the readers at the end of this chapter.

- **Hitting Time**: Let a variable $x^{(\tau)} = u$ denote that a random walker is at node u at step τ, and the *hitting time*-based closeness measure between users u and v can be represented as:

$$C_{HT}(u, v) = \mathbb{E}(\{\tau | x^{(\tau)} = v \wedge x^{(0)} = u\}), \tag{3.43}$$

where $\mathbb{E}(\cdot)$ denotes the expectation of the variable.

Considering a random walker can reach v from u via different paths. The above equation denotes the expected number of steps to reach v from u, which is also called the *average hitting time* [37, 67]. Generally, close friends in the online social networks will have a small *average hitting time*.

Another way to define the *hitting time* between nodes u and v is to count the minimum number of steps needed to reach v from u, which can be represented as:

$$C_{mHT}(u, v) = \min\{\tau | x^{(\tau)} = v \wedge x^{(0)} = u\}, \tag{3.44}$$

which is also called the *minimum hitting time* measure.

- **Commute Time**: According to the above definition of *hitting time*, we can see that the measure is actually asymmetric, i.e., $C_{HT}(u, v) \neq C_{HT}(v, u)$, especially when the networks are directed. Such an asymmetric property will cause some problems when applying the *hitting time* in measuring the closeness among users in the real-world social networks. To overcome such a problem, some new measures, like *Commute Time* [33, 67], have been proposed, which counts the *hitting time* between user pairs from both of the directions, i.e.,

$$C_{CT}(u, v) = C_{HT}(u, v) + C_{HT}(v, u). \tag{3.45}$$

Formally, based on the adjacency matrix $\mathbf{A}$, we can define its corresponding Laplace matrix as $\mathbf{L} = \mathbf{D} - \mathbf{A}$ ($\mathbf{D}$ is a diagonal matrix). The pseudo-inverse matrix of $\mathbf{L}$ can be represented as $\mathbf{L}^\dagger$, and the *commute time* for user pairs (u_i, u_j) can be represented as

$$C_{CT}(u_i, u_j) = 2|\mathcal{E}| \cdot (L^\dagger(i, i)$$
$$+ L^\dagger(j, j) - 2L^\dagger(i, j)). \tag{3.46}$$

The proof to the above equation will not be provided here, and more detailed information for the proof is available in [67].

- **Cosine Similarity**: With the pseudo-inverse matrix $\mathbf{L}^\dagger$, we can introduce a vector $\mathbf{z}_u = (\mathbf{L}^\dagger)^{\frac{1}{2}} \mathbf{e}_u$ and vector $\mathbf{e}_u$ is a binary vector of 0s except the entries corresponding to node u which is filled with 1. According to existing works, the closeness between users u and v can be defined based on the *cosine similarity* [23, 67] measure of vectors $\mathbf{z}_u$ and $\mathbf{z}_v$ as follows:

$$C_{CS}(u, v) = \frac{\mathbf{z}_u^\top \mathbf{z}_v}{\sqrt{(\mathbf{z}_u^\top \mathbf{z}_u)(\mathbf{z}_v^\top \mathbf{z}_u)}}. \tag{3.47}$$

Furthermore, based on the pseudo-inverse matrix $\mathbf{L}^\dagger$, the above cosine similarity can be represented as

$$C_{CS}(u_i, u_j) = \frac{L^\dagger(i, j)}{\sqrt{L^\dagger(i, i) \cdot L^\dagger(j, j)}}. \tag{3.48}$$

These above closeness measures are all defined based on the regular *random walk* model. Meanwhile, in recent years, several variant *random walk* models have been proposed, which allow the walker to jump back to the starting point with a certain chance. Based on the definition of random walk, if the walker is allowed to return to the starting point with a probability of $1-c$, where $c \in [0, 1]$, then the new random walk method is formally defined as *random walk with restart* (RWR) [41], whose updating equation is shown as follows:

$$\mathbf{x}_u^{(\tau+1)} = c\mathbf{P}^\top \mathbf{x}_u^{(\tau)} + (1-c)\mathbf{e}_u, \tag{3.49}$$

where vector $\mathbf{x}_u^{(\tau+1)}$ denotes the probability of the random walker at all the nodes in the network starting from u initially.

By keeping updating the vector $\mathbf{x}_u^{(\tau+1)}$ until convergence, if matrix $(\mathbf{I} - c\mathbf{P}^\top)$ is invertible, we can have the stationary distribution vector of the RWR model to be

$$\mathbf{x}_u = (1 - c)(\mathbf{I} - c\mathbf{P}^\top)^{-1}\mathbf{e}_u, \tag{3.50}$$

Furthermore, the closeness measure between user pairs u and v with the RWR model can be represented as

$$C_{RWR}(u, v) = \mathbf{x}_u(v), \tag{3.51}$$

where entry $\mathbf{x}_u(v)$ denotes the stationary probability of walking from u to v based on the RWR model.

3.3.4 Transitivity and Social Balance

The links in online social networks actually create various relationships among users. In this part, we will analyze several important properties about social networks based on the connections, which include *social transitivity* [19], *clustering coefficient* [5], and *social balance* [20, 57], respectively.

3.3.4.1 Social Transitivity
In discrete mathematics, a relation R on the domain $\mathcal{D}$ is a transitive relation iff $\forall u, v, w \in \mathcal{D}$ the following equation can hold:

$$R(u, v) \wedge R(v, w) \rightarrow R(u, w). \tag{3.52}$$

The transitive relation can also be used to describe the social connections among users in online social networks. In the real world, there is a social phenomenon that

> Friends of my friend can also be my friend.

Such a social phenomenon has been adopted in many friend recommender systems in online social networks for either recommendation or candidate pruning. Given three users $u, v, w \in \mathcal{V}$ in an online social network, if users u, v are friends, v, w are friends (i.e., links $(u, v), (v, w) \in \mathcal{E}$), and u, w also happen to be friends in the network, then we can observe a transitive friend relation among the three users. These three users together with the friendship connections among them will form a triangle. Therefore, to measure the transitivity of a network, the number of triangles existing in the network can be an important signal.

3.3.4.2 Clustering Coefficient
For a network with denser connections, there tend to be more triangles formed by the users in the network. We can measure how close a network compared to a complete network (i.e., a network with all node pairs connected) with the *clustering coefficient* concept. Formally, the network *clustering coefficient* [5] denotes among any three user nodes in the network, given that there exist two links connecting them already, how many of them will form triangles.

Formally, let set $\mathcal{P}^2 = \{(u, v, w)|u, v, w \in \mathcal{V} \wedge (u, v) \in \mathcal{E} \wedge (v, w) \in \mathcal{E}\}$ denote the node triples forming paths of length 2, and $\mathcal{T} = \{(u, v, w)|u, v, w \in \mathcal{V} \wedge (u, v) \in \mathcal{E} \wedge (v, w) \in \mathcal{E} \wedge (u, w) \in \mathcal{E}\}$

Fig. 3.8 An input graph for network clustering coefficient calculation

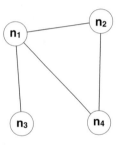

represent the set of node triples forming a triangle. We can represent the *clustering coefficient* of the network structure as follows:

$$CC = \frac{|\mathcal{T}|}{|\mathcal{P}^2|}. \tag{3.53}$$

Since in each triangle, there exist six different closed paths of length 2 and 2 different connected node triples in a path of length 2, the above equation can also be rewritten as follows:

$$CC = \frac{\text{Number of triangles} \times 6}{|\mathcal{P}^2|}$$
$$= \frac{\text{Number of triangles} \times 6}{\text{Number of connected triples of nodes} \times 2}, \tag{3.54}$$

which can make the counting works simpler.

Example 3.12 In Fig. 3.8, we show an input graph with four nodes and four links. Among all these nodes, there exists one single triangle structure, i.e., the triangle involving n_1, n_2, and n_4. Meanwhile, there are five different paths of length 2, i.e., $n_1 - n_2 - n_4$, $n_2 - n_4 - n_1$, $n_4 - n_1 - n_2$, $n_3 - n_1 - n_2$, and $n_3 - n_1 - n_4$. Therefore, according to the above definition, we can calculate the *clustering coefficient* score of the network to be $\frac{1 \times 6}{5 \times 2} = \frac{3}{5}$.

3.3.4.3 Social Balance

Another concept strongly correlated with *transitivity* is *social balance* [20,57], which denotes whether a triangle social structure is balanced or not especially in *signed networks* [32,77]. A *signed network* denotes a social network, where the links are associated with polarities (either positive or negative). Depending on the specific network settings, the polarities attached to the links will have different physical meanings, like *trust* vs. *distrust* [63], *friend* vs. *enemy* [59], and *good attitude* vs. *bad attitude* [64].

The *social balance* theory describes the consistency of the signed connections among users. Some informal cases of *social balanced* structures in networks include:

Friends of my friend can be my friend,
Friends of my enemy can be my enemy,
Enemies of my friend can be my enemy,
Enemies of my enemies can be my friend.

Given three users $u, v, w \in \mathcal{V}$ in a network, we can represent the signs of relationships among them as $s_{u,v}$, $s_{v,w}$ and $s_{u,w}$, respectively. For instance, sign $s_{u,v} = +1$ denotes that users u and v are friends, while sign $s_{u,v} = -1$ denotes that users u and v are enemies. The relationships among these

Fig. 3.9 Examples of structures based on the social balance theory

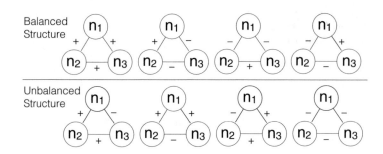

three users in the above four cases will form the *balanced structures*, and all the remaining structures among these three users are all called *unbalanced structure*.

Example 3.13   For instance, in Fig. 3.9, we provide an example about the *balanced* and *unbalanced* social structures formed by three users (i.e., n_1, n_2, and n_3). Among three users in a triangle, there can exist eight different social structure formed by them with signed links, which are shown in Fig. 3.9. In these eight cases, four of them are *balanced* (as shown at the top) and four are *unbalanced* (as shown at the bottom).

Actually, there exists a very simple method to determine whether a social structure is *balanced* or *unbalanced*. Based on the sign notations, the triangle formed by users u, v, w is a *balanced structure*, iff

$$s_{u,v} \cdot s_{v,w} \cdot s_{u,w} \geq 0. \tag{3.55}$$

Otherwise, the structure is said to be *unbalanced*.

3.4   Network Categories

The network concept introduced in the previous section can be used to model various types of network structured datasets available in the real world, including *online social networks* [38], *bibliographical networks* [52], *transportation networks* [6], and *computer networks* [10]. For instance, when we use the concept to define the *online social networks*, those various types of information entities in the social networks can be represented as the nodes, while the connections among the information entities are denoted as the links. Different online social networks are usually of different properties, and the corresponding network representations will have different kinds of characteristics as well.

For example, in some online social networks, the social connections among users can be (1) either *directed* (e.g., the social connections are the uni-directional follow links) or undirected (e.g., the social connections denote the bi-directional friendship links); (2) either *weighted* (e.g., users have different closeness scores with their friends) or *unweighted* (i.e., no closeness information is indicated in defining the social links); and (3) either *signed* (e.g., friendship links have different physical meanings actually and the link polarities denote different social attitudes) or *unsigned* (no social attitude information is provided in defining the social links).

Given a network $G = (\mathcal{V}, \mathcal{E})$, the nodes and links involved in it usually belong to different categories. Formally, we can represent the sets of node and link types involved in the network as

$\mathcal{N}$ and $\mathcal{R}$, respectively. Meanwhile, the corresponding network definition can be updated by adding the mappings indicating the node and link type information.

Definition 3.10 (Network) Formally, a network structured data can be represented as $G = (\mathcal{V}, \mathcal{E}, \phi, \psi)$, where $\mathcal{V}, \mathcal{E}$ are the sets of nodes and links in the network, and mappings $\phi : \mathcal{V} \to \mathcal{N}$, $\psi : \mathcal{E} \to \mathcal{R}$ project the nodes and links to their specific types, respectively. In many cases, the mappings ϕ, ψ are omitted assuming that the node and link types are known by default.

In this section, depending on the categories of information involved in the networks, we propose to categorize the network data into three groups: *homogeneous networks* [48], *heterogeneous networks* [51], and *multiple aligned heterogeneous networks* [29, 71–73], which will be introduced as follows, respectively.

3.4.1 Homogeneous Network

Definition 3.11 (Homogeneous Network) For a network $G = (\mathcal{V}, \mathcal{E}, \phi, \psi)$, if there exists one single type of nodes and one single type of links in the network (i.e., $|\mathcal{N}| = |\mathcal{R}| = 1$), then the network is called a *homogeneous network*.

Many different types of network structures can be represented as the *homogeneous networks* actually, like online social networks [38] involving users and friendship links only, company internal organizational network [74, 76, 78] involving employees and the management relationships, and computer networks [10] involving PCs and the internet connections. *Homogeneous networks* are one of the simplest network structures, analysis of which can provide many fundamental knowledge about networks with more complex structures. In the following part, we will introduce several common *homogeneous network* structures first.

3.4.1.1 Friendship Networks

Friendship network is one of the most common homogeneous social network structures, and they can be represented as the graph $G = (\mathcal{V}, \mathcal{E})$ defined before, where $\mathcal{V}$ represents the set of individuals while $\mathcal{E}$ denotes the set of social relationships among these individuals. Depending on whether the links in G are directed or undirected, the social links can denote either the *follow* links or *friendship* links among the individuals. Given an individual $u \in \mathcal{V}$ in an undirected friendship social network, the set of individuals connected to u can be represented as the friends of user u in the network G, denoted as $\Gamma(u) \subset \mathcal{V} = \{v | (u, v) \in \mathcal{E}\}$. The number of friends that user u has in the network is also called the degree of node u, i.e., $|\Gamma(u)|$.

Meanwhile, in a directed network G, the set of individuals followed by u (i.e., $\Gamma_{out}(u) = \{v | (u, v) \in \mathcal{E}\}$) are called the followees of u; and the set of individuals that follow u (i.e., $\Gamma_{in}(u) = \{v | (v, u) \in \mathcal{E}\}$) are called the followers of u. The number of users who follow u is called the in-degree of u, and the number of users followed by u is called the out-degree of u in the network. For the users with large out-degrees, they are called the *hubs* [27] in the network; while those with large in-degrees, they are called the *authorities* [27] in the network.

Example 3.14 In Fig. 3.10, we provide two examples of *friendship networks*, where plot (a) involves an undirected network and plot (b) contains a directed network. The links in plot (a) denote the friendship links, while those in plot (b) represent the follow links. Among all the users in plot (b),

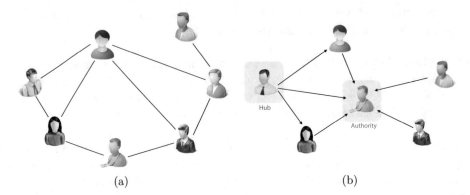

Fig. 3.10 Examples of friendship networks: (**a**) Undirected friendship network, (**b**) directed friendship network

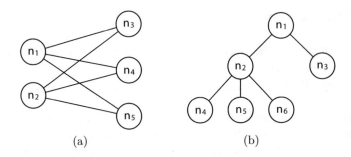

Fig. 3.11 Examples of homogeneous networks: (**a**) Bipartite network, (**b**) tree

we can identify one *authority user*, i.e., the one in blue square box with lots of in-links, and one *hub user*, i.e., the one in red square box with many out-links.

3.4.1.2 Computer Network

For the computer networks, like a local area network (LAN) or a wide area network (WAN), involving a set of computers and the access relationships among the computers, they can also be represented as the homogeneous networks as well. Generally, in a computer web, depending on the roles, the computers in the web network can serve as either the servers or the PCs. The PCs are the regular computers used by the end users, while the servers usually host some websites. The PCs can access the servers by visiting the websites or connecting with them via secure shell (SSH). If we don't consider the access relationships among the PCs and servers, respectively, then the computers together with their access relationships will form a *bipartite network* [65].

Definition 3.12 (Bipartite Computer Network) Formally, a *bipartite network* can be represented as $G = (\mathcal{V}_L \cup \mathcal{V}_R, \mathcal{E})$, where $\mathcal{V}_L$ and $\mathcal{V}_R$ denote the nodes on the left and right sides in the network and $\mathcal{E} \subset \mathcal{V}_L \times \mathcal{V}_R$ represents the access relationships between nodes on the left and right sides.

Example 3.15 An example of a *bipartite computer network* is shown in Fig. 3.11a, which involve five different nodes (two on the left and three on the right) and six links, where all the nodes on the left side are connected with the nodes on the right side. According to the above definition, the *bipartite network* can be formally represented as $G = (\{n_1, n_2\} \cup \{n_3, n_4, n_5\}, \{n_1, n_2\} \times \{n_3, n_4, n_5\})$.

3.4.1.3 Company Organizational Network

In many cases, the network structure or the sub-network structure of interest is a tree-structured diagram. Formally, in mathematics and computer science, *tree* is a special type of connected graph with no cycles formed by the nodes. As shown in Fig. 3.11b, for the nodes in *trees*, those with degree 1 are called the *leaf nodes* (i.e., the ones at the bottom) and the remaining ones are called *internal nodes*. The *tree* structured networks have several important properties, like *every tree has at least one edge and at least two nodes*, and *every tree with n nodes has exactly n − 1 links*. We will not provide the formal proof of these statements here. Tree is an important concept in networks representations, and many important network structures can be represented as a *tree* formally, like the *company organizational chart* as discussed in [74, 76, 78].

Definition 3.13 (Company Organizational Chart) Formally, a company management structure can be represented as a *rooted tree* $T = (\mathcal{V}, \mathcal{E}, root)$, where $\mathcal{V}$ and $\mathcal{E}$ denote the employees and management relationships among the employees in the company. Node $root \in \mathcal{V}$ usually denotes the CEO of the company.

Example 3.16 An example of the *company organizational chart* is shown in Fig. 3.12. As shown in the figure, in the *company organizational chart*, all the employees will have their managers except the CEO (i.e., Adam in the plot). The employees who are not in a management position (i.e., the leaf nodes) are named as the *base employees*. Different from the regular social networks, there generally exist no cycles in terms of management relationships in the *company organizational chart*. It is very important for companies, as a clear outline of the positions and responsibilities of the employees can avoid management confusion and chaos. What's more, in the *company organizational chart*, employees at higher levels can be connected to multiple lower-level employees, i.e., the *subordinates*, at the same time. Meanwhile, each employee at lower levels will be connected to one single employee at higher level, i.e., the *manager*. In other words, managers can manage multiple employees simultaneously, while each employee reports to one single manager.

Besides the *company organizational network*, many other networks can also be represented as tree structured diagram, like *ontologies* [18] outlining the relationships among different *categories of beings*, and the *cascades* [26] in information diffusion indicating how information propagates from the source users to the other users in the network.

Fig. 3.12 An example of company organizational chart

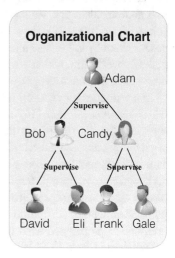

3.4.2   Heterogeneous Network

Definition 3.14 (Heterogeneous Network) For a network $G = (\mathcal{V}, \mathcal{E}, \phi, \psi)$, if there exist multiple types of nodes or links in the network (i.e., $|\mathcal{N}| > 1$, or $|\mathcal{R}| > 1$), then the network is called a *heterogeneous network*.

Most of the network structured data in the real world may contain very complex information involving multiple types of nodes and connections, which can be represented as the *heterogeneous networks* [51] formally. Representative examples include *heterogeneous social networks* [29, 54, 73] involving users, posts, check-ins, words, and timestamps, as well as the friendship links, write links, and the other links among these nodes; *bibliographic networks* [52] including authors, papers, conferences, and the write, cite, and publish-in links among them; and *movie knowledge libraries* [34] containing movies, casts, reviewers, review comments, as well as the complex links among these nodes. Many of the concepts introduced before for the *homogeneous networks* can also be applied to the *heterogeneous networks* as well.

3.4.2.1 Online Social Networks

The *online social networks* [29, 54, 73] usually allow the users to perform different social activities, like *make friends with other users*, *write posts online*, and *check-in at some places*, which will generate different kinds of information entities and very complex connections among these information entities. Formally, an *online social network* involving these diverse information entities and complex links is called a *heterogeneous social network*.

Example 3.17 In Fig. 3.13, we illustrate an example of a *heterogeneous social network*. Formally, according to the heterogeneous network definition, it can be represented as $G = (\mathcal{V}, \mathcal{E})$ (the mappings are not provided), where the node set $\mathcal{V}$ can be divided into several subsets $\mathcal{V} = \mathcal{U} \cup \mathcal{P} \cup \mathcal{L} \cup \mathcal{T}$ representing the *user*, *post*, *location*, and *timestamp* nodes, respectively. Meanwhile, depending on the node types that the links are connected to, the links in $\mathcal{E}$ can also have different physical meanings and can be further divided into subsets $\mathcal{E} = \mathcal{E}_{u,u} \cup \mathcal{E}_{u,p} \cup \mathcal{E}_{u,l} \cup \mathcal{E}_{p,t}$, which correspond to the friendship links among users, and the links between users and posts, locations and timestamps, respectively.

In the *heterogeneous social networks*, each node can be connected with a set of nodes belonging to different categories via various type of connections. For example, given a user $u \in \mathcal{U}$, the set of user node incident to u via the friendship links can be represented as the online friends of u, i.e., set $\{v | v \in \mathcal{U}, (u, v) \in \mathcal{E}_{u,u}\}$; the set of post node incident to u via the write links can be represented as the posts written by u, i.e., set $\{w | w \in \mathcal{P}, (u, w) \in \mathcal{E}_{u,p}\}$. It is very similar for the location and timestamp nodes as well, from which we can achieve the set of locations visited by users and the collection of timestamps that the users perform the social actions.

Many interesting research problems have been studied based on the *online social networks*, like *friend recommendation* [56, 61, 71, 72], *social community detection* [62, 68], *social information diffusion* [26, 75, 76, 79] via the connections among users. *Friend recommendation* problems aim at recommending online friends for users in the social networks, which can be formulated either as a ranking problem or as a link prediction problem. *Community detection* problem focuses on dividing the users into different social groups, where users who frequently interact with each other tend to appear in the same group. *Information diffusion* problems aim at modeling how information propagates within the online social networks, and when the users can be activated by certain information propagated from their friends. These problems mentioned here will also be covered in the following Chaps. 7–11 in great detail.

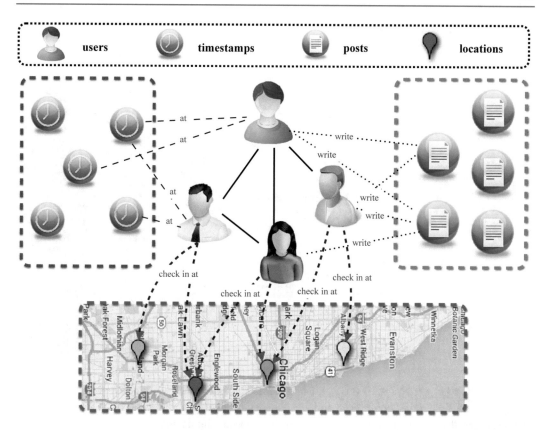

Fig. 3.13 An example of heterogeneous online social network

3.4.2.2 Bibliographic Networks

Another type of heterogeneous network well studied in research is called the *bibliographic networks* [52], which denote the academic networks depicting the paper authorship, paper citation, and paper publishing venues. Generally, the *bibliographic networks* may involve multiple types of information entities, like authors, papers, conferences/journals, and very complex connections among these information entities, which can be represented as a heterogeneous network as well.

Example 3.18 As shown in Fig. 3.14, a *bibliographic network* can be represented as graph $G = (\mathcal{V}, \mathcal{E})$, where $\mathcal{V} = \mathcal{A} \cup \mathcal{P} \cup \mathcal{V}$ containing the authors, papers, and venues (i.e., conferences or journals), and $\mathcal{E} = \mathcal{E}_{a,p} \cup \mathcal{E}_{p,p} \cup \mathcal{E}_{p,v}$ involving the authorship links between authors and papers, citation links among the papers, and publishing links between papers and venues. In the example, MLI [73], MNA [29], and MFC [79] are the model names proposed by the authors in their papers.

In many cases, the information entities in a *bibliographic network* may also be associated with a set of attributes indicating their properties, like expertise/skills about the authors, the title, abstract, keywords, categories of papers, and the year, categories (like data mining, machine learning) of the publication venues. Via the papers, the authors can get correlated with each other. For instance, given a paper $p \in \mathcal{P}$, we can obtain the set of authors who are involved in writing p as $\{a|a \in \mathcal{A}, (a, p) \in \mathcal{E}_{a,p}\}$. For any author pairs $a_1, a_2 \in \{a|a \in \mathcal{A}, (a, p) \in \mathcal{E}_{a,p}\}$, they will be the co-authors on paper p. Similarly, from the *bibliographic network*, we can also obtain the set of papers published at certain

Fig. 3.14 An example of heterogeneous bibliographic network

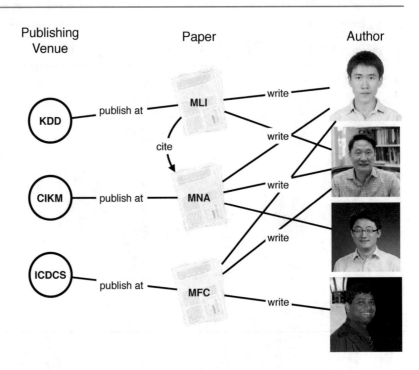

venue $v \in \mathcal{V}$ as $\{p | p \in \mathcal{P}, (p, v) \in \mathcal{E}_{p,v}\}$. Many other interesting information, like authors who have ever published at a similar conference and conferences frequently participated in by certain authors, can be analyzed with the meta path concept to be introduced later as well.

Many interesting problems can be studied in the *bibliographic networks*, like co-author recommendation [52], rankings of authors, papers and venues [53], project team formation [30, 78]. Co-author recommendation is an important problem for academia, as it will help researchers find their collaborators to carry out the projects. The researchers, papers, and publishing venues are usually of different quality, some of which are highly ranked but some are of lower ranks. An effective ranking of the researchers, papers, and venues will make it easier for people to find qualified collaborators, related works, and publishing venues. Meanwhile, in the real world, great researchers tend to write innovative research papers and get them published at top-tier publishing venues. The ranking problems of the authors, papers, and venues are usually strongly correlated. To finish certain research projects, the project leader may need to build a team of researchers with different kinds of required skills. Team formation problems aim at identifying the team members for projects. In this book, we will mainly focus on the *social networks*, and these aforementioned problems for bibliographic networks will not be covered in this book. However, the readers are recommended to read the referred articles, if you are interested in these problems.

3.4.2.3 Movie Knowledge Libraries

For the online movie review sites, like IMDB[1] and Douban,[2] they involve very complex information and can be represented as heterogeneous networks as well [34]. Generally, in these sites, users can post review comments and ratings for the movies to express their favor regarding some movies. Meanwhile, for the movies, we can obtain the cast involved in producing the movies, like the writers, directors,

---

[1] http://www.imdb.com/.

[2] https://www.douban.com/.

Fig. 3.15 An example of heterogeneous movie network

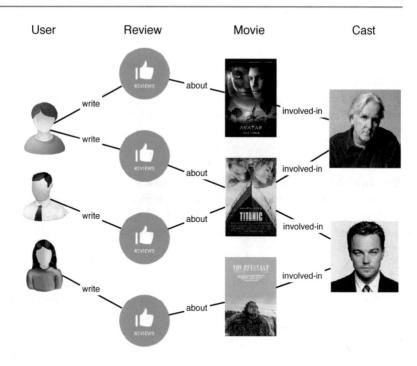

User   Review   Movie   Cast

actors, and actress. A set of attributes can be obtained for the movies and casts as well, like movie title, story outline, movie genres, and cast profile information.

Example 3.19 In Fig. 3.15, we illustrate an example of the *heterogeneous movie knowledge library*, which can be represented as a graph $G = (\mathcal{V}, \mathcal{E})$, where the node set $\mathcal{V} = \mathcal{M} \cup \mathcal{C} \cup \mathcal{U} \cup \mathcal{P}$ involves the movie nodes, cast nodes, user nodes and review post nodes, and link set $\mathcal{E} = \mathcal{E}_{m,c} \cup \mathcal{E}_{u,p} \cup \mathcal{E}_{p,m}$ contains the links between movies and casts, users, and review posts, and those between the review posts and movies.

Via the heterogeneous links in these online *movie knowledge libraries* [34], the nodes are extensively connected with each other and lots of interesting knowledge can be discovered from the online *online knowledge libraries*. For example, given a movie $m \in \mathcal{M}$, we can obtain the set of reviews posted for it as set $\{p | p \in \mathcal{P}, (p, m) \in \mathcal{E}_{p,m}\}$, based on the review comments and ratings contained by these review posts, we can analyze the sentiment and favor of audiences about the movie.

Based on the online *movie knowledge libraries*, research problems like *movie recommendation* [7], *movie box-office analysis* [3, 34], and *movie planning problem* [34, 45] can be studied. The movie recommendation problems aim at recommending movies for users based on their movie rating historical records, and inferring their potential ratings for the recommended movies. From the investors' perspective, they generally want to invest their money on promising movies that can achieve a good box-office, while the movie box-office depends on various factors, like movie genre, movie storyline, and movie cast. Given a movie basic profile information, inferring the potential box-office can be obtained by them is an important problem. The movie planning problem is studying the correlation between movie profile information and box-office in a reverse direction, which aims at designing the optimal movie configurations within the provided budget to achieve the largest movie box-office.

3.4.3   Aligned Heterogeneous Networks

In the real world, about the same information entities, e.g., social media users, researchers in academia, and the imported foreign movies, a large amount of information can actually be collected from various sources. These sources are usually of different varieties, like Facebook and Twitter, data mining and machine learning research areas, the USA and China online movie libraries. Generally, these multiple information sources sharing some common information entities can be modeled as *multiple aligned heterogeneous networks* [29, 68, 72, 73].

Definition 3.15 (Multiple Aligned Heterogeneous Networks)   Formally, the *multiple aligned heterogeneous networks* involving n networks can be defined as $\mathcal{G} = ((G^{(1)}, G^{(2)}, \ldots, G^{(n)}), (\mathcal{A}^{(1,2)}, \mathcal{A}^{(1,3)}, \ldots, \mathcal{A}^{(n-1,n)}))$, where the networks $G^{(1)}, G^{(2)}, \ldots, G^{(n)}$ denote these n heterogeneous networks and $\mathcal{A}^{(1,2)}, \mathcal{A}^{(1,3)}, \ldots, \mathcal{A}^{(n-1,n)}$ represent the sets of undirected *anchor links* aligning these networks.

In the above definition, the *anchor links* [29, 73] refer to the mappings of information entities across different sources, which actually correspond to the same information entity in the real world, e.g., users shared between online social networks, authors involved in multiple bibliographic networks, and the common movies shared in different movie libraries. As proposed in [29, 69, 70], *anchor links* are usually subject to the *one-to-one* cardinality constraint, which can be formally defined as follows.

Definition 3.16 (Anchor Link)   Given two heterogeneous networks $G^{(i)}$ and $G^{(j)}$ which share some common information entities, the set of *anchor links* connecting $G^{(i)}$ and $G^{(j)}$ can be represented as set $\mathcal{A}^{(i,j)} = \{(u_m^{(i)}, u_n^{(j)}) | u_m^{(i)} \in \mathcal{V}^{(i)} \wedge u_n^{(j)} \in \mathcal{V}^{(j)} \wedge u_m^{(i)}, u_n^{(j)} \text{denote the same information entity}\}$.

Example 3.20   In Fig. 3.16, we provide an example of *multiple aligned heterogeneous social networks*, which involve two heterogeneous networks Foursquare and Twitter, respectively. Both Foursquare and Twitter have very complex information, which can both be represented as the heterogeneous networks. Between these two networks, they share five common users, who are connected by the red dashed anchor links across networks.

Anchor links mainly exist between pairwise networks, when it comes to multiple (more than 2) aligned networks, there will exist a specific set of anchor links between any network pairs. The *anchor links* depict a transitive relationship among the information entities across different networks. Given three information entities $u_m^{(i)}, u_n^{(j)}, u_o^{(k)}$ from networks $G^{(i)}, G^{(j)}$, and $G^{(k)}$ respectively, if $u_m^{(i)}, u_n^{(j)}$ are connected by an anchor link and $u_n^{(j)}, u_o^{(k)}$ are connected by an anchor link, then the user pair $u_m^{(i)}$, $u_o^{(k)}$ will be connected by an anchor link by default.

For the information entities which are connected by the anchor links, they are named as the *anchor information entities*, like *anchor users* [73] in social networks, *anchor authors* in bibliographic networks, *anchor movies* in movie knowledge libraries. Meanwhile the remaining information entities are called the *non-anchor information entities*.

Definition 3.17 (Anchor Information Entities)   Given a pair of heterogeneous networks $G^{(i)}$ and $G^{(j)}$, the anchor links $\mathcal{A}^{(i,j)}$ aligning them, and the information entity sets $\mathcal{V}^{(i)}$ and $\mathcal{V}^{(j)}$ involved in them, respectively, the set of *anchor information entities* in $G^{(i)}$ can be represented as $\mathcal{V}_a^{(i),(i,j)} = \{u_m^{(i)} | u_m^{(i)} \in \mathcal{V}^{(i)}, \exists u_n^{(j)} \in \mathcal{V}^{(j)}, (u_m^{(i)}, u_n^{(j)}) \in \mathcal{A}^{(i,j)}\}$. Similarly, we can also represent the set of anchor information entities in $G^{(j)}$ as $\mathcal{V}_a^{(j),(i,j)} \subset \mathcal{V}^{(j)}$.

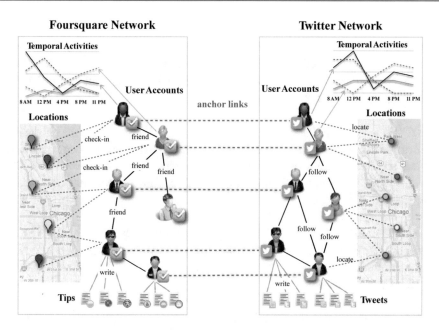

Fig. 3.16 An example of multiple aligned heterogeneous social networks

Definition 3.18 (Non-anchor Information Entities) Given a pair of heterogeneous networks $G^{(i)}$ and $G^{(j)}$, the anchor links $\mathcal{A}^{(i,j)}$ aligning them, and the information entity sets $\mathcal{V}^{(i)}$ and $\mathcal{V}^{(j)}$ involved in them, respectively, the set of *non-anchor information entities* in $G^{(i)}$ can be represented as $\mathcal{V}_{non\text{-}a}^{(i),(i,j)} = \{u_m^{(i)}|u_m^{(i)} \in \mathcal{V}^{(i)}, \forall u_n^{(j)} \in \mathcal{V}^{(j)}, (u_m^{(i)}, u_n^{(j)}) \notin \mathcal{A}^{(i,j)}\} = \mathcal{V}^{(i)} \setminus \mathcal{V}_a^{(i),(i,j)}$. In a similar way, we can represent the set of non-anchor information entities in network $G^{(j)}$ as well, which can be denoted as $\mathcal{V}_{non\text{-}a}^{(j),(i,j)}$.

The *anchor information entities* and *non-anchor information entities* concepts are defined based on the provided network pairs, which will be different (e.g., those in network $G^{(i)}$) as the network pair changes. For instance, the set of *anchor information entities* and *non-anchor information entities* in $G^{(i)}$ between network pairs $G^{(i)}$ and $G^{(j)}$ will be different from those in $G^{(i)}$ between network pairs $G^{(i)}$ and $G^{(k)}$. Furthermore, depending on the availability of *anchor information entities* and *non-anchor information entities*, the networks can be either *fully aligned*, *partially aligned*, and *non-aligned*, respectively.

Definition 3.19 (Full Alignment) Given a pair of heterogeneous networks $G^{(i)}$ and $G^{(j)}$ with non-anchor information entity sets $\mathcal{V}_{non\text{-}a}^{(i),(i,j)}$ and $\mathcal{V}_{non\text{-}a}^{(j),(i,j)}$, respectively. $G^{(i)}$ is said to be fully aligned with $G^{(j)}$ iff $\mathcal{V}_{non\text{-}a}^{(i),(i,j)} = \emptyset$, and $G^{(j)}$ is said to be fully aligned with $G^{(i)}$ iff $\mathcal{V}_{non\text{-}a}^{(j),(i,j)} = \emptyset$. $G^{(i)}$ and $G^{(j)}$ are said to be mutually fully aligned iff $\mathcal{V}_{non\text{-}a}^{(i),(i,j)} = \emptyset \wedge \mathcal{V}_{non\text{-}a}^{(j),(i,j)} = \emptyset$.

Network $G^{(i)}$ is said to be fully aligned with network $G^{(j)}$ if the information entities involved in $G^{(i)}$ are a subset of those involved in $G^{(j)}$, and vice versa. Networks $G^{(i)}$ and $G^{(j)}$ are mutually fully aligned if the information entities in $G^{(i)}$ and $G^{(j)}$ are actually identical. Fully aligned networks may exist in the real world, but a much common scenario will be *partial alignment* of networks instead.

Definition 3.20 (Partial Alignment) Given a pair of heterogeneous networks $G^{(i)}$ and $G^{(j)}$ with information entity sets $\mathcal{V}^{(i)}$ and $\mathcal{V}^{(j)}$ and anchor information entity sets $\mathcal{V}_a^{(i),(i,j)}$ and $\mathcal{V}_a^{(j),(i,j)}$, respectively. Network $G^{(i)}$ is partially aligned with network $G^{(j)}$ iff $\mathcal{V}_a^{(i),(i,j)} \neq \emptyset \wedge \mathcal{V}^{(i)} \neq \mathcal{V}_a^{(i),(i,j)}$, and vice versa. Networks $G^{(i)}$ and $G^{(j)}$ are said to be mutually partially aligned iff $\mathcal{V}_a^{(i),(i,j)} \neq \emptyset \wedge \mathcal{V}^{(i)} \neq \mathcal{V}_a^{(i),(i,j)}$ and $\mathcal{V}_a^{(j),(i,j)} \neq \emptyset \wedge \mathcal{V}^{(j)} \neq \mathcal{V}_a^{(j),(i,j)}$.

Network $G^{(i)}$ is said to be partially aligned with network $G^{(j)}$ if one part of the information entities in $G^{(i)}$ are involved in $G^{(j)}$. Both *full alignment* and *partial alignment* are not symmetric relationships. In the case that all the information entities in $G^{(i)}$ are also involved in $G^{(j)}$ while many information entities in $G^{(j)}$ are not involved in $G^{(i)}$, network $G^{(i)}$ will be fully aligned with $G^{(j)}$ but $G^{(i)}$ will be partially aligned with $G^{(i)}$ instead.

Definition 3.21 (Non-alignment) Given a pair of heterogeneous networks $G^{(i)}$ and $G^{(j)}$ with anchor information entity sets $\mathcal{V}_a^{(i),(i,j)}$ and $\mathcal{V}_a^{(j),(i,j)}$, respectively. Networks $G^{(i)}$ and $G^{(j)}$ are said to be non-aligned iff the information entities involved in two networks $G^{(i)}$ and $G^{(j)}$ are totally different, i.e., $\mathcal{V}_a^{(i),(i,j)} = \emptyset$ and $\mathcal{V}_a^{(j),(i,j)} = \emptyset$.

Different from *full alignment* and *partial alignment*, the *non-alignment* is a bi-directional relationships. In other words, if $G^{(i)}$ is non-aligned with $G^{(j)}$, then $G^{(j)}$ will be non-aligned with $G^{(i)}$ as well.

Lots of real-world network structures can actually share some common information entities, and can be represented as the *multiple aligned heterogeneous networks*. We will provide several examples as follows.

3.4.3.1 Multiple Aligned Heterogeneous Online Social Networks
To enjoy different kinds of social network services at the same time, users nowadays are usually involved in multiple online social networks simultaneously, e.g., Facebook, Twitter, Foursquare, and Google+. For the online social networks sharing common users, they can be represented as the *multiple aligned heterogeneous online social networks*.

Example 3.21 In Fig. 3.16, we have provided an example of two partially aligned heterogeneous online social networks: Foursquare and Twitter. Both Foursquare and Twitter can provide the users with different kinds of social network services, like make online friends with other users, write/like/comment on posts, check-in at some locations, and their online social activities are also associated with timestamps as well. Many users tend to join in Foursquare and Twitter at the same time, who are connected by the anchor links in the example.

In each of these two *aligned heterogeneous social networks*, we can have more data about the common users, which provides researchers and practitioners the opportunity to study users' social behaviors within these two networks. Moreover, the multiple aligned networks setting also allows the researchers to carry out a comparative study of users' social behaviors in different networks, which will provide a more comprehensive understanding about their social preferences and personal social behaviors.

3.4.3.2 Multiple Aligned Heterogeneous Bibliographic Networks
In the academia, the researchers are usually involved in various interdisciplinary projects and may collaborate with many researchers from other areas. For instance, the researchers of bioinformatics

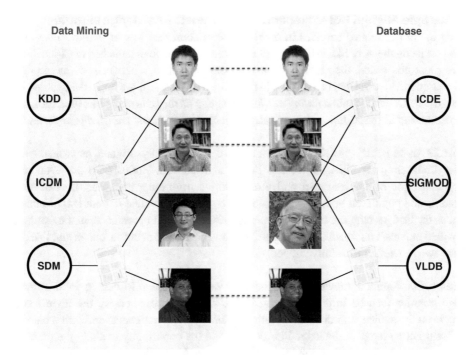

Fig. 3.17 An example of multiple aligned heterogeneous bibliographic networks

tend to have background in either computer science or biology; people working on data mining can publish works in either machine learning, data mining, or database; and the researchers working on neural networks can be experts on machine learning or neural science. Viewed in such a perspective, various closely related research areas may share lots of common researchers, and each researcher can also publish their works in different areas as well. Such complex relationships can be effectively modeled as the *multiple aligned heterogeneous bibliographic networks* formally.

In Fig. 3.17, we show an example of two partially aligned heterogeneous bibliographic networks in data mining and database. Between these two networks, there exist a large number of shared researchers, like Jiawei Zhang, Philip S. Yu, and Charu C. Aggarwal, who are active in both of these two areas and have published lots of academic papers in data mining and database conferences, like KDD, ICDM, SDM and ICDE, SIGMOD, VLDB. These two areas have different focuses in research actually, where data mining emphasizes more on knowledge discovery, while database is interested in data storage and management instead. Therefore, the researchers involved in these two areas are not exactly identical, and the shared researchers are indicated with the anchor links between them.

The multiple aligned heterogeneous bibliographic network setting allows us to study many interesting problems. In each of the networks, we can analyze the researchers' personal research interests, their preferred paper topics, frequently published conferences, which will be helpful to divide them into different research groups. Meanwhile, across the aligned bibliographic network, we can obtain their activities in different research areas. With the data about them across these different research domains, we can know their interdisciplinary research interest and activities, and it will provide extra information for us when studying researchers' personal cross-domain research interest shift, as well as their research progress in different domains.

3.4.3.3 Multiple Aligned Heterogeneous Online Movie Knowledge Libraries

To provide the movie related services in many different countries, lots of *online movie knowledge library* [34] exist on the web, like IMDB launched in the USA, Douban launched in China. Nowadays, to achieve more box-office, the movie import is a common practice between the movie markets in different countries. A movie can be on show in the USA first, and then get imported to show in China. Therefore, the IMDB and Douban *online movie knowledge library* can share lots of common movies, and can be modeled as the *multiple aligned heterogeneous online movie knowledge libraries* formally.

Example 3.22 In Fig. 3.18, we show an example of two partially aligned heterogeneous movie knowledge libraries in the USA and China: IMDB and Douban. Both IMDB and Douban have a very large collection of movies either native or imported from other countries. Lots of movies are very welcome and popular in both the USA and China, and are included in both IMDB and Douban, which act as the bridges aligning these different libraries together. For instance, in the example, these three provided movies, i.e., Avatar, Titanic, and The Revenant, exist in both Douban and IMDB, which make these two movie libraries fully aligned.

Generally, the common movie tend to have identical profile information in different movie knowledge libraries (can be in different languages), while they can receive the review comments and rating from the audience in different countries. These review comments and rating data obtained from different online movie knowledge libraries provide the opportunity to study the preferences of audiences from different countries about the shared movies. Moreover, many movies will be on show in the native countries first, and then get imported by other countries. Before these movies entering a new market, some prior knowledge about the movies in the original native country is available already, which will be very useful in scheduling the screenings in other countries, so as to maximize the revenue for theaters.

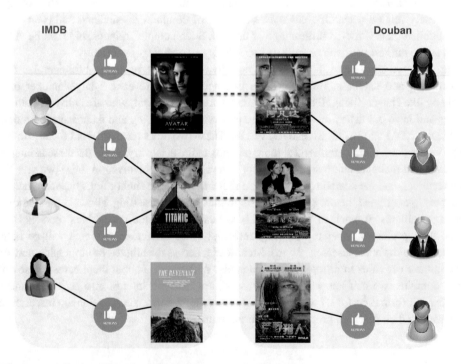

Fig. 3.18 An example of multiple aligned heterogeneous movie libraries

3.5 Meta Path

To deal with the social networks, especially those with heterogeneous information, a useful technique is the *meta path* [52, 53, 73]. *Meta path* is a concept defined based on the network schema, outlining the connections among nodes belonging to different categories. For the nodes which are not directly connected, their relationships can be depicted with the meta path concept. In this part, we will first introduce the meta path concept, and then talk about a set of meta paths within, as well as across real-world heterogeneous social networks.

3.5.1 Network Schema

Given a network $G = (\mathcal{V}, \mathcal{E})$, we can define its corresponding *network schema* [52, 53, 73] to describe the categories of nodes and links involved in G.

Definition 3.22 (Network Schema) Formally, the network schema of network $G = (\mathcal{V}, \mathcal{E}, \phi, \psi)$ can be represented as $S_G = (\mathcal{N}, \mathcal{R})$, where $\mathcal{N}$ and $\mathcal{R}$ denote the node type set and link type set of network G, respectively.

Network schema provides a meta level description of the network. Meanwhile, if a network G can be outlined by the network schema S_G, G is also called a *network instance* of the network schema. For a given node $u \in \mathcal{V}$, we can represent its corresponding node type as $\phi(u) = N \in \mathcal{N}$, and call u as an instance of node type N, which can also be represented as $u \in N$ for simplicity. Similarly, for a link (u, v), we can denote its link type as $\psi((u, v)) = R \in \mathcal{R}$. To represent that link (u, v) is an instance of the link type R, we can use the notations like $(u, v) \in R$, or $(u, v) \in S \xrightarrow{R} T$ for simplicity, where $\phi(u) = S \in \mathcal{N}$ and $\phi(v) = T \in \mathcal{N}$. The inverse relation type R^{-1} holds naturally for $T \xrightarrow{R^{-1}} S$, and R is generally not equal to R^{-1}, unless R is symmetric.

Example 3.23 In Fig. 3.19, we show the network schema of the heterogeneous social network on the left. According to the network structure, there exist four different node types, i.e., user, post, time, location, and four link types, i.e., follow, write, at, check-in at, in the network. These node types and link types together define the input network schema.

Meanwhile, in Figs. 3.20 and 3.21, we provide the network schemas of the input heterogeneous bibliographical network and the heterogeneous movie knowledge library, respectively. According to the bibliographical network structure, there exist three different node types and three link types, respectively. The movie knowledge library has a more complex structure, involving five different node types and four link types.

3.5.2 Meta Path in Heterogeneous Social Networks

Meta path [52, 53, 73] is a concept defined based on the network schema denoting the correlation of nodes based on the heterogeneous information (i.e., different types of nodes and links) in the networks.

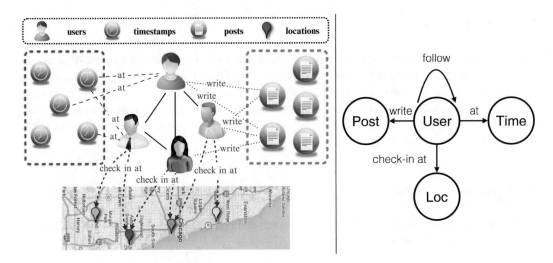

Fig. 3.19 An example of heterogeneous social network schema

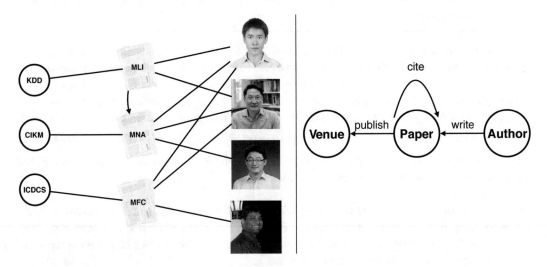

Fig. 3.20 An example of heterogeneous bibliographical network schema

Definition 3.23 (Meta Path) A meta path P defined based on the network schema $S_G = (\mathcal{N}, \mathcal{R})$ can be represented as $P = N_1 \xrightarrow{R_1} N_2 \xrightarrow{R_2} \cdots N_{k-1} \xrightarrow{R_{k-1}} N_k$, where $N_i \in \mathcal{N}, i \in \{1, 2, \ldots, k\}$ and $R_i \in \mathcal{R}, i \in \{1, 2, \ldots, k-1\}$.

Furthermore, depending on the categories of node and link types involved in the meta path, we can specify the meta path concept into two refined groups, like *homogeneous meta path* [73] and *heterogeneous meta path* [73].

Definition 3.24 (Homogeneous/Heterogeneous Meta Path) Let $P = N_1 \xrightarrow{R_1} N_2 \xrightarrow{R_2} \cdots N_{k-1} \xrightarrow{R_{k-1}} N_k$ denote a meta path defined based on the network schema $S_G = (\mathcal{N}, \mathcal{R})$. If

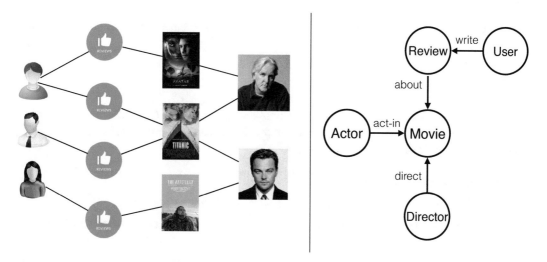

Fig. 3.21 An example of heterogeneous movie library schema

all the node types and link types involved in P are of the same category, P is called a *homogeneous meta path*; otherwise, P is called a *heterogeneous meta path*.

The meta paths connect any kinds of node type pairs, and specifically, for the meta paths starting and ending with the user node types within the same network, such a meta path is called the *social meta paths* [73].

Definition 3.25 (Social Meta Path) Let $P = N_1 \xrightarrow{R_1} N_2 \xrightarrow{R_2} \cdots N_{k-1} \xrightarrow{R_{k-1}} N_k$ denote a meta path defined based on the network schema $S_G = (\mathcal{N}, \mathcal{R})$. If the starting and ending node types N_1 and N_k are both the user node type, P is called a *social meta path*.

Users are usually the main focus in social network studies, and the *social meta paths* connecting the user node type will be frequently used in both research and real-world applications and services. If all the node types in the meta paths are user node type and the link types are also of an identical category, then the meta path is called the *homogeneous social meta path*. The number of path segments in the meta path is called the meta path length. For instance, the length of meta path $P = N_1 \xrightarrow{R_1} N_2 \xrightarrow{R_2} \cdots N_{k-1} \xrightarrow{R_{k-1}} N_k$ is $k - 1$. Meta paths can also been concatenated together with the *meta path composition operator* [52, 53, 73].

Definition 3.26 (Meta Path Composition) Meta paths $P^1 = N_1^1 \xrightarrow{R_1^1} N_2^1 \xrightarrow{R_2^1} \cdots N_{k-1}^1 \xrightarrow{R_{k-1}^1} N_k^1$, and $P^2 = N_1^2 \xrightarrow{R_1^2} N_2^2 \xrightarrow{R_2^2} \cdots N_{l-1}^2 \xrightarrow{R_{l-1}^2} N_l^2$ can be concatenated together to form a longer meta path $P = P^1 \circ P^2 = N_1^1 \xrightarrow{R_1^1} \cdots \xrightarrow{R_{k-1}^1} N_k^1 (\text{or } N_1^2) \xrightarrow{R_1^2} N_2^2 \xrightarrow{R_2^2} \cdots N_{l-1}^2 \xrightarrow{R_{l-1}^2} N_l^1$, if the ending node type of P^1 is the same as the starting node type of P^2, i.e., $N_k^1 = N_1^2$. The new composed meta path will be of length $k + l - 2$.

Meta path $P = N_1 \xrightarrow{R_1} N_2 \xrightarrow{R_2} \cdots N_{k-1} \xrightarrow{R_{k-1}} N_k$ can also been treated as the concatenation of simple meta paths $N_1 \xrightarrow{R_1} N_2$, $N_2 \xrightarrow{R_2} N_3$, ..., $N_{k-1} \xrightarrow{R_{k-1}} N_k$, which can be represented as $P = R_1 \circ R_2 \circ \cdots \circ R_{k-1} \circ R_k$. Here, we use the link type to denote the simplest meta paths of length 1.

Example 3.24 For instance, based on the network schemas shown in Figs. 3.19, 3.20, and 3.21, a group of meta paths can be defined. Here, we can provide a group of them as follows, which mainly connect the user/author/movie pairs specifically.

1. *Heterogeneous Social Network*
   - User $\xrightarrow{follow}$ User (or $U \to U$), which denotes a simple *follow* meta path.
   - User $\xleftarrow{follow}$ User $\xrightarrow{follow}$ User (or $U \leftarrow U \to U$), which denotes a *common follower* meta path.
   - User $\xrightarrow{follow}$ User $\xleftarrow{follow}$ User (or $U \to U \leftarrow U$), which denotes a *common followee* meta path.
   - User $\xrightarrow{check\text{-}in\ at}$ Location $\xleftarrow{check\text{-}in\ at}$ User (or $U \to L \leftarrow U$), which denotes a *common location check-in* meta path.

2. *Heterogeneous Bibliographic Network*
   - Author $\xrightarrow{write}$ Paper $\xleftarrow{write}$ User (or $A \to P \leftarrow A$), which denotes a *co-author* meta path.
   - Author $\xrightarrow{write}$ Paper $\xrightarrow{publish\ at}$ Venue $\xleftarrow{publish\ at}$ Paper $\xleftarrow{write}$ Author (or $A \to P \to V \leftarrow P \leftarrow A$), which denotes a *common publishing venue* meta path.
   - Author $\xrightarrow{write}$ Paper $\xrightarrow{cite}$ Paper $\xleftarrow{write}$ Author (or $A \to P \to P \leftarrow A$), which denotes a *citation* meta path.

3. *Heterogeneous Movie Library*
   - Movie $\xleftarrow{about}$ Review $\xleftarrow{write}$ User $\xrightarrow{write}$ Review $\xrightarrow{about}$ Movie (or $M \leftarrow R \leftarrow U \to R \to M$), which denotes a *shared review author* meta path.
   - Movie $\xleftarrow{direct}$ Director $\xrightarrow{direct}$ Movie (or $M \leftarrow D \to M$), which denotes a *shared director* meta path.
   - Movie $\xleftarrow{act\text{-}in}$ Actor $\xleftarrow{act\text{-}in}$ Movie (or $M \leftarrow A \to M$), which denotes a *shared actor* meta path.

   Besides these meta paths shown above, many other meta paths can also be defined based on the network schema structures, which will not be provided here and the readers can try to define some other useful meta paths on your own.

3.5.3   Meta Path Across Aligned Heterogeneous Social Networks

Besides the meta paths within a network, the meta paths can also be defined across multiple aligned heterogeneous networks via the *anchor meta path* [73] (or the anchor link type).

Definition 3.27 (Anchor Meta Path) Let $G^{(1)}$ and $G^{(2)}$ be two aligned heterogeneous networks sharing the common anchor information entity of types $N^{(1)} \in \mathcal{N}^{(1)}$ and $N^{(2)} \in \mathcal{N}^{(2)}$, respectively. The anchor meta path between the schemas of networks $G^{(1)}$ and $G^{(2)}$ can be represented as meta path $\Phi = N^{(1)} \xleftrightarrow{Anchor} N^{(2)}$ of length 1.

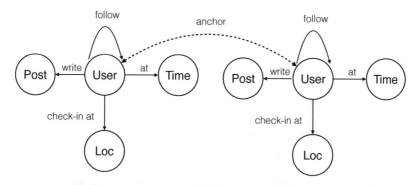

Fig. 3.22 An example of aligned heterogeneous social network schema

Formally, via the *anchor meta path*, given one pair of input *aligned heterogeneous social network* as shown in Fig. 3.16, we can formally represent the network schemas in Fig. 3.22. The *anchor meta path* is the simplest meta path across aligned networks, and a set of *inter-network meta paths* [73] can be defined based on the intra-network meta paths and the anchor meta path.

Definition 3.28 (Inter-Network Meta Path) Given a meta path $\Psi = N_1 \xrightarrow{R_1} N_2 \xrightarrow{R_2} \cdots N_{k-1} \xrightarrow{R_{k-1}} N_k$, Ψ is an *inter-network meta path* between networks $G^{(1)}$ and $G^{(2)}$ iff $\exists m \in \{1, 2, \ldots, k\}$, $R_m = Anchor$.

The *inter-network meta paths* can be viewed as a composition of *intra-network meta paths* and the *anchor meta path*. An *inter-network meta path* can be a meta path starting with an *anchor meta path* followed by the *intra-network meta paths*, or those with *anchor meta paths* in the middle and starting/ending with the *intra-network meta paths*. Here, we would like to introduce several categories *inter-network meta paths* involving the anchor meta paths at different positions as follows [73]:

- $\Psi(G^{(1)}, G^{(2)}) = \Phi(G^{(1)}, G^{(2)})$, which denotes the set of simplest *inter-network meta paths* composed of the anchor meta path only between networks $G^{(1)}$ and $G^{(2)}$.
- $\Psi(G^{(1)}, G^{(2)}) = \Phi(G^{(1)}, G^{(2)}) \circ P(G^{(2)})$, which denotes the set of *inter-network meta paths* starting with anchor meta path and followed by the intra-network meta path in network $G^{(2)}$ connected by an anchor meta path between networks $G^{(1)}$ and $G^{(2)}$.
- $\Psi(G^{(1)}, G^{(2)}) = P(G^{(1)}) \circ \Phi(G^{(1)}, G^{(2)})$, which denotes the set of *inter-network meta paths* starting with the intra-network meta path in network $G^{(1)}$ followed by an anchor meta path between networks $G^{(1)}$ and $G^{(2)}$.
- $\Psi(G^{(1)}, G^{(2)}) = P(G^{(1)}) \circ \Phi(G^{(1)}, G^{(2)}) \circ P(G^{(2)})$, which denotes the set of *inter-network meta paths* starting and ending with the intra-network meta path in networks $G^{(1)}$ and $G^{(2)}$, respectively, connected by an anchor meta path between networks $G^{(1)}$ and $G^{(2)}$.
- $\Psi(G^{(1)}, G^{(2)}) = P(G^{(1)}) \circ \Phi(G^{(1)}, G^{(2)}) \circ P(G^{(2)}) \circ \Phi(G^{(2)}, G^{(1)})$, which denotes the set of *inter-network meta paths* starting and ending with node types in network $G^{(1)}$ and traverse across the networks twice via the anchor meta path.
- $\Psi(G^{(1)}, G^{(2)}) = P(G^{(1)}) \circ \Phi(G^{(1)}, G^{(2)}) \circ P(G^{(2)}) \circ \Phi(G^{(2)}, G^{(1)}) \circ P(G^{(1)})$, which denotes the set of *inter-network meta paths* starting and ending with the *intra-network meta paths* in network $G^{(1)}$ and traverse across the networks twice via the anchor meta path between them.

Example 3.25 Based on the above descriptions, we can also represent several examples of *inter-network meta paths* across social networks as follows:

- $\text{User}^{(1)} \xleftrightarrow{anchor} \text{User}^{(2)} \xrightarrow{follow} \text{User}^{(2)}$ (or $U^{(1)} \leftrightarrow U^{(2)} \to U^{(2)}$).
- $\text{User}^{(1)} \xrightarrow{follow} \text{User}^{(1)} \xleftrightarrow{anchor} \text{User}^{(2)} \xleftarrow{follow} \text{User}^{(2)}$ (or $U^{(1)} \to U^{(1)} \leftrightarrow U^{(2)} \leftarrow U^{(2)}$).
- $\text{User}^{(1)} \xleftarrow{follow} \text{User}^{(1)} \xleftrightarrow{anchor} \text{User}^{(2)} \xrightarrow{follow} \text{User}^{(2)}$ (or $U^{(1)} \leftarrow U^{(1)} \leftrightarrow U^{(2)} \to U^{(2)}$).
- $\text{User}^{(1)} \xrightarrow{check\text{-}in\ at} \text{Location}^{(1)} \xleftarrow{check\text{-}in\ at} \text{User}^{(1)} \xleftrightarrow{anchor} \text{User}^{(2)} \xrightarrow{follow} \text{User}^{(2)}$ (or $U^{(1)} \to L^{(1)} \leftarrow U^{(1)} \leftrightarrow U^{(2)} \to U^{(2)}$).

Generally, shorter meta paths may convey more concrete physical meanings compared with the long meta paths. Due to the extensive connections among nodes in networks, extremely long meta paths will may not be useful, since almost all the node pairs in the network can be connected by such meta path instances. In the following parts, we will introduce several meta path-based network measures about node degree, node centrality, and node pair closeness, respectively.

3.5.4 Meta Path-Based Network Measures

The meta path concept introduced above provides a meta level description of information available within and across networks, and they can be used to compute various node and link measures based on the heterogeneous social networks. All the degree, centrality, and closeness measures introduced in the previous subsections are mainly based on the direct social links among users in homogeneous networks. In this part, we will extend these measures to the multiple aligned heterogeneous networks scenario based on the meta path concept specifically.

3.5.4.1 Meta Path-Based Node Degree

Via the meta paths, nodes in the networks which are not directly connected can be extensively correlated with each other. In this part, we will take the user node as an example, and try to study how the users are connected with each other via the meta paths. Let $\mathcal{U}^{(1)}$ be a user set in network $G^{(1)}$, and $\mathcal{P}$ be the set of various meta paths starting and ending with the user node type in network $G^{(1)}$ (which can be either intra-network or inter-network meta paths).

For each user pair in network $G^{(1)}$, e.g., $u, v \in \mathcal{U}^{(1)}$, based on one specific meta path $P_k \in \mathcal{P}$, we can denote the set of concrete meta path instances connecting u and v as set $P_k(u, v)$. The number of user nodes that u is connected with, i.e., its degree, based on meta path $P_k \in \mathcal{P}$ can be denoted as

$$D_{P_k}(u) = \sum_{v \in \mathcal{U}^{(1)}} |P_k(u, v)|. \tag{3.56}$$

Furthermore, for all these meta paths in set $\mathcal{P}$, we can represent the degree vector of user u as a $|\mathcal{P}|$-dimensional degree distribution vector

$$\mathbf{D}_{\mathcal{P}}(u) = [D_{P_1}(u), D_{P_2}(u), \ldots, D_{P_{|\mathcal{P}|}}(u)]^{\top}. \tag{3.57}$$

Generally, for these different meta paths in the set $\mathcal{P}$, they are usually of different weights. For instance, in some scenarios, shorter meta paths can denote stronger connections among users than longer meta paths; meta paths among the distinguishable node types (i.e., those only a small number of node types will be connected with them) will represent a more effective correlation than those composed of indistinguishable ones (i.e., those all the node types can be connected with them). By taking the meta path differences into consideration, we can represent the weighted meta path-based node degree as

$$D(u) = \sum_{P_i \in \mathcal{P}} w_{P_i} \cdot D_{P_i}(u),$$ (3.58)

where vector $[w_{P_1}, w_{P_2}, \ldots, w_{P_{|\mathcal{P}|}}]^{\top}$ $\left(\sum_{P_i \in \mathcal{P}} w_{P_i} = 1 \right)$ represents the weight parameters corresponding to the different meta paths.

3.5.4.2 Meta Path-Based Node Centrality and Closeness

Given the user node type and the set of meta paths $\mathcal{P}$ in $G^{(1)}$, based on each of the meta path $P_i \in \mathcal{P}$, the connections among users can be organized as homogeneous (weighted) graph $G_{P_i} = (\mathcal{U}, \mathcal{E}_{P_i}, w_{P_i})$, where the link set $\mathcal{E}_{P_i} = \{(u, v) | u, v \in \mathcal{U}, P_i(u, v) \neq \emptyset\}$. Mapping $w_{P_i} : \mathcal{E}_{P_i} \to \mathbb{R}$ denotes the weight of links in $\mathcal{E}_{P_i}$, where $w_{P_i}((u, v)) = |P_i(u, v)|$ represents the number of meta path instances of P_i connecting u and v. If we don't care about the link weights, the weight mapping is optional in the graph definition and can be discarded. In other words, based on the meta path concept, we can transform a *heterogeneous social network* into a group of *homogeneous networks* instead, where the edge weight equals to the meta path instance number.

Based on graph G_{P_i}, we can define the *centrality* measure of user $u \in \mathcal{U}$ as $C_{P_i}(u)$, which denotes either degree centrality, eigen-centrality, Katz centrality, pagerank centrality, or betweenness centrality that we have introduced before. Similar to the meta path-based degree concept introduced before, different meta path can play a different role in defining the users' centrality. One way to define the centrality measure of user u based on all the meta paths can be represented as

$$C(u) = \sum_{P_i \in \mathcal{P}} w_{P_i} \cdot C_{P_i}(u),$$ (3.59)

where w_{P_i} denotes the weight of the centrality measure based on meta path P_i.

In a similar way, the closeness measure among the user node pairs in the networks can be represented as

$$C(u, v) = \sum_{P_i \in \mathcal{P}} w_{P_i} \cdot C_{P_i}(u, v),$$ (3.60)

where $C_{P_i}(u, v)$ represents the closeness, e.g., *common neighbor* or *Jaccard's coefficient*, between users u and v in the network computed with meta path P_i.

3.6 Network Models

We have covered the basic knowledge about graphs, network measures, network category, and meta path already in this chapter. Before we end this chapter, we would like to introduce several models proposed for networks specifically. To model the link formation process in online social networks, several different models have been introduced, which can simulate how these networks are formed about the users. In this part, we will discuss several well-known network models, and analyze the properties, like degree distribution, clustering coefficient, and average path length, of networks generated by these models.

3.6.1 Random Graph Model

In the random graph model [14], the links among the nodes are assumed to be formed randomly, and each link will form with an equal chance. Based on such a simple assumption, the random graph model greatly simplifies the process of link formation in the real-world networks. Several different random graph models have been proposed already, and in this part, we will use the random graph model proposed by Gilbert [17] and Solomonoff and Rapoport [49] as an example.

In the random graph model, given a fixed number of nodes, e.g., n, the links among these nodes are formed independently with probability p. Formally, we denote the graph formed by following such a process as $G(n, p)$.

Theorem 3.3 *In graph $G(n, p)$, the number of links is not certain and the expected link number is $\frac{1}{2}\binom{n}{2}p$ (if the links are undirected).*

Proof We can represent the link number in the formed graph as m, and we have

$$
\begin{aligned}
m &= \frac{1}{2} \sum_{u,v \in \mathcal{V}, u \neq v} p((u, v)) \times 1 + (1 - p((u, v))) \cdot 0 \\
&= \frac{1}{2} \sum_{u,v \in \mathcal{V}, u \neq v} p((u, v)) \\
&= \frac{1}{2}\binom{n}{2}p.
\end{aligned}
\tag{3.61}
$$

Meanwhile, given a graph $G(n, p)$, we can also infer the probability of forming m links in $G(n, p)$ according to the following theorem.

Theorem 3.4 *In graph $G(n, p)$, the probability of forming m links is $\binom{\frac{\binom{n}{2}}{2}}{2} p^m (1 - p)^{\frac{\binom{n}{2}}{2} - m}$.*

Proof In graph $G(n, p)$, there exist $\frac{\binom{n}{2}}{2}$ potential links to be formed among these n nodes. Among these potential links, the probabilities that $0 \leq m \leq \frac{\binom{n}{2}}{2}$ of them are formed and the remaining are not formed can be denoted as p^m and $(1 - p)^{\frac{\binom{n}{2}}{2} - m}$, respectively. Therefore, the final probability that m out of these $\frac{\binom{n}{2}}{2}$ potential links are formed can be represented as

$$
P(m) = \binom{\frac{\binom{n}{2}}{2}}{2} p^m (1 - p)^{\frac{\binom{n}{2}}{2} - m}.
\tag{3.62}
$$

Theorem 3.5 *In graph $G(n, p)$, the expected degree of nodes is $(n - 1)p$.*

Proof For a node u in graph $G(n, p)$, it can be connected with the remaining $n - 1$ nodes all with probability p. Therefore, the expected degree of the node u in $G(n, p)$ can be represented as

$$
\begin{aligned}
\mathbb{E}(D(u)) &= \sum_{v \in \mathcal{V} \setminus \{u\}} p \cdot 1 + (1 - p) \cdot 0 \\
&= (n - 1)p.
\end{aligned}
\tag{3.63}
$$

Theorem 3.6 *In graph $G(n, p)$, the probability that a node has a degree of d is $\binom{n-1}{d} p^d (1-p)^{n-1-d}$.*

Proof Among these $n - 1$ potential neighbors of a given node, e.g., u, the node has $\binom{n-1}{d}$ different choices to select d neighbors for u to get connected with. Meanwhile, the probability of merely forming links with these selected d neighbors is $p^d (1 - p)^{n-1-d}$. In other words, the probability for a node u to have degree d will be

$$P(D(u) = d) = \binom{n-1}{d} p^d (1 - p)^{n-1-d}. \tag{3.64}$$

Theorem 3.7 *The global clustering coefficient of a random graph $G(n, p)$ is p.*

Proof According to the definition of clustering coefficient, we have

$$C(G(n, p)) = \frac{|\mathcal{T}|}{|\mathcal{P}|}, \tag{3.65}$$

where set $\mathcal{T}$ denotes the node triples which form triangles and set $\mathcal{P}$ denotes the node triples forming a path of length 2.

In the random graph $G(n, p)$, given three nodes $u, v, w \in \mathcal{V}$, the probability that they will form a path $u - v - w$ (where u and w can be either connected or unconnected) is p^2. Meanwhile, the probability that these three nodes will form a triangle will be p^3. Therefore, we have the sizes of sets $\mathcal{T}$ and $\mathcal{P}$ will be $\sum_{u,v,w \in \mathcal{V}, u \neq v \neq w} p^3$ and $\sum_{u,v,w \in \mathcal{V}, u \neq v \neq w} p^2$, respectively, and the global clustering coefficient is

$$C(G(n, p)) = \frac{\sum_{u,v,w \in \mathcal{V}, u \neq v \neq w} p^3}{\sum_{u,v,w \in \mathcal{V}, u \neq v \neq w} p^2} = p. \tag{3.66}$$

Theorem 3.8 *In graph $G(n, p)$, given two nodes $u, v \in \mathcal{V}$, the probability that there exists a path of length k connecting u and v is $\binom{n-2}{k-1} p^k$.*

Proof Between u and v, if there exists a path of length k connecting them, we can denote such a path as $P = u \rightarrow u_1 \rightarrow \cdots, u_{k-1} \rightarrow v$, where the intermediate nodes $u_1, u_2, \ldots, u_{k-1} \in \mathcal{V} \backslash \{u, v\}$. There exist $\binom{n-2}{k-1}$ different choices of these $k - 1$ nodes. Meanwhile, among these $k - 1$ selected nodes, the probability that they will form a path connecting u and v will be p^k. Therefore, we have the probability that there exists a path of length k connecting nodes u and v can be represented as

$$P(u, v, k) = \binom{n-2}{k-1} p^k. \tag{3.67}$$

Given the node number n, some properties of the random graph $G(n, p)$ will change as parameter p increases from 0 to 1. In the case that $p = 0$, the random graph $G(n, 0)$ will only involve n isolated nodes without any connections. In such a graph, the graph diameter will be 0 and the size of the largest connected component will contain merely 1 node and the average path length is 0 as no path exists among these nodes. As p increases, some links will be formed among the nodes, and the diameter of the graph $G(n, p)$ will increase which can also be greater than 1. At the same time, the size of the largest component increases, while the average path length will also increase and can be greater than 1. Meanwhile, in the case that $p = 1$, the graph will be a complete graph involving n nodes and $\frac{n(n-1)}{2}$ links with diameter 1, where all the nodes will be incorporated into one single connected component.

All the nodes will be connected, and the average path length in $G(n, 1)$ will be 1. Formally, the point where the diameter increases first and starts to shrink is called the *phase transition* [9] point.

Theorem 3.9 *The phase transition happens at* $p = \frac{1}{n-1}$ *in the random graph model.*

Proof The proof of the above theorem is left as an exercise for the readers.

In the random graph model, the formation of all the links is assumed to be independent with identical probabilities. However, in the real-world social networks, such an assumption cannot hold. For instance, in the socialization among users, people tend to form a small community involving connections with a very limited number of people, like friends, family members, and colleagues. Many other models, like the *small-world model* [28, 39, 58] can be used to model the formation process of such a phenomenon better.

3.6.2   Preferential Attachment Model

When making friends, generally the people with a large neighborhood can attract the connections more easily. For instance, in the real-world online social networks, the celebrities, like the politicians and super stars, are well known and they are usually among the top candidates that we choose to follow. A well-established method to model such an observation in network formation is called the *preferential attachment* model [4].

In the *preferential attachment* model, at the very beginning, there exist n_0 node in the network and new nodes will be added to form connections with these existing nodes. The new node will connected $n \leq n_0$ other existing nodes. Formally, we can represent the degrees of nodes in the existing graph, e.g., u, as $D(u)$, and new nodes are more likely to establish connections with the active nodes, i.e., those with a large degree. The probability for a new node to get connected with u can be represented as $P(u) = \frac{D(u)}{\sum_{v \in \mathcal{V}} D(v)}$.

Theorem 3.10 *The degree distribution of the graph generated by the preferential attachment model follows the power-law distribution with an exponent* $b = 3$.

Proof According to the introduction, the probability for the newly added node to connected with an existing node u is

$$P(u) = \frac{D(u)}{\sum_{v \in \mathcal{V}} D(v)}. \tag{3.68}$$

Meanwhile, at each step t, the expected increase of u's degree is proportional to $D(u)$, which can be modeled with a mean-field setting,

$$\frac{\mathrm{d}D(u)}{\mathrm{d}t} = nP(u)$$

$$= \frac{nD(u)}{\sum_{v \in \mathcal{V}} D(v)}$$

$$= \frac{nD(u)}{2nt}$$

$$= \frac{D(u)}{2t}. \tag{3.69}$$

In each step, n links will be added, and after t steps, the total node degree will be equal to $\sum_{v \in \mathcal{V}} D(v) = 2nt$. By solving such a partial differential equation, we can get

$$D(u) = n \left(\frac{t}{t_u} \right)^{\frac{1}{2}}, \tag{3.70}$$

where t_u denotes the step that u is added into the network.

The probability that $D(u)$ is less than d can be represented as

$$P(D(u) < d) = P \left(t_u > \frac{n^2 t}{d^2} \right) = 1 - P \left(t_u \leq \frac{n^2 t}{d^2} \right). \tag{3.71}$$

If we assume that $t_u \sim \text{Uniform}(0, t)$, we have

$$P(D(u) < d) = 1 - P \left(t_u \leq \frac{n^2 t}{d^2} \right) = 1 - \frac{n^2}{d^2} \frac{1}{n_0 + t}. \tag{3.72}$$

Let the node degree distribution density function to be $P(D)$, we have

$$P(d) = \frac{\partial P(D(u) < d)}{\partial d} = \frac{2n^2 t}{d^3 (t + n_0)} \approx_{t \to \infty} \frac{2n^2}{d^3}. \tag{3.73}$$

Theorem 3.11 *Based on the preferential attachment model, by using the mean-field analysis, the expected clustering coefficient of the generated network is*

$$C = \frac{n_0 - 1}{8} \frac{(\ln t)^2}{t}. \tag{3.74}$$

Theorem 3.12 *Based on the preferential attachment model, the average path length of nodes in the generated network is*

$$l \approx \frac{\ln |\mathcal{V}|}{\ln(\ln(|\mathcal{V}|))}. \tag{3.75}$$

The proofs of the above two theorems are out of the scope of this book, which will not be introduced here. For the readers who are interested in the proof, please refer to [65].

3.7   Summary

In this chapter, we provided an overview about the essential knowledge of online social networks, which can generally be represented as graphs involving nodes and connections among the nodes. Some basic information about graphs were provided at the beginning of this chapter, covering the different graph representation methods, e.g., adjacency matrix and adjacency list, and graph connectivity concepts, e.g., adjacent neighbors, incident links, walk, trail, tour, path, cycle, as well as node reachability and connect component.

We introduced the various measures for networks in this chapter, including degree, centrality, closeness, transitivity, and social balance. We talked about the node degree concept as well as the node degree distribution, which provide the basic information about the network connectivity structures. To denote the importance of node roles in the network, several different node centrality measures were introduced. The closeness between the node pairs in the networks can be computed with various closeness measures based on the local network structures, global paths, and random walks. We introduced the concepts of social transitivity, clustering coefficient, and social balance to analyze various social connection-based network properties.

Depending on the network structures and the involved information, the networks could be divided into various categories, e.g., homogeneous network, heterogeneous network, and aligned heterogeneous networks. The representative examples of homogeneous networks include the friendship network, computer network, and company organizational chart; the examples of heterogeneous networks cover the online social network, bibliographic network, and movie knowledge library, while the aligned heterogeneous networks concept provides the opportunity to model the information across multi-platforms.

To depict the diverse information inside the networks, meta path can be a very useful methods, which can outline the potential connections among the nodes. In the provided definition of the meta path concept based on the network schema, meta paths can be represented as the sequences of node types connected by the link types. Besides the meta paths within one single heterogeneous network, we also introduced the meta path across heterogeneous networks via the anchor meta path. Various network measures, e.g., degree, centrality, and closeness, were defined based on the meta path concept as well.

We concluded this chapter with several network models, including the random graph model and the preferential attachment model. A brief introduction and analysis about these two models were provided, which can also be applied to study various social network learning problems to be introduced in the following chapters as well.

3.8   Bibliography Notes

Studying online social networks and other related network structured data have been one of the most important research topics in the academia of machine learning and data mining in recent years, since lots of real-world data can be modeled as the networks [40]. There exist some survey articles on social networks [21], heterogeneous information networks [48, 51], and aligned social networks [66] published in recent years already, which can serve as the road map to study these related areas for the readers.

If the readers are interested in learning more knowledge about graph theory, you are very recommended to read the textbook *"Graph Theory and Complex Networks: An Introduction"* [55], which is well-written and well-organized book and covers a very broad topic about graphs. The recent *"Social Media Mining: An Introduction"* textbook [65] also provides a brief introduction to the graph related essential background knowledge, and the readers can take a look at that book as well.

Node degree distribution usually follows the power-law distribution [13], where the majority of the nodes only have a very small degree, while a very small number of the nodes can have a very large degree instead. Node centrality metric can measure the importance of nodes based on their positions inside the network, and a systematic overview of existing centrality measures is available in [11]. As to the node closeness, the readers can take a look at the recent survey article [67], which introduces various closeness measures as potential link predictors. The network transitivity, clustering coefficient, and social balance concepts are covered in [19, 20, 57], respectively.

A comprehensive survey about the network categories and existing network mining problems has been provided in [66], which also covers one section on network fusion and learning specifically. For the heterogeneous information network research works, the readers are suggested to read the lecture synthesis book [51], which covers the ranking, search, classification, and clustering problems on heterogeneous information networks. About the aligned heterogeneous social network alignment and mining problems, the readers are suggested to read the latest survey paper [48], which covers the alignment, link prediction, clustering, information diffusion, and embedding problems.

The meta path concept was initially introduced by Sun et al. in [53], and lately extended by Zhang et al. to the cross-network scenario in [73], which serves as an important tool for handling the heterogeneous network structures. Based on the assumptions that networks are generated randomly, the random graph models [14,17,49] can depict the generation process of graphs and certain properties that these generated graphs can have. Meanwhile, the preferential attachment model can depict the addition of new nodes into graphs, whose detailed description is available in [4].

3.9   Exercises

1. (Easy) Please compute the *diameter* of the graph shown in Fig. 3.23, and provide the maximum *shortest path*.
2. (Easy) Please compute the *betweenness centrality* and the *normalized betweenness centrality* of all the nodes in the input graph shown in Fig. 3.23.
3. (Easy) Please draw the *degree distribution* plot for the graph shown in Fig. 3.23.
4. (Easy) Please compute the *closeness* scores for all potential node pairs in Fig. 3.23 based on *common neighbor*, *Jaccard's coefficient*, and *Adamic/Adar*, respectively.
5. (Medium) Besides the heterogeneous network examples provided in this chapter, please think about some other data in the real world, which can be represented as a *heterogeneous network*. Please also provide its *network schema*, and list some *meta path* examples based on the schema.
6. (Medium) Based on the network schema, we can define a large number of meta paths. However, in many applications, extremely long meta paths (e.g., longer than 10) are not very useful. Please think why and write down the potential reasons.
7. (Medium) In Sect. 3.3.4.2, we show that the *clustering coefficient* equals to

$$CC = \frac{\text{Number of triangles} \times 6}{|\mathcal{P}^2|}. \tag{3.76}$$

Please also prove that the following equation also holds for computing the network *clustering coefficient*:

$$CC = \frac{\text{Number of triangles} \times 6}{\text{Number of connected triples of nodes}}. \tag{3.77}$$

8. (Hard) Please try to prove Theorem 3.9 regarding the *phase transition point* in the *random graph model*.

Fig. 3.23 An input graph example

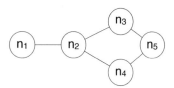

9. (Hard) Please prove that if a Markov chain is *irreducible* and *aperiodic* then the largest eigenvalue of the transition matrix **P** will be equal to 1 and all the other eigenvalues will be strictly less than 1, as introduced in Sect. 3.3.3.3.
10. (Hard) Please try to prove Theorems 3.11 and 3.12 about the *preferential attachment model*.

References

1. L. Adamic, E. Adar, Friends and neighbors on the Web. Soc. Netw. **25**(3), 211–230 (2003)
2. L. Adamic, R. Lukose, A. Puniyani, B. Huberman, Search in power-law networks. Phys. Rev. E **64**, 046135 (2001)
3. M. Bagella, L. Becchetti, The determinants of motion picture box office performance: evidence from movies produced in Italy. J. Cult. Econ. **23**(4), 237–256 (1999)
4. A. Barabasi, R. Albert, Emergence of scaling in random networks. Science **286**(5439), 509–512 (1999)
5. A. Barrat, M. Barthélemy, R. Pastor-Satorras, A. Vespignani, The architecture of complex weighted networks. Proc. Natl. Acad. Sci. **101**(11), 3747–3752 (2004)
6. H. Bast, D. Delling, A. Goldberg, M. Müller-Hannemann, T. Pajor, P. Sanders, D. Wagner, R. Werneck, Route planning in transportation networks (2015). arXiv:1504.05140
7. M. Berry, S. Dumais, G. O'Brien, Using linear algebra for intelligent information retrieval. SIAM Rev. **37**(4), 573–595 (1995)
8. C. Bettstetter, On the minimum node degree and connectivity of a wireless multihop network, in *Proceedings of the 3rd ACM International Symposium on Mobile Ad Hoc Networking and Computing* (ACM, New York, 2002)
9. B. Bollobás, S. Janson, O. Riordan, The phase transition in inhomogeneous random graphs. Random Struct. Algoritm. **31**(1), 3–122 (2007)
10. O. Bonaventure, *Computer Networking: Principles, Protocols, and Practice* (The Saylor Foundation, Washington, 2011)
11. S. Borgatti, M. Everett, A graph-theoretic perspective on centrality. Soc. Netw. **28**(4), 466–484 (2006)
12. U. Brandes, T. Erlebach, *Network Analysis: Methodological Foundations*. Lecture Notes in Computer Science (Springer, Berlin, 2005)
13. A. Clauset, C. Shalizi, M. Newman, Power-law distributions in empirical data. SIAM Rev. **51**(4), 661–703 (2009)
14. P. Erdos, A. Renyi, On the evolution of random graphs. Publ. Math. Inst. Hung. Acad. Sci. **5**(1), 17–60 (1960)
15. F. Fouss, A. Pirotte, J. Renders, M. Saerens, Random-walk computation of similarities between nodes of a graph with application to collaborative recommendation. IEEE Trans. Knowl. Data Eng. **19**, 355–369 (2007)
16. L. Freeman, A set of measures of centrality based on betweenness. Sociometry **40**, 35–41 (1977)
17. E. Gilbert, Random graphs. Ann. Math. Stat. **30**(4), 1141–1144 (1959)
18. T. Gruber, Toward principles for the design of ontologies used for knowledge sharing. Int. J. Hum. Comput. Stud. **43**(5–6), 907–928 (1995)
19. F. Harary, H. Kommel, Matrix measures for transitivity and balance. J. Math. Sociol. **6**(2), 199–210 (1979)
20. J. Harmon, The psychology of interpersonal relations. Soc. Forces **37**(3), 272–273 (1959)
21. J. Heidemann, M. Klier, F. Probst, Online social networks: a survey of a global phenomenon. Comput. Netw. **56**(18), 3866–3878 (2012)
22. D. Horton, R. Wohl, Mass communication and para-social interaction. Psychiatry **19**(3), 215–229 (1956)
23. A. Huang, Similarity measures for text document clustering, in *Proceedings of the Sixth New Zealand Computer Science Research Student Conference (NZCSRSC2008)* (Christchurch, 2008), pp. 49–56
24. P. Jaccard, Étude comparative de la distribution florale dans une portion des alpes et des jura. Bull. Soc. Vaud. Sci. Nat. **37**(142), 547–579 (1901)
25. L. Katz, A new status index derived from sociometric analysis. Psychometrika **18**, 39–43 (1953)
26. D. Kempe, J. Kleinberg, É. Tardos, Maximizing the spread of influence through a social network, in *Proceedings of the Ninth ACM SIGKDD International Conference on Knowledge Discovery and Data Mining* (ACM, New York, 2003), pp. 137–146
27. J. Kleinberg, Authoritative sources in a hyperlinked environment. J. ACM **46**(5), 604–632 (1999)
28. J. Kleinberg, The small-world phenomenon: an algorithmic perspective, in *Proceedings of the Thirty-Second Annual ACM Symposium on Theory of Computing* (ACM, New York, 2000), pp. 163–170
29. X. Kong, J. Zhang, P. Yu, Inferring anchor links across multiple heterogeneous social networks, in *Proceedings of the 22nd ACM International Conference on Information and Knowledge Management* (ACM, New York, 2013), pp. 179–188
30. T. Lappas, K. Liu, E. Terzi, Finding a team of experts in social networks, in *Proceedings of the 15th ACM SIGKDD International Conference on Knowledge Discovery and Data Mining* (ACM, New York, 2009), pp. 467–476
31. E. Leicht, P. Holme, M. Newman, Vertex similarity in networks. Phys. Rev. E **73**, 026120 (2006)

32. J. Leskovec, D. Huttenlocher, J. Kleinberg, Signed networks in social media, in *Proceedings of the SIGCHI Conference on Human Factors in Computing Systems* (ACM, New York, 2010), pp. 1361–1370
33. D. Liben-Nowell, J. Kleinberg, The link-prediction problem for social networks. J. Am. Soc. Inf. Sci. Technol. **58**(7), 1019–1031 (2007)
34. Y. Liu, J. Zhang, C. Zhang, P. Yu, Data-driven blockbuster planning on online movie knowledge library, in *2018 IEEE International Conference on Big Data* (IEEE, Piscataway, 2018)
35. F. Lorrain, H. White, Structural equivalence of individuals in social networks. J. Math. Sociol. **1**, 49–80 (1971)
36. L. Lovász, Random walks on graphs: a survey, in *Combinatorics, Paul Erdős is Eighty* (1996)
37. Q. Mei, D. Zhou, K. Church, Query suggestion using hitting time, in *Proceedings of the 17th ACM conference on Information and Knowledge Management* (ACM, New York, 2008)
38. A. Mislove, M. Marcon, K. Gummadi, P. Druschel, B. Bhattacharjee, Measurement and analysis of online social networks, in *Proceedings of the 7th ACM SIGCOMM Conference on Internet Measurement* (ACM, New York, 2007), pp. 29–42
39. M. Newman, Models of the small world. J. Stat. Phys. **101**(3–4), 819–841 (2000)
40. M. Newman, *Networks: An Introduction* (Oxford University Press, New York, 2010)
41. J. Pan, H. Yang, C. Faloutsos, P. Duygulu, Automatic multimedia cross-modal correlation discovery, in *Proceedings of the Tenth ACM SIGKDD International Conference on Knowledge Discovery and Data Mining* (ACM, New York, 2004), pp. 653–658
42. S. Pillai, T. Suel, S. Cha, The Perron-Frobenius theorem: some of its applications. IEEE Signal Process. Mag. **22**(2), 62–75 (2005)
43. A. Pothen, H. Simon, K. Liou, Partitioning sparse matrices with eigenvectors of graphs. SIAM J. Matrix Anal. Appl. **11**(3), 430–452 (1990)
44. E. Ravasz, A. Somera, D. Mongru, Z. Oltvai, A. Barabási, Hierarchical organization of modularity in metabolic networks. Science **297**(5586), 1551–1555 (2002)
45. M. Riedl, M. Young, Narrative planning: balancing plot and character. J. Artif. Int. Res. **39**(1), 217–267 (2010)
46. B. Ruhnau, Eigenvector-centrality a node-centrality? Soc. Netw. **22**(4), 357–365 (2000)
47. P. Savalle, E. Richard, N. Vayatis, Estimation of simultaneously sparse and low rank matrices, in *Proceedings of the 29th International Conference on Machine Learning* (2012)
48. C. Shi, Y. Li, J. Zhang, Y. Sun, P. S. Yu, A survey of heterogeneous information network analysis. IEEE Trans. Knowl. Data Eng. **29**, 17–37 (2017)
49. R. Solomonoff, A. Rapoport, Connectivity of random nets. Bull. Math. Biol. **13**, 107–117 (1951)
50. T. Sørensen, A method of establishing groups of equal amplitude in plant sociology based on similarity of species and its application to analyses of the vegetation on Danish commons. Biol. Skr. **5**, 1–34 (1948)
51. Y. Sun, J. Han, *Mining Heterogeneous Information Networks: Principles and Methodologies* (Morgan & Claypool Publishers, 2012)
52. Y. Sun, R. Barber, M. Gupta, C. C. Aggarwal, J. Han, Co-author relationship prediction in heterogeneous bibliographic networks, in *2011 International Conference on Advances in Social Networks Analysis and Mining* (IEEE, Piscataway, 2011)
53. Y. Sun, J. Han, X. Yan, P. Yu, T. Wu, Pathsim: meta path-based top-k similarity search in heterogeneous information networks. Proc. VLDB Endowment **4**(11), 992–1003 (2011)
54. J. Tang, T. Lou, J. Kleinberg, Inferring social ties across heterogeneous networks, in *Proceedings of the Fifth ACM International Conference on Web Search and Data Mining* (ACM, New York, 2012), pp. 743–752
55. M. van Steen, *Graph Theory and Complex Networks: An Introduction* (Maarten van Steen, Lexington, 2010)
56. Z. Wang, J. Liao, Q. Cao, H. Qi, Z. Wang, Friendbook: a semantic-based friend recommendation system for social networks. IEEE Trans. Mob. Comput. **14**(3), 538–551 (2015)
57. S. Wasserman, K. Faust, *Social Network Analysis: Methods and Applications* (Cambridge University Press, Cambridge, 1994)
58. D. Watts, S. Strogatz, Collective dynamics of small-world networks. Nature **393**, 440–442 (1998)
59. K. Wilcox, A.T. Stephen, Are close friends the enemy? Online social networks, self-esteem, and self-control. J. Consum. Res. **40**(1), 90–103 (2012)
60. R. Wilson, *Introduction to Graph Theory* (Wiley, London, 1986)
61. X. Xie, Potential friend recommendation in online social network, in *2010 IEEE/ACM International Conference on Green Computing and Communications & International Conference on Cyber, Physical and Social Computing* (IEEE, Piscataway, 2010). https://ieeexplore.ieee.org/abstract/document/5724926
62. J. Yang, J. McAuley, J. Leskovec, Community detection in networks with node attributes, in *2013 IEEE 13th International Conference on Data Mining (ICDM)* (IEEE, Piscataway, 2013)
63. Y. Yao, H. Tong, X. Yan, F. Xu, J. Lu, Matri: a multi-aspect and transitive trust inference model, in *Proceedings of the 22nd International Conference on World Wide Web* (ACM, New York, 2013), pp. 1467–1476

64. J. Ye, H. Cheng, Z. Zhu, M. Chen, Predicting positive and negative links in signed social networks by transfer learning, in *Proceedings of the 22nd International Conference on World Wide Web* (ACM, New York, 2013), pp. 1477–1488
65. R. Zafarani, M. Abbasi, H. Liu, *Social Media Mining: An Introduction* (Cambridge University Press, New York, 2014)
66. J. Zhang, Social network fusion and mining: a survey (2018). arXiv preprint. arXiv:1804.09874
67. J. Zhang, P. Yu, *Link Prediction Across Heterogeneous Social Networks: A Survey* (University of Illinois, Chicago, 2014)
68. J. Zhang, P. Yu, Community detection for emerging networks, in *Proceedings of the 2015 SIAM International Conference on Data Mining* (Society for Industrial and Applied Mathematics, Philadelphia, 2015), pp. 127–135
69. J. Zhang, P. Yu, Multiple anonymized social networks alignment, in *2015 IEEE International Conference on Data Mining* (IEEE, Piscataway, 2015)
70. J. Zhang, P. Yu, PCT: partial co-alignment of social networks, in *Proceedings of the 25th International Conference on World Wide Web* (International World Wide Web Conferences Steering Committee, Geneva, 2016), pp. 749–759
71. J. Zhang, X. Kong, P. Yu, Predicting social links for new users across aligned heterogeneous social networks (2013). arXiv preprint. arXiv:1310.3492
72. J. Zhang, X. Kong, P. Yu, Transferring heterogeneous links across location-based social networks, in *Proceedings of the 7th ACM International Conference on Web Search and Data Mining* (ACM, New York, 2014), pp. 303–312
73. J. Zhang, P. Yu, Z. Zhou, Meta-path based multi-network collective link prediction, in *Proceedings of the 20th ACM SIGKDD International Conference on Knowledge Discovery and Data Mining* (ACM, New York, 2014), pp. 1286–1295
74. J. Zhang, P. Yu, Y. Lv, Organizational chart inference, in *Proceedings of the 21st ACM SIGKDD International Conference on Knowledge Discovery and Data Mining* (ACM, New York, 2015), pp. 1435–1444
75. J. Zhang, S. Wang, Q. Zhan, P. Yu, Intertwined viral marketing in social networks, in *2016 IEEE/ACM International Conference on Advances in Social Networks Analysis and Mining (ASONAM)* (IEEE, Piscataway, 2016)
76. J. Zhang, P. Yu, Y. Lv, Q. Zhan, Information diffusion at workplace, in *Proceedings of the 25th ACM International on Conference on Information and Knowledge Management* (ACM, New York, 2016), pp. 1673–1682
77. J. Zhang, Q. Zhan, L. He, C. Aggarwal, P. Yu, Trust hole identification in signed networks, in *Joint European Conference on Machine Learning and Knowledge Discovery in Databases* (Springer, Berlin, 2016), pp. 697–713
78. J. Zhang, P. Yu, Y. Lv, Enterprise employee training via project team formation, in *Proceedings of the Tenth ACM International Conference on Web Search and Data Mining* (ACM, New York, 2017), pp. 3–12
79. J. Zhang, C. Aggarwal, P. Yu, Rumor initiator detection in infected signed networks, in *2017 IEEE 37th International Conference on Distributed Computing Systems (ICDCS)* (IEEE, Piscataway, 2017)
80. T. Zhou, L. Lü, Y. Zhang, Predicting missing links via local information. Eur. Phys. J. B **71**(4), 623–630 (2009)

Part II

Information Fusion: Social Network Alignment

Supervised Network Alignment

4

4.1 Overview

Online social networks, such as Facebook,[1] Twitter,[2] Foursquare,[3] and LinkedIn,[4] have become more and more popular in recent years. Each social network can be represented as a heterogeneous network containing abundant information about: who, where, when, and what, i.e., who the users are, where they have been to, what they have done, and when they did these activities. Different online social networks can provide unique social network services for the users. For instance, Facebook is a general public social sharing site, Twitter is a micro blogging social site mainly about short posts, Foursquare is a location based social network, and LinkedIn is a business oriented professional social network site.

Nowadays, to enjoy the social network services from multiple sites at the same time, people are usually getting involved in more and more different kinds of social networks simultaneously. For example, people usually share reviews or tips about different locations or places with their friends using Foursquare. At the same time, they may also share the latest news using Twitter, and share photos using Facebook. Thus, each user often has multiple separate accounts in different social networks. However, these accounts of the same user are mostly isolated without any connections or correspondence relationships to each other.

Discovering the correspondence between accounts of the same user is a crucial prerequisite for many interesting inter-network applications, such as *friend recommendation* [25, 26, 31, 35, 36], *social community detection* [27, 31, 32], and *social information diffusion* [10, 31, 39, 40, 42] using information from multiple networks simultaneously. For example, in the Foursquare network, the social connections and activities of new users can be very sparse. The friend and location recommendations for users are very hard merely using the Foursquare network. However, if the user's Twitter account is also known, his/her social connections and location check-in data in Twitter network will be used to improve the recommendation services in the Foursquare network.

In this book, we focus on studying the common users shared by different social networks, and the correspondence relationship between the shared users across social networks are called the *anchor links* [12, 31, 37]. Formally, inferring the common users shared by different social network sites is

---

[1] https://www.facebook.com.

[2] https://twitter.com.

[3] https://foursquare.com.

[4] https://www.linkedin.com.

© Springer Nature Switzerland AG 2019
J. Zhang, P. S. Yu, *Broad Learning Through Fusions*,
https://doi.org/10.1007/978-3-030-12528-8_4

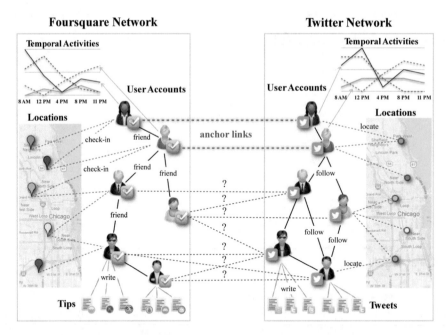

Fig. 4.1 An example of supervised network alignment problem

called the *network alignment* problem [12, 31, 33, 34, 38, 41], which can also be called the *anchor link prediction* problem [12, 31]. In this book, these two terms are used interchangeably when referring to the problem. In the real-world, the anchor links connecting the common information entities across different social networks are extremely hard to identify. Manual labeling of the anchor links between networks can be a tedious and complicated task. Depending on whether the pre-labeled training data is available or not, in this chapter as well as the following Chaps. 5–6, the network alignment problem will be introduced based on different learning settings, including *supervised learning setting*, *unsupervised learning setting*, and *semi-supervised learning setting*.

In this chapter, as illustrated in Fig. 4.1, the network alignment problem is studied based on the *supervised learning setting* specifically, assuming that we can obtain a set of labeled anchor links as the training set in advance. We will start this section with the *supervised network alignment* problem definition. Subject to the *one-to-one* constraint [12, 33, 34], two different network alignment models for *full network alignment* [12, 31, 33, 34] and *partial network alignment* [31, 38] models will be introduced afterwards, respectively, which both use the classification method as the base model. To incorporate the *one-to-one* cardinality constraint into the problem formulation, the *anchor link prediction with cardinality constraint* framework will be introduced, which is a generalized link prediction model and can be used in inferring other types of links subject to any cardinality constraints as well.

4.2   Supervised Network Alignment Problem Definition

Example 4.1 In Fig. 4.3, we show an example of two heterogeneous social networks (Twitter and Foursquare) with six users. Each user has two accounts in these two networks separately. In each network, users are connected with each other through the social links. Moreover, each user is also connected with a set of locations, timestamps, and text contents created by their online social activities. Note that the top two users in Fig. 4.3 also have another type of link, which connects the same user's accounts between two networks. We call these links the *anchor links* [12, 37] as introduced in Sect. 3.4.3. Each anchor link indicates a pair of accounts that belong to the same user. The task of network alignment problem is to discover which account pairs, as shown with question marks in Fig. 4.3, should be connected by the anchor links in the real world.

Both the Foursquare and Twitter networks shown in the example can be represented as heterogeneous social networks defined in the previous chapter. For instance, we can represent the Foursquare network as $G^{(1)} = (\mathcal{V}^{(1)}, \mathcal{E}^{(1)})$, where $\mathcal{V}^{(1)}$ and $\mathcal{E}^{(1)}$ denote the node and link sets, respectively. As introduced before, the Foursquare involves different types of information entities, and the node set $\mathcal{V}^{(1)} = \mathcal{U}^{(1)} \cup \mathcal{P}^{(1)} \cup \mathcal{L}^{(1)} \cup \mathcal{W}^{(1)} \cup \mathcal{T}^{(1)}$ contains the user, post, location check-in, word, and timestamp nodes, respectively. The node subset $\mathcal{U}^{(1)} = \{u_1^{(1)}, u_2^{(1)}, \ldots, u_{n^{(1)}}^{(1)}\}$ denotes the set of users in Foursquare. The posts written by the users are represented as the set $\mathcal{P}^{(1)}$, which may involve a set of location check-ins $\mathcal{L}^{(1)}$, text words $\mathcal{W}^{(1)}$, and timestamps $\mathcal{T}^{(1)}$. Meanwhile, the link set $\mathcal{E}^{(1)} = \mathcal{E}_{u,u}^{(1)} \cup \mathcal{E}_{u,p}^{(1)} \cup \mathcal{E}_{p,l}^{(1)} \cup \mathcal{E}_{p,w}^{(1)} \cup \mathcal{E}_{p,t}^{(1)}$ involves the links among users, between users and posts, as well as those between posts and locations check-ins, words, and timestamps. Similarly, the Twitter network can be represented as a heterogeneous social network $G^{(2)}$ with a similar structure, the user node set involved in which can be denoted as $\mathcal{U}^{(2)} = \{u_1^{(1)}, u_2^{(1)}, \ldots, u_{n^{(2)}}^{(1)}\}$.

In the network alignment problem settings, the anchor links are assumed to be subject to the *one-to-one* constraint, where a user from one network can be connected by at most one anchor link with users from another network. In the supervised network alignment setting, a set of existing and non-existing anchor links can be pre-identified and labeled as the (positive and negative) training set $\mathcal{A}_{train} \subset \mathcal{U}^{(1)} \times \mathcal{U}^{(2)}$. And the supervised network alignment problem aims at inferring the existence of the remaining potential anchor links, i.e., $\mathcal{A}_{test} = \mathcal{U}^{(1)} \times \mathcal{U}^{(2)} \setminus \mathcal{A}_{train}$, among users between networks $G^{(1)}$ and $G^{(2)}$.

In the supervised network alignment problem, a set of features will be extracted for the anchor links across networks with the heterogeneous information available in the networks. Meanwhile, the existing and non-existing anchor links will be labeled as positive and negative instances, respectively. Based on the training set $\mathcal{A}_{train}$, the feature vectors and labels of links in the set can be represented as tuples $\{(\mathbf{x}_l, y_l)\}_{l \in \mathcal{A}_{train}}$, where $\mathbf{x}_l$ represents the feature vector extracted for anchor link l and $y_l \in \{-1, +1\}$ denotes its label. Based on the training set, the *supervised network alignment* problem aims at building a mapping $f : \mathcal{A}_{test} \to \{-1, +1\}$ to determine the labels of the anchor links in the testing set.

4.3 Supervised Full Network Alignment

Depending on how many users in the networks are the shared anchor users, the supervised network alignment can be categorized into the *supervised full network alignment* [12] and *supervised (mutually) partial network alignment* [38] problems. In the *full network alignment* problem, the networks to be studied are fully aligned and each user will be connected by an anchor link in the results; while in the *partial network alignment* problem, the networks are partially aligned and many users should stay isolated without any matching partners across networks. In this section, we will introduce the model proposed to solve the *supervised full network alignment*, where the users in networks $G^{(1)}$ and $G^{(2)}$ are all anchor users and will get connected by the anchor links.

As introduced before, in this section, we will introduce a two-phase approach, namely MNA [12], to address the *full network alignment* problem. The first phase of MNA tackles the feature extraction problem, while the second phase takes care of the one-to-one constrained anchor link prediction.

The phase of feature extraction mainly explores two kinds of ideas on multiple heterogeneous social networks. First, we exploit social links in each network and the labeled anchor links across the two networks to extract social features for anchor link prediction. Second, we exploit the heterogeneous information in both networks to extract three sets of heterogeneous features for anchor link prediction, which correspond to aggregated patterns of users on spatial distribution, temporal activity distribution, and textual usage behavior separately. All these extracted features and the pairs of accounts with known labels will be used to learn a binary SVM for anchor link prediction. Since the label predictions of SVM don't satisfy the one-to-one constraint, the real-value scores of the SVM are used as the input for the second phase, and derive the anchor link predictions collectively according to the one-to-one constraint.

4.3.1 Feature Extraction for Anchor Links

There exist different types of information in the networks, including the social connections, location check-ins, text words, and timestamps, which can all indicate the correspondence relationships among users across networks.

Example 4.2 We show a case study to demonstrate these heterogeneous information from two networks are useful for identifying the anchor links. In Fig. 4.2, we show a case of five real-world users who have both Twitter and Foursquare accounts. These five users are socially connected in both networks, as shown in Fig. 4.2a. By considering this social information, we can significantly shrink the search space for anchor links if one or some of these users' accounts in both networks have already been labeled by anchor links. In Fig. 4.2b, we show the spatial distributions of different users in both networks. We can see that the spatial distribution of the same user is pretty similar to each other. Michelle is mainly located in the central states of the USA, when sending tweets and foursquare tips. The spatial distributions of her foursquare account and twitter account are pretty similar. In Fig. 4.2c, we show the temporal distributions of the users. We can see that Tristan's temporal activities across both Twitter account and Foursquare account are very consistent, and his distribution is very different from Lisa's temporal activity pattern. In Fig. 4.2d, we show some frequently used words by the users, where the choices of words of the same user can be pretty consistent. For example, Andrew seems to prefer to use "awsm" instead of "awesome" when writing tweets and tips.

Most existing features for link prediction, such as numbers of common neighbors and other social closeness measures introduced in the previous chapter, mainly focus on one single network setting, and the target links are assumed to be subject to the many-to-many cardinality constraint. These features cannot be directly used for the anchor link prediction task across multiple networks. Based on the above example, in the following part, we will introduce the features that can be extracted for the anchor links between networks with such heterogeneous social information.

4.3.1.1 Social Connection Based Features

Users often have similar social links in different social networks, such as Twitter and Facebook, because such social links usually indicate the user's social ties in real life. In other words, the social similarity between two user accounts from different social networks can be exploited to help locate the same user.

Our goal is to extract discriminative social features for a pair of user accounts in two disjoint social networks. Intuitively, the social neighbors of each user account can only involve user accounts from the same social network. For example, the neighbors for a Facebook account can only involve

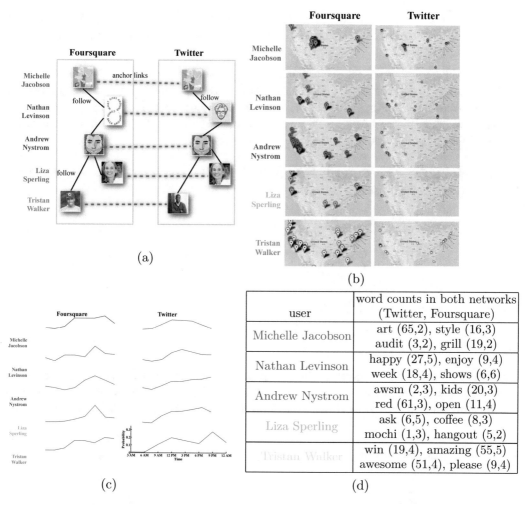

Fig. 4.2 Case study: five real-world users with their social, spatial, temporal, and text distributions. (**a**) Social. (**b**) Spatial. (**c**) Temporal. (**d**) Text

Facebook accounts instead of Twitter accounts. However, in anchor link prediction problem, we need to extract a set of features about a pair of user accounts in two different networks separately. The social neighbors for two user accounts are two disjoint sets of user accounts in two separate networks. There cannot exist any shared nodes among the neighbors of the pair of user accounts. In the following part, we will introduce the extension of several social features to multi-network settings.

Let $(u_i^{(1)}, u_j^{(2)})$ be a potential anchor link between these two networks, and $\mathcal{A}_{train}^+$ be the set of existing anchor links in the training set. Here we extend the definitions of some commonly used social closeness features in link prediction, i.e., "common neighbors," "Jaccard's coefficient," and "Adamic/Adar measure," to the *inter-network* scenarios for anchor links specifically.

- **Extended Common Neighbor**: The *extended common neighbor* measure $ECN(u_i^{(1)}, u_j^{(2)})$ represents the number of "common" neighbors between $u_i^{(1)}$ in network $G^{(1)}$ and $u_j^{(2)}$ in network $G^{(2)}$. We can denote the neighbors of $u_i^{(1)}$ in network $G^{(1)}$ as $\Gamma(u_i^{(1)})$, and the neighbors of $u_j^{(2)}$ in network $G^{(2)}$ as $\Gamma(u_j^{(2)})$. It is easy to identify that the sets $\Gamma(u_i^{(1)})$ and $\Gamma(u_j^{(2)})$ contain the users from two different networks, respectively, which are isolated without any common entries, i.e., $\Gamma(u_i^{(1)}) \cap \Gamma(u_j^{(2)}) = \emptyset$.

Meanwhile, based on the existing anchor links $\mathcal{A}_{train}^+$, some of the users in $\Gamma(u_i^{(1)})$ and $\Gamma(u_j^{(2)})$ can correspond to the accounts of the same users in these two networks, who are actually connected by the anchor links in $\mathcal{A}_{train}^+$. Therefore, based on the anchor links in set $\mathcal{A}_{train}^+$, the *extended common neighbor* measure between these two users can be defined as the number of shared anchor users in their neighbor sets, respectively.

Definition 4.1 (Extended Common Neighbor) The measure of *extended common neighbor* is defined as the number of shared users between $\Gamma(u_i^{(1)})$ and $\Gamma(u_j^{(2)})$.

$$
\begin{aligned}
&ECN(u_i^{(1)}, u_j^{(2)}) \\
&= \left| \{(u_p^{(1)}, u_q^{(2)}) | (u_p^{(1)}, u_q^{(2)}) \in \mathcal{A}_{train}^+, u_p^{(1)} \in \Gamma(u_i^{(1)}), u_q^{(2)} \in \Gamma(u_j^{(2)})\} \right| \\
&= \left| \Gamma(u_i^{(1)}) \bigcap_{\mathcal{A}_{train}^+} \Gamma(u_j^{(2)}) \right|.
\end{aligned}
\tag{4.1}
$$

Example 4.3 For instance, given an input network in Fig. 4.3 involving six users, there exist two known anchor links and the existing anchor link set $\mathcal{A}_{train}^+ = \{(Alice_F, Alice_T), (Cindy_F, Cindy_T)\}$, where the subscript denotes the network identifier (F: Foursquare, T: Twitter). Based

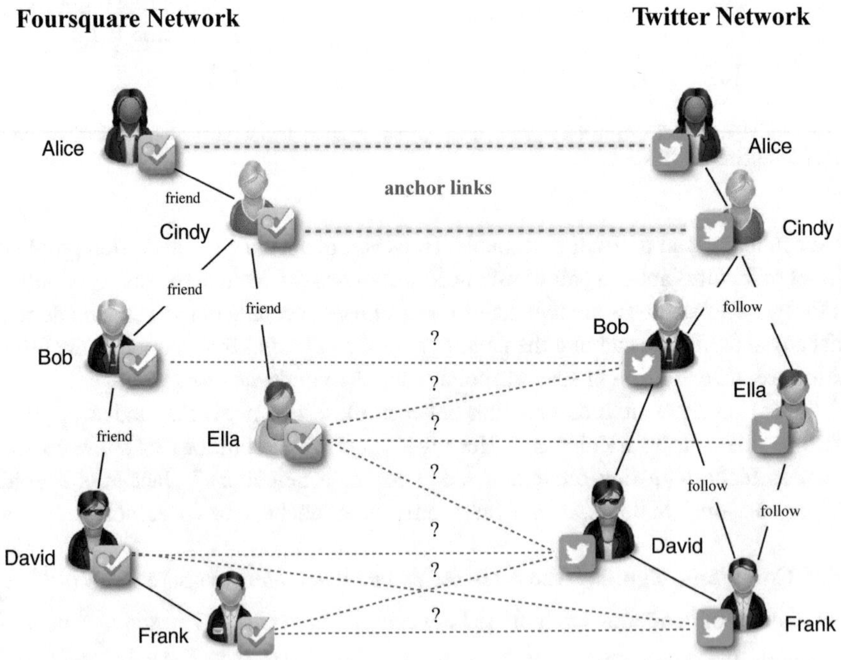

Fig. 4.3 An example of aligned social networks for feature extraction

on the input networks, the *extended common neighbor* feature between two different user pairs (Bob_F, Bob_T) and $(David_F, David_T)$ can be computed as follows:

(1) (Bob_F, Bob_T): We have the neighborhood sets of users Bob_F and Bob_T in the Foursquare and Twitter networks to be sets $\Gamma(Bob_F) = \{Cindy_F, David_F\}$ and $\Gamma(Bob_T) = \{Cindy_T, David_T, Frank_T\}$. We have

$$ECN(Bob_F, Bob_T) = \left| \Gamma(Bob_F) \bigcap_{\mathcal{A}^+_{train}} \Gamma(Bob_T) \right|$$

$$= |\{(Cindy_F, Cindy_T)\}|$$

$$= 1. \qquad (4.2)$$

(2) $(David_F, David_T)$: We can represent the neighborhood sets of $David_F$ and $David_T$ in Foursquare and Twitter to be $\Gamma(David_F) = \{Bob_F, Frank_F\}$ and $\Gamma(David_T) = \{Bob_T, Frank_T\}$. According to the existing anchor link set, no shared users exist in the neighbor sets of $David_F$ and $David_T$. In other words, we have

$$ECN(David_F, David_T) = 0. \qquad (4.3)$$

- **Extended Jaccard's Coefficient**: The measure of Jaccard's coefficient can also be extended to the multi-network setting for anchor links using a similar method of extending common neighbor. The *extended Jaccard's Coefficient* measure $EJC(u_i^{(1)}, u_j^{(2)})$ is a normalized version of the *extended common neighbor*, i.e., $ECN(u_i^{(1)}, u_j^{(2)})$ divided by the total number of distinct users in $\Gamma(u_i^{(1)})$ and $\Gamma(u_j^{(2)})$.

Definition 4.2 (Extended Jaccard's Coefficient) Given the neighborhood sets of users $u_i^{(1)}$ and $u_j^{(2)}$ in networks $G^{(1)}$ and $G^{(2)}$, respectively, the *Extended Jaccard's Coefficient* of user pair $u_i^{(1)}$ and $u_j^{(2)}$ can be represented as

$$EJC(u_i^{(1)}, u_j^{(2)}) = \frac{\left| \Gamma(u_i^{(1)}) \bigcap_{\mathcal{A}^+_{train}} \Gamma(u_j^{(2)}) \right|}{\left| \Gamma(u_i^{(1)}) \bigcup_{\mathcal{A}^+_{train}} \Gamma(u_j^{(2)}) \right|}, \qquad (4.4)$$

where

$$\left| \Gamma(u_i^{(1)}) \bigcup_{\mathcal{A}^+_{train}} \Gamma(u_j^{(2)}) \right| = |\Gamma(u_i^{(1)})| + |\Gamma(u_j^{(2)})| - \left| \Gamma(u_i^{(1)}) \bigcap_{\mathcal{A}^+_{train}} \Gamma(u_j^{(2)}) \right|. \qquad (4.5)$$

Example 4.4 For example, based on the aligned social networks in Fig. 4.3, we can compute the *extended Jaccard's Coefficient* of user pair (Bob_F, Bob_T) to be

$$EJC(Bob_F, Bob_T) = \frac{\left| \Gamma(Bob_F) \bigcap_{\mathcal{A}^+_{train}} \Gamma(Bob_T) \right|}{\left| \Gamma(Bob_F) \bigcup_{\mathcal{A}^+_{train}} \Gamma(Bob_T) \right|}$$

$$= \frac{1}{4}. \qquad (4.6)$$

- **Extended Adamic/Adar Index**: In addition, the Adamic/Adar measure can also be extended to the multi-network setting, where the common neighbors are weighted by their average degrees in both social networks.

Definition 4.3 (Extended Adamic/Adar Index) The *Extended Adamic/Adar Index* of the user pairs $u_i^{(1)}$ and $u_j^{(2)}$ across networks can be represented as

$$EAA(u_i^{(1)}, u_j^{(2)})$$

$$= \sum_{\forall (u_p^{(1)}, u_q^{(2)}) \in \Gamma(u_i^{(1)}) \cap_{\mathcal{A}_{train}^+} \Gamma(u_j^{(2)})} \log^{-1}\left(\frac{|\Gamma(u_p^{(1)})| + |\Gamma(u_q^{(2)})|}{2}\right). \tag{4.7}$$

In the EAA definition, for the common neighbors shared by $u_i^{(1)}$ and $u_j^{(2)}$, their degrees are defined as the average of their degrees in networks $G^{(1)}$ and $G^{(2)}$, respectively. Considering that different networks are of different scales, like Twitter is far larger than Foursquare, the node degree measure can be dominated by the degree of the larger networks. Some other weighted forms of the degree measure, like $\alpha \cdot |\Gamma(u_p^{(1)})| + (1 - \alpha) \cdot |\Gamma(u_q^{(2)})|$ ($\alpha \in [0, 1]$), can be applied to replace $\frac{|\Gamma(u_p^{(1)})| + |\Gamma(u_q^{(2)})|}{2}$ in the definition.

In addition to the social features mentioned above, heterogeneous social networks also involve abundant information about: where, when, and what. In the following part, we will introduce how to exploit the spatial, temporal, and text content information of different user accounts to facilitate anchor link prediction.

4.3.1.2 Spatial Check-In Distribution Features

Besides the social connection information, users' activities in the offline world may also provide important signals for inferring the anchor links across the networks as indicated in Fig. 4.2b. We notice that users in different social networks usually check-in at similar locations in real life, such as their home, working places, traveling spots, etc. The similarity between the spatial distributions of two user accounts from different social networks can also be used to help locate the same user.

Each location can be represented as a pair of (latitude, longitude) $= \ell \in \mathcal{L}^{(1)}$ (or $\ell \in \mathcal{L}^{(2)}$). Three different measures have been introduced in [12] to evaluate the similarity between the spatial distributions of two users accounts. Given a user $u_i^{(1)}$ in network $G^{(1)}$, we can represent the locations she/he has ever visited as set $L(u_i^{(1)}) \subset \mathcal{L}^{(1)}$. In a similar way, we can also represent the set of location visited by user $u_j^{(2)}$ as set $L(u_j^{(2)}) \subset \mathcal{L}^{(2)}$.

- **Number of Shared Locations**: Formally, we introduce a notation $l_1 = l_2$ to denote that these two locations are the same location, i.e., sharing common latitude and longitude. The number of shared locations which have been visited by users $u_i^{(1)}$ and $u_j^{(2)}$ can be denoted as

$$\left| \{ (l^{(1)}, l^{(2)}) | l^{(1)} \in L(u_i^{(1)}), l^{(2)} \in L(u_j^{(2)}), l^{(1)} = l^{(2)} \} \right|. \tag{4.8}$$

- **Cosine Similarity Between Location Vectors**: Let $\mathcal{L} = \mathcal{L}^{(1)} \cup \mathcal{L}^{(2)}$ be the set of all locations in both network $G^{(1)}$ and network $G^{(2)}$, for user $u_i^{(1)}$ (or $u_j^{(2)}$) in these two networks, the locations visited by them can be organized as a binary vector of length $|\mathcal{L}|$, i.e., $\mathbf{l}(u_i^{(1)}) \in \mathbb{R}^{|\mathcal{L}|}$ (or $\mathbf{l}(u_j^{(2)}) \in \mathbb{R}^{|\mathcal{L}|}$), the entries in which denote the visiting times for users at these locations. The similarity of the location visiting records between the users $u_i^{(1)}$ and $u_j^{(2)}$ can be denoted as the cosine similarity of the location-visiting record vectors $\mathbf{l}(u_i^{(1)})$ and $\mathbf{l}(u_j^{(2)})$, which can be used as another feature based on the location check-in records

$$\frac{\mathbf{l}(u_i^{(1)})^\top \mathbf{l}(u_j^{(2)})}{\sqrt{\left(\mathbf{l}(u_i^{(1)})^\top \mathbf{l}(u_i^{(1)})\right)\left(\mathbf{l}(u_j^{(2)})^\top \mathbf{l}(u_j^{(2)})\right)}}. \tag{4.9}$$

- **Average Distance of Visited Locations**: The physical distance between the regions that users pairs are active in can indicate their potential similarity from the geographical perspective. The average distance between the locations visited by the users $u_i^{(1)}$ and $u_j^{(2)}$ can be calculated as another feature based on the check-in records

$$D(L(u_i^{(1)}), L(u_j^{(2)})) = \frac{\sum_{l^{(1)} \in L(u_i^{(1)})} \sum_{l^{(2)} \in L(u_j^{(2)})} D(l^{(1)}, l^{(2)})}{|L(u_i^{(1)})||L(u_j^{(2)})|}, \tag{4.10}$$

where term $D(l^{(1)}, l^{(2)})$ represents the distance between locations $l^{(1)}$ and $l^{(2)}$ (Manhattan distance can be used here).

4.3.1.3 Temporal Distribution Features

We also notice that users in different social networks usually publish posts at similar time slots in real life, e.g., hours after work and weekends, etc. Such temporal distribution can effectively indicate the user's online activity patterns. For example, some users may like to send tweets at night, while other users may like to write tweets at commuting time on the bus or train. The temporal distribution of different user accounts can also help us find the anchor links between two networks.

Users' online social activities can be organized into a temporal vector of length 24, where each entry denotes the ratio of activities in each of the time bin. For instance, the temporal activity vector of user $u_i^{(1)}$ can be represented as a vector $\mathbf{t}(u_i^{(1)}) \in \mathbb{R}^{24}$. Similarly, we can also obtain the temporal activity similarity of users from different networks by calculating either the inner product or the cosine similarity of these two temporal activity vectors as follows:

$$(1) \text{ inner product: } \mathbf{t}(u_i^{(1)})^\top \mathbf{t}(u_j^{(2)}), \tag{4.11}$$

$$(2) \text{ cosine similarity: } \frac{\mathbf{t}(u_i^{(1)})^\top \mathbf{t}(u_j^{(2)})}{\sqrt{\left(\mathbf{t}(u_i^{(1)})^\top \mathbf{t}(u_i^{(1)})\right)\left(\mathbf{t}(u_j^{(2)})^\top \mathbf{t}(u_j^{(2)})\right)}}. \tag{4.12}$$

4.3.1.4 Textual Content Features

According to the textual content analysis provided in the table in Fig. 4.2d, the textual content of posts written by users in different social networks can also be a great hint for the anchor links, because different users may have different choices of words in their posts. The words used by the users can reveal either their personal habits or their word usage patterns.

The post contents written by each user can be converted into a bag-of-words vector weighted by TF-IDF. Let $\mathcal{W}$ denote the set of words used by all the users in networks $G^{(1)}$ and $G^{(2)}$. The word usage record vectors of users $u_i^{(1)}$ and $u_j^{(2)}$ in these two networks can be represented as vectors $\mathbf{w}(u_i^{(1)}) \in \mathbb{R}^{|\mathcal{W}|}$ and $\mathbf{w}(u_j^{(2)}) \in \mathbb{R}^{|\mathcal{W}|}$, respectively. Similar to the temporal activities information, the inner product and cosine similarity can be applied to these two vectors to denote how similar these users are in the textual word usage patterns:

$$(1)\ \text{inner product: } \mathbf{w}(u_i^{(1)})^\top \mathbf{w}(u_j^{(2)}), \tag{4.13}$$

$$(2)\ \text{cosine similarity: } \frac{\mathbf{w}(u_i^{(1)})^\top \mathbf{w}(u_j^{(2)})}{\sqrt{\left(\mathbf{w}(u_i^{(1)})^\top \mathbf{w}(u_i^{(1)})\right)\left(\mathbf{w}(u_j^{(2)})^\top \mathbf{w}(u_j^{(2)})\right)}}. \tag{4.14}$$

4.3.2 Supervised Anchor Link Prediction Model

With the features introduced in the previous subsection, in this part, we will introduce the supervised anchor link prediction model. We will provide the general description of the model architecture, and then use an example to show how to build the model.

Given the multiple aligned social networks, via manual labeling, we can identify a set of existing anchor links, denoted as set $\mathcal{A}_{train}^+$, and a set of non-existing anchor links, denoted as set $\mathcal{A}_{train}^-$. The anchor links in sets $\mathcal{A}_{train}^+$ and $\mathcal{A}_{train}^-$ are assigned with the positive and negative labels, respectively, i.e., $\{-1, +1\}$, depending on whether they exist or not. For instance, given a link $l \in \mathcal{A}_{train}^+$, it will be associated with a positive label, i.e., $y_l = +1$, while if link $l \in \mathcal{A}_{train}^-$, it will be associated with a negative label, i.e., $y_l = -1$. With the information across these two aligned heterogeneous social networks, a set of features introduced in the previous subsection can be extracted for the links in sets $\mathcal{A}_{train}^+$ and $\mathcal{A}_{train}^-$. For instance, for a link l in the training set $\mathcal{A}_{train}^+$ (or $\mathcal{A}_{train}^-$), we can represent its feature vector as $\mathbf{x}_l$, which will be called an *anchor link instance* and each feature is an *attribute* of the anchor link (more information about *instance*, *attribute*, and *label* concepts is available in Chap. 2.2). With these anchor link instances and their labels, a supervised learning model can be trained, like the classification models SVM (support vector machine), Decision Tree, or neural networks.

In the test procedure, for each link l in the test set $\mathcal{A}_{test}$, we can represent it with a similar set of features (or attributes) and denote its feature vector as $\mathbf{x}_l$. However, we don't know its label, and we may want to determine whether it exists or not (its label is positive or negative). By applying the trained model to the feature vector of the anchor link, we will obtain a prediction label, which will be returned as the final result.

Example 4.5 In Fig. 4.4, we show an example of the supervised anchor link prediction model architecture. Given the input aligned networks, we can identify a set of existing and non-existing anchor links between them. These links can be used as the training set with the existing anchor links as the positive instances and non-existing ones as the negative instances. Formally, we can denote the training set as $\mathcal{A}_{train}$, where the sets $\mathcal{A}_{train}^+ = \{(A_F, A_T), (C_F, C_T)\}$ and $\mathcal{A}_{train}^- = \{(A_F, C_T), (B_F, C_T), \ldots\}$ (the subscripts denotes the networks they belong to, i.e., F: Foursquare; T: Twitter).

These anchor links are labeled as the positive and negative instances, respectively, where the existing anchor links are assigned with the positive label and the non-existing anchor links are assigned with the negative label. Based on the heterogeneous information inside these two networks, a group of features (i.e., the features that we have introduced in the previous part) can be extracted for these links in the training sets. With the training data, a supervised learning model can be built, and further be applied to the feature vectors of some unknown anchor links, e.g., (B_F, B_T), which will output either a label or a score indicating its existence confidence scores.

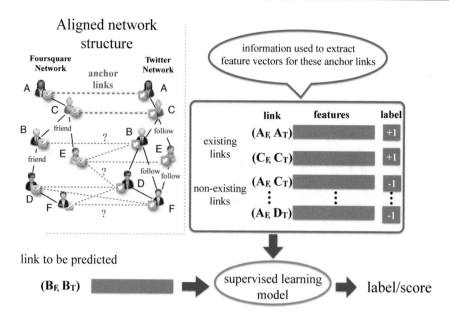

Fig. 4.4 An example of supervised anchor link prediction framework

Meanwhile, due to the *one-to-one* cardinality constraint on the anchor links, among all the potential anchor links incident to each user, only one of them will be positive and the remaining ones will be all negative. In other words, the negative training set is usually much larger than the positive set in the anchor link prediction model. In this part, we will not handle such a challenging problem, which will be taken care of in Sect. 4.4 instead.

4.3.3 Stable Matching

After extracting all the four types of heterogeneous features, we can train a binary classifier, such as SVM or logistic regression, for anchor link prediction. However, in the inference process, the predictions of the binary classifier cannot be directly used as anchor links due to the following issues:

- The inference of conventional classifiers are designed for constraint-free settings, and the one-to-one constraint may not necessarily hold in the label prediction of the classifier (e.g., SVM).
- Most classifiers also produce output scores, which can be used to rank the data points in the test set. However, these ranking scores are uncalibrated in scale to anchor link prediction task. Existing classifier calibration methods [29] can only be applied to classification problems without any constraint.

In order to tackle the above issues, we will introduce an inference process, called MNA (Multi-Network Anchoring) as proposed in [12], to infer anchor links based upon the ranking scores of the classifier. This model is motivated by the *stable marriage problem* [5] in mathematics.

Example 4.6 We first use a toy example in Fig. 4.5 to illustrate the main idea of MNA. Suppose in Fig. 4.5a, we are given the ranking scores from the classifiers, between the four user pairs across

Fig. 4.5 An example of anchor link inference by different methods. (**a**) is the input, ranking scores. (**b**)–(**d**) are the results of different methods for anchor link inference

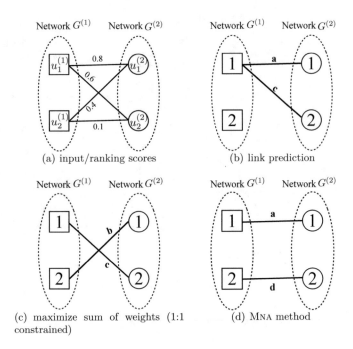

(a) input/ranking scores

(b) link prediction

(c) maximize sum of weights (1:1 constrained)

(d) MNA method

two networks (i.e., network $G^{(1)}$ and network $G^{(2)}$). We can see in Fig. 4.5b that link prediction methods with a fixed threshold may not be able to predict well, because the predicted links do not satisfy the constraint of one-to-one relationship. Thus one user account in network $G^{(1)}$ can be linked with multiple accounts in network $G^{(2)}$. In Fig. 4.5c, *weighted maximum matching* methods can find a set of links with the maximum sum of weights. However, it is worth noting that the input scores are uncalibrated, so the maximum weight matching may not be a good solution to the anchor link prediction problems. The input scores only indicate the ranking of different user pairs, i.e., the preference relationship among different user pairs.

Here we say "node x prefers node y over node z," if the score of pair (x, y) is larger than the score of pair (x, z). For example, in Fig. 4.5c, the weight of pair a, i.e., Score(a) = 0.8, is larger than Score(c) = 0.6. It shows that user $u_1^{(1)}$ (the first user in network $G^{(1)}$) *prefers* $u_1^{(2)}$ over $u_2^{(2)}$. The problem with the prediction result in Fig. 4.5c is that the pair $(u_1^{(1)}, u_1^{(2)})$ should be more likely to be an anchor link due to the following reasons: (1) $u_1^{(1)}$ prefers $u_1^{(2)}$ over $u_2^{(2)}$; (2) $u_1^{(2)}$ also prefers $u_1^{(1)}$ over $u_2^{(1)}$.

By following such an intuition, we can obtain the final stable matching result in Fig. 4.5d, where anchor links $(u_1^{(1)}, u_1^{(2)})$ and $(u_2^{(1)}, u_2^{(2)})$ are selected in the matching process.

Definition 4.4 (Matching) Mapping $\mu : \mathcal{U}^{(1)} \cup \mathcal{U}^{(2)} \to \mathcal{U}^{(1)} \cup \mathcal{U}^{(2)}$ is defined to be a *matching* iff (1) $|\mu(u_i^{(1)})| = 1, \forall u_i^{(1)} \in \mathcal{U}^{(1)}$ and $\mu(u_i^{(1)}) \in \mathcal{U}^{(2)}$; (2) $|\mu(u_j^{(2)})| = 1, \forall u_j^{(2)} \in \mathcal{U}^{(2)}$ and $\mu(u_j^{(2)}) \in \mathcal{U}^{(1)}$; and (3) $\mu(u_i^{(1)}) = u_j^{(2)}$ iff $\mu(u_j^{(2)}) = u_i^{(1)}$.

Definition 4.5 (Blocking Pair) A pair $(u_i^{(1)}, u_j^{(2)})$ is a blocking pair iff $u_i^{(1)}$ and $u_j^{(2)}$ both prefer each other over their current assignments, respectively in the predicted set of anchor links $\mathcal{A}'$.

Definition 4.6 (Stable Matching) An inferred anchor link set $\mathcal{A}'$ is stable if there is no blocking pair.

Based on the result from the previous step, MNA formulates the anchor link pruning problem as a stable matching problem between user accounts in network $G^{(1)}$ and accounts in network $G^{(2)}$. Assume that we have two sets of unlabeled user accounts, i.e., $\mathcal{U}^{(1)}$ in network $G^{(1)}$ and $\mathcal{U}^{(2)}$ in network $G^{(2)}$. Each user $u_i^{(1)}$ has a ranking list or preference list $P(u_i^{(1)})$ over all the user accounts in network $G^{(2)}$ ($u_j^{(2)} \in \mathcal{U}^{(2)}$) based upon the input scores of different pairs.

Example 4.7 For example, in Fig. 4.5a, the preference list of node $u_1^{(1)}$ is $P(u_1^{(1)}) = (u_1^{(2)} > u_2^{(2)})$, indicating that node $u_1^{(2)}$ is preferred by $u_1^{(1)}$ over $u_2^{(2)}$. The preference list of node $u_2^{(1)}$ is also $P(u_2^{(1)}) = (u_1^{(2)} > u_2^{(2)})$. Similarly, a preference list for each user account in network $G^{(2)}$ can also be built. In Fig. 4.5a, $P(u_1^{(2)}) = P(u_2^{(2)}) = (u_1^{(1)} > u_2^{(1)})$.

The proposed MNA method for anchor link prediction is shown in Algorithm 1. In each iteration, MNA first randomly selects a free user account $u_i^{(1)}$ from network $G^{(1)}$. Then MNA gets the most preferred user node $u_j^{(2)}$ by $u_i^{(1)}$ in its preference list $P(u_i^{(1)})$, and removes $u_j^{(2)}$ from the preference list, i.e., $P(u_i^{(1)}) = P(u_i^{(1)}) \setminus u_j^{(2)}$. If $u_j^{(2)}$ is also a free account, MNA adds the pair of accounts $(u_i^{(1)}, u_j^{(2)})$ into the current solution set $\mathcal{A}'$. Otherwise, $u_j^{(2)}$ is already occupied with $u_p^{(1)}$ in $\mathcal{A}'$. MNA then examines the preference of $u_j^{(2)}$. If $u_j^{(2)}$ also prefers $u_i^{(1)}$ over $u_p^{(1)}$, it means that the pair $(u_i^{(1)}, u_j^{(2)})$ is a blocking pair. MNA removes the blocking pair by replacing the pair $(u_p^{(1)}, u_j^{(2)})$ in the solution set $\mathcal{A}'$ with the pair $(u_i^{(1)}, u_j^{(2)})$. Otherwise, if $u_j^{(2)}$ prefers $u_p^{(1)}$ over $u_i^{(1)}$, MNA starts the next iteration to reach out the next free node in network $G^{(1)}$. The algorithm stops when all the users

---

Algorithm 1 Multi-network stable matching

Require: two heterogeneous social networks, $\mathcal{G}^{(1)}$ and $\mathcal{G}^{(2)}$.
        a set of known anchor links $\mathcal{A}$
Ensure: a set of inferred anchor links $\mathcal{A}'$
 1: Construct a training set of user account pairs with known labels using $\mathcal{A}$.
 2: For each pair $(u_i^{(1)}, u_j^{(2)})$, extract four types of features.
 3: Training classification model C on the training set.
 4: Perform classification using model C on the test set.
 5: For each unlabeled user account, sort the ranking scores into a preference list of the matching accounts.
 6: Initialize all unlabeled $u_i^{(1)}$ in $\mathcal{G}^{(1)}$ and $u_j^{(2)}$ in $\mathcal{G}^{(2)}$ as free
 7: $\mathcal{A}' = \emptyset$
 8: **while** $\exists$ free $u_i^{(1)}$ in $\mathcal{G}^{(1)}$ and $u_i^{(1)}$'s preference list is non-empty **do**
 9:     Remove the top-ranked account $u_j^{(2)}$ from $u_i^{(1)}$'s preference list
10:     **if** $u_j^{(2)}$ is free **then**
11:         $\mathcal{A}' = \mathcal{A}' \cup \{(u_i^{(1)}, u_j^{(2)})\}$
12:         Set $u_i^{(1)}$ and $u_j^{(2)}$ as occupied
13:     **else**
14:         $\exists u_p^{(1)}$ that $u_j^{(2)}$ is occupied with.
15:         **if** $u_j^{(2)}$ prefers $u_i^{(1)}$ to $u_p^{(1)}$ **then**
16:             $\mathcal{A}' = (\mathcal{A}' \setminus \{(u_p^{(1)}, u_j^{(2)})\}) \cup \{(u_i^{(1)}, u_j^{(2)})\}$
17:             Set $u_p^{(1)}$ as free and $u_i^{(1)}$ as occupied
18:         **end if**
19:     **end if**
20: **end while**

---

in network $G^{(1)}$ are occupied, or all the preference lists of free accounts in network $G^{(1)}$ are empty. Finally, the selected anchor links in set $\mathcal{A}'$ will be returned as the final network alignment result.

4.4   Supervised Partial Network Alignment

The method MNA introduced in the previous section assumes that the online social networks are fully aligned, and all the users in the networks are *anchor users*, who will all be connected by the anchor links in the final results. However, in the real world, such a strong assumption can hardly hold. These online social networks generally contain different groups of users, which can be highly likely to be partially aligned actually. For instance, As pointed out in [6], by the end of 2013, about 42% of online adults are using multiple social sites at the same time. Meanwhile, 93% of Instagram users are involved in Facebook concurrently and 53% Twitter users are using Instagram as well [16]. In other words, there still exist a large number of users who merely use Instagram or Twitter but are not involved in Facebook, which will make these online social networks partially aligned [37, 38] instead.

4.4.1   Partial Network Alignment Description

Different from the "*supervised full network alignment*" problem, we will study a more general *partial network alignment* problem in this section. There exist several significant differences between these two problems, which are provided as follows to help the readers distinguish these two works. Firstly, the networks studied in this section are partially aligned [37], which contain a large number of anchor and non-anchor users [37] at the same time. Secondly, the networks studied here are not confined to the Foursquare and Twitter social networks, and a more general feature extraction method will be needed. We hope a minor revision of the "*partial network alignment*" problem can be mapped to many other existing tough problems, e.g., large biology network alignment [2], entity resolution in database integration [3], ontology matching [8], and various types of entity matching in online social networks [19]. Thirdly, due to the cardinality constraints on the anchor links as mentioned at the end of Sect. 4.3.2, the number of *positive* and *negative* anchor link instances across networks can be highly imbalanced, i.e., the negative training set is far larger than the positive training set. Such a class imbalance problem will make most of the existing supervised classification model fail to work. Finally, many of the users will stay isolated in the alignment results, since they can be non-anchor users actually. The constraint on *anchor links* is updated to "*one-to-one$_\leq$*" (called "*one-to-at-most-one*"), i.e., each user in one network can be mapped to at most one user in another network. Across partially aligned networks, only anchor users can be connected by anchor links. Therefore, identifying the non-anchor users from networks and pruning all the predicted potential anchor links connected to them is a novel yet challenging problem. The "*one-to-one$_\leq$*" constraint on anchor links can distinguish the "*partial network alignment*" problem from most existing link prediction problems. For example, in traditional link prediction and link transfer problems [20, 21, 37], the constraint on links is "*many-to-many*," while in the "*anchor link inference*" problem [12] across fully aligned networks, the constraint on *anchor links* is strictly "*one-to-one*."

The *supervised partial network alignment* problem studied in this section follows the identical definition as that introduced in Sect. 4.2, except that the anchor links will follow the "*one-to-one$_\leq$*" constraint instead. To address such a problem, in this section, we will introduce the PNA method proposed in [38], which contains three phases: (1) *general feature extraction based on meta paths*, (2) *class-imbalance anchor link classification*, and (3) *generic stable matching* to preserve the "*one-to-one$_\leq$*" constraint on anchor links.

4.4.2   Inter-Network Meta Path Based Feature Extraction

Different from the diverse features extracted for the Twitter and Foursquare networks specifically as introduced in Sect. 4.3.1, in this part, we will introduce two different general feature extraction methods across online social networks, including both the *meta path* [37] based explicit feature extraction and *tensor* [11] based latent feature extraction.

4.4.2.1 Inter-Network Meta Paths

The *inter-network meta path* has been introduced in the previous chapter already (detailed information is available in Sect. 3.5). Via the instances of *inter-network meta paths*, users across *aligned social networks* can be extensively connected to each other. In these two partially aligned online social networks (e.g., $\mathcal{G} = ((G^{(1)}, G^{(2)}), (\mathcal{A}^{(1,2)}))$) studied here, we can represent their network schemas in Fig. 4.6, where the *Word*, *Location*, and *Timestamp* node types are attached to the *Post* node type. For the network schema of network $G^{(1)}$, users may also create several lists, which can contain a bunch of location check-ins. Various *inter-network meta paths* between $G^{(1)}$ (i.e., Foursquare) and $G^{(2)}$ (i.e., Twitter) can be defined as follows:

- *Common Out Neighbor Inter-Network Meta Path* (Ψ_1): $User^{(1)} \xrightarrow{follow} User^{(1)} \xleftrightarrow{Anchor} User^{(2)} \xleftarrow{follow} User^{(2)}$ or "$\mathcal{U}^{(1)} \to \mathcal{U}^{(1)} \leftrightarrow \mathcal{U}^{(2)} \leftarrow \mathcal{U}^{(2)}$" for short.
- *Common In Neighbor Inter-Network Meta Path* (Ψ_2): $User^{(1)} \xleftarrow{follow} User^{(1)} \xleftrightarrow{Anchor} User^{(2)} \xrightarrow{follow} User^{(2)}$ or "$\mathcal{U}^{(1)} \leftarrow \mathcal{U}^{(1)} \leftrightarrow \mathcal{U}^{(2)} \to \mathcal{U}^{(2)}$."
- *Common Out In Neighbor Inter-Network Meta Path* (Ψ_3): $User^{(1)} \xrightarrow{follow} User^{(1)} \xleftrightarrow{Anchor} User^{(2)} \xrightarrow{follow} User^{(2)}$ or "$\mathcal{U}^{(1)} \to \mathcal{U}^{(1)} \leftrightarrow \mathcal{U}^{(2)} \to \mathcal{U}^{(2)}$."
- *Common In Out Neighbor Inter-Network Meta Path* (Ψ_4): $User^{(1)} \xleftarrow{follow} User^{(1)} \xleftrightarrow{Anchor} User^{(2)} \xleftarrow{follow} User^{(2)}$ or "$\mathcal{U}^{(1)} \leftarrow \mathcal{U}^{(1)} \leftrightarrow \mathcal{U}^{(2)} \leftarrow \mathcal{U}^{(2)}$."

Besides the users who can be shared across the online social networks, many other nodes can be shared across networks as well by capturing the identical physical information, like words, location latitude-longitude pairs, and timestamps, which are defined as the *bridge nodes* [38] across the aligned social networks.

Fig. 4.6 Schema of aligned heterogeneous network

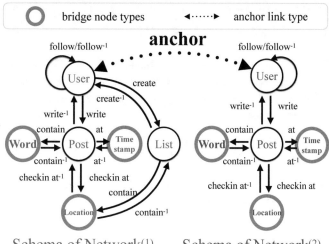

Schema of Network$^{(1)}$ Schema of Network$^{(2)}$

Definition 4.7 (Bridge Nodes) The *bridge nodes* shared between $G^{(1)}$ and $G^{(2)}$ can be represented as $\mathcal{B}^{(1,2)} = \{v | (v \in \mathcal{V}^{(1)} \setminus \mathcal{U}^{(1)}) \wedge (v \in \mathcal{V}^{(2)} \setminus \mathcal{U}^{(2)})\}$.

The *bridge node* concept is different from the *anchor node* concept defined before. Different from the concrete user information entities denoted by *anchor node*, the *bridge node* can get shared across networks merely because of their identical physical meanings (e.g., the words) or physical representations (e.g., the POI latitude-longitude pairs or timestamps).

For instance, in the network schema as shown in Fig. 4.6, the word node type, location node type, and timestamp node type are the bridge node types shared by the networks $G^{(1)}$ and $G^{(2)}$. These *inter-network meta paths* defined above all involve one single node type, i.e., the "User" node type, across the *partially aligned social networks*. In addition to these meta paths, there can exist many other *inter-network meta paths* consisting of user node type and other *bridge node* types from Foursquare to Twitter, e.g., Location, Word, and Timestamp.

- *Common Location Check-in Inter-Network Meta Path 1 (Ψ_5)*: $User^{(1)} \xrightarrow{write} Post^{(1)} \xrightarrow{check-in\ at} Location \xleftarrow{check-in\ at} Post^{(2)} \xleftarrow{write} User^{(2)}$ or "$\mathcal{U}^{(1)} \to \mathcal{P}^{(1)} \to \mathcal{L} \leftarrow \mathcal{P}^{(2)} \leftarrow \mathcal{U}^{(2)}$."
- *Common Location Check-in Inter-Network Meta Path 2 (Ψ_6)*: $User^{(1)} \xrightarrow{create} List^{(1)} \xrightarrow{contain} Location \xleftarrow{check-in\ at} Post^{(2)} \xleftarrow{write} User^{(2)}$ or "$\mathcal{U}^{(1)} \to \mathcal{I}^{(1)} \to \mathcal{L} \leftarrow \mathcal{P}^{(2)} \leftarrow \mathcal{U}^{(2)}$."
- *Common Timestamps Inter-Network Meta Path (Ψ_7)*: $User^{(1)} \xrightarrow{write} Post^{(1)} \xrightarrow{at} Time \xleftarrow{at} Post^{(2)} \xleftarrow{write} User^{(2)}$ or "$\mathcal{U}^{(1)} \to \mathcal{P}^{(1)} \to \mathcal{T} \leftarrow \mathcal{P}^{(2)} \leftarrow \mathcal{U}^{(2)}$."
- *Common Word Usage Inter-Network Meta Path (Ψ_8)*: $User^{(1)} \xrightarrow{write} Post^{(1)} \xrightarrow{contain} Word \xleftarrow{contain} Post^{(2)} \xleftarrow{write} User^{(2)}$ or "$\mathcal{U}^{(1)} \to \mathcal{P}^{(1)} \to \mathcal{W} \leftarrow \mathcal{P}^{(2)} \leftarrow \mathcal{U}^{(2)}$."

4.4.2.2 Explicit Inter-Network Adjacency Features Extraction

Based on the above defined *inter-network meta paths*, different kinds of inter-network meta path based adjacency relationship can be extracted from the network. Formally, a new concept of *inter-network adjacency score* has been defined to describe such relationships among users across *partially aligned social networks*.

Definition 4.8 (Inter-Network Meta Path Instance) Based on *inter-network meta path* $\Psi_i = T_1 \xrightarrow{R_1} T_2 \xrightarrow{R_2} \cdots \xrightarrow{R_{k-1}} T_k$, a concrete path $\psi = n_1 - n_2 - \ldots - n_{k-1} - n_k$ is an instance of Ψ_i iff n_j is an instance of node type T_j, $\forall j \in \{1, 2, \ldots, k\}$ and (n_j, n_{j+1}) is an instance of link type $T_j \xrightarrow{R_j} T_{j+1}, \forall j \in \{1, 2, \ldots, k-1\}$.

Definition 4.9 (Inter-Network Adjacency Score) The *inter-network adjacency* score is quantified as the number of concrete path instances of various *inter-network meta paths* connecting users across networks. The *inter-network adjacency score* between $u^{(1)} \in \mathcal{U}^{(1)}$ and $v^{(2)} \in \mathcal{U}^{(2)}$ based on meta path Ψ_i is defined as:

$$\text{score}_{\Psi_i}(u^{(1)}, v^{(2)}) = \left| \{\psi | (\psi \in \Psi_i) \wedge (u^{(1)} \in T_1) \wedge (v^{(2)} \in T_k)\} \right|, \tag{4.15}$$

where path ψ starts and ends with node types T_1 and T_k, respectively and $\psi \in \Psi_i$ denotes that ψ is a path instance of meta path Ψ_i.

Fig. 4.7 An example of input aligned social networks

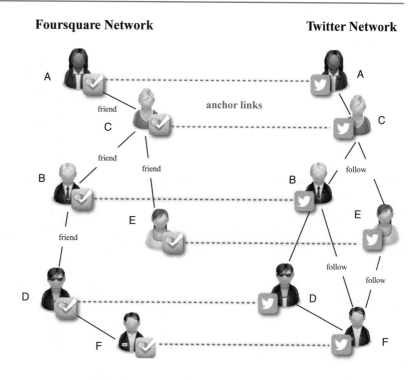

Example 4.8 For instance, given a pair of input aligned social networks as shown in Fig. 4.7, where the Foursquare network is $G^{(1)}$ and the Twitter network is $G^{(2)}$, two meta paths can be defined as follows:

1. $\mathcal{U}^{(1)} \to \mathcal{U}^{(1)} \leftrightarrow \mathcal{U}^{(2)} \leftarrow \mathcal{U}^{(2)}$
2. $\mathcal{U}^{(1)} \to \mathcal{U}^{(1)} \to \mathcal{U}^{(1)} \leftrightarrow \mathcal{U}^{(2)} \leftarrow \mathcal{U}^{(2)}$

For a random user pair, like (B_F, E_T), based on meta path 1, we identify one single path $B_F \to C_F \leftrightarrow C_T \leftarrow E_T$ connecting these two users. Meanwhile, based on meta path 2, there will exist another path connecting them, which includes $B_F \to D_F \to F_F \leftrightarrow F_T \leftarrow E_T$. In other words, the *inter-network adjacency score* between B_F and E_T based on meta paths 1 and 2 are both 1 actually.

The *anchor adjacency scores* among all users across *partially aligned networks* can be stored in the *anchor adjacency matrix* as follows:

Definition 4.10 (Inter-Network Adjacency Matrix) Given a certain *anchor meta path*, e.g., Ψ, the *anchor adjacency matrix* between networks $G^{(1)}$ and $G^{(2)}$ can be defined as $\mathbf{A}_\Psi \in \mathbb{N}^{|\mathcal{U}^{(1)}| \times |\mathcal{U}^{(2)}|}$ and $A_\Psi(l, m) = score_\Psi(u_l^{(1)}, u_m^{(2)})$ for users $u_l^{(1)} \in \mathcal{U}^{(1)}, u_m^{(2)} \in \mathcal{U}^{(2)}$.

Multiple *anchor adjacency matrix* can be grouped together to form a *high-order tensor*. A *tensor* [11] is a multidimensional array and an N-order *tensor* is an element of the tensor product of N vector spaces, each of which can have its own coordinate system. As a result, a 1-order *tensor* is a vector, a 2-order *tensor* is a matrix and *tensors* of three or higher order are called the *higher-order tensor* [11, 18].

Fig. 4.8 Tensor
decomposition

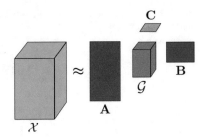

Definition 4.11 (Inter-Network Adjacency Tensor) Based on the eight defined *inter-network meta paths* $\{\Psi_1, \Psi_2, \ldots, \Psi_8\}$, a set of *anchor adjacency matrices* between users in two *partially aligned networks* can be obtained to be $\{\mathbf{A}_{\Psi_1}, \mathbf{A}_{\Psi_2}, \ldots, \mathbf{A}_{\Psi_8}\}$. With $\{\mathbf{A}_{\Psi_1}, \mathbf{A}_{\Psi_2}, \ldots, \mathbf{A}_{\Psi_8}\}$, a 3-order *anchor adjacency tensor* $\mathcal{X} \in \mathbb{R}^{|\mathcal{U}^{(1)}| \times |\mathcal{U}^{(2)}| \times 8}$ can be constructed, where the ith layer of $\mathcal{X}$ is the *anchor adjacency matrix* based on *anchor meta path* Ψ_i, i.e., $\mathcal{X}(:, :, i) = \mathbf{A}_{\Psi_i}, i \in \{1, 2, \ldots, 8\}$.

Based on the *anchor adjacency tensor*, a set of *explicit anchor adjacency features* can be extracted for *anchor links* across *partially aligned social networks*. For a certain *anchor link* $(u_l^{(1)}, u_m^{(2)})$, the *explicit anchor adjacency feature vectors* extracted based on the *anchor adjacency tensor* $\mathcal{X}$ can be represented as $\mathbf{x} = [x_1, x_2, \ldots, x_8]$ (i.e., the *anchor adjacency scores* between $u_l^{(1)}$ and $u_m^{(2)}$ based on these 8 different *anchor meta paths*), where $x_k = \mathcal{X}(l, m, k), k \in \{1, 2, \ldots, 8\}$.

4.4.2.3 Latent Topological Feature Vectors Extraction

Explicit anchor adjacency features can express manifest properties of the connections across *partially aligned networks* and are the *explicit topological features*. Besides these explicit topological connections, there can also exist some hidden common connection patterns [28] across *partially aligned networks*. In [38], a group of *latent topological feature vectors* are extracted from the *anchor adjacency tensor*.

As proposed in [11, 18], a *higher-order tensor* can be decomposed into a *core tensor*, e.g., $\mathcal{G}$, multiplied by a matrix along each mode, e.g., $\mathbf{A}, \mathbf{B}, \ldots, \mathbf{Z}$, with various *tensor decomposition methods*, e.g., Tucker decomposition [11]. For example, in Fig. 4.8, the *3-order anchor adjacency tensor* $\mathcal{X}$ can be decomposed into three matrices $\mathbf{A} \in \mathbb{R}^{|\mathcal{U}^{(1)}| \times P}$, $\mathbf{B} \in \mathbb{R}^{|\mathcal{U}^{(2)}| \times Q}$, and $\mathbf{C} \in \mathbb{R}^{8 \times R}$ and a core tensor $\mathcal{G} \in \mathbb{R}^{P \times Q \times R}$, where P, Q, R are the number of columns of matrices $\mathbf{A}, \mathbf{B}, \mathbf{C}$ [11]:

$$\mathcal{X} = \sum_{p=1}^{P} \sum_{q=1}^{Q} \sum_{r=1}^{R} g_{pqr} \mathbf{a}_p \circ \mathbf{b}_q \circ \mathbf{c}_r = [\mathcal{G}; \mathbf{A}, \mathbf{B}, \mathbf{C}], \tag{4.16}$$

where $\mathbf{a}_p \circ \mathbf{b}_q$ denotes the vector outer product of $\mathbf{a}_p$ and $\mathbf{b}_q$. Each row of $\mathbf{A}$ and $\mathbf{B}$ represents a *latent topological feature vector* of users in $\mathcal{U}^{(1)}$ and $\mathcal{U}^{(2)}$, respectively [18].

4.4.3 Class-Imbalance Classification Model

Based on the *one-to-one*$_\leq$ cardinality constraint, the number of non-existing anchor links will be far more than the existing anchor links. In other words, the negative instances will be far more than that of the positive instances in both the training set and testing set of the supervised link prediction model. To overcome such a disadvantage, in this part, we will introduce several existing techniques that can be applied to sample or prune the training/testing sets.

4.4.3.1 Training Set Sampling

Based on the anchor adjacency scores calculated according to various anchor meta paths in previous section, various supervised link prediction models [12,36,37] can be built to infer the potential anchor links across networks. As proposed in [15, 17], conventional supervised link prediction methods [22] can suffer from the *class imbalance* problem a lot. To address the problem, two effective methods (*down sampling* [14] and *over sampling* [4]) can be applied.

Down sampling methods aim at deleting the *unreliable negative instances* from the training set. In Fig. 4.9, we show the distribution of training instances in the feature space, where negative instances can be generally divided into 4 different categories [14]:

- *noisy instances*: instances mixed in the positive instances;
- *borderline instances*: instances close to the decision boundary;
- *redundant instances*: instances which are too far away from the decision boundary in the negative region;
- *safe instances*: instances which are helpful for determining the classification boundary.

Different heuristics have been proposed to remove the *noisy instances* and *borderline instances*, like the *Tomek links* method proposed in [14, 23]. For any two given instances $\mathbf{x}_1$ and $\mathbf{x}_2$ of different labels, pair $(\mathbf{x}_1, \mathbf{x}_2)$ is called a *tomek link* if there exists no other instances, e.g., $\mathbf{z}$, such that $d(\mathbf{x}_1, \mathbf{z}) < d(\mathbf{x}_1, \mathbf{x}_2)$ and $d(\mathbf{x}_2, \mathbf{z}) < d(\mathbf{x}_1, \mathbf{x}_2)$. Examples that participate in *Tomek links* are either borderline or noisy instances [14, 23]. As to the *redundant instances*, they will not harm correct classifications as their existence will not change the classification boundary but they can lead to extra classification costs. To remove the *redundant instances*, a *consistent subset* $\mathcal{C}$ of the training set will be created, e.g., $\mathcal{S}$ [14]. Subset $\mathcal{C}$ is *consistent* with $\mathcal{S}$ if classifiers built with $\mathcal{C}$ can correctly classify instances in $\mathcal{S}$. Initially, $\mathcal{C}$ consists of all positive instances and one randomly selected negative instances. A classifier, e.g., kNN, built with $\mathcal{C}$ is applied to $\mathcal{S}$, where instances that are misclassified will be added into $\mathcal{C}$. The final set $\mathcal{C}$ only contains the *safe links*.

Another method to overcome the *class imbalance* problem in the training set is to *over sample* the *minority class*. Many *over sampling* methods have been proposed, e.g., *over sampling with replacement, over sampling with "synthetic" instances* [4]. Among them, the *over sampling with "synthetic" instances* is frequently used due to its effectiveness and wide usage in many scenarios [4]. The minority class is over sampled by introducing new "synthetic" examples along the line segment joining m of the k nearest minority class neighbors for each minority class instances. The value of parameter m can be determined according to the ratio to *over sample* the minority class. For example, if the minority class need to be *over sampled* by 200%, then $m = 2$. The instance to be created between a certain example $\mathbf{x}$ and one of its nearest neighbor $\mathbf{y}$ can be denoted as $\mathbf{x} + \boldsymbol{\theta}^\top (\mathbf{x} - \mathbf{y})$,

Fig. 4.9 Instance distribution in feature space

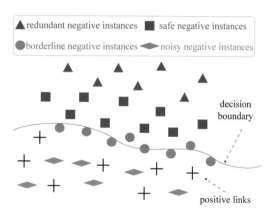

where $\mathbf{x}$ and $\mathbf{y}$ are the feature vectors of two instances and $\boldsymbol{\theta}^\top$ is the transpose of a coefficient vector containing random numbers in range $[0, 1]$.

4.4.3.2 Test Set Pre-pruning

Across two *partially aligned social networks*, users in a certain network can have a large number of potential *anchor link candidates* in the other network, which can lead to great time and space costs in predicting the anchor links. The problem can be even worse when the networks are of large scales, e.g., containing millions or even billions of users, which can make the *partial network alignment* problem unsolvable. To shrink size of the candidate set, a *candidate pre-pruning* step of links in the test set will be conducted before applying the *class imbalance link prediction model*, i.e., $\mathcal{M}$, to links in the test set.

Each user in one network can have millions of potential *anchor link candidates* in another network. To address such a problem, the *candidate pre-pruning* step can be conducted on the test set $\mathcal{L}$ before applying the built model $\mathcal{M}$ to predict these links in $\mathcal{L}$. The *pre-pruning* method adopted here includes (1) *profile pre-pruning* and (2) *inter-network adjacency score pre-pruning*.

- *profile pre-pruning*: the profile information of users shared across *partially aligned social networks*, e.g., Foursquare and Twitter, can include username and hometown [30]. Given an anchor link $(u_l^{(1)}, u_m^{(2)}) \in \mathcal{A}_{test}$, if the username and hometown of $u_l^{(1)}$ and $u_m^{(2)}$ are totally different, e.g., cosine similarity scores are 0, then link $(u_l^{(1)}, u_m^{(2)})$ will be pruned from the testing set $\mathcal{A}_{test}$.
- *inter-network adjacency score pruning*: based on the *explicit inter-network adjacency tensor* $\mathcal{X}$ introduced in the previous sections, for a given link $(u_l^{(1)}, u_m^{(2)}) \in \mathcal{A}_{test}$, if its extracted *explicit inter-network adjacency features* are all 0, i.e., $\mathcal{X}(l, m, x) = 0, x \in \{1, 2, \dots, 8\}$, then link $(u_l^{(1)}, u_m^{(2)})$ will be pruned from the testing set $\mathcal{A}_{test}$.

Example 4.9 For example, in Fig. 4.10, we give 6 users in two different networks together with all the 9 potential *anchor links* between them in the test set, where "*William*" and "*Wm*" are the same

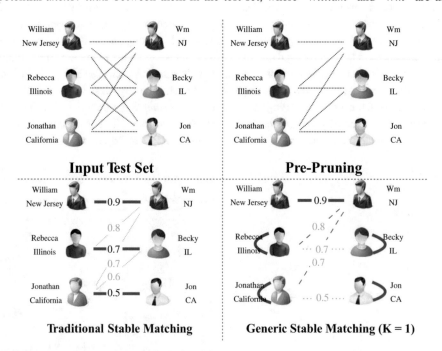

Fig. 4.10 Partial network alignment with pruning

user in these two networks and the remaining users are all different users. Users' profile information (i.e., names and hometowns) is given in the figure. By applying the *profile pre-pruning* method, we can remove 3 *anchor links* from the test set. The advantages of the *pre-pruning* will be more significant when being applied to very large-scale real-world *partially aligned social networks*.

4.4.4 Generic Stable Matching

Given the user sets $\mathcal{U}^{(1)}$ and $\mathcal{U}^{(2)}$ of two *partially aligned social networks* $G^{(1)}$ and $G^{(2)}$, each user in $\mathcal{U}^{(1)}$(or $\mathcal{U}^{(2)}$) has his preference over users in $\mathcal{U}^{(2)}$(or $\mathcal{U}^{(1)}$). Term $v_j P_{u_i}^{(1)} v_k$ can be used to denote that $u_i \in \mathcal{U}^{(1)}$ prefers v_j to v_k for simplicity, where $v_j, v_k \in \mathcal{U}^{(2)}$ and $P_{u_i}^{(1)}$ is the preference operator of $u_i \in \mathcal{U}^{(1)}$. Similarly, we can use term $u_i P_{v_j}^{(2)} u_k$ to denote that $v_j \in \mathcal{U}^{(2)}$ prefers u_i to u_k in $\mathcal{U}^{(1)}$ as well.

Stable matching based method [12] introduced for MNA in Sect. 4.3.3 can only work well in *fully aligned social networks*. However, in the real world, few social networks are fully aligned and lots of users in social networks are involved in one network only, i.e., *non-anchor users*, and they should not be connected by any anchor links. However, traditional *stable matching* method cannot identify these *non-anchor users* and remove the predicted *potential anchor links* connected with them. To overcome such a problem, the *generic stable matching* has been introduced in [38] to identify the *non-anchor users* and prune the anchor link results to meet the *one-to-one$_{\leq}$* constraint.

In partial network matching method PNA [38], a novel concept, *self-matching*, has been introduced, which allows users to be mapped to themselves if they are discovered to be *non-anchor users*. In other words, the *non-anchor users* will be identified as those who are mapped to themselves in the final matching results.

Definition 4.12 (Self-matching) For the given two partially aligned networks $G^{(1)}$ and $G^{(2)}$, user $u_i \in \mathcal{U}^{(1)}$ can have his preference $P_{u_i}^{(1)}$ over users in $\mathcal{U}^{(2)} \cup \{u_i\}$ and u_i preferring u_i himself denotes that u_i is a *non-anchor user* and prefers to stay unconnected, which is formally defined as *self-matching*.

Users in one social network will be matched with either partners in other social networks or themselves according to their preference lists (i.e., from high preference scores to low preference scores). Only partners that users prefer over themselves will be *accepted* finally, otherwise users will be matched with themselves instead.

Definition 4.13 (Acceptable Partner) For a given *matching* $\mu : \mathcal{U}^{(1)} \cup \mathcal{U}^{(2)} \to \mathcal{U}^{(1)} \cup \mathcal{U}^{(2)}$, the mapped partner of users $u_i \in \mathcal{U}^{(1)}$, i.e., $\mu(u_i)$, is *acceptable* to u_i iff $\mu(u_i) P_{u_i}^{(1)} u_i$.

To cut off the partners with very low *preference scores*, the *partial matching strategy* is proposed to obtain the promising partners, who will participate in the matching finally.

Definition 4.14 (Partial Matching Strategy) The *partial matching strategy* of user $u_i \in \mathcal{U}^{(1)}$, i.e., $Q_{u_i}^{(1)}$, consists of the first K the *acceptable partners* in u_i's preference list $P_{u_i}^{(1)}$, which are in the same order as those in $P_{u_i}^{(1)}$, and u_i in the $(K+1)$th entry of $Q_{u_i}^{(1)}$. Parameter K is called the *partial matching rate* as introduced in [38].

Fig. 4.11 An example of
partial matching strategy
($K = 2$)

Preference List

$(K+1)_{th}$ entry

Partial Matching Strategy

Example 4.10 An example is given in Fig. 4.11, where to get the top two promising partners for the user, the user himself is placed at the third cell in the preference list. All the remaining potential partners will be cut off and only the top three users will participate in the final matching.

Based on the concepts of *self-matching* and *partial matching strategy*, the concepts of *partial stable matching* and *generic stable matching* can be defined as follows:

Definition 4.15 (Partial Stable Matching) For a given *matching* μ, μ is (1) *rational* if $\mu(u_i) Q_{u_i}^{(1)} u_i, \forall u_i \in \mathcal{U}^{(1)}$ and $\mu(v_j) Q_{v_j}^{(2)} v_j, \forall v_j \in \mathcal{U}^{(2)}$, (2) *pairwise stable* if there exist no *blocking pairs* in the matching results, and (3) *stable* if it is both *rational* and *pairwise stable*.

Definition 4.16 (Generic Stable Matching) For a given *matching* μ, μ is a *generic stable matching* iff μ is a *self-matching* or μ is a *partial stable matching*.

Example 4.11 An example of *generic stable matching* is shown in the bottom two plots of Fig. 4.10. *Traditional stable matching* can prune most non-existing anchor links and make sure the results can meet the *one-to-one* constraint. However, it preserves the anchor links (Rebecca, Becky) and (Jonathan, Jon), which are connecting *non-anchor users* actually. In the *generic stable matching* with parameter $K = 1$, users will be either connected with their most preferred partner or stay *unconnected*. Users "William" and "Wm" are matched as link (William, Wm) with the highest score. "Rebecca" and "Jonathan" will prefer to stay *unconnected* as their most preferred partner "Wm" is connected with "William" already. Furthermore, "Becky" and "Jon" will stay *unconnected* as their most preferred partners "Rebecca" and "Jonathan" prefer to stay *unconnected*. In this way, *generic stable matching* can further prune the non-existing anchor links (Rebecca, Becky) and (Jonathan, Jon).

The *generic stable matching* results can be achieved with the *Generic Gale-Shapley* algorithm, whose pseudo-code is available in Algorithm 2.

Algorithm 2 Generic Gale-Shapley algorithm

Require: user sets of aligned networks: $\mathcal{U}^{(1)}$ and $\mathcal{U}^{(2)}$.
        classification results of potential anchor links in $\mathcal{L}$
        known anchor links in $\mathcal{A}^{(1,2)}$
        truncation rate K
Ensure: a set of inferred anchor links $\mathcal{L}'$
 1: Initialize the preference lists of users in $\mathcal{U}^{(1)}$ and $\mathcal{U}^{(2)}$ with predicted existence probabilities of links in $\mathcal{L}$ and known anchor links in $\mathcal{A}^{(1,2)}$, whose existence probabilities are 1.0
 2: construct the truncated strategies from the preference lists
 3: Initialize all users in $\mathcal{U}^{(1)}$ and $\mathcal{U}^{(2)}$ as *free*
 4: $\mathcal{L}' = \emptyset$
 5: **while** $\exists$ *free* $u_i^{(1)}$ in $\mathcal{U}^{(1)}$ and $u_i^{(1)}$'s truncated strategy is non-empty **do**
 6:     Remove the top-ranked account $u_j^{(2)}$ from $u_i^{(1)}$'s truncated strategy
 7:     **if** $u_j^{(2)} == u_i^{(1)}$ **then**
 8:         $\mathcal{L}' = \mathcal{L}' \cup \{(u_i^{(1)}, u_i^{(1)})\}$
 9:         Set $u_i^{(1)}$ as *stay unconnected*
10:     **else**
11:         **if** $u_j^{(2)}$ is *free* **then**
12:             $\mathcal{L}' = \mathcal{L}' \cup \{(u_i^{(1)}, u_j^{(2)})\}$
13:             Set $u_i^{(1)}$ and $u_j^{(2)}$ as *occupied*
14:         **else**
15:             $\exists u_p^{(1)}$ that $u_j^{(2)}$ is occupied with.
16:             **if** $u_j^{(2)}$ prefers $u_i^{(1)}$ to $u_p^{(1)}$ **then**
17:                 $\mathcal{L}' = (\mathcal{L}' - \{(u_p^{(1)}, u_j^{(2)})\}) \cup \{(u_i^{(1)}, u_j^{(2)})\}$
18:                 Set $u_p^{(1)}$ as *free* and $u_i^{(1)}$ as *occupied*
19:             **end if**
20:         **end if**
21:     **end if**
22: **end while**

4.5   Anchor Link Inference with Cardinality Constraint

In the previous two sections, the anchor links are assumed to be subject to the *one-to-one* and "*one-to-one$_\leq$*" constraints, respectively. Besides these two cases, users can also have multiple accounts in one social network. For instance, some people will create multiple accounts in Facebook, and tend to use different accounts to socialize with different groups of online friends. In such a scenario, the *cardinality constraint* on the anchor links will become *many-to-many* or *one-to-many*. In this section, we will introduce a model ITERCLIPS [43], which can infer the anchor links with a general *cardinality constraint*, covering *one-to-one*, *one-to-many*, and *many-to-many* simultaneously. The problem definition is identical to that introduced in Sect. 4.2, except that the positive and negative instances labels become $+1$ and 0, respectively. Actually, the ITERCLIPS model to be introduced here can not only infer the anchor links, but also be applied to solve many other types of link prediction tasks.

4.5.1   Loss Function for Anchor Link Prediction

Let set $\mathcal{L} = \mathcal{U}^{(1)} \times \mathcal{U}^{(2)}$ denote all the potential anchor links between networks $G^{(1)}$ and $G^{(2)}$, where $\mathcal{L} = \mathcal{A}_{train} \cup \mathcal{A}_{test}$. Based on the whole link set $\mathcal{L}$, as introduced in the previous sections, a set of features can be extracted for these links with the information available across the social networks, which can be represented as set $\mathcal{X} = \{\mathbf{x}_l\}_{l \in \mathcal{L}}$ ($\mathbf{x}_l \in \mathbb{R}^m, \forall l \in \mathcal{L}$). Given the link existence label set $\mathcal{Y} = \{0, 1\}$, the objective of the anchor link prediction problem is to achieve a general link inference function $f : \mathcal{X} \to \mathcal{Y}$ to map the link feature vectors to their corresponding labels. Here, 0 denotes the label of the negative class, and $+1$ denotes the label of the positive class. Depending on the specific application settings and information available in the networks, the feature vectors extracted for links in $\mathcal{L}$ can be very diverse. Various explicit and latent features introduced in the previous sections can be applied, and we will not repeat them here.

Formally, the loss introduced in the mapping $f(\cdot)$ can be represented as function $L : \mathcal{X} \times \mathcal{Y} \to \mathbb{R}$ over the link feature vector/label pairs. Meanwhile, for one certain input feature vector $\mathbf{x}_l$ of link $l \in \mathcal{L}$, its inferred label introducing the minimum loss can be denoted as:

$$\hat{y}_l = \arg \min_{y_l \in \mathcal{Y}, \mathbf{w}} L(\mathbf{x}_l, y_l; \mathbf{w}), \tag{4.17}$$

where vector $\mathbf{w}$ represents the parameters involved in the mapping function $f(\cdot)$.

Therefore, given the pre-defined loss function $L(\cdot)$, the general form of the objective mapping $f : \mathcal{X} \to \mathcal{Y}$ parameterized by vector $\mathbf{w}$ can be represented as:

$$f(\mathbf{x}; \mathbf{w}) = \arg \min_{y_l \in \mathcal{Y}} L(\mathbf{x}, y; \mathbf{w}). \tag{4.18}$$

In many cases (e.g., when the links are not linearly separable), the feature vector $\mathbf{x}_l$ of link l needs to be transformed as $g(\mathbf{x}_l) \in \mathbb{R}^k$ (k is the transformed feature number) and the transformation function $g(\cdot)$ can be different *kernel projections* depending on the separability of instances. Here, we can assume loss function $L(\cdot)$ to be linear based on some combined representation of the transformed link feature vector $g(\mathbf{x}_l)^\top$ and label y_l, i.e.,

$$L(\mathbf{x}_l, y_l; \mathbf{w}) = (\langle \mathbf{w}, g(\mathbf{x}_l) \rangle - y_l)^2 = (\mathbf{w}^\top g(\mathbf{x}_l) - y_l)^2. \tag{4.19}$$

Furthermore, based on all the links in $\mathcal{L}$, the extracted feature vectors for these links can be represented as matrix $\mathbf{X} = [g(\mathbf{x}_{l_1}), g(\mathbf{x}_{l_2}), \dots, g(\mathbf{x}_{l_{|\mathcal{L}|}})]^\top \in \mathbb{R}^{|\mathcal{L}| \times k}$ (for simplicity, linear kernel projection can used here, and $g(\mathbf{x}_l) = \mathbf{x}_l$). Meanwhile, their existence labels can be represented as vector $\mathbf{y} = [y_{l_1}, y_{l_2}, \dots, y_{l_{|\mathcal{L}|}}]^\top$, where $y_l \in \{0, 1\}, \forall l \in \mathcal{L}$. Specifically, for the existing links in $\mathcal{A}^+_{train}$, their labels are known to be positive in advance, i.e., $y_l = 1, \forall l \in \mathcal{A}^+_{train}$. According to the above loss function definition, based on $\mathbf{X}$ and $\mathbf{y}$, the loss introduced by all links in $\mathcal{L}$ can be represented to be

$$L(\mathbf{X}, \mathbf{y}; \mathbf{w}) = \|\mathbf{X}\mathbf{w} - \mathbf{y}\|_2^2. \tag{4.20}$$

To learn the parameter vector $\mathbf{w}$ and infer the potential label vector $\mathbf{y}$, the loss term introduced by all the links in $\mathcal{L}$ will be minimized. Meanwhile, to avoid overfitting the training set, besides minimizing the loss function $L(\mathbf{X}, \mathbf{y}; \mathbf{w})$, a regularization term $\|\mathbf{w}\|_2^2$ about the parameter vector $\mathbf{w}$ is added to the objective function:

$$\min_{\mathbf{w}, \mathbf{y}} \frac{1}{2} \|\mathbf{w}\|_2^2 + \frac{c}{2} \|\mathbf{X}\mathbf{w} - \mathbf{y}\|_2^2,$$

$$s.t. \ \mathbf{y} \in \{0, 1\}^{|\mathcal{L}| \times 1}, \text{ and } y_l = 1, \forall l \in \mathcal{A}^+_{train}, \tag{4.21}$$

where constant c denotes the weight of the loss term in the function.

4.5.2 Cardinality Constraint Description

The *cardinality constraints* define both the limit on link cardinality and the limit on node degrees that those links are incident to. To be general, the links studied can be either uni-directional or bi-directional, where uni-directional links are treated as bi-directional. For each node $u \in \mathcal{V}$ in the network, we can represent the potential links going-out from u as set $\Gamma^{out}(u) = \{l | l \in \mathcal{L}, \exists v \in \mathcal{V}, l = (u, v)\}$, and those going-into u as set $\Gamma^{in}(u) = \{l | l \in \mathcal{L}, \exists v \in \mathcal{V}, l = (v, u)\}$. Furthermore, with the link label variables $\{y_l\}_{l \in \mathcal{L}}$, we can represent the out-degree and in-degree of node $u \in \mathcal{V}$ as $degree^{out}(u) = \sum_{l \in \Gamma^{out}(u)} y_l$ and $degree^{in}(u) = \sum_{l \in \Gamma^{in}(u)} y_l$, respectively. Considering that the node degrees cannot be negative, besides the upper bounds introduced by the *cardinality constraints*, a lower bound "≥ 0" is also added to guarantee validity of node degrees by default.

One-to-One Cardinality Constraint
For the bi-directional anchor links with 1:1 *cardinality constraint*, the nodes in the information networks can be attached with at most one such kinds of link. In other words, for all the nodes (e.g., $u \in \mathcal{V}$) in the network, their in-degree and out-degree cannot exceed 1, i.e.,

$$0 \leq \sum_{l \in \Gamma^{out}(u)} y_l \leq 1, \forall u \in \mathcal{V}, \text{ and } 0 \leq \sum_{l \in \Gamma^{in}(u)} y_l \leq 1, \forall u \in \mathcal{V}. \tag{4.22}$$

One-to-Many Cardinality Constraint
Meanwhile, when aligning two network structured data sources, where the cardinality constraint on *anchor links* is *one-to-many*. For instance, in some social networks, they may require the SSN or ID number from users in registration and each user can only contain one single account. The alignment of such a social network with the general social networks without such a limitation will be subject to the *one-to-many* cardinality constraint instead. In such a case, for all the nodes (e.g., $u \in \mathcal{V}$) in the network, their *out-degree* cannot exceed N and the *in-degree* should be exactly 1, i.e.,

$$0 \leq \sum_{l \in \Gamma^{out}(u)} y_l \leq N, \forall u \in \mathcal{V}, \text{ and } 1 \leq \sum_{l \in \Gamma^{in}(u)} y_l \leq 1, \forall u \in \mathcal{V}. \tag{4.23}$$

Many-to-Many Cardinality Constraint
In many cases, there usually exist no specific *cardinality constraints* on the anchor links, and nodes can be connected with each other freely. Simply, we assume the node *in-degrees* and *out-degrees* to be limited by the maximum degree parameter $N = |\mathcal{V}| - 1$, i.e.,

$$0 \leq \sum_{l \in \Gamma^{out}(u)} y_l \leq N, \forall u \in \mathcal{V}, \text{ and } 0 \leq \sum_{l \in \Gamma^{in}(u)} y_l \leq N, \forall u \in \mathcal{V}. \tag{4.24}$$

General Cardinality Constraint Representation
The *cardinality constraint* on links can be generally represented with the linear algebra equations. The relationship between nodes $\mathcal{V}$ and links $\mathcal{L}$ can actually be represented as matrices $\mathbf{T}^{out} \in \{0, 1\}^{|\mathcal{V}| \times |\mathcal{L}|}$ and $\mathbf{T}^{in} \in \{0, 1\}^{|\mathcal{V}| \times |\mathcal{L}|}$, where entry $\mathbf{T}^{out}(u, l) = 1$ iff $l \in \Gamma^{out}(u)$ and $\mathbf{T}^{in}(u, l) = 1$ iff $l \in \Gamma^{in}(u)$. Based on the link label vector $\mathbf{y}$, the node out-degrees and in-degrees can be represented as vectors $\mathbf{T}^{out} \cdot \mathbf{y}$ and $\mathbf{T}^{in} \cdot \mathbf{y}$, respectively. The general representation of the *cardinality constraints* introduced above can be rewritten as follows:

$$\underline{\mathbf{b}}^{out} \preccurlyeq \mathbf{T}^{out} \cdot \mathbf{y} \preccurlyeq \overline{\mathbf{b}}^{out}, \text{ and } \underline{\mathbf{b}}^{in} \preccurlyeq \mathbf{T}^{in} \cdot \mathbf{y} \preccurlyeq \overline{\mathbf{b}}^{in}, \tag{4.25}$$

where vectors $\underline{\mathbf{b}}^{out}$, $\overline{\mathbf{b}}^{out}$, $\underline{\mathbf{b}}^{in}$, and $\overline{\mathbf{b}}^{in}$ can take different values depending on the cardinality constraint on the links (e.g., for the 1:1 constraint, the bounds will have values $\underline{\mathbf{b}}^{out} = \underline{\mathbf{b}}^{in} = \mathbf{0}$ and $\overline{\mathbf{b}}^{out} = \overline{\mathbf{b}}^{in} = \mathbf{1}$).

4.5.3 Joint Optimization Function

Based on the above remarks, the constrained optimization objective function of the problem can be represented as

$$\min_{\mathbf{w},\mathbf{y}} \frac{1}{2}\|\mathbf{w}\|_2^2 + \frac{c}{2}\|\mathbf{X}\mathbf{w} - \mathbf{y}\|_2^2,$$

$$s.t. \ \ \mathbf{y} \in \{0,1\}^{|\mathcal{L}|\times 1}, \ y_l = 1, \forall l \in \mathcal{E},$$

$$\underline{\mathbf{b}}^{out} \preccurlyeq \mathbf{T}^{out} \cdot \mathbf{y} \preccurlyeq \overline{\mathbf{b}}^{out}, \underline{\mathbf{b}}^{in} \preccurlyeq \mathbf{T}^{in} \cdot \mathbf{y} \preccurlyeq \overline{\mathbf{b}}^{in}. \tag{4.26}$$

The above objective function involves variables $\mathbf{w}$ and $\mathbf{y}$ at the same time, which is actually not jointly convex and can be very challenging to solve. In [43], the proposed ITERCLIPS model addresses the function with an alternative updating framework by fixing one variable and updating the other one iteratively. The framework involves two steps:

Step 1: Fix $\mathbf{y}$ and Update $\mathbf{w}$

By fixing $\mathbf{y}$ (i.e., treating $\mathbf{y}$ as a constant vector), the objective function about $\mathbf{w}$ can be simplified as

$$\min_{\mathbf{w}} \frac{1}{2}\|\mathbf{w}\|_2^2 + \frac{c}{2}\|\mathbf{X}\mathbf{w} - \mathbf{y}\|_2^2. \tag{4.27}$$

Let $h(\mathbf{w}) = \frac{1}{2}\|\mathbf{w}\|_2^2 + \frac{c}{2}\|\mathbf{X}\mathbf{w} - \mathbf{y}\|_2^2$. By taking the derivative of the function $h(\mathbf{w})$ regarding $\mathbf{w}$ we can have

$$\frac{dh(\mathbf{w})}{d\mathbf{w}} = \mathbf{w} + c\mathbf{X}\mathbf{w}\mathbf{X}^\top - c\mathbf{y}\mathbf{X}^\top. \tag{4.28}$$

By making the derivation to be zero, the optimal vector $\mathbf{w}$ can be represented to be

$$\mathbf{w} = c(\mathbf{I} + c\mathbf{X}^\top\mathbf{X})^{-1}\mathbf{X}^\top\mathbf{y}, \tag{4.29}$$

and the minimum value of the function will be

$$\frac{c}{2}\mathbf{y}^\top\mathbf{y} - \frac{c^2}{2}\mathbf{y}^\top\mathbf{X}(\mathbf{I} + c\mathbf{X}^\top\mathbf{X})^{-1}\mathbf{X}^\top\mathbf{y}. \tag{4.30}$$

Step 2: Fix $\mathbf{w}$ and Update $\mathbf{y}$

When fixing $\mathbf{w}$ and treating it as a constant vector, the objective function about $\mathbf{y}$ can be represented as

$$\min_{\mathbf{y}} \frac{c}{2}\|\hat{\mathbf{y}} - \mathbf{y}\|_2^2,$$

$$s.t. \ \ \mathbf{y} \in \{0,1\}^{|\mathcal{L}|\times 1}, \ y_l = 1, \forall l \in \mathcal{E},$$

$$\underline{\mathbf{b}}^{out} \preccurlyeq \mathbf{T}^{out} \cdot \mathbf{y} \preccurlyeq \overline{\mathbf{b}}^{out}, \underline{\mathbf{b}}^{in} \preccurlyeq \mathbf{T}^{in} \cdot \mathbf{y} \preccurlyeq \overline{\mathbf{b}}^{in}, \tag{4.31}$$

Algorithm 3 Greedy link selection

Require: link estimate result $\hat{\mathbf{y}}$, parameter k
Ensure: link label vector $\mathbf{y}$
 1: initialize link label vector $\mathbf{y} = \mathbf{0}$
 2: **for** $l \in \mathcal{E}$ **do**
 3:     $y_l = 1$
 4: **end for**
 5: **for** $l \in \mathcal{L} \setminus \mathcal{E}$ and $\hat{y}_l < 0.5$ **do**
 6:     $y_l = 0$
 7: **end for**
 8: Let $\tilde{\mathcal{L}} = \{l | l \in \mathcal{L} \setminus \mathcal{E}, \hat{y}_l \geq 0.5\}$
 9: **while** $\tilde{\mathcal{L}} \neq \emptyset$ **do**
10:     select $l \in \tilde{\mathcal{L}}$ with the highest estimation score
11:     **if** add l as positive instance violates the *cardinality constraint* or more than k links have been selected **then**
12:         $y_l = 0$
13:     **else**
14:         $y_l = 1$
15:     **end if**
16: **end while**
17: **return y**

Algorithm 4 Cardinality constrained anchor link prediction framework

---

Require: link feature vector $\mathbf{X}$
        weight parameter c
Ensure: parameter vector $\mathbf{w}$, link label vector $\mathbf{y}$
 1: Initialize label vector $\mathbf{y} = \frac{1}{2} \cdot \mathbf{1}$
 2: For links in $\mathcal{E}$, assign their label as 1
 3: Initialize parameter vector $\mathbf{w} = \mathbf{0}$
 4: Initialize convergence-tag = False
 5: **while** convergence-tag == False **do**
 6:     Update vector $\mathbf{w}$ with equation $\mathbf{w} = c(\mathbf{I} + c\mathbf{X}^\top \mathbf{X})^{-1}\mathbf{X}^\top \mathbf{y}$
 7:     Calculate link estimation result $\hat{\mathbf{y}} = \mathbf{X}\mathbf{w}$
 8:     Update vector $\mathbf{y}$ with Algorithm Greedy($\hat{\mathbf{y}}$)
 9:     **if** $\mathbf{w}$ and $\mathbf{y}$ both converge **then**
10:         convergence-tag = True
11:     **end if**
12: **end while**

---

where $\hat{\mathbf{y}} = \mathbf{X}\mathbf{w}$ denotes the inference results of the links in $\mathcal{L}$ with the updated parameter vector $\mathbf{w}$ from Step 1. The objective function is a constrained non-linear integer programming problem about variable $\mathbf{y}$. Formally, the above optimization sub-problem is named as the CLS (Cardinality Constrained Link Selection) problem [43]. The CLS problem is shown to be NP-hard (we will analyze it in the next subsection), and achieving the optimal solution to it is very time consuming. To preserve the *cardinality constraints* on the variables and minimize the loss term, one brute-force way to achieve the optimal solution $\mathbf{y}$ is to enumerate all the feasible combination of links candidates to be selected as the positive instances, which will lead to very high time complexity. In [43], a greedy link selection algorithm is proposed to resolve the problem, and the pseudo-code of the greedy link selection method is available in Algorithm 3. Meanwhile, the framework is illustrated with the pseudo-code available in Algorithm 4. The framework updates vectors $\mathbf{w}$ and $\mathbf{y}$ alternatively until both of them converge.

4.5.4   Problem and Algorithm Analysis

In this part, we will show the CLS problem with M:N *cardinality constraints* can be reduced to the *k-maximum weighted matching problem* [7], which is NP-hard and not solvable in polynomial time. In addition, we will also prove that the greedy method can actually achieve a $\frac{1}{2}$-approximation of the optimal result of the CLS problem.

In the CLS problem, for all the existing links in $\mathcal{A}^+_{train}$, we know their label should be 1 in advance. For all the links in $\mathcal{L} \setminus \mathcal{A}^+_{train}$ with estimation score (i.e., $\hat{y}$) lower than 0.5, assigning their label with value 0 will introduce less loss and has no impact on the cardinality constraints. Therefore, in Algorithm 3, these links are handled in advance to simplify the problem. For the remaining links, we need to select those with high scores to assign with label 1 (so as to minimize the loss term), and preserve the *cardinality constraints* at the same time. For the links selection of which violate the cardinality constraints, they will be assigned with label 0 instead.

Formally, we can represent the unlabeled links with confidence scores greater than 0.5 as set $\tilde{\mathcal{L}} = \{l | l \in \mathcal{L} \setminus \mathcal{A}^+_{train}, \hat{y}_l > 0.5\}$. For all the links in set $\tilde{\mathcal{L}}$, the introduced loss term can be represented as

$$\sum_{l \in \tilde{\mathcal{L}}} (\hat{y}_l - y_l)^2 = \sum_{l \in \tilde{\mathcal{L}}} \hat{y}_l^2 + \sum_{l \in \tilde{\mathcal{L}}} y_l^2 - \sum_{l \in \tilde{\mathcal{L}}} 2\hat{y}_l \cdot y_l, \tag{4.32}$$

where term $\sum_{l \in \tilde{\mathcal{L}}} \hat{y}_l^2$ is a constant, term $\sum_{l \in \tilde{\mathcal{L}}} y_l^2$ denotes the number of selected links, and $\sum_{l \in \tilde{\mathcal{L}}} 2\hat{y}_l \cdot y_l$ represents the confidence scores of the selected links. Let's assume k links are selected finally, i.e., $\sum_{l \in \tilde{\mathcal{L}}} y_l^2 = k$, the optimal k links which can minimize the loss term can be achieved by maximizing the confidence scores of the selected links:

$$\max \sum_{l \in \tilde{\mathcal{L}}} \hat{y}_l y_l$$

$$s.t. \ \ y_l \in \{0, 1\}, \forall l \in \tilde{\mathcal{L}}, \sum_{l \in \tilde{\mathcal{L}}} y_l = k,$$

$$\underline{\mathbf{b}}^{out} \preccurlyeq \mathbf{T}^{out} \cdot \mathbf{y} \preccurlyeq \overline{\mathbf{b}}^{out}, \underline{\mathbf{b}}^{in} \preccurlyeq \mathbf{T}^{in} \cdot \mathbf{y} \preccurlyeq \overline{\mathbf{b}}^{in}. \tag{4.33}$$

By enumerating different k values in range $[1, |\tilde{\mathcal{L}}|]$, the optimal link set can be identified for the CLS problem.

Theorem 4.1 *The k-maximum weighted matching problem can be reduced to the above optimization problem with a general M:N cardinality constraint.*

Proof The above optimization problem with 1:1 *cardinality constraint* is actually identical to the *k-maximum weighted matching problem* studied in the existing works [7], and the reduction is trivial. Meanwhile, for the above optimization problem with N:1 *cardinality constraints* on the links, we can have vectors $\overline{\mathbf{b}}^{in} = \underline{\mathbf{b}}^{in} = [1, 1, \ldots, 1]^\top$, $\overline{\mathbf{b}}^{out} = [N, N, \ldots, N]^\top$, and $\underline{\mathbf{b}}^{out} = \mathbf{0}$. Given the information network G with $1 : N$ cardinality constraints, N dummy nodes can be constructed for each the nodes with out-going links. The constructed dummy nodes are connected to the original nodes to indicate the belonging relationships. For each original link, e.g., (u, v), a dummy directed link connecting the dummy node created for u with node v will be added, whose weight is identical to the weight of the original link (u, v). Given the *k-maximum weighted matching* result on the constructed dummy network, the optimal solution to the above optimization problem on network G can be obtained by replacing all the created dummy nodes with the original nodes corresponding to them. Meanwhile, for any solution to the above optimization problem on network G, the solution to the *k-maximum weighted matching* problem can be obtained on the constructed dummy network. In other words, the *k-maximum weighted matching problem* can be reduced to the above optimization problem with N:1 *cardinality constraints* via the constructed dummy network. Meanwhile, for the networks with general M:N constraint, dummy nodes can be created for both nodes with out-going links and in-coming links at the same time, whose reduction to the *k-maximum weighted matching problem* is not provided due to the limited space. In addition, in the M:N case, to avoid the case that solutions pick links connecting the more than one link connecting the dummy nodes corresponding the common node pairs (e.g., both (u', v') and (u'', v'') are selected, where u', u'' and v', v'' are the dummy nodes of u and v, respectively), more constraints will be added to the objective function of the *k-maximum weighted matching problem*.

According to the existing works [9], the *k-maximum weighted matching problem* is actually NP-hard. To address the problem efficiently, a greedy link selection method is applied as introduced in the previous subsection. As shown in Algorithm 3, among all the remaining links in $\tilde{\mathcal{L}}$, the greedy link selection method picks the links with the highest confidence scores $\hat{y}_l$. If the selection of a link doesn't violate the *cardinality constraint*, the greedy method will add it to the final result. We will show that the method can actually achieve a $\frac{1}{2}$-approximation of the optimal result.

Theorem 4.2 *The greedy method can achieve a $\frac{1}{2}$-approximation of the optimal solution to the* CLS *problem.*

Proof Formally, let $\mathcal{C}$ be the set of links selected by the greedy method to assign with label $+1$, while the optimal solution to the CLS problem can be represented as OPT. Every time, when the method selects the links with the highest confidence score (e.g., $l = (u, v)$) to add to $\mathcal{C}$, the degrees of nodes u and v will get increased by 1 and some other links incident to u, v will no longer get added to $\mathcal{C}$ due to the degree limit (introduced by the *cardinality constraint*). At most two links incident to u and v can get removed due to the selection of (u, v), since (u, v) occupies the degree space of u and v by one, respectively. Formally, the set of links incident to l can be represented as set $\Gamma(l) = \Gamma^{out}(u) \cup \Gamma^{in}(v)$. Depending on whether link $l \in \mathcal{C}$ is in OPT or not and the number of links in the optimal solution but are removed in $\mathcal{C}$ due to the selection of l (i.e., links in $\Gamma(l) \cap (OPT \setminus \mathcal{C})$), there exist three cases:

1. $l \in OPT$: Link l also belongs to the optimal result, and adding l into $\mathcal{C}$ will not affect the selection of other links.
2. $l \notin OPT$ and $\Gamma(l) \cap (OPT \setminus \mathcal{C}) = \{l_1\}$: Link l is not in the optimal solution, and adding l to the result $\mathcal{C}$ will occupy the degree space and make the optimal link $l_1 \in OPT$ (incident to either u or v) fail to be selected. Meanwhile, since l is the link with the highest score at selection, if l_1 is not selected ahead of l, it is easy to show that $\hat{y}_l > \hat{y}_{l_1} > \frac{1}{2}\hat{y}_{l_1}$.
3. $l \notin OPT$ and $\Gamma(l) \cap (OPT \setminus \mathcal{C}) = \{l_1, l_2\}$: Link l is not in the optimal solution, and adding of $l = (u, v)$ will occupy the degrees of nodes u and v by 1 and make links $l_1, l_2 \in OPT$ incident to u and v, respectively to be removed. Since l has the highest score, if links l_1 and l_2 are not selected ahead of l, it is easy to show that $\hat{y}_l > \hat{y}_{l_1}$ and $\hat{y}_l > \hat{y}_{l_2}$. Therefore, we have $\hat{y}_l > \frac{1}{2}(\hat{y}_{l_1} + \hat{y}_{l_2})$.

Based on the above remarks, for all the selected links in $\mathcal{C}$, we have

$$
\begin{aligned}
\hat{y}(\mathcal{C}) &= \hat{y}\left((\mathcal{C} \cap OPT) \cup (\mathcal{C} \setminus OPT)\right) \\
&= \hat{y}(\mathcal{C} \cap OPT) + \hat{y}(\mathcal{C} \setminus OPT) \\
&= \hat{y}(\mathcal{C} \cap OPT) + \sum_{l \in \mathcal{C} \setminus OPT} \hat{y}_l \\
&> \frac{1}{2}\hat{y}(OPT \cap \mathcal{C}) + \frac{1}{2} \sum_{l \in OPT \setminus \mathcal{C}} \hat{y}_l \\
&= \frac{1}{2}\hat{y}(OPT).
\end{aligned}
\tag{4.34}
$$

where $\hat{y}(\mathcal{C}) = \sum_{l \in \mathcal{C}} \hat{y}_l$ denotes the score sum of the links in $\mathcal{C}$.

Therefore, the greedy anchor link selection algorithm can achieve a $\frac{1}{2}$ approximation of the optimal solution for the CLS problem with $M{:}N$ link *cardinality constraint*, and the time complexity of the greedy method is $O(|\tilde{\mathcal{L}}|)$.

4.5.5   Distributed Algorithm

Meanwhile, for large-scale networks involving billions of nodes and links, the complete network data can hardly be stored in one single machine and the learning framework may suffer from the high computing cost problem a lot. In this section, we will introduce a scalable version of the ITERCLIPS model introduced before based on distributed computational platforms proposed in [43]. The framework involves two iterative steps actually. In the first step, it updates vector $\mathbf{w}$ to calculate the confidence vector $\hat{\mathbf{y}} = \mathbf{X}\mathbf{w} = c\mathbf{X}(\mathbf{I} + c\mathbf{X}^{\top}\mathbf{X})^{-1}\mathbf{X}^{\top}\mathbf{y}$, where matrix $c\mathbf{X}(\mathbf{I} + c\mathbf{X}^{\top}\mathbf{X})^{-1}\mathbf{X}^{\top}$ can actually be pre-computed and divided into blocks to be stored in different slaves (i.e., worker nodes in a cluster). For instance, in Spark, the matrix can be divided into rows, where each row can be saved as an RDD (resilient distributed dataset) in one slave, and each entry in vector $\hat{\mathbf{y}}$ can be updated independently in different slaves simultaneously. The updated values in $\hat{\mathbf{y}}$ can be exchanged among the slaves with very low communication costs. Meanwhile, for the second step in framework, how to generalize the greedy method to the distributed version is not very straightforward, which will be the focus in the following part of this subsection.

According to Theorem 4.1, the *k-maximum weighted matching problem* can be reduced to the objective function of k-CLS with general $M{:}N$ cardinality constraint in polynomial time. Therefore, next we will talk about the distributed version of the greedy method for the CLS with 1:1 constraint specifically (which can be applied for the general $M{:}N$ cardinality constraint as well). Before diving into details of the distributed greedy algorithm, we first provide some intuitive ideas about how the distributed method works. In the distributed weighted link selection, each process representing one node in the graph knows its neighbor nodes as well as their inferred confidence scores $\hat{y}$. These processes can also communicate with each other by sending and receiving messages. Via the communication among processes, links with locally highest confidence scores can be identified concurrently. Intuitively, by running the algorithm on all the nodes simultaneously, the same matching result can be obtained as the greedy algorithm based on the stand-alone mode.

Algorithm 5 Distributed Greedy algorithm with 1:1 cardinality constraint

---

Require: node u, neighbor set $\Gamma(u)$
 1: Initialize neighborhood set $N = \Gamma(u)$
 2: Initialize matching candidate set $C = \emptyset$
 3: Select candidate $c = candidate(u, N)$
 4: **if** $c \neq null$ **then**
 5:      Send $\langle invite \rangle$ message to c
 6: **end if**
 7: **while** $N \neq \emptyset$ **do**
 8:      Receive a message m from neighbor v
 9:      **if** $m == \langle invite \rangle$ **then**
10:         $C = C \cup \{v\}$
11:      **end if**
12:      **if** $m == \langle remove \rangle$ **then**
13:         $N = N \setminus \{v\}$
14:         $C = C \setminus \{v\}$
15:         **if** $v == c$ **then**
16:            Select new candidate $c = candidate(u, N)$
17:            **if** $c \neq null$ **then**
18:               Send $\langle invite \rangle$ message to c
19:            **end if**
20:         **end if**
21:      **end if**
22:      **if** $c \neq null \wedge c \in C$ **then**
23:         **for** $w \in C \setminus \{c\}$ **do**
24:            Send $\langle remove \rangle$ message to w
25:         **end for**
26:         $C = \emptyset$
27:      **end if**
28: **end while**

Formally, the pseudo-code of the distributed greedy algorithm is available in Algorithm 5. According to the algorithm, for each node u, its neighbor set N can be initialized as $\Gamma(u)$ (N will change dynamically in the algorithm). Function $c = candidate(u, N)$ returns the candidate of u, whose link with u is of the highest confidence score, i.e.,

$$c = candidate(u, N) = \arg_{v \in N} \max \hat{y}_{(u,v)}. \tag{4.35}$$

Initially, node u will send the *invite* message to the candidate c, and receive messages from all the neighbors in set N. If the message received from neighbor v is also an *invite* (i.e., u is the most promising candidate of v), u will add v to its *matching candidate set* C. Meanwhile, if the message is *remove*, it denotes v has already found its partner and link between them has already been selected. Node v will be removed from u's *neighbor set* and *matching candidate set*. What's more, if v happens to be candidate c, node u will retrieve the next most promising candidate c and send a new *invite* message again. Finally, if candidate c invites u and u also invites c, the link between whom will be of the highest score and selected finally.

Lemma 4.1 *In the distributed greedy algorithm, each process (node) sends out at most one message over each incident edge.*

Proof In the algorithm, for each node u, it will send an *invite* message to the first candidate as well as other candidate if the previous candidates send a *remove* message to u. Therefore, for each potential candidate c obtained via function $candidate(u, N)$, u sends at most one *invite* message to c.

Meanwhile, the *remove* messages are merely sent to the other neighbors in N (excluding candidates c) only once. In other words, all the neighbors in $\Gamma(u)$ only receive exactly one message (either *invite* or *remove*) from u in the whole process.

According to the analysis in Lemma 4.1, we can prove the time complexity of the distributed greedy Algorithm to be $O(|\tilde{\mathcal{L}}|)$.

4.6   Summary

In this chapter, we focused on the supervised network alignment problem. Based on a set of labeled anchor links, the supervised network alignment problem aims at learning a mapping to infer the potential labels of the unlabeled anchor links. To address such a problem, three different supervised network alignment approaches have been introduced in this chapter, including the full network alignment method MNA, partial network alignment method PNA, and the general network alignment method ITERCLIPS with different cardinality constraints.

Based on a detailed data analysis, we illustrated that the heterogeneous information available in the online social networks can be utilized to extract some useful features for the anchor links across networks, which include the social connections, textual contents and spatial check-ins and temporal activity distribution. Based on these features, we introduced the MNA model, which covers two phases: (1) anchor link classification, and (2) social network matching for anchor link pruning. In MNA, the studied social networks are assumed to be fully aligned, i.e., all the involved users will be connected by an anchor link, and the anchor links are assumed to be subject to the one-to-one cardinality constraint.

Instead of studying the network alignment problem based on two specific online social network, e.g., Foursquare and Twitter, we introduced another general network alignment method PNA. Based on the heterogeneous information across the social networks, PNA introduces a general feature extraction approach based on meta path and tensor decomposition. The social networks to be aligned by PNA are partially aligned instead, where a bunch of the users are the non-anchor users. In other words, the anchor links are subject to the one-to-one $\le$ cardinality constraint instead. To pruning the non-existing anchor links across the social networks, PNA adopts a generic stable matching algorithm to extract the final mapping of users across networks.

Finally, in the last section of this chapter, we generalized the cardinality constraint on anchor links, and introduced the network alignment approach ITERCLIPS. Model ITERCLIPS can handle the network alignment problem very well, where the cardinality constraint on anchor links can be either *one-to-one*, *one-to-many*, or *many-to-many*, respectively. ITERCLIPS models the cardinality constraint on anchor links as the mathematical constraint on node degrees instead, and addresses the network alignment problem as an optimization problem. ITERCLIPS adopts a greedy search approach to pick the anchor links across networks, which can achieve a $\frac{1}{2}$-approximation of the optimal solution.

4.7   Bibliography Notes

In recent years, witnessing the rapid growth of online social networks, researchers start to shift their attention to align multiple online social networks. Homogeneous network alignment was studied in [24], enlightened by which the problem of aligning two bipartite networks is studied by Koutra [13], where a fast alignment algorithm which can be applied to large-scale networks is introduced. Users can have various types of attribute information in social networks generated by their social activities, based on which Zafarani et al. study the cross-network user matching problem in [30]. These proposed approaches are mostly proposed based on some simple heuristics and assumptions.

The supervised network alignment problem initially proposed in [12] has become one of the most important research problems in social network studies. By extending the traditional supervised link prediction approach [1] to the inter-network scenario, Kong et al. [12] introduces a two-phase approach to address the network alignment via anchor link prediction and network matching. A set of useful inter-network features extracted for anchor links have also been provided in Kong et al. [12] as well, which can capture very useful signals about the anchor links.

Supervised partial network alignment is introduced in [38], which allows users to stay isolated without connections to anchor links in the network alignment process. A detailed description about the tensor and related tensor decomposition approaches is available in [11]. Class imbalance is a serious problem in traditional machine learning, and the frequently used techniques proposed to handle such a problem include both *down sampling* [14] and *over sampling* [4]. The readers may refer to these papers for more information when reading Sect. 4.4.

Graph matching has been an important research problem in graph studies for a very long time, and Jack Edmonds introduced an efficient algorithm to address the maximum matching problem based on graphs in [7]. As proposed in the existing works [9], the *k-maximum weighted matching problem* is actually NP-hard. The network alignment approach ITERCLIPS with the general cardinality constraint was initially introduced in [43], which unifies the prediction tasks of links subject to different cardinality constraints into the one framework.

4.8 Exercises

1. (Easy) Given two partially aligned networks shown in Fig. 4.12, please compute the *extended common neighbor* between user pair $C^{(1)}$ and $C^{(2)}$ across the networks, respectively.

2. (Easy) Given two partially aligned networks shown in Fig. 4.12, please compute the *extended Jaccard's Coefficient* between user pair $C^{(1)}$ and $C^{(2)}$ across the networks, respectively.

3. (Easy) Given two partially aligned networks shown in Fig. 4.12, please compute the *extended Adamic/Adar Index* between user pair $C^{(1)}$ and $C^{(2)}$ across the networks, respectively.

4. (Easy) Please compute the number of meta path instances connecting B_F and D_T across the networks shown in Fig. 4.12 ($G^{(1)}$: Foursquare, $G^{(2)}$: Twitter) based on meta path $U^{(1)} \rightarrow U^{(1)} \rightarrow U^{(1)} \leftrightarrow U^{(2)} \leftarrow U^{(2)}$.

5. (Easy) Please compute the number of meta path instances connecting B_F and E_T across the networks shown in Fig. 4.12 ($G^{(1)}$: Foursquare, $G^{(2)}$: Twitter) based on meta path $U^{(1)} \rightarrow U^{(1)} \rightarrow U^{(1)} \leftrightarrow U^{(2)} \leftarrow U^{(2)} \leftarrow U^{(2)}$.

6. (Medium) Please identify the *stable matching* of the networks as shown in Fig. 4.13 with Algorithm 1.

7. (Medium) Please identify the *generic stable matching* of the networks as shown in Fig. 4.13 with Algorithm 2 (with $K = 1$).

8. (Medium) Please identify the *greedy matching* of the networks subject to *one-to-one* cardinality constraint as shown in Fig. 4.13 with the *greedy link selection* method introduced in Algorithm 3.

9. (Hard) Please try to implement the *stable matching* algorithm (i.e., Algorithm 1) and the *generic stable matching* algorithm (i.e., Algorithm 2) in a programming language you prefer. You can also input the aligned network structures shown in Fig. 4.13 as the input to test the correctness of your implementation.

10. (Hard) Please try to implement the *greedy link selection* algorithm (i.e., Algorithm 3) and use the networks in Fig. 4.13 to text the implementation correctness.

Fig. 4.12 Multiple
aligned social networks
input

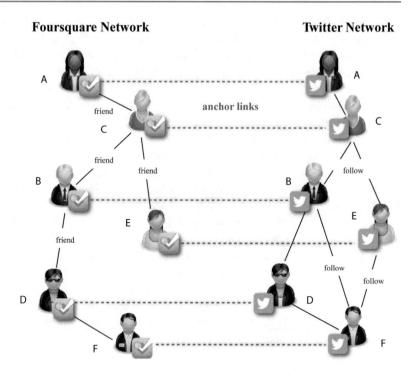

Fig. 4.13 Aligned social
networks with inference
confidence scores

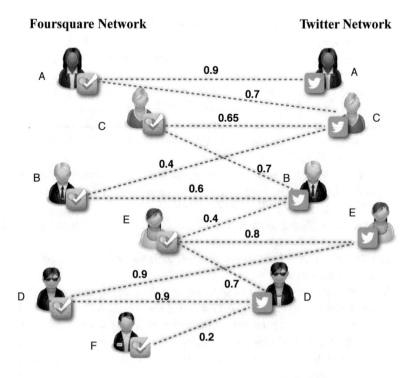

References

1. M. Al Hasan, V. Chaoji, S. Salem, M. Zaki, Link prediction using supervised learning, in *Workshop on Link Analysis, Counterterrorism and Security (SDM 06)* (2006)
2. M. Bayati, M. Gerritsen, D. Gleich, A. Saberi, Y. Wang, Algorithms for large, sparse network alignment problems, in *2009 Ninth IEEE International Conference on Data Mining* (2009)
3. I. Bhattacharya, L. Getoor, Collective entity resolution in relational data. ACM Trans. Knowl. Discov. Data **1**(1), 5 (2007)
4. N. Chawla, K. Bowyer, L. Hall, P. Kegelmeyer, Smote: synthetic minority over-sampling technique. J. Artif. Int. Res. **16**, 321–357 (2002)
5. L. Dubins, D. Freedman, Machiavelli and the Gale-Shapley algorithm. Am. Math. Mon. **88**(7), 485–494 (1981)
6. M. Duggan, A. Smith, Social media update 2013 (2013). Report available at http://www.pewinternet.org/2013/12/30/social-media-update-2013/
7. J. Edmonds, Maximum matching and a polyhedron with 0, 1 vertices. J. Res. Natl. Bur. Stand. **69**(125–130), 55–56 (1965)
8. J. Euzenat, P. Shvaiko, *Ontology Matching* (Springer, Secaucus, 2007)
9. M. Garey, D. Johnson, *Computers and Intractability; A Guide to the Theory of NP-Completeness* (W. H. Freeman & Co., New York, 1990)
10. D. Kempe, J. Kleinberg, É. Tardos, Maximizing the spread of influence through a social network, in *Proceedings of the Ninth ACM SIGKDD International Conference on Knowledge Discovery and Data Mining (KDD '03)* (2003)
11. T. Kolda, B. Bader, Tensor decompositions and applications. SIAM Rev. **51**(3), 455–500 (2009)
12. X. Kong, J. Zhang, P. Yu, Inferring anchor links across multiple heterogeneous social networks, in *Proceedings of the 22nd ACM International Conference on Information & Knowledge Management (CIKM '13)* (2013)
13. D. Koutra, H. Tong, D. Lubensky, Big-align: fast bipartite graph alignment, in *2013 IEEE 13th International Conference on Data Mining* (2013)
14. M. Kubat, S. Matwin, Addressing the curse of imbalanced training sets: one-sided selection, in *Proceedings of the Fourteenth International Conference on Machine Learning* (1997)
15. R. Lichtenwalter, J. Lussier, N. Chawla, New perspectives and methods in link prediction, in *Proceedings of the 16th ACM SIGKDD International Conference on Knowledge Discovery and Data Mining (KDD '10)* (2010)
16. MarketingCharts, Majority of twitter users also use Instagram (2014). Report available at http://www.marketingcharts.com/wp/online/majority-of-twitter-users-also-use-instagram-38941/
17. A. Menon, C. Elkan, Link prediction via matrix factorization, in *Machine Learning and Knowledge Discovery in Databases (ECML PKDD 2011)* (2011)
18. S. Moghaddam, M. Jamali, M. Ester, ETF: extended tensor factorization model for personalizing prediction of review helpfulness, in *Proceedings of the Fifth ACM International Conference on Web Search and Data Mining (WSDM '12)* (2012)
19. O. Peled, M. Fire, L. Rokach, Y. Elovici, Entity matching in online social networks, in *2013 International Conference on Social Computing* (2013)
20. Y. Sun, R. Barber, M. Gupta, C. Aggarwal, J. Han, Co-author relationship prediction in heterogeneous bibliographic networks, in *2011 International Conference on Advances in Social Networks Analysis and Mining* (2011)
21. Y. Sun, J. Han, C. Aggarwal, N. Chawla, When will it happen?: relationship prediction in heterogeneous information networks, in *Proceedings of the 5th ACM International Conference on Web Search and Data Mining (WSDM 2012)* (2012)
22. J. Tang, H. Gao, X. Hu, H. Liu, Exploiting homophily effect for trust prediction, in *Proceedings of the Sixth ACM International Conference on Web Search and Data Mining (WSDM '13)* (2013)
23. I. Tomek, Two modifications of CNN, IEEE Trans. Syst. Man Cybern. **6**, 769–772 (1976)
24. S. Umeyama, An eigendecomposition approach to weighted graph matching problems, IEEE Trans. Pattern Anal. Mach. Intell. **10**(5), 695–703 (1988)
25. Z. Wang, J. Liao, Q. Cao, H. Qi, Z. Wang, Friendbook: a semantic-based friend recommendation system for social networks. IEEE Trans. Mob. Comput. **14**(3), 538–551 (2015)
26. X. Xie, Potential friend recommendation in online social network, in *2010 IEEE/ACM International Conference on Green Computing and Communications International Conference on Cyber, Physical and Social Computing* (2010)
27. J. Yang, J. McAuley, J. Leskovec, Community detection in networks with node attributes, CoRR, abs/1401.7267 (2014)
28. J. Ye, H. Cheng, Z. Zhu, M. Chen, Predicting positive and negative links in signed social networks by transfer learning, in *Proceedings of the 22nd International Conference on World Wide Web (WWW '13)* (2013)

29. B. Zadrozny, C. Elkan, Transforming classifier scores into accurate multiclass probability estimates, in *Proceedings of the Eighth ACM SIGKDD International Conference on Knowledge Discovery and Data Mining (KDD '02)* (2002)
30. R. Zafarani, H. Liu, Connecting users across social media sites: a behavioral-modeling approach, in *Proceedings of the 19th ACM SIGKDD International Conference on Knowledge Discovery and Data Mining (KDD '13)* (2013)
31. J. Zhang, Social network fusion and mining: a survey (2018). arXiv preprint. arXiv:1804.09874
32. J. Zhang, P. Yu, Community detection for emerging networks, in *Proceedings of the 2015 SIAM International Conference on Data Mining* (2015)
33. J. Zhang, P. Yu, Multiple anonymized social networks alignment, in *2015 IEEE International Conference on Data Mining* (2015)
34. J. Zhang, P. Yu, PCT: partial co-alignment of social networks, in *Proceedings of the 25th International Conference on World Wide Web (WWW '16)* (2016)
35. J. Zhang, X. Kong, P. Yu, Predicting social links for new users across aligned heterogeneous social networks, in *2013 IEEE 13th International Conference on Data Mining* (2013)
36. J. Zhang, X. Kong, P. Yu, Transferring heterogeneous links across location-based social networks, in *Proceedings of the 7th ACM International Conference on Web Search and Data Mining (WSDM '14)* (2014)
37. J. Zhang, P. Yu, Z. Zhou, Meta-path based multi-network collective link prediction, in *Proceedings of the 20th ACM SIGKDD International Conference on Knowledge Discovery and Data Mining (KDD '14)* (2014)
38. J. Zhang, W. Shao, S. Wang, X. Kong, P. Yu, Pna: partial network alignment with generic stable matching, in *2015 IEEE International Conference on Information Reuse and Integration* (2015)
39. J. Zhang, S. Wang, Q. Zhan, P. Yu, Intertwined viral marketing in social networks, in *2016 IEEE/ACM International Conference on Advances in Social Networks Analysis and Mining (ASONAM)* (2016)
40. J. Zhang, P. Yu, Y. Lv, Q. Zhan, Information diffusion at workplace, in *Proceedings of the 25th ACM International on Conference on Information and Knowledge Management (CIKM '16)* (2016)
41. J. Zhang, Q. Zhan, P. Yu, Concurrent alignment of multiple anonymized social networks with generic stable matching, in *Information Reuse and Integration* (2016)
42. J. Zhang, C. Aggarwal, P. Yu, Rumor initiator detection in infected signed networks, in *2017 IEEE 37th International Conference on Distributed Computing Systems (ICDCS)* (2017)
43. J. Zhang, J. Chen, J. Zhu, Y. Chang, P. Yu, Link prediction with cardinality constraints, in *Proceedings of the Tenth ACM International Conference on Web Search and Data Mining (WSDM '17)* (2017)

Unsupervised Network Alignment

5

5.1 Overview

Identifying the common users shared by different online social sites is a very hard task even for humans. Manually labeling of the anchor links can be extremely challenging, expensive (in human efforts, time, and money costs), and tedious, and the scale of the real-world online social networks involving millions even billions of users also renders the training data labeling much more difficult. In this chapter, we will introduce several approaches to resolve the *network alignment* problem based on the unsupervised learning setting instead, where no labeled training data will be needed in model building.

Given two heterogeneous online social networks, which can be represented as $G^{(1)} = (\mathcal{V}^{(1)}, \mathcal{E}^{(1)})$ and $G^{(2)} = (\mathcal{V}^{(2)}, \mathcal{E}^{(2)})$, respectively, the *unsupervised network alignment* problem aims at inferring the set of potential anchor links connecting the shared users between networks $G^{(1)}$ and $G^{(2)}$. Let $\mathcal{U}^{(1)} \subset \mathcal{V}^{(1)}$ and $\mathcal{U}^{(2)} \subset \mathcal{V}^{(2)}$ be the user sets in these two networks, respectively, we can represent the set of potential anchor links between networks $G^{(1)}$ and $G^{(2)}$ as $\mathcal{A} = \mathcal{U}^{(1)} \times \mathcal{U}^{(2)}$. In the *unsupervised network alignment* problem, among all the potential anchor links in set $\mathcal{A} = \mathcal{U}^{(1)} \times \mathcal{U}^{(2)}$, we want to infer which one in set $\mathcal{A}$ should exist in the real world.

In this chapter, we will study the network alignment problem based on the unsupervised learning setting with no training data at all. We will first introduce several unsupervised heuristics for measuring the similarities of user accounts across networks, which can serve as the basic predictors of anchor links across networks. After that we will introduce the unsupervised network alignment problem of homogeneous networks especially, and talk about the existing approaches proposed to address the problem. To handle the heterogeneous social networks involving both complex structures and different types of attribute information, a state-of-the-art unsupervised heterogeneous network alignment algorithm will be introduced afterwards. In the real-world social networks, besides the users, many other types of information entities can also be shared across different networks, like products, videos, and POIs (points of interest). Identifying multiple types of common information entities shared across social networks simultaneously is called the *network co-alignment* problem. A novel unsupervised network co-alignment model will be introduced to solve the problem. Besides the pairwise networks, aligning multiple (more than 2) networks simultaneously is called the *multiple network alignment* problem. In the *multiple network alignment* problem, preserving the consistency of the alignment results among all these networks is necessary, which will introduce more constraints on

© Springer Nature Switzerland AG 2019
J. Zhang, P. S. Yu, *Broad Learning Through Fusions*,
https://doi.org/10.1007/978-3-030-12528-8_5

the unsupervised alignment model at the same time. Finally, we will introduce a novel unsupervised *multiple network alignment* algorithm at the end of this chapter.

5.2 Heuristics Based Unsupervised Network Alignment

To infer the correspondence relationships of the shared users between different social networks, many different types of unsupervised heuristics can be applied. Besides the similarity measures (or features) calculated based on the social network structures, location check-ins, textual contents, and temporal activities of users introduced in Sect. 4.3.1, in this section we will introduce two other categories of unsupervised network alignment heuristics based on the user name and user profile information, respectively.

5.2.1 User Names Based Network Alignment Heuristics

Formally, given a user pair $u_i^{(1)}$ and $u_j^{(2)}$ from networks $G^{(1)}$ and $G^{(2)}$, respectively, we can represent their names in these two networks as name$(u_i^{(1)})$ and name$(u_j^{(2)})$. Generally, user names can be denoted as strings involving one or several words. To help their friends identify them, users tend to use their real names in many online social networks, like Facebook, Twitter, and LinkedIn. Several heuristics can be proposed to infer the user anchor links across networks with user names. Exact matching of user names is the most intuitive idea, which is based on the assumption that users tend to use exactly the same names in different sites. For instance, if the names of $u_i^{(1)}$ and $u_j^{(2)}$ are exactly the same (i.e., name$(u_i^{(1)})$ = name$(u_j^{(2)})$), then $u_i^{(1)}$ and $u_j^{(2)}$ are highly likely to be the same person.

However, in real-world social networks, such an assumption can be violated with certain degrees and this method may not perform well due to several reasons:

- *Abbreviated Name*: For some reasons, users tend to abbreviate their names in some sites. For instance, name "Anne-Marie Slaughter" can be written as "A.-M. Slaughter" by using the initials of the first name instead.
- *Order-Reversed Name*: Depending on the sites, user names can display in different ways. For instance, names "Anne-Marie Slaughter" and "Slaughter, Anne-Marie" may refer to the same user, but the string representations have flipped the first and last name, which may create many problems for exact name matching.
- *Alias Names*: Some people can have their alias names, like the nick name. For example, "Michael" and "Mike" may refer to the same person, but they are written in totally different ways.
- *Duplicated Names*: For many common names, lots of users may have exactly the same name, e.g., "James," "Michael," "Robert," "Mary," "Maria," and "David." Simply name matching may not be sufficient to identify the exact shared users accounts.

To overcome these aforementioned problems (except *duplicated names*), in this part, we will introduce the unsupervised network alignment heuristics based on the user names to compute the user similarities. For the *duplicated name* case, besides the user name information, we will introduce several other network alignment heuristics by using the user profile information in the following subsection.

5.2.1.1 Name Similarity Metrics

Instead of using the exact name matching, *name similarity metrics* are usually applied to calculate how likely the users are the same person based on their names. Currently, the *name similarity metrics* can be categorized into three groups: *character based similarity metrics*, *token based similarity metrics*, and *phonetic similarity metrics*.

Character Based Similarity Metrics *Character based similarity metrics* are usually used to handle typographical differences of user name strings, and the well-known *character based similarity metrics* include

- *Edit Distance*: Given two strings, the *edit distance* [31, 34, 41] denotes the minimal number of *insertion*, *deletion*, and *substitution* edit operations of characters needed to transform one string into the other one. Given two user names, if the *edit distance* between them is short, they will be similar to each other.
- *Damerau-Levenshtein Distance*: *Damerau-Levenshtein distance* [28] is a variation of *edit distance*. Besides the basic *insertion*, *deletion*, and *substitution* edit operations, it also takes the character *transposition* as another elementary edit operation.
- *Jaro Distance*: *Jaro distance* [22] is also a type of edit distance computed based on the number of shared characters between strings and the number of transpositions operations. Formally, given two strings s_1 and s_2, the *Jaro Distance* between them is defined as

$$d_j(s_1, s_2) = \begin{cases} 0, & \text{if } m = 0, \\ \frac{1}{3}\left(\frac{m}{|s_1|} + \frac{m}{|s_2|} + \frac{m-t}{m} \right), & \text{otherwise,} \end{cases} \tag{5.1}$$

where m denotes the number of matching characters between s_1 and s_2, t denotes half the number of transpositions to change one shared character sequence into the other, and $|s_1|$, $|s_2|$ denote the length of s_1 and s_2, respectively. Two characters from s_1 and s_2, respectively, are considered matching only if they are the same and their indexes are not farther than $\left\lfloor \frac{\max(|s_1|,|s_2|)}{2} \right\rfloor - 1$ way. We will introduce more information about the *Jaro distance* in Example 5.2.

- *Jaro-Winkler Distance*: *Jaro-Winkler distance* [42] is a variant of *Jaro distance*, and it uses a prefix scale p to give more favorable ratings to strings that match from the beginning for a set prefix length. Given two strings s_1 and s_2, the *Jaro-Winkler distance* between them is defined as

$$d_w(s_1, s_2) = d_j(s_1, s_2) + l \cdot p \cdot (1 - d_j(s_1, s_2)), \tag{5.2}$$

where l is the length of common prefix at the start of the string up to a maximum of four characters, and p is a constant scaling factor for how much the score is adjusted upward for having common prefixes (whose standard value is $p = 0.1$).

- *Longest Common Sub-string (LCS)*: LCS [21] repeatedly identifies and removes the longest common sub-string from two names, whose total length can denote the similarity between the user names. For instance, the total length of shared sub-string (involving more than 1 character) by input names "Michael Jordan" and "Yordan Michelin" is 11 (the shared sub-strings include "mich," "el," and "ordan").

Example 5.1 For instance, given two user name strings "Michael Jordan" and "Mike Jordan," we can convert "Mike Jordan" into "Michael Jordan" with 4 edits: 1 *substitution* and 3 *insertion*:

- *substitute "k" with "c"*: with this operation, we can convert "Mike Jordan" to "Mice Jordan";
- *insert "h" after "c"*: with this operation, we will be able to convert "Mice Jordan" to "Miche Jordan";
- *insert "a" after "h"*: with this operation, we can convert "Miche Jordan" to "Michae Jordan";
- *insert "l" after "e"*: with this operation, we will be able to successfully convert "Michae Jordan" into "Michael Jordan."

Example 5.2 Given two strings "TRACE" and "CRATE," we can identify the matching characters between them are "R," "A," and "E" and $m = 3$. Here, characters "T" and "C" are shared by these two strings, but they are not considered to be matching characters since their index distance is 3, which is father than $\left\lfloor \frac{\max(5,5)}{2} \right\rfloor - 1 = 1$ away. Considering that the matching characters "R-A-E" are already in the same order in both of these two strings, so no transpositions are needed, and we have $t = 0$.

Therefore, the *Jaro distance* between "TRACE" and "CRATE" is

$$
\begin{aligned}
\frac{1}{3} &\left(\frac{m}{|s_1|} + \frac{m}{|s_2|} + \frac{m-t}{m} \right) \\
&= \frac{1}{3} \left(\frac{3}{5} + \frac{3}{5} + \frac{3-0}{3} \right) \\
&= \frac{11}{15}
\end{aligned}
\tag{5.3}
$$

Token Based Similarity Metrics The *character based similarity metrics* may fail to handle the names with first and last name flipped. To handle such a case, a set of *token based similarity metrics* are proposed, which divide name strings into a set of tokens (i.e., words) instead.

- *Common Token*: *Common token* metric counts the number of common words shared by different user names. Given two user names s_1 and s_2, we can transform them into two sets of tokens in advance, which can be denoted as set(s_1) and set(s_2), respectively. The number of *common tokens* [44] shared by them can be denoted as

$$
\text{CT}(\text{set}(s_1), \text{set}(s_2)) = |\text{set}(s_1) \cap \text{set}(s_2)|.
\tag{5.4}
$$

- *Token based Jaccard's coefficient*: In some cases, people can have very long names, like "Jose Arcadio Buendia" and "Jose Arcadio," which share two common tokens but actually refer to different people in *"One Hundred Years of Solitude."* *Token based Jaccard's coefficient* proposes to penalize the long names, and the score calculated for names set(s_1) and set(s_2) can be denoted as

$$
\text{TJC}(\text{set}(s_1), \text{set}(s_2)) = \frac{|\text{set}(s_1) \cap \text{set}(s_2)|}{|\text{set}(s_1) \cup \text{set}(s_2)|}.
\tag{5.5}
$$

- *Token based Cosine Similarity*: *Cosine Similarity* [5] expresses strings as term vectors, where each word denotes a dimension in the vector representation. Formally, given two user names s_1 and s_2, we can represent their token vectors as $\mathbf{s}_1$ and $\mathbf{s}_2$, respectively, and the *token based cosine similarity* score of the user names can be represented as

$$\text{TSC}(\mathbf{s}_1, \mathbf{s}_2) = \frac{\mathbf{s}_1^\top \cdot \mathbf{s}_2}{\|\mathbf{s}_1\| \cdot \|\mathbf{s}_2\|}. \tag{5.6}$$

- *TFIDF-TCS*: *TFIDF* denotes *"Term Frequency/Inverted Document Frequency"* [35], which counts frequency of a word but also penalizes it with its occurrence in other documents. *TFIDF* can be applied to weight the name's feature vectors before using them to compute the *Token based Cosine Similarity* measure. Formally, the *Token based Cosine Similarity* measure subject to the *TFIDF* weighted user name vectors can be denoted as *TFIDF-TCS*.

As to the partition of names into tokens, different techniques can be applied, including specific delimiter based partition, n-gram [8] based partition, etc. We will not introduce these name partition methods here since it will be out of the scope of this book.

Example 5.3 For instance, given two user names s_1 ="Michael Jordan" and s_2="Jordan, Michael," based on the character based similarity metrics, e.g., *edit distance*, we can compute their edit distance to be 12. Meanwhile, by dividing the user names into tokens by separator " " (i.e., the space between tokens) or ", ", we can transform the name strings into two unit token sets $set(s_1) = \{Michael, Jordan\}$ and $set(s_1) = \{Jordan, Michael\}$.

According to the *common token, token based Jaccard's coefficient* or *token based cosine similarity* metrics, we can compute their similarity scores all to be 1.0. Viewed in such a perspective, these *token based similarity metrics* can effectively amend the disadvantages of the *character based similarity metrics*.

Phonetic Similarity Metrics These metrics introduced before are mostly based on the string representation of the names. In many cases, people like to use abbreviated names based on the pronunciation instead of the textual representations. For instance, people like to use "Mike" as a nick name of "Michael," which is pronounced in a very similar way but is very different in their writings. Some *phonetic similarity metrics* [37] have been proposed to measure the phonetic similarity between strings, which can be applied to compute the user name similarities as well.

- *Soundex Matching*: *Soundex* [39] is a phonetic algorithm for indexing names by sound. The Soundex code [39] for a name consists of a letter followed by three numerical digits: the letter is the first letter of the name, and the digits encode the remaining consonants. Given two name strings s_1 and s_2, if their *Soundex codes* are the same, their *Soundex matching* will be 1; otherwise, it will be 0.
- *Soundex Similarity*: Instead of exact matching the *Soundex codes*, *Soundex similarity* proposes to count the shared characters in the *Soundex codes* of input strings, which is normalized by the returned *Soundex code* length.

5.2.2 Profile Based Network Alignment Heuristics

In the real-world online social networks, many users will share the same name (especially the regular names like "Mike," "David," etc.) and the above name based similarity metrics will fail to differentiate these users in the alignment process. To help improve the alignment results, a set of *profile based network alignment heuristics* will be introduced in this part based on the user profile information, including *hometown*, *employment history*, *educational records*, *gender*, *birthday*, and other personal information. Generally, these information are represented as strings as well, and it seems the string similarity measures introduced before can also be applied here. Meanwhile, slightly different from names, the profile information usually consists of several components and is much longer than names, which renders the similarity metrics to be used here different from those introduced before.

- *Location Distance*: Given the hometown of two users from different online social networks, the hometown locations are usually represented as strings and we can apply the character/token based string similarity metrics introduced before to measure how close the users are in terms of geograph-ical locations. Meanwhile, in many cases, there exist various different string representations for the same location, e.g., "PA," "Penn," and "Pennsylvania," all denote the Pennsylvania state in the USA, which will make the introduced metrics fail to work. A better way to handle the hometown information is using the online Map APIs to transform the hometown strings to the (latitude, longitude) coordinate pairs, and calculate the geographical distance between the coordinate pairs as the metric instead. Several different Map APIs are available online, like Google Maps,[1] Bing Maps,[2] Apple Maps.[3]
- *Birthday*: Online social networks may ask for users' birthday, and will offer some special services for users on their birthday by either sending notifications to friends online or generating some celebration posts on their homepage timeline. User's birthday is usually represented as a string containing the year, month, and day. Besides the string similarity metrics, these birthday strings can be transformed into a date object, based on which we can calculate the birthday closeness by computing how many days the users' birthdays are apart from each other. The open source toolkits that can handle the date related strings include Python datetime[4] and Java Date (java.util.Date[5]).
- *Vector Space Model*: For the remaining information represented in strings, they can be handled with either exact matching or *character based similarity metrics* for the *gender* information, *token based similarity metrics* and *phonetic similarity metrics* for the employment and educational records. To deal with such long string information together, we can treat user profile as a document and the *vector space model* [36] can be applied here. Vector space model or term vector model is an algebraic model for representing textual documents as vectors of identifiers. Given two users' profile information, one of user's profile can be treated as the query "profile document" and used to retrieve the similar "profile document" from another network. The similarity metric frequently used in the *vector space model* is *cosine similarity*, and TFIDF can also be applied to weight the feature vector representations prior to the similarity score computation.

5.3   Pairwise Homogeneous Network Alignment

Given two homogeneous networks $G^{(1)}$ and $G^{(2)}$, matching the nodes between them is an extremely challenging task, which is also called the *graph isomorphism* problem [12, 30]. By this context so far, no efficient algorithm exists that can address the problem in polynomial time. In this part, we will

---

[1]https://developers.google.com/maps/.

[2]https://www.bingmapsportal.com.

[3]https://developer.apple.com/maps/.

[4]https://docs.python.org/2/library/datetime.html.

[5]https://www.tutorialspoint.com/java/util/java_util_date.htm.

introduce several heuristics based network alignment models, and a mapping matrix inference model to solve the *pairwise homogeneous network alignment* problem.

5.3.1 Heuristics Based Network Alignment Model

The information generated by users' online social activities can indicate their personal characteristics. Besides the features extracted in Sect. 4.3.1 and user name/profile based heuristics introduced in Sect. 5.2, in this part, we will introduce a new category of measures, *Relative Centrality Difference* (RCD) [47, 48], which computes the similarity of user node centrality scores in different networks as the alignment results. Based on the assumption that "users with similar *centrality* scores are more likely to be the same user," the *relative centrality difference* can also be applied to solve the *unsupervised network alignment* problem.

The *centrality* concept introduced in Chap. 3.3.2 denotes the importance of nodes in the network structured data. Here, we assume that important users in one social network (like celebrities, movie stars and politicians) will be important as well in other networks. Viewed in such a perspective, the *centrality* of users in different networks can be an important signal for inferring the anchor links across networks.

Definition 5.1 (Relative Centrality Difference) Given two users $u_i^{(1)}$, $u_j^{(2)}$ from networks $G^{(1)}$ and $G^{(2)}$, respectively, let $C(u_i^{(1)})$ and $C(u_j^{(2)})$ denote the *centrality* scores of these two users, we can define the *relative centrality difference* (RCD) between them as

$$RCD\left(u_i^{(1)}, u_j^{(2)}\right) = \left(1 + \frac{\left|C\left(u_i^{(1)}\right) - C\left(u_j^{(2)}\right)\right|}{\left(C\left(u_i^{(1)}\right) + C\left(u_j^{(2)}\right)\right)/2}\right)^{-1}. \tag{5.7}$$

Depending on the *centrality* measures applied here, different types of *relative centrality difference* measures can be defined. For instance, if we use node degree [1, 6] as the *centrality* measure, the *relative degree difference* can be represented as

$$RDD\left(u_i^{(1)}, u_j^{(2)}\right) = \left(1 + \frac{\left|D\left(u_i^{(1)}\right) - D\left(u_j^{(2)}\right)\right|}{\left(D\left(u_i^{(1)}\right) + D\left(u_j^{(2)}\right)\right)/2}\right)^{-1}. \tag{5.8}$$

Meanwhile, if the *eigen-centrality* [6, 7] definition is adopted here, we can represent the nodes *relative eigen-centrality difference* measure as

$$RECD\left(u_i^{(1)}, u_j^{(2)}\right) = \left(1 + \frac{\left|C_{eigen}\left(u_i^{(1)}\right) - C_{eigen}\left(u_j^{(2)}\right)\right|}{\left(C_{eigen}\left(u_i^{(1)}\right) + C_{eigen}\left(u_j^{(2)}\right)\right)/2}\right)^{-1}. \tag{5.9}$$

Example 5.4 Based on the input homogeneous network structures shown in Fig. 5.1, if the *relative centrality difference* (with *degree* as the *centrality* measure) is adopted, we can compute the RDD

Fig. 5.1 An example of input pairwise homogeneous networks

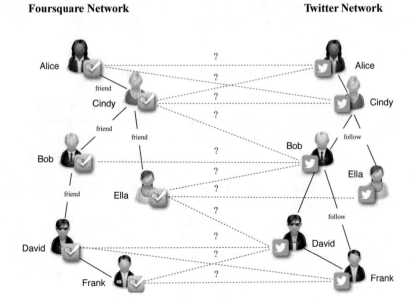

scores for the three potential anchor links incident to $Ella_F$ as follows:

$$RDD(Ella_F, Bob_T) = \left(1 + \frac{|D(Ella_F) - D(Bob_T)|}{(D(Ella_F) + D(Bob_T))/2}\right)^{-1}$$

$$= \left(1 + \frac{|1 - 3|}{(1 + 3)/2}\right)^{-1}$$

$$= \frac{1}{2} \tag{5.10}$$

$$RDD(Ella_F, Ella_T) = \left(1 + \frac{|D(Ella_F) - D(Ella_T)|}{(D(Ella_F) + D(Ella_T))/2}\right)^{-1}$$

$$= \left(1 + \frac{|1 - 1|}{(1 + 1)/2}\right)^{-1}$$

$$= 1 \tag{5.11}$$

$$RDD(Ella_F, David_T) = \left(1 + \frac{|D(Ella_F) - D(David_T)|}{(D(Ella_F) + D(David_T))/2}\right)^{-1}$$

$$= \left(1 + \frac{|1 - 2|}{(1 + 2)/2}\right)^{-1}$$

$$= \frac{3}{5} \tag{5.12}$$

Therefore, compared with Bob_T and $David_T$, $Ella_F$ is more likely to be connected by the anchor link with $Ella_T$, since they have a closer node degree (both with degree 1) in the input network structures.

5.3.2 IsoRank

Another model called IsoRank [38] initially proposed to align the biomedical networks can also be applied to align the *homogeneous social networks*. The IsoRank algorithm has two stages. It first associates a score with each possible anchor link between user nodes across networks. For instance, we can denote $r(u_i^{(1)}, u_j^{(2)})$ as the reliability score of a potential anchor link $(u_i^{(1)}, u_j^{(2)})$ between the networks $G^{(1)}$ and $G^{(2)}$, and all such scores can be organized into a vector $\mathbf{r}$ of length $|\mathcal{U}^{(1)}| \times |\mathcal{U}^{(2)}|$. The second stage of IsoRank is to construct the mapping for the networks by extracting information from the vector $\mathbf{r}$.

Definition 5.2 (Reliability Score) The *reliability score* $r(u_i^{(1)}, u_j^{(2)})$ of the anchor link $(u_i^{(1)}, u_j^{(2)})$ is highly correlated with the support provided by the neighbors of users $u_i^{(1)}$ and $u_j^{(2)}$. Therefore, we can define score $r(u_i^{(1)}, u_j^{(2)})$ as

$$r\left(u_i^{(1)}, u_j^{(2)}\right) = \sum_{u_m^{(1)} \in \Gamma\left(u_i^{(1)}\right)} \sum_{u_n^{(2)} \in \Gamma\left(u_i^{(2)}\right)} \frac{1}{\left|\Gamma\left(u_i^{(1)}\right)\right| \left|\Gamma\left(u_j^{(2)}\right)\right|} r\left(u_m^{(1)}, u_n^{(2)}\right), \qquad (5.13)$$

where sets $\Gamma(u_i^{(1)})$ and $\Gamma(u_i^{(2)})$ represent the neighbors of users $u_i^{(1)}$ and $u_i^{(1)}$, respectively, in networks $G^{(1)}$ and $G^{(2)}$.

The above definition is for the regular *unweighted networks*. Meanwhile, if the studied networks are weighted, and all the intra-network connections like $(u_i^{(1)}, u_m^{(1)})$ are associated with a weight $w(u_i^{(1)}, u_m^{(1)})$, we can represent the *reliability* measure of the weighted network as

$$r\left(u_i^{(1)}, u_j^{(2)}\right)$$

$$= \sum_{u_m^{(1)} \in \Gamma\left(u_i^{(1)}\right)} \sum_{u_n^{(2)} \in \Gamma\left(u_j^{(2)}\right)} \frac{w\left(u_i^{(1)}, u_m^{(1)}\right) w\left(u_j^{(2)}, u_n^{(2)}\right) \cdot r\left(u_m^{(1)}, u_n^{(2)}\right)}{\left(\sum_{u_p^{(1)} \in \Gamma\left(u_i^{(1)}\right)} w\left(u_i^{(1)}, u_p^{(1)}\right)\right) \left(\sum_{u_q^{(2)} \in \Gamma\left(u_j^{(2)}\right)} w\left(u_j^{(2)}, u_q^{(2)}\right)\right)},$$

$$(5.14)$$

As we can see, Eq. (5.13) is a special case of Eq. (5.14) with link weight $w(u_i^{(1)}, u_j^{(1)}) = 1$ for $u_i^{(1)} \in \mathcal{U}^{(1)}$ and $u_j^{(2)} \in \mathcal{U}^{(2)}$.

To simplify the problem settings, we will take the *unweighted social networks* as an example in the following analysis, and Eq. (5.13) can also be rewritten with a linear algebra as follows:

$$\mathbf{r} = \mathbf{A}\mathbf{r}, \qquad (5.15)$$

where the matrix $\mathbf{A}$ is of dimension $(|\mathcal{U}^{(1)}||\mathcal{U}^{(2)}|) \times (|\mathcal{U}^{(1)}||\mathcal{U}^{(2)}|)$, and the row and column indexes correspond to different potential anchor links across the networks. The matrix entry

$$A(i, j)(p, q) = \begin{cases} \dfrac{1}{|\Gamma(u_i^{(1)})||\Gamma(u_j^{(2)})|}, & \text{if } \left(u_i^{(1)}, u_p^{(1)}\right) \in \mathcal{E}^{(1)}, \left(u_j^{(2)}, u_q^{(2)}\right) \in \mathcal{E}^{(2)}, \\ 0, & \text{otherwise,} \end{cases} \tag{5.16}$$

which corresponds the anchor links $(u_i^{(1)}, u_j^{(2)})$ and $(u_p^{(1)}, u_q^{(2)})$. As we can see, the above equation has a very similar representation to the stationary equation of *random walk*, whose solutions correspond to the principal eigenvector of the matrix $\mathbf{A}$ corresponding to the eigenvalue 1. For more information about the *random walk* model, please refer to Chap. 3.3.3.3.

Example 5.5 Here, we will use an example from [38] to illustrate the IsoRank algorithm. Formally, given the two input graphs as shown in Fig. 5.2, we can represent the *reliability* score of the potential anchor links across the networks with the following equations:

$$r_{aa'} = \frac{1}{4}r_{bb'}, \tag{5.17}$$

$$r_{bb'} = \frac{1}{3}r_{ac'} + \frac{1}{3}r_{a'c} + r_{aa'} + \frac{1}{9}r_{cc'}, \tag{5.18}$$

$$r_{cc'} = \frac{1}{4}r_{bb'} + \frac{1}{2}r_{be'} + \frac{1}{2}r_{bd'} + \frac{1}{2}r_{eb'} + \frac{1}{2}r_{db'} + r_{ee'} + r_{ed'} + r_{de'} + r_{dd'}, \tag{5.19}$$

$$r_{dd'} = \frac{1}{9}r_{cc'}, \tag{5.20}$$

$$r_{ee'} = \frac{1}{9}r_{cc'}. \tag{5.21}$$

By assigning the *reliability* scores of all these anchor links with random initial values and updating the scores according to the above equations, we can achieve the stationary score values of all the links as shown in the right matrix (the *reliability* score vector $\mathbf{r}$ is reshaped into a matrix for ease of viewing). For the entries without values, they are filled with the 0 by default.

IsoRank adopts a greedy strategy to select the anchor links from the *reliability* score matrix subject to the *one-to-one* cardinality constraint, where the anchor links corresponding to the largest *reliability* score will be selected first and no other anchor links incident to these selected nodes will be selected in the following rounds. For instance, among all the potential anchor links in this example, the

Fig. 5.2 An example of IsoRank

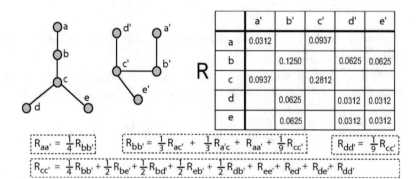

	a'	b'	c'	d'	e'
a	0.0312		0.0937		
b		0.1250		0.0625	0.0625
c	0.0937		0.2812		
d		0.0625		0.0312	0.0312
e		0.0625		0.0312	0.0312

$R_{aa'} = \frac{1}{4}R_{bb'}$ $R_{bb'} = \frac{1}{3}R_{ac'} + \frac{1}{3}R_{a'c} + R_{aa'} + \frac{1}{9}R_{cc'}$ $R_{dd'} = \frac{1}{9}R_{cc'}$

$R_{cc'} = \frac{1}{4}R_{bb'} + \frac{1}{2}R_{be'} + \frac{1}{2}R_{bd'} + \frac{1}{2}R_{eb'} + \frac{1}{2}R_{db'} + R_{ee'} + R_{ed'} + R_{de'} + R_{dd'}$

anchor links (c, c'), (b, b'), (a, a') will be selected first. As indicated in [38], by using sequence information, the ambiguities in the remaining anchor links (d, d'), (d, e'), (e, d'), and (e, e') with the same *reliability* scores can be resolved successfully.

5.3.3 IsoRankN

IsoRankN algorithm introduced in [29] is an extension to IsoRank, which uses a different method of spectral clustering on the induced graph of pairwise alignment scores. The new approach provides significant advantages not only over the original IsoRank algorithm but also over other methods. IsoRankN has four main steps: (1) initial network alignment with IsoRank, (2) star spread, (3) spectral partition, and (4) star merging. Next, we will introduce these four steps in detail.

5.3.3.1 Initial Network Alignment

Given k networks $G^{(1)}, G^{(2)}, \ldots, G^{(k)}$, IsoRankN computes the local alignment scores of node pairs across networks with the IsoRank algorithm. For instance, if the networks are unweighted, the alignment score between nodes $u_l^{(i)}$ and $u_m^{(j)}$ in networks $G^{(i)}, G^{(j)}$ can be denoted as:

$$r\left(u_l^{(i)}, u_m^{(j)}\right) = \sum_{u_p^{(i)} \in \Gamma\left(u_l^{(i)}\right)} \sum_{u_q^{(j)} \in \Gamma\left(u_m^{(j)}\right)} \frac{1}{|\Gamma\left(u_l^{(i)}\right)| |\Gamma\left(u_m^{(j)}\right)|} r\left(u_p^{(i)}, u_q^{(j)}\right), \qquad (5.22)$$

It will lead to a weighted k-partite graph, where the links denote the anchor links across networks weighted by the *reliability* scores calculated above. If the networks $G^{(1)}, \ldots G^{(k)}$ are all complete graphs, the alignment results will be the maximum weighted cliques. However, in the real world, such an assumption can hardly hold, and IsoRankN proposes to use a technique called "Star Spread" to select a subgraph with high weights.

5.3.3.2 Star Spread

For each node in a network, e.g., $u_l^{(i)}$ in network $G^{(i)}$, the set of nodes from all the other networks connected with $u_l^{(i)}$ via potential anchor links can be denoted as set $\Gamma(u_l^{(i)})$. The nodes in $\Gamma(u_l^{(i)})$ can be further pruned by removing those connected with *weak* anchor links. Here, the term "weak" denotes the anchor links with low *reliability* scores calculated with IsoRank. Formally, among all the nodes in $\Gamma(u_l^{(i)})$, we can denote the node connected to $u_l^{(i)}$ with the strongest link as $v^* = \arg_{v \in \Gamma(u_l^{(i)})} \max r(u_l^{(i)}, v)$. For all the nodes with weights lower than $\beta \cdot r(u_l^{(i)}, v^*)$ will be removed from $\Gamma(u_l^{(i)})$ (where β is a threshold parameter), and the remaining nodes together with $u_l^{(i)}$ will form a star structured graph $S_{u_l^{(i)}}$.

5.3.3.3 Spectral Partition

For each node $u_l^{(i)}$, IsoRankN aims at selecting a subgraph $S_{u_l^{(i)}}^*$ from the star-structured graph $S_{u_l^{(i)}}$, which contains the highly weighted neighbors of $u_l^{(i)}$. To achieve such an objective, IsoRankN proposes to identify a subgraph with a low *conductance* from $S_{u_l^{(i)}}$ instead.

Formally, given a network $G = (\mathcal{V}, \mathcal{E})$, let $\mathcal{S} \subset \mathcal{V}$ denote a node subset of G. The *conductance* of the subgraph involving $\mathcal{S}$ can be formally represented as

$$\phi(\mathcal{S}) = \frac{\sum_{u \in \mathcal{S}} \sum_{v \in \bar{\mathcal{S}}} r_{u,v}}{\min(\text{vol}(\mathcal{S}), \text{vol}(\bar{\mathcal{S}}))}, \tag{5.23}$$

where $\bar{\mathcal{S}} = \mathcal{V} \setminus \mathcal{S}$, and $\text{vol}(\mathcal{S}) = \sum_{u \in \mathcal{S}} \sum_{v \in \mathcal{V}} r_{u,v}$.

IsoRankN points out that a node subset $\mathcal{S}$ containing node $u_l^{(i)}$ can be computed effectively and efficiently with the personalized PageRank algorithm starting from node $u_l^{(i)}$, which will not be introduced here.

5.3.3.4 Star Merging

Considering that links in the selected star graph $S^*_{u_l^{(i)}}$ are all the anchor links across networks, there exist no intra-network links at all, e.g., the links in network $G^{(i)}$ only. However, in many cases, there may exist multiple nodes corresponding to the same entity inside the network as well. To solve such a problem, IsoRankN proposes a star merging step to combine several star graphs together, e.g., $S^*_{u_l^{(i)}}$ and $S^*_{u_m^{(j)}}$.

Formally, given two star graphs $S^*_{u_l^{(i)}}$ and $S^*_{u_m^{(j)}}$ if the following conditions both hold, $S^*_{u_l^{(i)}}$ and $S^*_{u_m^{(j)}}$ can be merged into one star graph.

$$\forall v \in S^*_{u_m^{(j)}} \setminus \{u_m^{(j)}\}, r\left(v, u_l^{(i)}\right) \geq \beta \cdot \max_{v' \in \Gamma\left(u_l^{(i)}\right)} r\left(v', u_l^{(i)}\right), \tag{5.24}$$

$$\forall v \in S^*_{u_l^{(i)}} \setminus \{u_l^{(i)}\}, r\left(v, u_m^{(j)}\right) \geq \beta \cdot \max_{v' \in \Gamma\left(u_m^{(j)}\right)} r\left(v', u_m^{(j)}\right). \tag{5.25}$$

Via these aforementioned four steps, IsoRankN will be able to compute the alignment among multiple input networks, where steps (3) and (4) will repeat until all the nodes are assigned to a cluster.

5.3.4 Matrix Inference Based Network Alignment

For homogeneous network alignment, we will introduce another algorithm based on matrix inference at the end of this section. Formally, the set of *anchor links* actually define a mapping of nodes across networks, which is subject to the *one-to-one* constraint.

Formally, given a homogeneous network $G^{(1)} = (\mathcal{V}^{(1)}, \mathcal{E}^{(1)})$, its structure can be organized as the adjacency matrix $\mathbf{A}_{G^{(1)}} \in \mathbb{R}^{|\mathcal{V}^{(1)}| \times |\mathcal{V}^{(1)}|}$. If network $G^{(1)}$ is unweighted, then matrix $\mathbf{A}_{G^{(1)}}$ will be a binary matrix and entry $A_{G^{(1)}}(i, p) = 1$ (or $A_{G^{(1)}}(u_i^{(1)}, u_p^{(1)}) = 1$) iff the correspond social link $(u_i^{(1)}, u_p^{(1)})$ exists in the link set $\mathcal{E}^{(1)}$. In the case that the network is weighted, the entries, e.g., $A_{G^{(1)}}(i, p)$, will denote the weight of link $(u_i^{(1)}, u_p^{(1)})$ and the entry will be 0 if the link $(u_i^{(1)}, u_p^{(1)})$ doesn't exist. The concept of *adjacency matrix* has been introduced when talking about the network representations in Sect. 3.2.1. In a similar way, we can also represent the social adjacency matrix $\mathbf{A}_{G^{(2)}}$ for network $G^{(2)}$ as well.

The network alignment problem actually aims at inferring a *one-to-one* node mappings, that can project nodes from one network to the other network. For instance, we can denote the mapping between networks $G^{(1)}$ to $G^{(2)}$ as $f : \mathcal{V}^{(1)} \to \mathcal{V}^{(2)}$. Via the mapping f, besides the nodes, the network structure can be projected across networks as well. For instance, given a social connection $(u_i^{(1)}, u_p^{(1)})$ in $G^{(1)}$, we can represent its corresponding projected connection in $G^{(2)}$ as $(f(u_i^{(1)}), f(u_p^{(1)}))$.

Based on the assumption that the mapped nodes should have similar network connection patterns across different networks, via the mapping f, we can denote the projection error as the network structure differences introduced by the projection between $G^{(1)}$ and $G^{(2)}$, i.e.,

$$L(G^{(1)}, G^{(2)}, f) = \sum_{u_i^{(1)} \in \mathcal{V}^{(1)}} \sum_{u_p^{(1)} \in \mathcal{V}^{(1)}} \left(A_{G^{(2)}} \left(u_i^{(1)}, u_p^{(1)} \right) - A_{G^{(1)}} \left(f \left(u_i^{(1)} \right), f \left(u_p^{(1)} \right) \right) \right)^2. \quad (5.26)$$

Formally, such a node projection can be represented as a matrix $\mathbf{P} \in \{0, 1\}^{|\mathcal{V}^{(1)}| \times |\mathcal{V}^{(2)}|}$ as well, where entry $P(i, j) = 1$ iff anchor link $(u_i^{(1)}, u_j^{(2)})$ exists between networks $G^{(1)}$ and $G^{(2)}$. Via the matrix $\mathbf{P}$, we can represent the above loss term as

$$L(\mathbf{A}_{G^{(1)}}, \mathbf{A}_{G^{(2)}}, \mathbf{P}) = \left\| \mathbf{P}^\top \mathbf{A}_{G^{(1)}} \mathbf{P} - \mathbf{A}_{G^{(2)}} \right\|^2. \quad (5.27)$$

If there exists a perfect mapping of users across networks, we can obtain a mapping matrix $\mathbf{P}$ introducing zero loss in the above function, i.e., $L(\mathbf{A}_{G^{(1)}}, \mathbf{A}_{G^{(2)}}, \mathbf{P}) = 0$. However, in the real-world scenarios, few networks can be perfectly aligned together. Inferring the optimal mapping matrix $\mathbf{P}$ which can introduce the minimum loss can be represented as the following objective function

$$\mathbf{P}^* = \arg\min_{\mathbf{P}} \left\| \mathbf{P}^\top \mathbf{A}_{G^{(1)}} \mathbf{P} - \mathbf{A}_{G^{(2)}} \right\|^2, \quad (5.28)$$

where the matrix $\mathbf{P}$ is usually subject to some constraints, like $\mathbf{P}$ is binary and each row and column should contain at most one entry being filled with value 1.

In general, it is not easy to find the optimal solution to the above objective function, as it is a purely combinatorial optimization problem. Identifying the optimal solution requires the enumeration of all the potential user mapping across different networks. In [40], Umeyama provides an algorithm that can solve the function with a nearly optimal solution based on eigen-decomposition. If the readers are interested in the solution, please refer to [40] for more detailed information. In the following sections, we will introduce an approximation method to address a similar optimization problem by relaxing the hard binary constraint to real values instead.

5.4 Multiple Homogeneous Network Alignment with Transitivity Penalty

The works introduced in the previous section are all about pairwise network alignment, which focus on the alignment of two networks only. However, in the real world, people are normally involved in multiple (usually more than two) social networks simultaneously. In this section, we will focus on the simultaneous alignment problem of multiple (more than two) networks, which is called the M-NASA problem formally [47].

5.4.1   Multiple Network Alignment Problem Description

Example 5.6 To help illustrate the M-NASA problem more clearly, we also give an example in Fig. 5.3, which involves three different social networks (i.e., networks I, II, and III). Users in these three networks are all anonymized and their names are replaced with randomly generated meaningless identifiers. Each pair of these three anonymized networks can actually share some common users, e.g., "David" participates in both networks I and II simultaneously, "Bob" is using networks I and III concurrently, and "Charles" is involved in all these three networks at the same time. Besides these shared anchor users, in these three partially aligned networks, some users are involved in one single network only (i.e., the non-anchor users [49]), e.g., "Alice" in network I, "Eva" in network II, and "Frank" in network III. The M-NASA problem studied in this section aims at discovering the anchor links (i.e., the dashed bi-directional red lines) connecting anchor users across these three social networks.

The significant difference of M-NASA from existing *pairwise* network alignment problems is due to the *"transitivity law"* that anchor links follow. In traditional set theory [26], a relation $\mathcal{R}$ is defined to be a *transitive relation* in domain $\mathcal{X}$ iff $\forall a, b, c \in \mathcal{X}, (a, b) \in \mathcal{R} \wedge (b, c) \in \mathcal{R} \rightarrow (a, c) \in \mathcal{R}$. If we treat the union of user account sets of all these social networks as the target domain $\mathcal{X}$ and treat anchor links as the relation $\mathcal{R}$, then anchor links depict a *"transitive relation"* among users across networks. We can take the networks shown in Fig. 5.3 as an example. Let u be a user involved in networks I, II, and III simultaneously, whose accounts in these networks are u^I, u^{II}, and u^{III}, respectively. If anchor links (u^I, u^{II}) and (u^{II}, u^{III}) are identified in aligning networks (I, II) and networks (II, III), respectively (i.e., u^I, u^{II}, and u^{III} are discovered to be the same user), then anchor link (u^I, u^{III}) should also exist in the alignment result of networks (I, III) as well. In the M-NASA problem, we need to guarantee the inferred anchor links can meet the *transitivity law*.

Formally, the M-NASA problem can be formally defined as follows. Given the n isolated anonymized social networks $\{G^{(1)}, G^{(2)}, \ldots, G^{(n)}\}$, the M-NASA problem aims at discovering the anchor links among these n networks, i.e., the undirected anchor link sets $\mathcal{A}^{(1,2)}, \mathcal{A}^{(1,3)}, \ldots, \mathcal{A}^{(n-1,n)}$. Networks $G^{(1)}, G^{(2)}, \ldots, G^{(n)}$ are partially aligned and the constraint on anchor links in $\mathcal{A}^{(1,2)}, \mathcal{A}^{(1,3)}, \ldots, \mathcal{A}^{(n-1,n)}$ is *one-to-one*, which also follow the *transitivity law*.

Fig. 5.3 An example of multiple anonymized partially aligned social networks

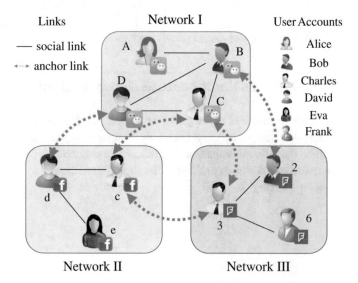

To solve the M-NASA problem, a novel network alignment framework UMA (Unsupervised Multi-network Alignment) proposed in [47] will be introduced in this section. UMA addresses the M-NASA problem with two steps: (1) unsupervised transitive anchor link inference across multi-networks, and (2) transitive multi-network matching to maintain the *one-to-one constraint*. In step (1), UMA infers a set of potential anchor links with unsupervised learning techniques by minimizing the *friendship inconsistency* and preserving the *alignment transitivity* property across networks. In step (2), UMA keeps the one-to-one constraint on anchor links by selecting those which can maximize the overall existence probabilities while maintaining the *matching transitivity* property at the same time. Next, we will introduce these two steps in great detail.

5.4.2 Unsupervised Multiple Network Alignment

In this part, we will introduce the first step of the UMA framework. We will first introduce the problem setting and objective function for pairwise network alignment based on matrix inference, which will be extended to the *multiple network* settings subject to the *transitivity law*. The integrated objective function can be addressed by relaxing the hard binary constraints, and the approximated solution will serve as the input for the second step of UMA to be introduced in Sect. 5.4.3.

5.4.2.1 Unsupervised Pairwise Network Alignment

Anchor links between any two given networks $G^{(i)}$ and $G^{(j)}$ actually define a *one-to-one* mapping (of users and social links) between $G^{(i)}$ and $G^{(j)}$. To evaluate the quality of different inferred mappings (i.e., the inferred anchor links), two new concepts of cross-network *Friendship Consistency/Inconsistency* will be introduced here. The optimal inferred anchor links are those which can maximize the *Friendship Consistency* (or minimize the *Friendship Inconsistency*) across networks.

As introduced in Sect. 5.3.4, given two partially aligned social networks $G^{(i)} = (\mathcal{U}^{(i)}, \mathcal{E}^{(i)})$ and $G^{(j)} = (\mathcal{U}^{(j)}, \mathcal{E}^{(j)})$, we can represent their corresponding *social adjacency* matrices to be $\mathbf{S}^{(i)} \in \mathbb{R}^{|\mathcal{U}^{(i)}| \times |\mathcal{U}^{(i)}|}$ and $\mathbf{S}^{(j)} \in \mathbb{R}^{|\mathcal{U}^{(j)}| \times |\mathcal{U}^{(j)}|}$, respectively.

Meanwhile, given the anchor link set $\mathcal{A}^{(i,j)} \subset \mathcal{U}^{(i)} \times \mathcal{U}^{(j)}$ between networks $G^{(i)}$ and $G^{(j)}$, the *binary transitional matrix* defined based on $\mathcal{A}^{(i,j)}$ can be represented as $\mathbf{T}^{(i,j)} \in \{0, 1\}^{|\mathcal{U}^{(i)}| \times |\mathcal{U}^{(j)}|}$, where $\mathbf{T}^{(i,j)}(l, m) = 1$ iff link $(u_l^{(i)}, u_m^{(j)}) \in \mathcal{A}^{(i,j)}$, $u_l^{(i)} \in \mathcal{U}^{(i)}$, $u_m^{(j)} \in \mathcal{U}^{(j)}$. The *binary transitional matrix* from $G^{(j)}$ to $G^{(i)}$ can be defined in a similar way, which can be represented as $\mathbf{T}^{(j,i)} \in \{0, 1\}^{|\mathcal{U}^{(j)}| \times |\mathcal{U}^{(i)}|}$, where $(\mathbf{T}^{(i,j)})^\top = \mathbf{T}^{(j,i)}$ as the anchor links between $G^{(i)}$ and $G^{(j)}$ are undirected. Considering that anchor links have an inherent *one-to-one* constraint, each row and each column of the *binary transitional matrices* $\mathbf{T}^{(i,j)}$ and $\mathbf{T}^{(j,i)}$ should have at most one entry filled with 1, which will pose a constraint on the inference space of potential *binary transitional matrices* $\mathbf{T}^{(i,j)}$ and $\mathbf{T}^{(j,i)}$.

Meanwhile, the *friendship inconsistency* can be defined as the number of non-shared social links between those mapped from $G^{(i)}$ and the original links in $G^{(j)}$. Based on the inferred *anchor transitional matrix* $\mathbf{T}^{(i,j)}$, the introduced *friendship inconsistency* between networks $G^{(i)}$ and $G^{(j)}$ can be represented as:

$$\left\| (\mathbf{T}^{(i,j)})^\top \mathbf{S}^{(i)} \mathbf{T}^{(i,j)} - \mathbf{S}^{(j)} \right\|_F^2, \tag{5.29}$$

where $\|\cdot\|_F$ denotes the Frobenius norm. And the optimal *binary transitional matrix* $\bar{\mathbf{T}}^{(i,j)}$, which can lead to the minimum *friendship inconsistency* can be represented as follows:

$$\bar{\mathbf{T}}^{(i,j)} = \arg\min_{\mathbf{T}^{(i,j)}} \left\| (\mathbf{T}^{(i,j)})^\top \mathbf{S}^{(i)} \mathbf{T}^{(i,j)} - \mathbf{S}^{(j)} \right\|_F^2,$$

$$s.t. \quad \mathbf{T}^{(i,j)} \in \{0,1\}^{|\mathcal{U}^{(i)}| \times |\mathcal{U}^{(j)}|},$$

$$\mathbf{T}^{(i,j)} \mathbf{1}^{|\mathcal{U}^{(j)}| \times 1} \preccurlyeq \mathbf{1}^{|\mathcal{U}^{(i)}| \times 1},$$

$$(\mathbf{T}^{(i,j)})^\top \mathbf{1}^{|\mathcal{U}^{(i)}| \times 1} \preccurlyeq \mathbf{1}^{|\mathcal{U}^{(j)}| \times 1}, \tag{5.30}$$

where the last two equations are added to maintain the *one-to-one* constraint on anchor links. The inequality $\mathbf{X} \preccurlyeq \mathbf{Y}$ holds iff $\mathbf{X}$ is of the same dimensions as $\mathbf{Y}$ and every entry in $\mathbf{X}$ is no greater than the corresponding entry in $\mathbf{Y}$.

5.4.2.2 Multiple Network Alignment with Transitivity Constraint

Isolated network alignment can work well in addressing the alignment problem of two social networks. However, in the M-NASA problem studied in this section, multiple social networks (more than two) social networks are to be aligned simultaneously. Besides minimizing the *friendship inconsistency* between each pair of networks, the *transitivity* property of anchor links also needs to be preserved in the transitional matrices inference.

The *transitivity* property should hold for the alignment of any n networks, where the minimum of n is 3. To help illustrate the *transitivity property* more clearly, we will use three network alignment as an example to introduce the M-NASA problem, which can be easily generalized to the case of n networks alignment. Let $G^{(i)}$, $G^{(j)}$, and $G^{(k)}$ be three social networks to be aligned concurrently. To accommodate the alignment results and preserve the *transitivity* property, a new *alignment transitivity penalty* is introduced as follows:

Definition 5.3 (Alignment Transitivity Penalty) Let $\mathbf{T}^{(i,j)}$, $\mathbf{T}^{(j,k)}$, and $\mathbf{T}^{(i,k)}$ be the inferred binary transitional matrices from $G^{(i)}$ to $G^{(j)}$, from $G^{(j)}$ to $G^{(k)}$ and from $G^{(i)}$ to $G^{(k)}$, respectively, among these three networks. The *alignment transitivity penalty* $C(\{G^{(i)}, G^{(j)}, G^{(k)}\})$ introduced by the inferred transitional matrices can be quantified as the number of inconsistent social links being mapped from $G^{(i)}$ to $G^{(k)}$ via two different alignment paths $G^{(i)} \to G^{(j)} \to G^{(k)}$ and $G^{(i)} \to G^{(k)}$, i.e.,

$$C(\{G^{(i)}, G^{(j)}, G^{(k)}\})$$

$$= \left\| (\mathbf{T}^{(j,k)})^\top (\mathbf{T}^{(i,j)})^\top \mathbf{S}^{(i)} \mathbf{T}^{(i,j)} \mathbf{T}^{(j,k)} - (\mathbf{T}^{(i,k)})^\top \mathbf{S}^{(i)} \mathbf{T}^{(i,k)} \right\|_F^2. \tag{5.31}$$

Alignment transitivity penalty is a general penalty concept and can be applied to n networks $\{G^{(1)}, G^{(2)}, \ldots, G^{(n)}\}$, $n \geq 3$ as well, which can be defined as the summation of penalty introduced by any three networks in the set, i.e.,

$$C(\{G^{(1)}, G^{(2)}, \ldots, G^{(n)}\})$$

$$= \sum_{\forall \{G^{(i)}, G^{(j)}, G^{(k)}\} \subset \{G^{(1)}, G^{(2)}, \ldots, G^{(n)}\}} C(\{G^{(i)}, G^{(j)}, G^{(k)}\}). \tag{5.32}$$

The optimal *binary transitional matrices* $\bar{\mathbf{T}}^{(i,j)}$, $\bar{\mathbf{T}}^{(j,k)}$, and $\bar{\mathbf{T}}^{(k,i)}$ which can minimize friendship inconsistency and the *alignment transitivity penalty* at the same time will be

$$\bar{\mathbf{T}}^{(i,j)}, \bar{\mathbf{T}}^{(j,k)}, \bar{\mathbf{T}}^{(k,i)}$$

$$= \arg\min\nolimits_{\mathbf{T}^{(i,j)}, \mathbf{T}^{(j,k)}, \mathbf{T}^{(k,i)}} \left\| (\mathbf{T}^{(i,j)})^\top \mathbf{S}^{(i)} \mathbf{T}^{(i,j)} - \mathbf{S}^{(j)} \right\|_F^2$$

$$+ \left\| (\mathbf{T}^{(j,k)})^\top \mathbf{S}^{(j)} \mathbf{T}^{(j,k)} - \mathbf{S}^{(k)} \right\|_F^2 + \left\| (\mathbf{T}^{(k,i)})^\top \mathbf{S}^{(k)} \mathbf{T}^{(k,i)} - \mathbf{S}^{(i)} \right\|_F^2$$

$$+ \alpha \cdot \left\| (\mathbf{T}^{(j,k)})^\top (\mathbf{T}^{(i,j)})^\top \mathbf{S}^{(i)} \mathbf{T}^{(i,j)} \mathbf{T}^{(j,k)} - \mathbf{T}^{(k,i)} \mathbf{S}^{(i)} (\mathbf{T}^{(k,i)})^\top \right\|_F^2$$

$$s.t. \ \mathbf{T}^{(i,j)} \in \{0, 1\}^{|\mathcal{U}^{(i)}| \times |\mathcal{U}^{(j)}|}, \ \mathbf{T}^{(j,k)} \in \{0, 1\}^{|\mathcal{U}^{(j)}| \times |\mathcal{U}^{(k)}|}$$

$$\mathbf{T}^{(k,i)} \in \{0, 1\}^{|\mathcal{U}^{(k)}| \times |\mathcal{U}^{(i)}|}$$

$$\mathbf{T}^{(i,j)} \mathbf{1}^{|\mathcal{U}^{(j)}| \times 1} \preccurlyeq \mathbf{1}^{|\mathcal{U}^{(i)}| \times 1}, (\mathbf{T}^{(i,j)})^\top \mathbf{1}^{|\mathcal{U}^{(i)}| \times 1} \preccurlyeq \mathbf{1}^{|\mathcal{U}^{(j)}| \times 1},$$

$$\mathbf{T}^{(j,k)} \mathbf{1}^{|\mathcal{U}^{(k)}| \times 1} \preccurlyeq \mathbf{1}^{|\mathcal{U}^{(j)}| \times 1}, (\mathbf{T}^{(j,k)})^\top \mathbf{1}^{|\mathcal{U}^{(j)}| \times 1} \preccurlyeq \mathbf{1}^{|\mathcal{U}^{(k)}| \times 1},$$

$$\mathbf{T}^{(k,i)} \mathbf{1}^{|\mathcal{U}^{(i)}| \times 1} \preccurlyeq \mathbf{1}^{|\mathcal{U}^{(k)}| \times 1}, (\mathbf{T}^{(k,i)})^\top \mathbf{1}^{|\mathcal{U}^{(k)}| \times 1} \preccurlyeq \mathbf{1}^{|\mathcal{U}^{(i)}| \times 1}, \quad (5.33)$$

where parameter α denotes the weight of the alignment transitivity penalty term.

5.4.2.3 Relaxation of the Optimization Problem

The objective function introduced in the previous subsection aims at obtaining the *hard* mappings among users across different networks and entries in all these *transitional matrices* are binary, which can lead to a fatal drawback: *hard assignment* can be neither possible nor realistic for networks with star structures as proposed in [25] and the hard subgraph isomorphism [27] is NP-hard.

To overcome such a problem, the UMA framework proposes to relax the binary constraint of entries in transitional matrices to allow them to be real values within range [0, 1]. Each entry in the transitional matrix represents a probability, denoting the confidence of certain user-user mapping across networks. Such a relaxation can make the *one-to-one* constraint no longer hold (which will be addressed with transitive network matching in the next subsection) as multiple entries in rows/columns of the transitional matrix can have non-zero values. To limit the existence of non-zero entries in the transitional matrices, we replace the one-to-one constraint, e.g.,

$$\mathbf{T}^{(k,i)} \mathbf{1}^{|\mathcal{U}^{(i)}| \times 1} \preccurlyeq \mathbf{1}^{|\mathcal{U}^{(k)}| \times 1}, (\mathbf{T}^{(k,i)})^\top \mathbf{1}^{|\mathcal{U}^{(k)}| \times 1} \preccurlyeq \mathbf{1}^{|\mathcal{U}^{(i)}| \times 1} \quad (5.34)$$

with *sparsity constraints*

$$\left\| \mathbf{T}^{(k,i)} \right\|_0 \leq t \quad (5.35)$$

instead, where term $\|\mathbf{T}\|_0$ denotes the L_0 norm of matrix $\mathbf{T}$, i.e., the number of non-zero entries in $\mathbf{T}$, and t is a small positive number to limit the non-zero entries in the matrix (i.e., the sparsity). Furthermore, the framework UMA adds term $\|\mathbf{T}\|_0$ to the minimization objective function, as it can be hard to determine the value of t in the constraint.

Based on the above relaxations, we can obtain the updated objective function to be

$$\bar{\mathbf{T}}^{(i,j)}, \bar{\mathbf{T}}^{(j,k)}, \bar{\mathbf{T}}^{(k,i)}$$

$$= \arg\min_{\mathbf{T}^{(i,j)}, \mathbf{T}^{(j,k)}, \mathbf{T}^{(k,i)}} \left\| (\mathbf{T}^{(i,j)})^\top \mathbf{S}^{(i)} \mathbf{T}^{(i,j)} - \mathbf{S}^{(j)} \right\|_F^2$$

$$+ \left\| (\mathbf{T}^{(j,k)})^\top \mathbf{S}^{(j)} \mathbf{T}^{(j,k)} - \mathbf{S}^{(k)} \right\|_F^2 + \left\| (\mathbf{T}^{(k,i)})^\top \mathbf{S}^{(k)} \mathbf{T}^{(k,i)} - \mathbf{S}^{(i)} \right\|_F^2$$

$$+ \alpha \cdot \left\| (\mathbf{T}^{(j,k)})^\top (\mathbf{T}^{(i,j)})^\top \mathbf{S}^{(i)} \mathbf{T}^{(i,j)} \mathbf{T}^{(j,k)} - \mathbf{T}^{(k,i)} \mathbf{S}^{(i)} (\mathbf{T}^{(k,i)})^\top \right\|_F^2$$

$$+ \beta \cdot \left\| \mathbf{T}^{(i,j)} \right\|_0 + \gamma \cdot \left\| \mathbf{T}^{(j,k)} \right\|_0 + \theta \cdot \left\| \mathbf{T}^{(k,i)} \right\|_0$$

$$s.t.\ \mathbf{0}^{|\mathcal{U}^{(i)}| \times |\mathcal{U}^{(j)}|} \preccurlyeq \mathbf{T}^{(i,j)} \preccurlyeq \mathbf{1}^{|\mathcal{U}^{(i)}| \times |\mathcal{U}^{(j)}|},$$

$$\mathbf{0}^{|\mathcal{U}^{(j)}| \times |\mathcal{U}^{(k)}|} \preccurlyeq \mathbf{T}^{(j,k)} \preccurlyeq \mathbf{1}^{|\mathcal{U}^{(j)}| \times |\mathcal{U}^{(k)}|},$$

$$\mathbf{0}^{|\mathcal{U}^{(k)}| \times |\mathcal{U}^{(i)}|} \preccurlyeq \mathbf{T}^{(k,i)} \preccurlyeq \mathbf{1}^{|\mathcal{U}^{(k)}| \times |\mathcal{U}^{(i)}|}, \tag{5.36}$$

which involves three variables $\mathbf{T}^{(i,j)}$, $\mathbf{T}^{(j,k)}$, and $\mathbf{T}^{(k,i)}$ simultaneously, obtaining the joint optimal solution for which at the same time is very hard and time consuming. The UMA framework proposes to address the above objective function by fixing two variables and updating the other variable alternatively with the gradient descent method [3]. As proposed in [25], if during the alternating updating steps, the entries of the transitional matrices become invalid (i.e., values less than 0 or greater than 1), UMA applies the projection technique introduced in [25] to adjust negative entries to 0 and change entries greater than 1 to 1 instead. With such a process, the updating equations of matrices $\mathbf{T}^{(i,j)}$, $\mathbf{T}^{(j,k)}$, $\mathbf{T}^{(k,i)}$ at step $t+1$ are given as follows:

$$\mathbf{T}^{(i,j)}(t+1)$$

$$= \mathbf{T}^{(i,j)}(t) - \eta^{(i,j)} \frac{\partial \mathcal{L}\left(\mathbf{T}^{(i,j)}(t), \mathbf{T}^{(j,k)}(t), \mathbf{T}^{(k,i)}(t), \beta, \gamma, \theta\right)}{\partial \mathbf{T}^{(i,j)}}, \tag{5.37}$$

$$\mathbf{T}^{(j,k)}(t+1)$$

$$= \mathbf{T}^{(j,k)}(t) - \eta^{(j,k)} \frac{\partial \mathcal{L}\left(\mathbf{T}^{(i,j)}(t+1), \mathbf{T}^{(j,k)}(t), \mathbf{T}^{(k,i)}(t), \beta, \gamma, \theta\right)}{\partial \mathbf{T}^{(j,k)}}, \tag{5.38}$$

$$\mathbf{T}^{(k,i)}(t+1)$$

$$= \mathbf{T}^{(k,i)}(t) - \eta^{(k,i)} \frac{\partial \mathcal{L}\left(\mathbf{T}^{(i,j)}(t+1), \mathbf{T}^{(j,k)}(t+1), \mathbf{T}^{(k,i)}(t), \beta, \gamma, \theta\right)}{\partial \mathbf{T}^{(k,i)}}. \tag{5.39}$$

In the updating equations, $\eta^{(i,j)}$, $\eta^{(j,k)}$, and $\eta^{(k,i)}$ are the learning steps in updating $\mathbf{T}^{(i,j)}$, $\mathbf{T}^{(j,k)}$, and $\mathbf{T}^{(k,i)}$, respectively. The Lagrangian function of the objective function can be represented as

$$
\begin{aligned}
\mathcal{L}(\mathbf{T}^{(i,j)}, \mathbf{T}^{(j,k)}, \mathbf{T}^{(k,i)}, \beta, \gamma, \theta) &= \left\| (\mathbf{T}^{(i,j)})^\top \mathbf{S}^{(i)} \mathbf{T}^{(i,j)} - \mathbf{S}^{(j)} \right\|_F^2 \\
&+ \left\| (\mathbf{T}^{(j,k)})^\top \mathbf{S}^{(j)} \mathbf{T}^{(j,k)} - \mathbf{S}^{(k)} \right\|_F^2 + \left\| (\mathbf{T}^{(k,i)})^\top \mathbf{S}^{(k)} \mathbf{T}^{(k,i)} - \mathbf{S}^{(i)} \right\|_F^2 \\
&+ \alpha \cdot \left\| (\mathbf{T}^{(j,k)})^\top (\mathbf{T}^{(i,j)})^\top \mathbf{S}^{(i)} \mathbf{T}^{(i,j)} \mathbf{T}^{(j,k)} - \mathbf{T}^{(k,i)} \mathbf{S}^{(i)} (\mathbf{T}^{(k,i)})^\top \right\|_F^2 \\
&+ \beta \cdot \left\| \mathbf{T}^{(i,j)} \right\|_0 + \gamma \cdot \left\| \mathbf{T}^{(j,k)} \right\|_0 + \theta \cdot \left\| \mathbf{T}^{(k,i)} \right\|_0.
\end{aligned}
\tag{5.40}
$$

Meanwhile, considering that $\|\cdot\|_0$ is not differentiable because of its discrete values [43], we will replace the $\|\cdot\|_0$ with the $\|\cdot\|_1$ instead (i.e., the sum of absolute values of all entries). Furthermore, as all the negative entries will be projected to 0, the L_1 norm of transitional matrix $\mathbf{T}$ can be represented as $\left\| \mathbf{T}^{(k,i)} \right\|_1 = \mathbf{1}^\top \mathbf{T}^{(k,i)} \mathbf{1}$ (i.e., the sum of all entries in the matrix). In addition, the Frobenius norm $\|\mathbf{X}\|_F^2$ can be represented with trace $\mathrm{Tr}(\mathbf{X}\mathbf{X}^\top)$.

The partial derivatives of function $\mathcal{L}$ with regard to $\mathbf{T}^{(i,j)}$, $\mathbf{T}^{(j,k)}$, and $\mathbf{T}^{(k,i)}$ will be:

$$
\begin{aligned}
(1) \ &\frac{\partial \mathcal{L}\left(\mathbf{T}^{(i,j)}, \mathbf{T}^{(j,k)}, \mathbf{T}^{(k,i)}, \beta, \gamma, \theta\right)}{\partial \mathbf{T}^{(i,j)}} \\
&= 2 \cdot \mathbf{S}^{(i)} \mathbf{T}^{(i,j)} (\mathbf{T}^{(i,j)})^\top (\mathbf{S}^{(i)})^\top \mathbf{T}^{(i,j)} \\
&+ 2 \cdot (\mathbf{S}^{(i)})^\top \mathbf{T}^{(i,j)} (\mathbf{T}^{(i,j)})^\top \mathbf{S}^{(i)} \mathbf{T}^{(i,j)} \\
&+ 2\alpha \cdot \mathbf{S}^{(i)} \mathbf{T}^{(i,j)} \mathbf{T}^{(j,k)} (\mathbf{T}^{(j,k)})^\top (\mathbf{T}^{(i,j)})^\top (\mathbf{S}^{(i)})^\top \mathbf{T}^{(i,j)} \mathbf{T}^{(j,k)} (\mathbf{T}^{(j,k)})^\top \\
&+ 2\alpha \cdot (\mathbf{S}^{(i)})^\top \mathbf{T}^{(i,j)} \mathbf{T}^{(j,k)} (\mathbf{T}^{(j,k)})^\top (\mathbf{T}^{(i,j)})^\top \mathbf{S}^{(i)} \mathbf{T}^{(i,j)} \mathbf{T}^{(j,k)} (\mathbf{T}^{(j,k)})^\top \\
&- 2 \cdot \mathbf{S}^{(i)} \mathbf{T}^{(i,j)} (\mathbf{S}^{(j)})^\top - 2 \cdot (\mathbf{S}^{(i)})^\top \mathbf{T}^{(i,j)} \mathbf{S}^{(j)} \\
&- 2\alpha \cdot (\mathbf{S}^{(i)})^\top \mathbf{T}^{(i,j)} \mathbf{T}^{(j,k)} \mathbf{T}^{(k,i)} \mathbf{S}^{(i)} (\mathbf{T}^{(k,i)})^\top (\mathbf{T}^{(j,k)})^\top \\
&- 2\alpha \cdot \mathbf{S}^{(i)} \mathbf{T}^{(i,j)} \mathbf{T}^{(j,k)} \mathbf{T}^{(k,i)} (\mathbf{S}^{(i)})^\top (\mathbf{T}^{(k,i)})^\top (\mathbf{T}^{(j,k)})^\top - \beta \cdot \mathbf{1}\mathbf{1}^\top.
\end{aligned}
\tag{5.41}
$$

$$
\begin{aligned}
(2) \ &\frac{\partial \mathcal{L}\left(\mathbf{T}^{(i,j)}, \mathbf{T}^{(j,k)}, \mathbf{T}^{(k,i)}, \beta, \gamma, \theta\right)}{\partial \mathbf{T}^{(j,k)}} \\
&= 2 \cdot \mathbf{S}^{(j)} \mathbf{T}^{(j,k)} (\mathbf{T}^{(j,k)})^\top (\mathbf{S}^{(j)})^\top \mathbf{T}^{(j,k)} \\
&+ 2 \cdot (\mathbf{S}^{(j)})^\top \mathbf{T}^{(j,k)} (\mathbf{T}^{(j,k)})^\top \mathbf{S}^{(j)} \mathbf{T}^{(j,k)} \\
&+ 2\alpha \cdot (\mathbf{T}^{(i,j)})^\top \mathbf{S}^{(i)} \mathbf{T}^{(i,j)} \mathbf{T}^{(j,k)} (\mathbf{T}^{(j,k)})^\top (\mathbf{T}^{(i,j)})^\top (\mathbf{S}^{(i)})^\top \mathbf{T}^{(i,j)} \mathbf{T}^{(j,k)} \\
&+ 2\alpha \cdot (\mathbf{T}^{(i,j)})^\top (\mathbf{S}^{(i)})^\top \mathbf{T}^{(i,j)} \mathbf{T}^{(j,k)} (\mathbf{T}^{(j,k)})^\top (\mathbf{T}^{(i,j)})^\top \mathbf{S}^{(i)} \mathbf{T}^{(i,j)} \mathbf{T}^{(j,k)} \\
&- 2 \cdot \mathbf{S}^{(j)} \mathbf{T}^{(j,k)} (\mathbf{S}^{(k)})^\top - 2 \cdot (\mathbf{S}^{(j)})^\top \mathbf{T}^{(j,k)} \mathbf{S}^{(k)} \\
&- 2\alpha \cdot (\mathbf{T}^{(i,j)})^\top (\mathbf{S}^{(i)})^\top \mathbf{T}^{(i,j)} \mathbf{T}^{(j,k)} \mathbf{T}^{(k,i)} \mathbf{S}^{(i)} (\mathbf{T}^{(k,i)})^\top \\
&- 2\alpha \cdot (\mathbf{T}^{(i,j)})^\top \mathbf{S}^{(i)} \mathbf{T}^{(i,j)} \mathbf{T}^{(j,k)} \mathbf{T}^{(k,i)} (\mathbf{S}^{(i)})^\top (\mathbf{T}^{(k,i)})^\top - \gamma \cdot \mathbf{1}\mathbf{1}^\top.
\end{aligned}
\tag{5.42}
$$

$$(3) \frac{\partial \mathcal{L}\left(\mathbf{T}^{(i,j)}, \mathbf{T}^{(j,k)}, \mathbf{T}^{(k,i)}, \beta, \gamma, \theta\right)}{\partial \mathbf{T}^{(k,i)}}$$

$$= 2 \cdot \mathbf{S}^{(k)} \mathbf{T}^{(k,i)} (\mathbf{T}^{(k,i)})^\top (\mathbf{S}^{(k)})^\top \mathbf{T}^{(k,i)}$$

$$+ 2 \cdot (\mathbf{S}^{(k)})^\top \mathbf{T}^{(k,i)} (\mathbf{T}^{(k,i)})^\top \mathbf{S}^{(k)} \mathbf{T}^{(k,i)}$$

$$+ 2\alpha \mathbf{T}^{(k,i)} (\mathbf{S}^{(i)})^\top (\mathbf{T}^{(k,i)})^\top \mathbf{T}^{(k,i)} \mathbf{S}^{(i)}$$

$$+ 2\alpha \mathbf{T}^{(k,i)} \mathbf{S}^{(i)} (\mathbf{T}^{(k,i)})^\top \mathbf{T}^{(k,i)} (\mathbf{S}^{(i)})^\top$$

$$- 2 \cdot \mathbf{S}^{(k)} \mathbf{T}^{(k,i)} (\mathbf{S}^{(i)})^\top - 2 \cdot (\mathbf{S}^{(k)})^\top \mathbf{T}^{(k,i)} \mathbf{S}^{(i)}$$

$$- 2\alpha \cdot (\mathbf{T}^{(j,k)})^\top (\mathbf{T}^{(i,j)})^\top (\mathbf{S}^{(i)})^\top \mathbf{T}^{(i,j)} \mathbf{T}^{(j,k)} \mathbf{T}^{(k,i)} \mathbf{S}^{(i)}$$

$$- 2\alpha \cdot (\mathbf{T}^{(j,k)})^\top (\mathbf{T}^{(i,j)})^\top \mathbf{S}^{(i)} \mathbf{T}^{(i,j)} \mathbf{T}^{(j,k)} \mathbf{T}^{(k,i)} (\mathbf{S}^{(i)})^\top - \theta \cdot \mathbf{1} \mathbf{1}^\top. \qquad (5.43)$$

Such an iterative updating process will stop when all *transitional matrices* converge. The achieved variable matrices $\mathbf{T}^{(i,j)}$, $\mathbf{T}^{(j,k)}$, and $\mathbf{T}^{(k,i)}$ will serve as the input for the *transitive network matching* to be introduced in the following subsection to maintain both the *one-to-one* constraint and the *transitivity law* constraint.

5.4.3 Transitive Network Matching

The constraint relaxation in the previous section actually violates the *one-to-one* property on anchor links seriously, since each node in a network can be connected by multiple anchor links with different inferred scores in matrices $\mathbf{T}^{(i,j)}$, $\mathbf{T}^{(j,k)}$, and $\mathbf{T}^{(k,i)}$. To resolve such a problem, UMA adopts the *transitive network matching* step as proposed in [47] to prune the introduced redundant anchor links. The matching results (i.e., selected anchor links) need to meet both the *one-to-one* constraint and the *transitivity property*.

Formally, given two networks $G^{(i)}$ and $G^{(j)}$, each potential anchor link, e.g., $(u_l^{(i)}, u_m^{(j)})$, between $G^{(i)}$ and $G^{(j)}$ is associated with a binary variable $x_{l,m}^{(i,j)} \in \{0, 1\}$ to denote whether anchor link $(u_l^{(i)}, u_m^{(j)})$ is selected or not in the matching, where

$$x_{l,m}^{(i,j)} = \begin{cases} 1 & \text{if } \left(u_l^{(i)}, u_m^{(j)}\right) \text{ is selected,} \\ 0, & \text{otherwise.} \end{cases} \qquad (5.44)$$

For each user in network $G^{(i)}$, e.g., $u_l^{(i)} \in \mathcal{U}^{(i)}$, at most one potential anchor link attached to him/her will be selected in the final alignment result with another network, e.g., $G^{(j)}$ (or $G^{(k)}$). So, based on the introduced binary variables, the *one-to-one* constraint on anchor links between networks $G^{(i)}$ and $G^{(j)}$ as well as between networks $G^{(i)}$ and $G^{(k)}$ can be represented as follows:

$$\sum_{u_m^{(j)} \in \mathcal{U}^{(j)}} x_{l,m}^{(i,j)} \le 1, \quad \sum_{u_o^{(k)} \in \mathcal{U}^{(k)}} x_{l,o}^{(i,k)} \le 1, \forall u_l^{(i)} \in \mathcal{U}^{(i)}. \qquad (5.45)$$

Similarly, we can also define the binary variables $x_{m,o}^{(j,k)}, x_{o,l}^{(k,i)} \in \{0, 1\}$ and the corresponding *one-to-one* constraints for potential anchor links $(u_m^{(j)}, u_o^{(k)})$ and $(u_o^{(k)}, u_l^{(i)})$ between networks $G^{(j)}, G^{(k)}$ and between networks $G^{(k)}, G^{(i)}$, respectively, to represent whether these links are selected or not.

According to the definition of "transitivity law," if anchor links $(u_l^{(i)}, u_m^{(j)})$ and $(u_m^{(j)}, u_o^{(k)})$ are selected $\forall l \in \{1, 2, \ldots, |\mathcal{U}^{(i)}|\}, m \in \{1, 2, \ldots, |\mathcal{U}^{(j)}|\}, o \in \{1, 2, \ldots, \mathcal{U}^{(k)}|\}$ in matching networks $G^{(i)}, G^{(j)}$ and networks $G^{(j)}, G^{(k)}$, then anchor link $(u_o^{(k)}, u_l^{(i)})$ should be selected as well in the matching of networks $G^{(k)}, G^{(i)}$, i.e., $x_{o,l}^{(k,i)} = 1$. In other words, in three networks matching, the case that only two variables in $\{x_{l,m}^{(i,j)}, x_{m,o}^{(j,k)}, x_{o,l}^{(k,i)}\}$ are assigned with value 1 while the remaining one is 0 cannot hold in the final matching results, i.e.,

$$x_{l,m}^{(i,j)} + x_{m,o}^{(j,k)} + x_{o,l}^{(k,i)} \neq 2, \forall l \in \{1, 2, \ldots, |\mathcal{U}^{(i)}|\},$$

$$\forall m \in \{1, 2, \ldots, |\mathcal{U}^{(j)}|\}, \forall o \in \{1, 2, \ldots, \mathcal{U}^{(k)}|, \quad (5.46)$$

which is called the *matching transitivity constraint*. The *matching transitivity constraint* can be easily generalized to the case of matching n ($n \geq 3$) networks.

Definition 5.4 (Matching Transitivity Constraint) Let $\mathcal{G} = \{G^{(1)}, G^{(2)}, \ldots, G^{(n)}\}$ be a set of n networks, the *matching transitivity constraint* (MTC) for matching these n networks in $\mathcal{G}$ can be defined recursively as follows:

$$\text{MTC}(\mathcal{G}) = \left\{ \sum x_{\mathcal{G}} \neq |\mathcal{G}| - 1 \right\} \cup \left\{ \bigcup_{\mathcal{G}' \subset \mathcal{G}, |\mathcal{G}'| = |\mathcal{G}| - 1} \text{MTC}(\mathcal{G}') \right\}. \quad (5.47)$$

In the above equation, the variable summation term $\sum x_{\mathcal{G}} = x_{l,m}^{(1,2)} + x_{m,o}^{(2,3)} + \cdots + x_{p,l}^{(n,1)}, \forall l \in \{1, 2, \ldots, |\mathcal{U}^{(1)}|\}, \forall m \in \{1, 2, \ldots, |\mathcal{U}^{(2)}|\}, \ldots, \forall p \in \{1, 2, \ldots, |\mathcal{U}^{(n)}|\}$ represents the transitivity constraint involving all these n networks.

The final selected anchor links should be those with high confidence scores in the inferred *transitional matrices* but also can meet the *one-to-one matching* constraint and *matching transitivity* constraint simultaneously. The *transitive network matching* can be formulated as the following optimization problem:

$$\max_{\mathbf{x}^{(i,j)}, \mathbf{x}^{(j,k)}, \mathbf{x}^{(k,i)}} \sum_{l,m} x_{l,m}^{(i,j)} \mathbf{T}^{(i,j)}(l, m) + \sum_{l,m} x_{l,m}^{(i,j)} \mathbf{T}^{(i,j)}(l, m)$$

$$+ \sum_{l,m} x_{l,m}^{(i,j)} \mathbf{T}^{(i,j)}(l, m),$$

$$s.t. \sum_{u_m^{(j)} \in \mathcal{U}^{(j)}} x_{l,m}^{(i,j)} \leq 1, \sum_{u_o^{(k)} \in \mathcal{U}^{(k)}} x_{l,o}^{(i,k)} \leq 1, \forall u_l^{(i)} \in \mathcal{U}^{(i)},$$

$$\sum_{u_l^{(i)} \in \mathcal{U}^{(i)}} x_{m,l}^{(j,i)} \leq 1, \sum_{u_o^{(k)} \in \mathcal{U}^{(k)}} x_{m,o}^{(j,k)} \leq 1, \forall u_m^{(j)} \in \mathcal{U}^{(j)},$$

$$\sum_{u_l^{(i)} \in \mathcal{U}^{(i)}} x_{o,l}^{(k,i)} \leq 1, \quad \sum_{u_m^{(j)} \in \mathcal{U}^{(j)}} x_{o,m}^{(k,j)} \leq 1, \forall u_o^{(k)} \in \mathcal{U}^{(k)},$$

$$x_{l,m}^{(i,j)} + x_{m,o}^{(j,k)} + x_{o,l}^{(k,i)} \neq 2, \forall l \in \{1, 2, \ldots, |\mathcal{U}^{(i)}|\},$$

$$\forall m \in \{1, 2, \ldots, |\mathcal{U}^{(j)}|\}, \forall o \in \{1, 2, \ldots, \mathcal{U}^{(k)}|,$$

$$x_{l,m}^{(i,j)} \in \{0, 1\}, \forall u_l^{(i)} \in \mathcal{U}^{(i)}, u_m^{(j)} \in \mathcal{U}^{(j)}.$$

$$x_{m,o}^{(j,k)} \in \{0, 1\}, \forall u_m^{(j)} \in \mathcal{U}^{(j)}, u_o^{(k)} \in \mathcal{U}^{(k)}.$$

$$x_{o,l}^{(k,i)} \in \{0, 1\}, \forall u_o^{(k)} \in \mathcal{U}^{(k)}, u_l^{(i)} \in \mathcal{U}^{(i)}. \tag{5.48}$$

In the above objective function, the matching transitivity constraint $x_{l,m}^{(i,j)} + x_{m,o}^{(j,k)} + x_{o,l}^{(k,i)} \neq 2$ is actually non-convex, which can be another challenge in addressing the function. In [47], framework UMA proposes to (1) remove the matching transitivity constraint from the objective function, and (2) apply the matching transitivity constraint to post-process the solution (obtained from the objective function without the constraint).

The objective function (with the matching transitivity constraint removed) can be solved with open source optimization toolkit, e.g., Scipy.Optimization[6] and GLPK,[7] and we will not describe how to solve in detail due to the limited space. Among all the obtained solutions, we can check all the links whose corresponding variables meeting $x_{l,m}^{(i,j)} + x_{m,o}^{(j,k)} + x_{o,l}^{(k,i)} = 2$ and assign the variable with value 0 with 1 instead. For example, for three given variables $x_{l,m}^{(i,j)}$, $x_{m,o}^{(j,k)}$ and $x_{o,l}^{(k,i)}$, if $x_{l,m}^{(i,j)} = x_{m,o}^{(j,k)} = 1$ but $x_{o,l}^{(k,i)} = 0$, we will assign $x_{o,l}^{(k,i)}$ with new value 1 and $x_{o,x}^{(k,i)} = 0, \forall x \neq l$, $x_{x,l}^{(k,i)} = 0, \forall x \neq o$ to preserve the matching transitivity constraint.

5.5   Heterogeneous Network Co-alignment

The real-world social networks usually contain heterogeneous information, including both very complex network structures and different categories of attribute information, which can all provide extra signals for inferring the potential anchor links across networks. Besides these common users, social networks offering similar services can also share other common information entities, e.g., locations shared between Foursquare and Yelp, and products sold in both Amazon and Ebay. To distinguish the anchor links between different types of information entities, those aligning the common users are called the *user anchor links*, while those connecting locations (or products) are called the *location anchor links* (or *product anchor links*).

In this section, we will study the problem to infer different categories of anchor links connecting various anchor instances across social networks simultaneously, which is formally defined as the network "<u>P</u>artial <u>C</u>o-alignmen<u>T</u>" (PCT) problem. PCT is a general research problem and can be applied to different types of social networks, like Foursquare and Yelp, Amazon and eBay. In this section, we will take online social networks as an example, and will mainly focus on the partial co-alignment of location based social networks via shared users and locations with the various connection and attribute information available in the networks.

---

[6]http://docs.scipy.org/doc/scipy/reference/optimize.html.

[7]http://www.gnu.org/software/glpk/.

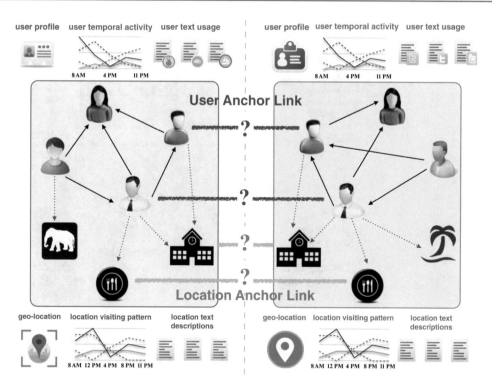

user profile user temporal activity user text usage   user profile user temporal activity user text usage

Fig. 5.4 An example of the PCT problem

5.5.1 Network Co-alignment Problem Description

Example 5.7 As shown in Fig. 5.4, we have two location based online social networks. In these social networks, users can check-in at the locations of their interest. Besides the connections among users and those between users and locations, both the users and locations are associated with different types of attribute information as well. For instance, for the users, we can obtain their profile information, the temporal activities information, and the posts written by them. As introduced before, these different types of attribute information can reveal the personal characteristics and is helpful for identifying the shared common users. Similarly, for the locations, we can have the profile information of them, and can also accumulate the timestamps and words from the posts attaching check-ins at these locations.

The timestamps and words of posts written at these places can reveal the characteristics of the locations as well. For instance, as illustrated in Fig. 5.5, we have two totally different locations: the Lincoln Park Zoo[8] and Scarlet Bar.[9] The Lincoln Park Zoo is the largest free zoo in Chicago and is open during 10:00 AM–5:00 PM. The Scarlet Bar is one of the most famous bars in Chicago, where people can drink with friends, dance to enjoy their night life, and it is open during 8:00 PM–2:00 AM.

We also have four online posts published by people at these two places in either Foursquare or Twitter. From the content of these posts, we find that people usually publish words about animals, pictures, and the scene at the Lincoln Park Zoo. However, people who visit the Scarlet Bar mainly talk about the atmosphere in the bar, the drinks, the dance floor, and the music there. Meanwhile, we can also accumulate the timestamps of posts published at these two places. The timestamps of posts

---

[8]http://www.lpzoo.org.

[9]http://www.scarletbarchicago.com.

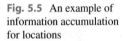

Fig. 5.5 An example of information accumulation for locations

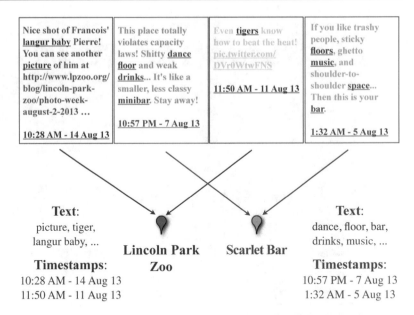

published at the Lincoln Park Zoo are mostly during the daytime, while those of posts published at the Scarlet Bar are at night. Such accumulated information can serve as the profile information of these locations, respectively.

Based on the above example and descriptions, we can define the PCT problem as follows. For any two given social networks $G^{(1)}$ and $G^{(2)}$, with the link and attribute information in both $G^{(1)}$ and $G^{(2)}$, the PCT problem aims at inferring the potential anchor links between users and locations across $G^{(1)}$ and $G^{(2)}$, respectively. To differentiate the user anchor links and location anchor links, we use sets $\mathcal{A}_u^{(1,2)}$ and $\mathcal{A}_l^{(1,2)}$ to represent these two types of anchor links between networks $G^{(1)}$ and $G^{(2)}$, respectively. To address the problem, in this section, we will introduce UNICOAT model proposed in [48].

5.5.2   Anchor Link Co-inference

As introduced in the problem definition, anchor set $\mathcal{A}_u^{(1,2)}$ between networks $G^{(1)}$ and $G^{(2)}$ actually maps users between networks $G^{(1)}$ and $G^{(2)}$. Considering that users in different social networks are associated with both links and attribute information, the quality of the inferred anchor links $\mathcal{A}_u^{(1,2)}$ can be measured by the costs introduced by such mappings calculated with users' link and attribute information, i.e.,

$$cost(\mathcal{A}_u^{(1,2)}) = \text{cost in links } (\mathcal{A}_u^{(1,2)}) + \alpha \cdot \text{cost in attributes}(\mathcal{A}_u^{(1,2)}), \qquad (5.49)$$

where α denotes the weight of the cost obtained from the attribute information (α is set as 1 in the experiments for simplicity, i.e., the link and attribute information is treated to be of the same importance). Considering that locations are also attached with link and attributes, similar cost function can be defined for the inferred location anchor links in $\mathcal{A}_l^{(1,2)}$:

$$cost(\mathcal{A}_l^{(1,2)}) = \text{cost in links } (\mathcal{A}_l^{(1,2)}) + \alpha \cdot \text{cost in attributes}(\mathcal{A}_l^{(1,2)}). \qquad (5.50)$$

The optimal user and location anchor links $(\mathcal{A}_u^{(1,2)})^*$ and $(\mathcal{A}_l^{(1,2)})^*$ to be inferred in the PCT problem that minimize the cost functions can be represented as

$$(\mathcal{A}_u^{(1,2)})^*, (\mathcal{A}_l^{(1,2)})^* = \arg \min_{\mathcal{A}_u^{(1,2)}, \mathcal{A}_l^{(1,2)}} cost(\mathcal{A}_u^{(1,2)}) + cost(\mathcal{A}_l^{(1,2)}). \tag{5.51}$$

To resolve the objective function, in the following parts of this section, we will introduce the isolated user anchor link inference in Sect. 5.5.2.1, the isolated location anchor link inference in Sect. 5.5.2.2, and the joint co-inference framework of user and location anchor links in Sect. 5.5.3.

5.5.2.1 User Anchor Links Inference

Social connections among users clearly illustrate the social community structures of users in online social networks. Meanwhile, attribute information (e.g., profile information, text usage patterns, temporal activities) can reveal users' unique personal characteristics. Common users in different networks tend to form similar community structures [46] and have very close personal characteristics [45]. As a result, link and attribute information about the users both play very important roles in inferring potential user anchor links across networks. In this part, we will introduce how to use such information to improve the user anchor link inference results.

User Anchor Link Inference with Link Information
Similar to the matrix inference based network alignment approach introduced in Sect. 5.3.4, based on the social links among users in both $G^{(1)}$ and $G^{(2)}$ (i.e., $\mathcal{E}_{u,u}^{(1)}$ and $\mathcal{E}_{u,u}^{(2)}$, respectively), we can construct the binary *social adjacency matrices* [32] $\mathbf{S}^{(1)} \in \mathbb{R}^{|\mathcal{U}^{(1)}| \times |\mathcal{U}^{(1)}|}$ and $\mathbf{S}^{(2)} \in \mathbb{R}^{|\mathcal{U}^{(2)}| \times |\mathcal{U}^{(2)}|}$ for networks $G^{(1)}$ and $G^{(2)}$, respectively. Meanwhile, via the inferred user anchor links $\mathcal{A}_u^{(1,2)}$, users as well as their social connections can be mapped between networks $G^{(1)}$ and $G^{(2)}$. We can represent the inferred user anchor links $\mathcal{A}_u^{(1,2)}$ with binary *user transitional matrix* $\mathbf{P} \in \mathbb{R}^{|\mathcal{U}^{(1)}| \times |\mathcal{U}^{(2)}|}$, where the ($i$th, lth) entry $P(i, l) = 1$ iff link $(u_i^{(1)}, u_l^{(2)}) \in \mathcal{A}_u^{(1,2)}$. Considering that the constraint on user anchor links is *one-to-one*, each column and each row of $\mathbf{P}$ can contain at most one entry being assigned with value 1, i.e.,

$$\mathbf{P}\mathbf{1}^{|\mathcal{U}^{(2)}| \times 1} \preccurlyeq \mathbf{1}^{|\mathcal{U}^{(1)}| \times 1}, \ \mathbf{P}^\top \mathbf{1}^{|\mathcal{U}^{(1)}| \times 1} \preccurlyeq \mathbf{1}^{|\mathcal{U}^{(2)}| \times 1}. \tag{5.52}$$

The optimal user anchor links are those which can minimize the inconsistency of mapped social links across networks and the cost introduced by the inferred user anchor link set $\mathcal{A}_u^{(1,2)}$ with the link information can be represented as

$$\text{cost in link}(\mathcal{A}_u^{(1,2)}) = \text{cost in link}(\mathbf{P}) = \left\| \mathbf{P}^\top \mathbf{S}^{(1)} \mathbf{P} - \mathbf{S}^{(2)} \right\|_F^2, \tag{5.53}$$

where $\|\cdot\|_F$ denotes the Frobenius norm of the corresponding matrix.

User Anchor Link Inference with Attribute Information
Besides social links, users in social networks can be associated with a set of attributes, which can provide extra hints for identifying the correspondence relationships about users across networks. In this part, we will introduce the method to infer the user anchor links with attribute information, which includes *username information*, *text usage patterns*, and *temporal activity information*.

Username that can differentiate users from each other in online social networks is like their online ID, which is an important factor in inferring potential anchor links. Let $(u_i^{(1)}, u_l^{(2)})$ be a potential anchor link between $G^{(1)}$ and $G^{(2)}$, the usernames of $u_i^{(1)}$ and $u_l^{(2)}$ can be represented as two sets of characters $n(u_i^{(1)})$ and $n(u_l^{(2)})$, respectively, based on which various metrics introduced in Sect. 5.2.1 can be applied to measure the similarity between $u_i^{(1)}$ and $u_l^{(2)}$. For instance, by adopting the *token based Jaccard's coefficient* measure, we can compute the user similarity score between users $u_i^{(1)}$ and $u_j^{(2)}$ as follows:

$$sim\left(n\left(u_i^{(1)}\right), n\left(u_l^{(2)}\right)\right) = \frac{|n\left(u_i^{(1)}\right) \cap n\left(u_l^{(2)}\right)|}{|n\left(u_i^{(1)}\right) \cup n\left(u_l^{(2)}\right)|}. \tag{5.54}$$

Users usually have their unique active temporal patterns in online social networks [24]. For example, some users like to socialize with their online friends in the early morning, but some may prefer to do so in the evening after work. Users' online active time can be extracted based on their post publishing timestamps effectively. Let $\mathbf{t}(u_i^{(1)})$ and $\mathbf{t}(u_l^{(2)})$ be the normalized temporal activity distribution vectors of users $u_i^{(1)}$ and $u_l^{(2)}$, which are both of length 24. Entries of $\mathbf{t}(u_i^{(1)})$ and $\mathbf{t}(u_l^{(2)})$ contain the ratios of posts being published at the corresponding hour in a day. For example, $\mathbf{t}(u_i^{(1)})(3)$ denotes the ratio of all posts written by $u_i^{(1)}$ at 3 AM. Based on vectors $\mathbf{t}(u_i^{(1)})$ and $\mathbf{t}(u_l^{(2)})$, we can calculate the inner product of the temporal distribution vectors [24] as the similarity scores between $u_i^{(1)}$ and $u_l^{(2)}$ in their temporal activity patterns, i.e.,

$$sim\left(\mathbf{t}\left(u_i^{(1)}\right), \mathbf{t}\left(u_l^{(2)}\right)\right) = \mathbf{t}\left(u_i^{(1)}\right)^{\top} \mathbf{t}\left(u_l^{(2)}\right). \tag{5.55}$$

Besides profile and online activity temporal distribution information, people normally have very different text usage habits online [45], which can reveal personal unique characteristics and can be applied in inferring the user anchor links across networks. We represent the text content used by users $u_i^{(1)}$ and $u_l^{(2)}$ as bag-of-words vectors [24], $\mathbf{w}(u_i^{(1)})$ and $\mathbf{w}(u_l^{(2)})$, weighted by TF-IDF [23], respectively. Commonly used text similarity measure: *Cosine similarity* [10] can be applied to measure the similarities in text usage patterns between $u_i^{(1)}$ and $u_l^{(2)}$, i.e.,

$$sim\left(\mathbf{w}\left(u_i^{(1)}\right), \mathbf{w}\left(u_l^{(2)}\right)\right) = \frac{\mathbf{w}\left(u_i^{(1)}\right)^{\top} \cdot \mathbf{w}\left(u_l^{(2)}\right)}{\left\|\mathbf{w}\left(u_i^{(1)}\right)\right\| \cdot \left\|\mathbf{w}\left(u_l^{(2)}\right)\right\|}. \tag{5.56}$$

With these different attribute information (i.e., username, temporal activity, and text content), we can calculate the similarities between users across networks $G^{(1)}$ and $G^{(2)}$. We represent such similarity matrix as $\mathbf{\Lambda} \in \mathbb{R}^{|\mathcal{U}^{(1)}| \times |\mathcal{U}^{(2)}|}$, where entry $\mathbf{\Lambda}(i, l)$ is the similarity between $u_i^{(1)}$ and $u_l^{(2)}$. $\mathbf{\Lambda}(i, l)$ can be represented as a combination of $sim(n(u_i^{(1)}), n(u_l^{(2)}))$, $sim(\mathbf{t}(u_i^{(1)}), \mathbf{t}(u_l^{(2)}))$, and $sim(\mathbf{w}(u_i^{(1)}), \mathbf{w}(u_l^{(2)}))$ and linear combination is used here due to its simplicity and wide usages. The optimal weights of similarity scores calculated with different attribute information can be learned from the data theoretically, but it will make the model too complicated. To focus on the co-alignment problem itself, we can assume they are all of the same importance and propose to assign them with the same weight for simplicity. In other words, we have

$$\mathbf{\Lambda}(i, l) = \frac{1}{3}\left(sim\left(n\left(u_i^{(1)}\right), n\left(u_l^{(2)}\right)\right) + sim\left(\mathbf{t}\left(u_i^{(1)}\right), \mathbf{t}\left(u_l^{(2)}\right)\right)\right.$$
$$\left. + sim\left(\mathbf{w}\left(u_i^{(1)}\right), \mathbf{w}\left(u_l^{(2)}\right)\right)\right). \tag{5.57}$$

Similar users across social networks are more likely to be the same user and user anchor links in $\mathcal{A}_u^{(1,2)}$ that align similar users together should lead to lower cost. The cost function introduced by the inferred user anchor links $\mathcal{A}_u^{(1,2)}$ with attribute information can be represented as

$$\text{cost in attribute}(\mathcal{A}_u^{(1,2)}) = \text{cost in attribute}(\mathbf{P}) = -\|\mathbf{P} \circ \mathbf{\Lambda}\|_1, \tag{5.58}$$

where $\|\cdot\|_1$ is the L_1 norm [33] of the corresponding matrix, entry $(\mathbf{P} \circ \mathbf{\Lambda})(i, l)$ can be represented as $\mathbf{P}(i, l) \cdot \mathbf{\Lambda}(i, l)$ and $\mathbf{P} \circ \mathbf{\Lambda}$ denotes the Hadamard product [9] of matrices $\mathbf{P}$ and $\mathbf{\Lambda}$.

User Anchor Link Inference with Link and Attribute Information
Both link and attribute information can be very important for user anchor link inference. By taking these two categories of information into considerations simultaneously, the cost introduced by the inferred user anchor link set $\mathcal{A}_u^{(1,2)}$ can be represented as

$$cost(\mathcal{A}_u^{(1,2)}) = \text{cost in link}(\mathcal{A}_u^{(1,2)}) + \alpha \cdot \text{cost in attribute}(\mathcal{A}_u^{(1,2)})$$
$$= \left\|\mathbf{P}^\top \mathbf{S}^{(1)} \mathbf{P} - \mathbf{S}^{(2)}\right\|_F^2 - \alpha \cdot \|\mathbf{P} \circ \mathbf{\Lambda}\|_1. \tag{5.59}$$

The optimal *user transitional matrix* $\mathbf{P}^*$ which can lead to the minimum cost will be achieved by addressing the following objective function:

$$\mathbf{P}^* = \arg\min_{\mathbf{P}} cost(\mathcal{A}_u^{(1,2)})$$
$$= \arg\min_{\mathbf{P}} \left\|\mathbf{P}^\top \mathbf{S}^{(1)} \mathbf{P} - \mathbf{S}^{(2)}\right\|_F^2 - \alpha \cdot \|\mathbf{P} \circ \mathbf{\Lambda}\|_1$$
$$s.t. \quad \mathbf{P} \in \{0, 1\}^{|\mathcal{U}^{(1)}| \times |\mathcal{U}^{(2)}|},$$
$$\mathbf{P}\mathbf{1}^{|\mathcal{U}^{(2)}| \times 1} \preccurlyeq \mathbf{1}^{|\mathcal{U}^{(1)}| \times 1}, \mathbf{P}^\top \mathbf{1}^{|\mathcal{U}^{(1)}| \times 1} \preccurlyeq \mathbf{1}^{|\mathcal{U}^{(2)}| \times 1}. \tag{5.60}$$

5.5.2.2 Location Anchor Links Inference
Similar to users, locations in online social networks are also associated with both link and attribute information (like the location links between users and locations, profile information and text descriptions about the locations, as well as the (longitude, latitude) coordinate information). The (longitude, latitude) pairs of the same location in different networks are usually not identical and various nearby locations can have very close coordinates, which pose great challenges in differentiating the locations from each other.

Location Anchor Link Inference with Link Information
Let $\mathcal{L}^{(1)}$ and $\mathcal{L}^{(2)}$ be the sets of locations in networks $G^{(1)}$ and $G^{(2)}$, respectively. Based on the location links between users and locations in networks $G^{(1)}$ and $G^{(2)}$ (i.e., $\mathcal{E}_{u,l}^{(1)}$ and $\mathcal{E}_{u,l}^{(2)}$), we can construct the binary *location adjacency matrices* $\mathbf{L}^{(1)} \in \mathbb{R}^{|\mathcal{U}^{(1)}| \times |\mathcal{L}^{(1)}|}$ and $\mathbf{L}^{(2)} \in \mathbb{R}^{|\mathcal{U}^{(2)}| \times |\mathcal{L}^{(2)}|}$ for networks $G^{(1)}$ and $G^{(2)}$, respectively. Entries in $\mathbf{L}^{(1)}$ and $\mathbf{L}^{(1)}$ $\left(\text{e.g., } \mathbf{L}^{(1)}(i, j) \text{ and } \mathbf{L}^{(2)}(l, m)\right)$ are filled with value 1 iff user $u_i^{(1)}$ has visited location $l_j^{(1)}$ in $G^{(1)}$ and user $u_i^{(2)}$ has visited location $l_m^{(2)}$ in $G^{(2)}$.

Besides the *user transitional matrix* $\mathbf{P}$ which maps users between $G^{(1)}$ and $G^{(2)}$, we can also construct the binary *location transitional matrix* $\mathbf{Q} \in \{0, 1\}^{|\mathcal{L}^{(1)}| \times |\mathcal{L}^{(2)}|}$ based on the inferred location anchor link set $\mathcal{A}_l^{(1,2)}$, which maps locations between $G^{(1)}$ and $G^{(2)}$. The cost introduced by the inferred location anchor link set $\mathcal{A}_l^{(1,2)}$ can be defined as the number of mis-mapped location links across networks, i.e.,

$$\text{cost in link}(\mathcal{A}_l^{(1,2)}) = \left\| \mathbf{P}^\top \mathbf{L}^{(1)} \mathbf{Q} - \mathbf{L}^{(2)} \right\|_F^2 . \tag{5.61}$$

Location Anchor Link Inference with Attribute Information
In location-based social networks, each location has its own profile page, which shows the name and all the review comments about the location. Similar to the similarity scores for user anchor links, for any two locations $l_i \in \mathcal{L}^{(1)}$ and $l_m \in \mathcal{L}^{(2)}$, based on the names of locations l_i and l_m, we can calculate the similarity scores between l_i and l_m to be

$$sim(n(l_i), n(l_m)) = \frac{|n(l_i) \cap n(l_m)|}{|n(l_i) \cup n(l_m)|}. \tag{5.62}$$

Users' review comments can summarize the unique features about locations, which are also very important hints for inferring potential location anchor links. Similarly, we represent users' review comments posted at locations l_i and l_m as bag-of-word vectors [19] weighted TF-IDF [35], $\mathbf{w}(l_i)$ and $\mathbf{w}(l_m)$. And the similarity between l_i and l_m based on the review comments can be represented as

$$sim(\mathbf{w}(l_i), \mathbf{w}(l_m)) = \mathbf{w}(l_i)^\top \cdot \mathbf{w}(l_i). \tag{5.63}$$

Closer locations are more likely to the same site than the ones which are far away. Based on the (latitude, longitude) information, the similarity score between locations l_i and l_m can be defined as follows:

$$sim(lat\text{-}long(l_i), lat\text{-}long(l_m))$$
$$= 1.0 - \frac{\sqrt{(lat(l_i) - lat(l_m))^2 + (long(l_i) - long(l_m))^2}}{\sqrt{(180 - (-180))^2 + (90 - (-90))^2}}. \tag{5.64}$$

Furthermore, we can also construct the similarity matrix between locations in $G^{(1)}$ and $G^{(2)}$ as $\Theta \in \mathbb{R}^{|\mathcal{L}^{(1)}| \times |\mathcal{L}^{(2)}|}$, where we have entry

$$\Theta(j, m) = \frac{1}{3}\Big(sim(n(l_i), n(l_m)) + sim(\mathbf{w}(l_i), \mathbf{w}(l_m))$$
$$+ sim(lat\text{-}long(l_i), lat\text{-}long(l_m))\Big). \tag{5.65}$$

The optimal *location transitional matrix* $\mathbf{Q}$ which can minimize the cost in attribute information can be represented as

$$\text{cost in attribute}(\mathcal{A}_l^{(1,2)}) = -\|\mathbf{Q} \circ \Theta\|_1 . \tag{5.66}$$

Location Anchor Link Inference with Link and Attribute Information
By considering the location links and attributes attached to locations simultaneously, the cost function of inferred location anchor links $\mathcal{A}_l^{(1,2)}$ can be represented as

$$\text{cost}(\mathcal{A}_l^{(1,2)}) = \text{cost in link}(\mathcal{A}_l^{(1,2)}) + \alpha \cdot \text{cost in attribute}(\mathcal{A}_l^{(1,2)})$$
$$= \left\| \mathbf{P}^\top \mathbf{L}^{(1)} \mathbf{Q} - \mathbf{L}^{(2)} \right\|_F^2 - \alpha \cdot \|\mathbf{Q} \circ \Theta\|_1 . \tag{5.67}$$

The optimal user and location transitional matrices $\mathbf{P}^*$ and $\mathbf{Q}^*$ that can minimize the mapping cost will be

$$\mathbf{P}^*, \mathbf{Q}^* = \arg\min_{\mathbf{P}, \mathbf{Q}} \mathrm{cost}(\mathcal{A}_l^{(1,2)})$$

$$= \arg\min_{\mathbf{P}, \mathbf{Q}} \left\| \mathbf{P}^\top \mathbf{L}^{(1)} \mathbf{Q} - \mathbf{L}^{(2)} \right\|_F^2 - \alpha \cdot \left\| \mathbf{Q} \circ \boldsymbol{\Theta} \right\|_1,$$

$$s.t. \quad \mathbf{Q} \in \{0, 1\}^{|\mathcal{L}^{(1)}| \times |\mathcal{L}^{(2)}|}, \mathbf{P} \in \{0, 1\}^{|\mathcal{U}^{(1)}| \times |\mathcal{U}^{(2)}|},$$

$$\mathbf{P} \mathbf{1}^{|\mathcal{U}^{(2)}| \times 1} \preccurlyeq \mathbf{1}^{|\mathcal{U}^{(1)}| \times 1}, \mathbf{P}^\top \mathbf{1}^{|\mathcal{U}^{(1)}| \times 1} \preccurlyeq \mathbf{1}^{|\mathcal{U}^{(2)}| \times 1}.$$

$$\mathbf{Q} \mathbf{1}^{|\mathcal{L}^{(2)}| \times 1} \preccurlyeq \mathbf{1}^{|\mathcal{L}^{(1)}| \times 1}, \mathbf{Q}^\top \mathbf{1}^{|\mathcal{L}^{(1)}| \times 1} \preccurlyeq \mathbf{1}^{|\mathcal{L}^{(2)}| \times 1}, \tag{5.68}$$

where *location anchor links* also have *one-to-one* constraint, and the last two equations are added to maintain such a constraint.

5.5.2.3 Co-inference of Anchor Links

User transitional matrix $\mathbf{P}$ is involved in the objective functions of inferring both *user anchor links* and *location anchor links*, and these two different anchor link inference tasks are strongly correlated (due to $\mathbf{P}$) and can be inferred simultaneously. By integrating the objective equations of anchor link inference for both users and locations, the optimal transitional matrices $\mathbf{P}^*$ and $\mathbf{Q}^*$ can be obtained simultaneously by solving the following objective function:

$$\mathbf{P}^*, \mathbf{Q}^* = \arg\min_{\mathbf{P}, \mathbf{Q}} \mathrm{cost}(\mathcal{A}_u^{(1,2)}) + \mathrm{cost}(\mathcal{A}_l^{(1,2)})$$

$$= \arg\min_{\mathbf{P}, \mathbf{Q}} \left\| \mathbf{P}^\top \mathbf{S}^{(1)} \mathbf{P} - \mathbf{S}^{(2)} \right\|_F^2 + \left\| \mathbf{P}^\top \mathbf{L}^{(1)} \mathbf{Q} - \mathbf{L}^{(2)} \right\|_F^2$$

$$- \alpha \cdot \left\| \mathbf{P} \circ \boldsymbol{\Lambda} \right\|_1 - \alpha \cdot \left\| \mathbf{Q} \circ \boldsymbol{\Theta} \right\|_1,$$

$$s.t. \quad \mathbf{P} \in \{0, 1\}^{|\mathcal{U}^{(1)}| \times |\mathcal{U}^{(2)}|}, \mathbf{Q} \in \{0, 1\}^{|\mathcal{L}^{(1)}| \times |\mathcal{L}^{(2)}|},$$

$$\mathbf{P} \mathbf{1}^{|\mathcal{U}^{(2)}| \times 1} \preccurlyeq \mathbf{1}^{|\mathcal{U}^{(1)}| \times 1}, \mathbf{P}^\top \mathbf{1}^{|\mathcal{U}^{(1)}| \times 1} \preccurlyeq \mathbf{1}^{|\mathcal{U}^{(2)}| \times 1},$$

$$\mathbf{Q} \mathbf{1}^{|\mathcal{L}^{(2)}| \times 1} \preccurlyeq \mathbf{1}^{|\mathcal{L}^{(1)}| \times 1}, \mathbf{Q}^\top \mathbf{1}^{|\mathcal{L}^{(1)}| \times 1} \preccurlyeq \mathbf{1}^{|\mathcal{L}^{(2)}| \times 1}. \tag{5.69}$$

The objective function is a constrained nonlinear integer programming problem, which is hard to address mathematically. Many relaxation algorithms have been proposed so far [2]. To solve the problem, as proposed in [48], the binary constraint of matrices $\mathbf{P}$ and $\mathbf{Q}$ can be relaxed to the real numbers in range [0, 1] and entries in $\mathbf{P}$ and $\mathbf{Q}$ will denote the existence probabilities/confidence scores of the corresponding anchor links. Redundant anchor links introduced by such a relaxation will be pruned with the co-matching algorithm to be introduced in the next section.

Meanwhile, the Hadamard product terms $\mathbf{P} \circ \boldsymbol{\Lambda}$ and $\mathbf{Q} \circ \boldsymbol{\Theta}$ can be very hard to deal with when solving the optimization problem. Considering that matrices $\mathbf{P}, \boldsymbol{\Lambda}, \mathbf{Q}$, and $\boldsymbol{\Theta}$ are all positive matrices, the L_1 norm of Hadamard product terms can be replaced according to the following Lemmas.

Lemma 5.1 *For any given matrix $\mathbf{A}$, the square of its Frobenius norm equals to the trace of $\mathbf{A}\mathbf{A}^\top$, i.e., $\|\mathbf{A}\|_F^2 = tr(\mathbf{A}\mathbf{A}^\top)$.*

The proof of the above lemma will be left as an exercise for the readers.

Lemma 5.2 *For two given positive matrices* $\mathbf{A}$ *and* $\mathbf{B}$ *of the same dimensions, the* L_1 *norm of the Hadamard product about* $\mathbf{A}$ *and* $\mathbf{B}$ *equals to the trace of* $\mathbf{A}^\top\mathbf{B}$ *or* $\mathbf{AB}^\top$, *i.e.,* $\|\mathbf{A} \circ \mathbf{B}\|_1 = tr(\mathbf{A}^\top\mathbf{B}) = tr(\mathbf{AB}^\top)$.

Proof According to the definitions of matrix trace, terms $tr(\mathbf{A}^\top\mathbf{B})$ and $tr(\mathbf{AB}^\top)$ equal to the Frobenius product [33] of matrices $\mathbf{A}$ and $\mathbf{B}$, i.e.,

$$tr(\mathbf{A}^\top\mathbf{B}) = tr(\mathbf{AB}^\top) = \sum_{i,j} \mathbf{A}(i,j)\mathbf{B}(i,j). \tag{5.70}$$

Meanwhile,

$$\|\mathbf{A} \circ \mathbf{B}\|_1 = \sum_{i,j} |(\mathbf{A} \circ \mathbf{B})(i,j)| = \sum_{i,j} |\mathbf{A}(i,j) \cdot \mathbf{B}(i,j)|. \tag{5.71}$$

Considering that both $\mathbf{A}$ and $\mathbf{B}$ are positive matrices, so the following equation can always hold:

$$\|\mathbf{A} \circ \mathbf{B}\|_1 = \sum_{i,j} \mathbf{A}(i,j) \cdot \mathbf{B}(i,j) = tr(\mathbf{A}^\top\mathbf{B}) = tr(\mathbf{AB}^\top). \tag{5.72}$$

To solve the objective function, the alternating projected gradient descent (APGD) method introduced in [25] can be applied here and the *one-to-one* constraint will be relaxed. The constraints $\mathbf{P1} \preccurlyeq \mathbf{1}, \mathbf{P}^\top\mathbf{1} \preccurlyeq \mathbf{1}$ are replaced with $\|\mathbf{P}\|_1 \leq t$ instead, where t is a small constant. Similarly, the *one-to-one* constraint on $\mathbf{Q}$ is also relaxed and replaced with $\|\mathbf{Q}\|_1 \leq t$. Furthermore, by incorporating terms $\|\mathbf{P}\|_1$ and $\|\mathbf{Q}\|_1$ into the minimization objective function, based on the relaxed constraints as well as Lemmas 5.1–5.2, the new objective function can be represented to be

$$\arg\min_{\mathbf{P},\mathbf{Q}} \; f(\mathbf{P},\mathbf{Q}) = tr\left((\mathbf{P}^\top\mathbf{S}^{(1)}\mathbf{P} - \mathbf{S}^{(2)})(\mathbf{P}^\top\mathbf{S}^{(1)}\mathbf{P} - \mathbf{S}^{(2)})^\top\right)$$

$$+ tr\left((\mathbf{P}^\top\mathbf{L}^{(1)}\mathbf{Q} - \mathbf{L}^{(2)})(\mathbf{P}^\top\mathbf{L}^{(1)}\mathbf{Q} - \mathbf{L}^{(2)})^\top\right)$$

$$- \alpha \cdot tr(\mathbf{P}\mathbf{\Lambda}^\top) - \alpha \cdot tr(\mathbf{Q}\mathbf{\Theta}^\top) + \gamma \cdot \|\mathbf{P}\|_1 + \mu \cdot \|\mathbf{Q}\|_1$$

$$s.t. \quad \mathbf{0}^{|\mathcal{U}^{(1)}|\times|\mathcal{U}^{(2)}|} \preccurlyeq \mathbf{P} \preccurlyeq \mathbf{1}^{|\mathcal{U}^{(1)}|\times|\mathcal{U}^{(2)}|},$$

$$\mathbf{0}^{|\mathcal{L}^{(1)}|\times|\mathcal{L}^{(2)}|} \preccurlyeq \mathbf{Q} \preccurlyeq \mathbf{1}^{|\mathcal{L}^{(1)}|\times|\mathcal{L}^{(2)}|}, \tag{5.73}$$

where γ and μ denote the weights on $\|\mathbf{P}\|_1$ and $\|\mathbf{Q}\|_1$, respectively.

As we can see, the objective function is with respect to $\mathbf{P}$ and $\mathbf{Q}$ and we cannot give a closed-form solution for the objective function. In [48], the optimal $\mathbf{P}$ and $\mathbf{Q}$ are learned with an alternative updating procedure based on the gradient descent algorithm: (1) fix $\mathbf{Q}$ and minimize the objective function *w.r.t.* $\mathbf{P}$; and (2) fix $\mathbf{P}$ and minimize the objective function *w.r.t.* $\mathbf{Q}$. If during these two updating procedures, entries in $\mathbf{P}$ or $\mathbf{Q}$ become invalid, we use a projection to guarantee the [0, 1] constraint: (1) if $\mathbf{P}(i,j) > 1$ or $\mathbf{Q}(i,j) > 1$, we project it to 1; and (2) if $\mathbf{P}(i,j) < 0$ or $\mathbf{Q}(i,j) < 0$, we project it to 0 [25]. The alternative updating equations of these two matrices are available as follows:

$$\mathbf{P}^{\tau} = \mathbf{P}^{\tau-1} - \eta_1 \cdot \frac{\partial \Gamma(\mathbf{P}^{\tau-1}, \mathbf{Q}^{\tau-1}, \gamma, \mu)}{\partial \mathbf{P}}$$

$$= \mathbf{P}^{\tau-1} - 2\eta_1 \cdot \Big(\mathbf{S}^{(1)}\mathbf{P}\mathbf{P}^{\top}(\mathbf{S}^{(1)})^{\top}\mathbf{P} + (\mathbf{S}^{(1)})^{\top}\mathbf{P}\mathbf{P}^{\top}\mathbf{S}^{(1)}\mathbf{P}$$

$$+ \mathbf{L}^{(1)}\mathbf{Q}\mathbf{Q}^{\top}(\mathbf{L}^{(1)})^{\top}\mathbf{P} - \mathbf{S}^{(1)}\mathbf{P}(\mathbf{S}^{(2)})^{\top} - (\mathbf{S}^{(1)})^{\top}\mathbf{P}\mathbf{S}^{(2)}$$

$$- \mathbf{L}^{(1)}\mathbf{Q}(\mathbf{L}^{(2)})^{\top} - \frac{1}{2}\alpha\mathbf{\Lambda} + \frac{1}{2}\gamma\mathbf{1}\mathbf{1}^{\top} \Big), \tag{5.74}$$

$$\mathbf{Q}^{\tau} = \mathbf{Q}^{\tau-1} - \eta_2 \cdot \frac{\partial \Gamma(\mathbf{P}^{\tau}, \mathbf{Q}^{\tau-1}, \gamma, \mu)}{\partial \mathbf{Q}}$$

$$= \mathbf{Q}^{\tau-1} - 2\eta_2 \cdot \Big((\mathbf{L}^{(1)})^{\top}\mathbf{P}\mathbf{P}^{\top}\mathbf{L}^{(1)}\mathbf{Q} - (\mathbf{L}^{(1)})^{\top}\mathbf{P}\mathbf{L}^{(2)}$$

$$- \frac{1}{2}\alpha\mathbf{\Theta} + \frac{1}{2}\mu\mathbf{1}\mathbf{1}^{\top} \Big), \tag{5.75}$$

where η_1 and η_2 are the *learning rate* in updating $\mathbf{P}$ and $\mathbf{Q}$, respectively. Such an updating process will continue until both $\mathbf{P}$ and $\mathbf{Q}$ converge.

5.5.3 Network Co-matching

To solve the objective function, the *one-to-one* constraint on both *user anchor links* and *location anchor links* are relaxed, which can take values in range [0, 1]. As a result, users and locations in each network can be connected by multiple user/location anchor links of various confidence scores across networks simultaneously and the *one-to-one* constraint can no longer hold any more. To maintain such a constraint on both user and location anchor links, the redundant ones introduced due to the relaxation with *network flow* will be pruned based on a network co-matching algorithm in this subsection.

Based on user sets $\mathcal{U}^{(1)}$ and $\mathcal{U}^{(2)}$, location sets $\mathcal{L}^{(1)}$ and $\mathcal{L}^{(2)}$, as well as the existence confidence scores of potential user and location anchor links between networks $G^{(1)}$ and $G^{(1)}$ (i.e., entries of $\mathbf{P}$ and $\mathbf{Q}$), we can construct the user and location preference bipartite graphs as shown in the left plots of Fig. 5.6.

User Preference Bipartite Graph
The *user preference bipartite graph* can be represented as $BG_{\mathcal{U}} = (\mathcal{U}^{(1)} \cup \mathcal{U}^{(2)}, \mathcal{U}^{(1)} \times \mathcal{U}^{(2)}, \mathcal{W}_{\mathcal{U}})$, where $\mathcal{U}^{(1)} \cup \mathcal{U}^{(2)}$ denotes the user nodes in $G^{(1)}$ and $G^{(2)}$, $\mathcal{U}^{(1)} \times \mathcal{U}^{(2)}$ contains all the potential user anchor links between $G^{(1)}$ and $G^{(2)}$, and $\mathcal{W}_{\mathcal{U}}$ will map links in $\mathcal{U}^{(1)} \times \mathcal{U}^{(2)}$ to their confidence scores (i.e., entries in $\mathbf{P}$) inferred in the previous section.

Location Preference Bipartite Graph
Similarly, we can also represent the *location preference bipartite graph* to be $BG_{\mathcal{L}} = (\mathcal{L}^{(1)} \cup \mathcal{L}^{(2)}, \mathcal{L}^{(1)} \times \mathcal{L}^{(2)}, \mathcal{W}_{\mathcal{L}})$, where the weight mapping of potential *location anchor links* (i.e., $\mathcal{W}_{\mathcal{L}}$) can be obtained from *location transitional matrix* $\mathbf{Q}$ in a similar way as introduced before.

Co-matching Network Flow Graph
To prune the non-existing anchor links, the traditional network flow algorithm can be employed to match users and locations across networks $G^{(1)}$ and $G^{(2)}$ simultaneously, which are grouped together in an integrated network flow model, named "co-matching network flow." As shown in the right plot

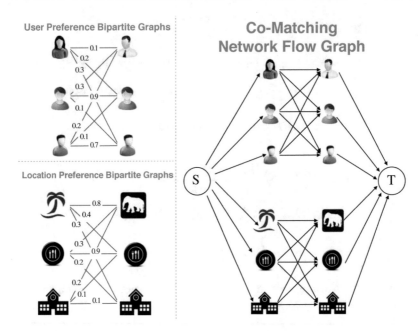

Fig. 5.6 User and location preference bipartite graphs and co-matching network flow graph

of Fig. 5.6, based on the *user preference bipartite graphs* and *location preference bipartite graphs*, the *co-matching network flow graph* can be constructed by adding (1) a source node S, (2) a sink node T, (3) links connecting node S and links in $\mathcal{U}^{(1)} \cup \mathcal{L}^{(1)}$ (i.e., $\{S\} \times (\mathcal{U}^{(1)} \cup \mathcal{L}^{(1)})$), and (4) links connecting nodes in $\mathcal{U}^{(2)} \cup \mathcal{L}^{(2)}$ and node T (i.e., $(\mathcal{U}^{(2)} \cup \mathcal{L}^{(2)}) \times \{T\}$).

Bound Constraint

In the network flow model, each link in the *co-matching network flow graph* is associated with an *upper bound* and *lower bound* to control the amount of flow going through it. For example, the upper and lower bounds of potential user anchor link $(u, v) \in \mathcal{U}^{(1)} \times \mathcal{U}^{(2)}$ in the *co-matching network flow graph* can be represented as

$$\underline{B}(u, v) \leq F(u, v) \leq \overline{B}(u, v), \tag{5.76}$$

where $F(u, v)$ denotes the flow amount going through link (u, v), $\underline{B}(u, v)$ and $\overline{B}(u, v)$ represent the *lower bound* and *upper bound* associated with link (u, v), respectively.

Considering that the constraint on both user and location anchor links is *one-to-one* and networks studied in this section are partially aligned, users in online social networks include both anchor and non-anchor users; so is the case for locations. In other words, each user and location in online social networks can be connected by at most one anchor links across networks, which can be achieved by adding the following upper and lower bound constraint on links $\{S\} \times (\mathcal{U}^{(1)} \cup \mathcal{L}^{(1)})$ and $(\mathcal{U}^{(2)} \cup \mathcal{L}^{(2)}) \times \{T\}$:

$$0 \leq F(u, v) \leq 1, \forall (u, v) \in \{S\} \times (\mathcal{U}^{(1)} \cup \mathcal{L}^{(1)}) \cup (\mathcal{U}^{(2)} \cup \mathcal{L}^{(2)}) \times \{T\}. \tag{5.77}$$

Among all the potential user anchor links in $\mathcal{U}^{(1)} \times \mathcal{U}^{(2)}$ and location anchor links in $\mathcal{L}^{(1)} \times \mathcal{L}^{(2)}$, only part of these links will be selected finally due to the *one-to-one* constraint. To represent whether a link (u, v) is selected or not, the flow amount going through links $\mathcal{U}^{(1)} \times \mathcal{U}^{(2)} \cup \mathcal{L}^{(1)} \times \mathcal{L}^{(2)}$ will be set as integers with upper and lower bounds to be 0 a 1 (1 denotes the link is selected, and 0 otherwise), respectively, i.e.,

$$F(u, v) \in \{0, 1\}, \forall (u, v) \in \mathcal{U}^{(1)} \times \mathcal{U}^{(2)} \cup \mathcal{L}^{(1)} \times \mathcal{L}^{(2)}. \tag{5.78}$$

Mass Balance Constraint

In addition, in network flow model, for each node in the graph (except the source and sink node), the amount of flow going through it should meet the *mass balance constraint*, i.e., for each node in the network, the amount of network flow going into it should equal to that going out from it:

$$\sum_{w \in \mathcal{N}_F, (w,u) \in \mathcal{L}_F} F(w, u) = \sum_{v \in \mathcal{N}_F, (u,v) \in \mathcal{L}_F} F(u, v), \tag{5.79}$$

where $\mathcal{N}_F = \{S\} \cup \mathcal{U}^{(1)} \cup \mathcal{U}^{(2)} \cup \mathcal{L}^{(1)} \cup \mathcal{L}^{(2)} \cup \{T\}$ denotes all the nodes in the *co-matching network flow graph* and $\mathcal{L}_F = \{S\} \times (\mathcal{U}^{(1)} \cup \mathcal{L}^{(1)}) \cup \mathcal{U}^{(1)} \times \mathcal{U}^{(2)} \cup \mathcal{L}^{(1)} \times \mathcal{L}^{(2)} \cup (\mathcal{U}^{(2)} \cup \mathcal{L}^{(2)}) \times \{T\}$ represents all the links in graph.

Maximum Confidence Objective Function

All the potential links connecting users and locations across networks are associated with certain costs in network flow model, where links with lower costs are more likely to be selected. The model can be modified a little to select the links introducing the maximum confidence scores instead from $\mathcal{U}^{(1)} \times \mathcal{U}^{(2)}$ and $\mathcal{L}^{(1)} \times \mathcal{L}^{(2)}$, respectively, which can be obtained with the following objective functions:

$$\max \sum_{(u,v) \in (\mathcal{U}^{(1)} \times \mathcal{U}^{(2)})} F(u, v) \cdot \mathcal{W}_{\mathcal{U}}(u, v), \tag{5.80}$$

$$\max \sum_{(m,n) \in (\mathcal{L}^{(1)} \times \mathcal{L}^{(2)})} F(m, n) \cdot \mathcal{W}_{\mathcal{L}}(m, n). \tag{5.81}$$

The final objective equation of simultaneous *co-matching* of users and locations across networks can be represented to be

$$\max \sum_{(u,v) \in (\mathcal{U}^{(1)} \times \mathcal{U}^{(2)})} F(u, v) \cdot \mathcal{W}_{\mathcal{U}}(u, v) + \sum_{(m,n) \in (\mathcal{L}^{(1)} \times \mathcal{L}^{(2)})} F(m, n) \cdot \mathcal{W}_{\mathcal{L}}(m, n),$$

$$s.t. \ 0 \le F(u, v) \le 1, \forall (u, v) \in \{S\} \times (\mathcal{U}^{(1)} \cup \mathcal{L}^{(1)}) \cup (\mathcal{U}^{(2)} \cup \mathcal{L}^{(2)}) \times \{T\},$$

$$F(u, v) \in \{0, 1\}, \forall (u, v) \in \mathcal{U}^{(1)} \times \mathcal{U}^{(2)} \cup \mathcal{L}^{(1)} \times \mathcal{L}^{(2)},$$

$$\sum_{w \in \mathcal{N}_F, (w,u) \in \mathcal{L}_F} F(w, u) = \sum_{v \in \mathcal{N}_F, (u,v) \in \mathcal{L}_F} F(u, v). \tag{5.82}$$

The above network flow objective function can be solved with open-source toolkits (e.g., Scipy.Optimization[10] and GLPK[11]). In the obtained solution, the flow amount variable of potential

---

[10]http://docs.scipy.org/doc/scipy/reference/optimize.html.
[11]http://www.gnu.org/software/glpk/.

user and location anchor links achieving value 1 are the selected ones which will be assigned with label $+1$, while the remaining (i.e., those achieving value 0) are not selected which are assigned with label -1. The matching results obtained from the above objective function will be outputted as the final network co-alignment result.

5.6   Summary

In this chapter, we introduced the network alignment problem based on the unsupervised learning setting, where no training data (i.e., labeled anchor links) is available or necessarily needed. Technically speaking, the unsupervised network alignment problem is very challenging to address, which can be identically modeled as the graph isomorphism problem. To resolve the problem, we talked about several heuristics based network alignment approaches and several matrix inference approaches at first.

Based on the users' names and profile information, we provided a detailed description about how to utilize such information to compute the similarity scores among users. Based on the assumption that similar users are more likely to be the same user, several different similarity metrics have been introduced in this chapter. With the user names, we can compute the similarity scores among users based on the characters, tokens, and phonetic representations. Meanwhile, with the profile information, we can compute the similarity scores of users in their hometown locations, birthday, and textual information.

The anchor links actually define a mapping of users across networks, and we also introduce several network alignment approaches via inferring the mapping matrix about anchor links. Via the anchor links, both the user nodes and the social connections can be mapped from one network to the other network. The good mappings should be able to minimize the projection inconsistency about the network structures. Furthermore, there also exists a hard binary and one-to-one cardinality constraint on the mapping matrix, which renders the inference process to be extremely challenging.

We introduced an approach to apply the matrix inference based network alignment method to infer the anchor links across multiple (more than two) networks, where the alignment results should also preserve the transitivity law. To resolve the objective function, we talked about a two-phase solution: matrix inference via constraint relaxation, and post-processing of the alignment results via transitive matching.

For the alignment of networks via multiple types of shared information entities simultaneously, we also introduced a network co-alignment approach, which learns the mapping matrices of multiple types of anchor links simultaneously. By extending the anchor link mapping matrix inference approach to the scenario with both network structure and diverse attribute information, the introduced method is able to infer the user anchor links and location anchor links across heterogeneous networks at the same time.

5.7   Bibliography Notes

Graph isomorphism problem is an extremely challenging research problem, which is one of few standard problems in computational complexity theory belonging to NP. But by this context so far, it is still not known whether it belongs to P or NP-complete. It is one of only two problems whose complexity remains unresolved as listed in [15], the other being integer factorization. There have been several research works proposing efficient algorithms to address the graph isomorphism problem in the past century [12, 30]. In 2015, László Babai claimed to have proven that the graph isomorphism problem is solvable in quasi-polynomial time [4], but the proof has not been vetted yet.

Entity resolution is a common problem in many areas, e.g., database, statistics, and artificial intelligence. Based on the entity names, [11] provides an introduction about several string distance metrics for name-matching tasks, including the edit distance like metrics, token based distance metrics, and hybrid distance metrics. Meanwhile, for a comprehensive survey about the text similarity approaches, the readers may refer to [18], which covers the similarity metrics between words, sentences, paragraphs, and documents.

The multiple network simultaneously alignment method is based on [47], where the transitivity law property on anchor links in alignment was initially pointed out in that work. Meanwhile, the heterogeneous network co-alignment approach via multiple types of shared information entities was introduced in [48]. The network matching procedure in [47, 48] are both formulated as the maximum network flow problem [20]. Over the years, many improved solutions to the problem have been discovered, e.g., the shortest augmenting path algorithm [14], the blocking flow algorithm [13], the push-relabel algorithm [17], and the binary blocking flow algorithm [16].

5.8 Exercises

1. (Easy) Please compute the *edit distance*, *Jaro distance*, and *Jaro-Winkler distance* of a pair of input strings "DIXON" and "DICKSONX."
2. (Easy) Please compute the *common token*, *token based Jaccard's coefficient*, and *token based Cosine Similarity* for the user names "Mike Jordan" and "Michael Jordan."
3. (Medium) Please identify the alignment results of the input network shown in Fig. 5.7 with the *relative degree difference* measure.

Fig. 5.7 A pair of aligned input networks

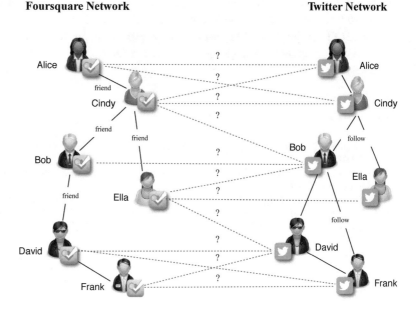

Fig. 5.8 Multiple aligned input networks

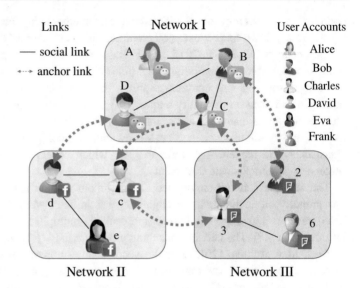

4. (Medium) Please identify the alignment results of the input network shown in Fig. 5.7 with the IsoRank algorithm.
5. (Medium) Please identify the alignment results of the input network shown in Fig. 5.8 with the IsoRankN algorithm.
6. (Medium) Please try to prove the Lemma 5.1.
7. (Hard) Please try to implement an algorithm with your preferred programming language to compute the *edit distance* of two input strings (Hint: You may consider to use the dynamic programming algorithm).
8. (Hard) Please try to implement the matrix inference based network alignment algorithm introduced in Sect. 5.3.4, and test it based on a small-sized synthetic aligned homogeneous network dataset.
9. (Hard) Please implement the UMA algorithm in your preferred programming language, and test it based on a small-sized synthetic aligned network dataset.
10. (Hard) Please implement the *network co-alignment* algorithm introduced in Sect. 5.5.2.2 your preferred programming language, and test it with a small-sized synthetic co-aligned network dataset.

References

1. L. Adamic, R. Lukose, A. Puniyani, B. Huberman, Search in power-law networks. Phys. Rev. E **64**, 046135 (2001). cs.NI/0103016
2. Y. Aflaloa, A. Bronsteinb, R. Kimmel, On convex relaxation of graph isomorphism. Proc. Natl. Acad. Sci. U S A **112**(10), 2942–2947 (2015)
3. M. Avriel, *Nonlinear Programming: Analysis and Methods* (Prentice-Hall, Englewood Cliffs, 1976)
4. L. Babai, Graph isomorphism in quasipolynomial time. CoRR, abs/1512.03547 (2015)
5. R. Baeza-Yates, B. Ribeiro-Neto, *Modern Information Retrieval* (Addison-Wesley Longman Publishing Co., Inc., Boston, 1999)
6. C. Bettstetter, On the minimum node degree and connectivity of a wireless multihop network, in *Proceedings of the 3rd ACM International Symposium on Mobile Ad Hoc Networking & Computing (MobiHoc '02)* (ACM, New York, 2002)
7. S. Borgatti, M. Everett, A graph-theoretic perspective on centrality. Soc. Net. **28**(4), 466–484 (2006)

8. W. Cavnar, J. Trenkle, N-gram-based text categorization, in *Proceedings of SDAIR-94, 3rd Annual Symposium on Document Analysis and Information Retrieval* (1994)
9. D. Chandler, The norm of the Schur product operation. Numer. Math. **4**(1), 343–344 (1962)
10. M. Charikar, Similarity estimation techniques from rounding algorithms, in *Proceedings of the Thirty-Fourth Annual ACM Symposium on Theory of Computing (STOC '02)* (ACM, New York, 2002)
11. W. Cohen, P. Ravikumar, S. Fienberg, A comparison of string distance metrics for name-matching tasks, in *Proceedings of the 2003 International Conference on Information Integration on the Web (IIWEB'03)* (AAAI Press, Palo Alto, 2003)
12. D. Corneil, C. Gotlieb, An efficient algorithm for graph isomorphism. J. ACM **17**(1), 51–64 (1970)
13. E. Dinic, Algorithm for solution of a problem of maximum flow in a network with power estimation. Sov. Math. Dokl. **11**, 1277–1280 (1970)
14. J. Edmonds, R. Karp, Theoretical improvements in algorithmic efficiency for network flow problems. J. ACM **19**(2), 248–264 (1972)
15. M. Garey, D. Johnson, *Computers and Intractability; A Guide to the Theory of NP-Completeness* (W. H. Freeman & Co., New York, 1990)
16. A.A. Goldberg, S. Rao, Beyond the flow decomposition barrier. J. ACM **45**(5), 783–797 (1998)
17. A. Goldberg, R. Tarjan, A new approach to the maximum-flow problem. J. ACM **35**(4), 921–940 (1988)
18. W. Gomaa, A. Fahmy, Article: a survey of text similarity approaches. Int. J. Comput. Appl. **68**(3), 13–18 (2013)
19. Z. Harris, Distributional structure. Word **10**(23), 146–162 (1954)
20. T. Harris, F. Ross, *Fundamentals of a Method for Evaluating Rail Net Capacities*. Research Memorandum (The RAND Corporation, Santa Monica, 1955)
21. D. Hirschberg, Algorithms for the longest common subsequence problem. J. ACM **24**(4), 664–675 (1977)
22. M. Jaro, Advances in record-linkage methodology as applied to matching the 1985 census of Tampa, Florida. J. Am. Stat. Assoc. **84**(406), 414–420 (1989)
23. T. Joachims, A probabilistic analysis of the Rocchio algorithm with TFIDF for text categorization, in *Proceedings of the Fourteenth International Conference on Machine Learning (ICML '97)* (Morgan Kaufmann Publishers Inc., San Francisco, 1997)
24. X. Kong, J. Zhang, P. Yu, Inferring anchor links across multiple heterogeneous social networks, in *Proceedings of the 22nd ACM International Conference on Information & Knowledge Management (CIKM '13)* (ACM, New York, 2013)
25. D. Koutra, H. Tong, D. Lubensky, Big-align: fast bipartite graph alignment, in *2013 IEEE 13th International Conference on Data Mining* (IEEE, Piscataway, 2013)
26. K. Kunen, *Set Theory* (Elsevier Science Publishers, Amsterdam, 1980)
27. J. Lee, W. Han, R. Kasperovics, J. Lee, An in-depth comparison of subgraph isomorphism algorithms in graph databases, in *Proceedings of the VLDB Endowment*. VLDB Endowment (2012)
28. V.I. Levenshtein, Binary codes capable of correcting deletions, insertions and reversals. Sov. Phy. Dok. **10**, 707 (1966)
29. C. Liao, K. Lu, M. Baym, R. Singh, B. Berger, IsoRankN: spectral methods for global alignment of multiple protein networks. Bioinformatics **25**(12), i253–i258 (2009)
30. B. McKay, Practical graph isomorphism. Congr. Numer. **30**, 45–87 (1981)
31. G. Navarro, A guided tour to approximate string matching. ACM Comput. Surv. **33**(1), 31–88 (2001)
32. M. Newman, Analysis of weighted networks. Phy. Rev. E **70**, 056131 (2004)
33. K. Petersen, M. Pedersen, *The Matrix Cookbook*. Technical report. Technical University of Denmark, Lyngby (2012)
34. E. Ristad, P. Yianilos, Learning string-edit distance. IEEE Trans. Pattern Anal. Mach. Intell. **20**(5), 522–532 (1998)
35. G. Salton, C. Buckley, Term-weighting approaches in automatic text retrieval. Inf. Process. Manage. **24**(5), 513–523 (1988)
36. G. Salton, A. Wong, C.S. Yang, A vector space model for automatic indexing. Commun. ACM **18**(11), 613–620 (1975)
37. T. Shi, S. Kasahara, T. Pongkittiphan, N. Minematsu, D. Saito, K. Hirose, A measure of phonetic similarity to quantify pronunciation variation by using ASR technology, in *18th International Congress of Phonetic Sciences (ICPhS 2015)* (University of Glasgow, Glasgow, 2015)
38. R. Singh, J. Xu, B. Berger, Global alignment of multiple protein interaction networks with application to functional orthology detection. Natl. Acad. Sci. **105**(35), 12763–12768 (2008)
39. The National Archives, The Soundex indexing system (2007)
40. S. Umeyama, An eigendecomposition approach to weighted graph matching problems. IEEE Trans. Pattern Anal. Mach. Intell. **10**(5), 695–703 (1988)
41. R. Wagner, M. Fischer, The string-to-string correction problem. J. ACM **21**(1), 168–173 (1974)
42. W. Winkler, String comparator metrics and enhanced decision rules in the Fellegi-Sunter model of record linkage, in *Proceedings of the Section on Survey Research* (1990)

43. D. Wipf, B. Rao, L0-norm minimization for basis selection, in *Proceedings of the 17th International Conference on Neural Information Processing Systems (NIPS'04)* (MIT Press, Cambridge, 2005)
44. M. Yu, G. Li, D. Deng, J. Feng, String similarity search and join: a survey. Front. Comput. Sci. **10**(3), 399–417 (2016)
45. R. Zafarani, H. Liu, Connecting users across social media sites: a behavioral-modeling approach, in *Proceedings of the 19th ACM SIGKDD International Conference on Knowledge Discovery and Data Mining (KDD '13)* (ACM, New York, 2013)
46. J. Zhang, P. Yu, MCD: mutual clustering across multiple social networks, in *2015 IEEE International Congress on Big Data* (IEEE, Piscataway, 2015)
47. J. Zhang, P. Yu, Multiple anonymized social networks alignment, in *2015 IEEE International Conference on Data Mining* (IEEE, Piscataway, 2015)
48. J. Zhang, P. Yu, PCT: partial co-alignment of social networks, in *Proceedings of the 25th International Conference on World Wide Web (WWW '16)* (International World Wide Web Conferences Steering Committee Republic and Canton of Geneva, Geneva, 2016)
49. J. Zhang, P. Yu, Z. Zhou, Meta-path based multi-network collective link prediction, in *Proceedings of the 20th ACM SIGKDD International Conference on Knowledge Discovery and Data Mining (KDD '14)* (ACM, New York, 2014)

Semi-supervised Network Alignment

6

6.1   Overview

As mentioned before, in the real-world online social networks, the anchor links are extremely difficult to label manually. The training set we can obtain is usually of a small size compared with the network scale, and most of the potential anchor links are unlabeled actually. For instance, given the Facebook and Twitter networks containing millions or billions of users, identifying a very small training set merely with hundreds of correct anchor links is however not an easy task. Therefore, it is not realistic to achieve a large set of labeled anchor links as required by the *supervised network alignment* models introduced in Chap. 4. On the other hand, completely ignoring the (small) set of labeled anchor links, just like the *unsupervised network alignment* models introduced in Chap. 5, may also create lots of problems, since these labeled anchor links can provide important signals for the network alignment model building. In this chapter, we will introduce another category of network alignment models based on the *semi-supervised learning* setting [8, 23], where both the (small) labeled and (large) unlabeled sets will be utilized in the model building process.

However, significantly different from the traditional semi-supervised learning problems, the anchor link instances studied in the network alignment problem are not independent. The *one-to-one* constraint on the anchor links actually limits the number of existing anchor links incident to each user node across networks, which can also introduce extra information for inferring the anchor links. For instance, given an identified anchor link $(u_i^{(1)}, u_j^{(2)})$ between networks $G^{(1)}$ and $G^{(2)}$, we can know that the remaining unlabeled anchor links incident to either $u_i^{(1)}$ or $u_j^{(2)}$ should not exist.

Given two heterogeneous online social networks $G^{(1)}$ and $G^{(2)}$, we can represent the small set of positively labeled anchor links and large number of unlabeled anchor links as $\mathcal{A}_{train}$ and $\mathcal{A}_{unlabeled} = \mathcal{U}^{(1)} \times \mathcal{U}^{(2)} \setminus \mathcal{A}_{train}$, respectively. The main objective of semi-supervised network alignment task is to build a model to infer the existence labels of these unlabeled anchor links with both sets $\mathcal{A}_{train}$ and $\mathcal{A}_{unlabeled}$. In our network alignment task, the test set is actually identical to the unlabeled set, i.e., $\mathcal{A}_{unlabeled} = \mathcal{A}_{test}$. The built model will be further applied to the test set to infer the potential labels of these anchor links.

In this chapter, we will focus on studying the network alignment problem based on the semi-supervised learning setting. This chapter will be organized as follows: At the very beginning, we will provide an introduction to the semi-supervised learning task, which is very different from the supervised and unsupervised learning tasks introduced before. With such a new learning setting, we

© Springer Nature Switzerland AG 2019
J. Zhang, P. S. Yu, *Broad Learning Through Fusions*,
https://doi.org/10.1007/978-3-030-12528-8_6

will introduce three different network alignment models based on the semi-supervised learning [8,23], active learning [16], and positive-unlabeled (PU) learning [12] settings, respectively.

6.2   Semi-supervised Learning: Overview

Semi-supervised learning [8, 23] is halfway between supervised and unsupervised learning. Besides the unlabeled data, the learning algorithms are also provided with some supervision information from a small set of labeled training data. In this section, we will provide a basic introduction to the classic semi-supervised learning task. In addition, the *semi-supervised learning* problem also has several special types, including *active learning* [16] and *positive-unlabeled learning* [12], which will be introduced in this section as well.

6.2.1   Semi-supervised Learning Problem Setting

Semi-supervised learning [8,23] is a new type of learning tasks which were not covered in the machine learning overview provided in Chap. 2. In semi-supervised learning tasks, besides the set of labeled training data instances $\{(\mathbf{x}_1, y_1), (\mathbf{x}_2, y_2), \ldots, (\mathbf{x}_l, y_l)\}$, there also exists a large-sized unlabeled data instance set $\{\mathbf{x}_{l+1}, \mathbf{x}_{l+2}, \ldots, \mathbf{x}_{l+u}\}$. Semi-supervised learning tasks attempt to make use of this combined information from both the labeled and unlabeled sets to surpass the performance that could be obtained by either the supervised learning merely with the labeled instances or unsupervised learning merely with the unlabeled instances.

Existing *semi-supervised learning tasks* can be generally divided into two main categories, i.e., *transductive semi-supervised learning* [23] and *inductive semi-supervised learning* [23]. The *transductive semi-supervised learning* tasks aim at inferring the correct labels of instances in the unlabeled data set $\{\mathbf{x}_{l+1}, \mathbf{x}_{l+2}, \ldots, \mathbf{x}_{l+u}\}$, while the *inductive semi-supervised learning* tasks want to infer the correct mapping from the feature space to the label space instead (not just limited to the data instances in the unlabeled set).

By reading here, a question may naturally arise in the readers' mind: "Is semi-supervised learning useful?" To answer the question from the mathematical perspective, the "semi-supervised learning is useful" iff the knowledge obtained from the unlabeled instance feature vector distribution $P(\mathbf{x})$ is helpful for the inference of posterior probability $P(y|\mathbf{x})$. If this is not the case, semi-supervised learning will not yield any improvement over supervised learning (merely with the labeled set). Otherwise, the involvement of the unlabeled instances will lead to a great improvement in learning the probability function $P(y|\mathbf{x})$, i.e., semi-supervised learning will be useful.

To ensure the effectiveness of semi-supervised learning, certain assumptions need to hold, like the *smoothness assumption* [8], *cluster assumption* [8], and *manifold assumption* [8]. These assumptions will provide the way to use the unlabeled instances in improving the learned models with the labeled instances. Subject to these different assumptions, various *semi-supervised learning* models have been proposed already, some of which will be introduced in this part as well.

6.2.1.1 Smoothness Assumption
The *smoothness assumption* [8] is the most popular assumption used in semi-supervised learning tasks, which goes as follows:

   "*Given the feature vectors of two instances, $\mathbf{x}_1$ and $\mathbf{x}_2$, if their feature vectors are close, so will be their corresponding labels.*"

The *smoothness assumption* implies that if two data instances are similar in their feature representations, they will be more likely to share common labels. Clearly, with such an assumption, we can generalize the finite training set by incorporating the unlabeled instances. The *smoothness assumption* can be applied in both semi-supervised classification and regression models.

6.2.1.2 Cluster Assumption

In the feature space, the data instances tend to form clusters, where the data instances with similar feature vectors tend to lie in the same cluster, while those which are different in feature representations will be partitioned into different clusters instead. In the semi-supervised learning, the *cluster assumption* [8] denotes

"For the data instances in the same cluster, they are more likely to have similar labels."

Based on the *cluster assumption*, the unlabeled data instances can be used to help identify the boundaries of the clusters in a more accurate way. We could run a clustering algorithm and use the labeled instances to assign a class label to each cluster. In this way, depending on the belonging relationships of the unlabeled instances, we can determine the potential labels of these unlabeled instances. Here, we also want to clarify that the *cluster assumption* doesn't imply that each class will only form one single cluster, and it only denotes that the instances within the same cluster will have the same label. For the same class, the data instances are also possible to form multiple clusters.

6.2.1.3 Manifold Assumption

Another frequently used assumption in *semi-supervised learning* is called the *manifold assumption* [8]:

"The high dimensional data instances lie on a lower-dimensional manifold."

In this case we can attempt to learn the manifold [8] using both the labeled and unlabeled data to avoid the *curse of dimensionality* [5]. The manifold assumption is practical when high-dimensional data are being generated by some processes that may be hard to model directly but only have a few degrees of freedom. If the data happen to lie on a low-dimensional manifold, however, then the learning algorithm can essentially operate in a space of the corresponding dimension, thus avoiding the curse of dimensionality.

6.2.2   Semi-supervised Learning Models

Several existing models can be adjusted to be applied in the semi-supervised learning tasks. In this part, we will introduce some existing semi-supervised learning models, which incorporate the unlabeled instances in the model training in different ways.

6.2.2.1 Semi-supervised Support Vector Machine (S3VM)

For the *smoothness assumption* introduced in the previous subsection, another revised version [4] is that

"Given the feature vectors of two instances, $\mathbf{x}_1$ and $\mathbf{x}_2$, if the feature vectors are close in the high-density regions, so will be their corresponding labels."

Generally, in semi-supervised learning, the decision boundary of learning models is assumed to be situated in a low-density region (in terms of unlabeled data). Let notation $f(\mathbf{x}) \in \mathcal{Y}$ ($\mathcal{Y} = \{-1, +1\}$) denote the inferred label of the data instance with feature vector representation $\mathbf{x}$, where $f(\cdot)$ is the built model and $\mathcal{Y} = \{-1, +1\}$ is the binary label space. The decision boundary inferred by the model $f(\cdot)$ can be denoted as $f(\mathbf{x}) = 0$. Furthermore, the loss function on an unlabeled instance $\mathbf{x}$ can be represented as the following loss function:

$$l(f, \mathbf{x}) = \max(1 - |f(\mathbf{x})|, 0), \tag{6.1}$$

where the loss term $l(f, \mathbf{x}) > 0$ when $-1 < f(\mathbf{x}) < +1$; otherwise, it will be 0.

By considering all the unlabeled instances in set $\{\mathbf{x}_{l+1}, \mathbf{x}_{l+2}, \ldots, \mathbf{x}_{l+u}\}$, the average loss term of these instance will become

$$l(f, \{\mathbf{x}_{l+1}, \mathbf{x}_{l+2}, \ldots, \mathbf{x}_{l+u}\}) = \frac{1}{u} \sum_{i=l+1}^{l+u} \max(1 - |f(\mathbf{x}_i)|, 0). \tag{6.2}$$

Generally, the average loss term counts the violations in the margin separation, which will lead to a ranking score of potential mapping $f \in \mathcal{F}$. The top ranked mapping f denotes the one whose decision boundary avoids most unlabeled instances by a large margin.

If we use support vector machine (SVM) as the base model, i.e., mapping $f(\mathbf{x}) = \mathbf{w}^\top \mathbf{x} + b$, where $\mathbf{w}$ and b are the variables involved in the model. By adding this loss term with the objective function of support vector machine (SVM), we can represent the joint optimization objective function as follows:

$$\min_{\mathbf{w}, b} \frac{1}{l} \sum_{i=1}^{l} \max(1 - y_i(\mathbf{w}^\top \mathbf{x}_i + b), 0) + c \|\mathbf{w}\|_2^2 + \lambda \cdot \frac{1}{u} \sum_{i=l+1}^{l+u} \max(1 - |\mathbf{w}^\top \mathbf{x}_i + b|, 0), \tag{6.3}$$

where the first term denotes the introduced loss on the labeled data instances, and $\|\mathbf{w}\|_2^2$ denotes the regularization on the model variables. The parameters c and λ are the weights of the last two terms, respectively.

By solving the objective function, the variables of the model can be learned. Actually, the loss term is non-convex, and the learning process of the objective function can be hard. Some existing algorithms, like *deterministic annealing* [15], *continuation method* [1], and *concave–convex procedure (CCCP)* [17], can be used to handle such a challenge. Formally, the above support vector machine model with the semi-supervised learning setting is also named as the S3VM model [4].

6.2.2.2 Semi-supervised Graph Based Model

Many of the semi-supervised learning models are based on graphs [24], where the data instances are represented as the nodes in the graph and the links denote the pairwise distance of the instances. For instance, given all the data instances, $\mathcal{V} = \mathcal{A}_{train} \cup \mathcal{A}_{unlabeled}$, we can represent it as a weighted graph $G = (\mathcal{V}, \mathcal{L}, w)$, where link set $\mathcal{L} \subset \mathcal{V} \times \mathcal{V} \setminus \{(u, u)\}_{u \in \mathcal{V}}$. The mapping function $w : \mathcal{L} \to \mathbb{R}$ denotes the weight of the links. For instance, given a link $e = (u, v) \in \mathcal{L}$, its weight $w(e)$ denotes how similar u and v are. If there exists no link between nodes u and v, then the weight of the potential link between them will be 0 instead.

The graph based semi-supervised learning models [24] can be viewed as estimating a projection function f to project the nodes in the graph to the label set. Generally, the projection function f needs to meet two requirements: (1) the inferred labels of the labeled nodes (i.e., data instances with labels) should be close to their true labels, and (2) it should be smooth on the whole graph. The first requirement can be viewed as minimizing the learning loss on labeled data, and the second requirement can be viewed as a regularization term about the model. So far, the existing various graph based semi-supervised models [24] mainly differ with each other in three aspects: (1) graph construction, (2) the loss function, and (3) the regularization term.

Graph construction is the key point of the graph based semi-supervised models, and it is still an open question to this context so far. Several different approaches have been proposed to construct the graphs already, which include:

- *Graph construction with domain knowledge*
- *Neighbor graph construction*
- *Graph construction with local fit*

More information about these graph construction approaches is provided in [24]. Besides graph construction, choosing different loss functions and regularization terms will lead to different semi-supervised models. We will introduce some of them as follows.

1. MinCut Model In the binary case, the classification of the data instances can be viewed as a cut problem to partition nodes in the graph into two disjoint subsets. We can treat the positive labels as the sources and the negative labels as the sinks. The objective of MinCut model [6] is to find a set of edges, removal of which can block all the flow between the source and sink nodes. In the cut result, nodes connected to the source nodes will be classified as the positive instances, and those connected to the sink nodes are classified as the negative instances.

Formally, the loss term introduced in the MinCut model can be represented as

$$
loss = \sum_{i=1}^{l} (f(\mathbf{x}_i) - y_i)^2,
\tag{6.4}
$$

where $f(\mathbf{x}_i)$ denotes the inferred label for the data instance $\mathbf{x}_i$.

Meanwhile, the regularization term in the *MinCut model* can be represented as

$$
reg = \frac{1}{2} \sum_{i,j=l+1, i \neq j}^{l+u} w(\mathbf{x}_i, \mathbf{x}_j) \cdot \left(f(\mathbf{x}_i) - f(\mathbf{x}_j) \right)^2.
\tag{6.5}
$$

The regularization terms can ensure the smoothness of the model. By minimizing the regularization term, the data instance pairs with closer representations, i.e., $w(\mathbf{x}_i, \mathbf{x}_j)$ is large, should have closer labels, i.e., $\left(f(\mathbf{x}_i) - f(\mathbf{x}_j) \right)^2$ will be small.

To ensure that the labeled instances are classified correctly, the loss term is usually assigned with a very large weight. The joint optimization function can be represented as

$$
\min \alpha \cdot \sum_{i=1}^{l} (f(\mathbf{x}_i) - y_i)^2 + \frac{1}{2} \sum_{i,j=l+1, i \neq j}^{l+u} w(\mathbf{x}_i, \mathbf{x}_j) \cdot (f(\mathbf{x}_i) - f(\mathbf{x}_j))^2
$$

$$
s.t. \ f(\mathbf{x}_i) \in \{+1, -1\}, \forall i \in \{1, 2, \ldots, l, l+1, \ldots, l+u\},
\tag{6.6}
$$

where α is assigned with a very large value, like $\alpha = \infty$. For simplicity, in the learning process, instead of learning the function $f(\cdot)$, we can treat $f(\mathbf{x}) = f_i$ as a variable instead, which takes values from the label space $\mathcal{Y} = \{+1, -1\}$. The learned variables will be outputted as the inferred labels for the data instances.

2. Local and Global Consistency Model The links in the weighted graph G can be represented as a weighted adjacency matrix $\mathbf{W} \in \mathbb{R}^{|\mathcal{V}| \times |\mathcal{V}|}$, where entry $W(i, j)$ denotes

$$W(i, j) = \begin{cases} w((u_i, u_j)), & \text{if } (u_i, u_j) \in \mathcal{L}, \\ 0, & \text{otherwise.} \end{cases} \tag{6.7}$$

Based on the weight matrix $\mathbf{W}$, we can define its corresponding normalized and unnormalized Laplacian matrix to be

$$\mathbf{L}_n = \mathbf{I} - \mathbf{D}^{-\frac{1}{2}} \mathbf{W} \mathbf{D}^{-\frac{1}{2}}, \text{ and } \mathbf{L} = \mathbf{D} - \mathbf{W}, \text{ respectively,} \tag{6.8}$$

where the diagonal matrix $\mathbf{D}$ has value $D(i, i) = \sum_j W(i, j)$ on its diagonal.

The *local and global consistency* model [22] uses the following loss function:

$$loss = \sum_{i=1}^{l} (f_i - y_i)^2, \tag{6.9}$$

where f_i denotes the inferred label of the input data instance featured by vector $\mathbf{x}_i$. Meanwhile, the regularization function used in the *local and global consistency* model can be represented as

$$reg = \mathbf{f}^\top \mathbf{L}_n \mathbf{f}. \tag{6.10}$$

Besides the regularization term used above, many other regularization functions, like Tikhonov regularizer $\mathbf{f}^\top \mathbf{L} \mathbf{f}$ [2], or manifold regularizer $\|\mathbf{f}\|_K^2 + \|\mathbf{f}\|_I^2$ [3], can all be used to define the objective function of the semi-supervised learning models, which will lead to different learning performance. In the equations, K denotes a base kernel, where $\|\mathbf{f}\|_K^2$ denotes an "intrinsic norm" on $\mathbf{f}$ in the reproducing Kernel Hilbert space (RKHS), and $\|\mathbf{f}\|_I^2 = \frac{1}{(l+k)^2} \mathbf{f}^\top \mathbf{L} \mathbf{f}$.

6.2.2.3 Semi-supervised Generative Model
In the case that the base model used is a generative model, like

$$f_{\boldsymbol{\theta}}(\mathbf{x}) = \arg\max_y P(y|\mathbf{x}, \boldsymbol{\theta}) = \arg\max_y \frac{P(\mathbf{x}, y|\boldsymbol{\theta})}{\sum_{y'} P(\mathbf{x}, y'|\boldsymbol{\theta})}, \tag{6.11}$$

where $\boldsymbol{\theta}$ denotes the parameter vector involved in the generative model. The term $P(\mathbf{x}, y|\boldsymbol{\theta})$ denotes the joint probability of instance's feature vector and label in the generative model.

For the unlabeled instances, the likelihood for them to fit in the model can be represented as

$$L(f_{\boldsymbol{\theta}}, \{\mathbf{x}_{l+1}, \mathbf{x}_{l+2}, \ldots, \mathbf{x}_{l+k}\}) = \sum_{i=l+1}^{l+u} \log \left(\sum_{y \in \mathcal{Y}} P(\mathbf{x}_i, y|\boldsymbol{\theta}) \right). \tag{6.12}$$

To learn the model that can fit well for both the labeled instances and unlabeled instances, we can represent the objective function as follows:

$$\arg\max_{\boldsymbol{\theta}} \log\left(\sum_{i=1}^{l} P(y_i|\mathbf{x}_i, \boldsymbol{\theta})\right) + \lambda \cdot \sum_{i=l+1}^{l+u} \log\left(\sum_{y\in\mathcal{Y}} P(\mathbf{x}_i, y|\boldsymbol{\theta})\right). \tag{6.13}$$

The parameter $\boldsymbol{\theta}$ can be learned with the expectation–maximization (EM) algorithm [9] and some existing numerical optimization methods.

Besides these three models introduced in this section, there also exist many other types of *semi-supervised learning* models, like *semi-supervised co-training models* [7]. For the readers who are interested in *semi-supervised learning* works, please refer to [8, 23] for more information. These models introduced in this part can all be applied to solve the network alignment problem to infer the anchor links between different social networks. By utilizing the unknown anchor links, more accurate decision boundary can be determined with a small number of labeled instances. More information about the network alignment method based on semi-supervised learning setting is available in Sect. 6.3.

6.2.3 Active Learning

Active learning [16] is a special case of semi-supervised learning tasks, in which a learning algorithm is able to interactively query an oracle (denoting an information source) to obtain the desired labels of some unlabeled data instances. In the situation where manual labeling of the data instances is extremely hard, if the learning algorithm can actively query for the labels of some data instances, the learning process will be more efficient and effective. In active learning, the number of required labeled instances will be much smaller than the number of labels required by the normal supervised learning models. Meanwhile, in the data instance label query process, choosing the most informative instances [16] will be crucial for active learning, which can also reduce the number of required queries significantly to determine a good decision boundary.

Compared with the classical semi-supervised learning tasks, both active learning and semi-supervised learning tasks aim at obtaining a good learning performance without demanding too many labeled instances. Meanwhile, there also exist some differences in the way they work. Semi-supervised learning focuses more on using the unlabeled data instances to assist the learning models to improve the learning results. However, the objective of active learning is to choose one part of unlabeled data instances to query for their labels, which will be involved in the model training process as the known instances instead.

Formally, let $\mathcal{T} = \{(\mathbf{x}_1, y_1), (\mathbf{x}_2, y_2), \ldots, (\mathbf{x}_l, y_l)\}$ denote the set of labeled data instances, and $\mathcal{U} = \{\mathbf{x}_{l+1}, \mathbf{x}_{l+2}, \ldots, \mathbf{x}_{l+u}\}$ represent the set of unlabeled instances. Active learning aims at partitioning the unlabeled set $\mathcal{U}$ into two disjoint subsets $\mathcal{U}_Q$ and $\mathcal{U}_U$, and will query for the labels of the instances in set $\mathcal{U}_Q$. Therefore, the crucial task in active learning is to choose the data instances for set $\mathcal{U}_Q$ from the unlabeled data instance set.

Different query strategies [16] have been proposed already to determine the instances to be selected for set $\mathcal{U}_Q$, which include

- *Uncertainty Sampling*: The *uncertainty sampling* strategy will select the data instances that the current model is least certain about what the output should be. Many measures can be adopted to measure the prediction *uncertainty*, e.g., *posterior prediction probability* for probabilistic models, prediction result *entropy* for classification tasks, and the prediction *loss* for regression tasks.

- *Query by Committee*: In this approach, a variety of models are built with the current labeled data instances, which will vote on the output labels for the unlabeled data instances. For the data instances that these current models disagree the most, they will be selected finally by the *query by committee* strategy.
- *Expected Model Change*: Labeling some unknown data instances and adding them to the training set will lead to changes of the current model. The *expected model change* strategy aims at selecting the data instances which can introduce the maximum model changes. Different metrics can be used for measuring the expected model changes, like the introduced *gradient* by the new data instance in the model loss function.
- *Expected Error Reduction*: Labeling the data instances and retraining the models with these newly labeled instances may reduce the model's generalization error. The *expected error reduction* strategy will choose to label the data instances that can lead to the maximum expected error reduction instead.
- *Variance Reduction*: Minimizing the expectation of a loss function directly is expensive, and in general this cannot be done in closed form. However, we can still reduce generalization error indirectly by minimizing the output variance, which sometimes does have a closed-form solution. The *variance reduction* strategy focuses on selecting the data instances that can lead to the maximum variance reduction in the result.
- *Balance Exploration and Exploitation*: Labeling the unlabeled data instances is seen as a dilemma between the exploration and the exploitation over the data space representation. Such a strategy manages this compromise by modeling the active learning problem as a contextual bandit problem instead.
- *Exponentiated Gradient Exploration*: This strategy uses a sequential algorithm named exponentiated gradient (EG)-active that can improve any active learning algorithm by an optimal random exploration.

Example 6.1 In Fig. 6.1, we show an example of active learning with both labeled and unlabeled data instances. The complete data distribution is provided in plot (a), where the green dots and red triangles denote the data instances belonging to two different classes, respectively. Given a few labeled data instances in the feature space as shown in plot (b), we can fit a model with these labeled data instances, whose decision boundary is denoted as the purple line. Generally, for the data instances which are far away from the decision boundary, we can know that they are more likely to be either the positive or negative instances. Meanwhile, for those lying near the decision boundary, we are less certain about their specific labels and obtaining the true labels of these instances will help to determine more correct decision boundary. For instance, in plot (c), we further query for some labels nearby the decision boundary. By adding the new labeled data instance into the labeled training set, the new decision boundary is updated, which can not only classify the queried data instances but also the remaining unlabeled data instances as well.

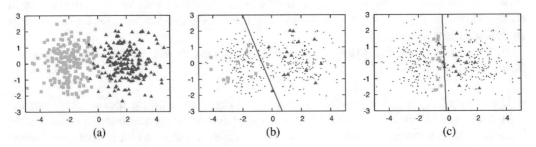

Fig. 6.1 An example of active learning

Active learning can also be applied to solve the network alignment problem when inferring the anchor links across networks. By keeping querying the labels of unlabeled instances, the model can refine the decision boundary with a very small training set. Different methods can be applied in selecting the unlabeled instances to query for their labels in the network alignment problem, which will be introduced in great detail in Sect. 6.4.

6.2.4 Positive and Unlabeled (PU) Learning

In some special case, the labeled training set may only involve the data instances belonging to one single class. For instance, in the e-commerce sites, when recommending products for the users, the training data available is merely the products that users have purchased in the past but no data about the products that the users will not purchase definitely. If we label the purchased products as the positive instances for the users, the training data available will only involve the positive instance only. Besides these positively labeled instances, there also exist a large number of instances in the site that we have no idea about whether the user is interested in or not. These remaining products will be the unlabeled instances on the other hand. Such examples are very common in the real world. On the web, the materials/contents that users are interested are relatively easy to obtain, while we have no idea about those that they dislike. Learning from these positively labeled and unlabeled data is called the *positive-unlabeled learning* (PU learning) task.

Definition 6.1 (Positive and Unlabeled (PU) Learning) Formally, the categories of learning tasks with a positive set $\mathcal{P}$ and unlabeled set $\mathcal{U}$ are called the *positive and unlabeled (PU) learning* tasks. With the $\mathcal{P}$ and $\mathcal{U}$ sets, PU learning aims at building a model to classify the unlabeled instances in $\mathcal{U}$ or some other future data.

The PU learning task is one type of the semi-supervised learning tasks as the unlabeled instances are involved in the model building. Different from classic semi-supervised learning tasks, the labeled instances in the PU learning tasks belong to one single type of class. Viewed in such a perspective, the PU learning task is also one type of one-class learning task [13], which is also known as the *unary learning* tasks aiming at identifying objects of one specific class among all the objects.

Generally speaking, it is "not learnable" merely with the positive instances. However, the addition of the unlabeled instances will make learning from the positive instances possible. Formally, let $(\mathbf{x}, y)$ be an instance tuple with feature vector $\mathbf{x}$ and label $y \in \{-1, +1\}$, the built model can be represented as a mapping $f : \mathbf{x} \to y$. We can rewrite the probability of achieving a wrong prediction as

$$P(f(\mathbf{x}) \neq y) = P(f(\mathbf{x}) = +1, y = -1) + P(f(\mathbf{x}) = -1, y = +1), \qquad (6.14)$$

which denotes the cases that $f(\cdot)$ misclassify the negative (or positive) instances to be positive (or negative).

On the other hand, we know that

$$
\begin{aligned}
P(f(\mathbf{x}) &= +1, y = -1) \\
&= P(f(\mathbf{x}) = +1) - P(f(\mathbf{x}) = +1, y = +1) \\
&= P(f(\mathbf{x}) = +1) - (P(y = +1) - P(f(\mathbf{x}) = -1, y = +1)).
\end{aligned}
\qquad (6.15)
$$

By plugging it into Eq. (6.14), we can have

$$P(f(\mathbf{x}) \neq y)$$
$$= P(f(\mathbf{x}) = +1) - P(y = +1) + 2P(f(\mathbf{x}) = -1, y = +1)$$
$$= P(f(\mathbf{x}) = +1) - P(y = +1) + 2P(f(\mathbf{x}) = -1|y = +1)P(y = +1). \qquad (6.16)$$

Noting that $P(y = +1)$ is a constant number, if we can also control $P(f(\mathbf{x}) = -1|y = +1)$ to be a small value, then the learning process (i.e., error minimization) is approximately the same as minimizing $P(f(\mathbf{x}) = +1)$. Meanwhile, holding $P(f(\mathbf{x}) = -1|y = +1)$ small is equivalent to ensuring $P(f(\mathbf{x}) = +1|y = +1)$ to be as large as possible while minimizing $P(f(\mathbf{x}) = +1, y = +1 \vee y = -1)$ at the same time. Here, the notation $P(f(\mathbf{x}) = +1|y = +1)$ denotes the probability of classifying positive instances correctly, and $P(f(\mathbf{x}) = +1, y = +1 \vee y = -1)$ represents the probability of classifying unlabeled instances as positive instances. Therefore, if we can ensure the positive instances are correctly classified, while the unlabeled are less likely to be classified as positive instances, the error of the model will be relatively low. Several different techniques have been proposed to ensure the low loss of the learned model, like the *spy techniques* [12] and *bridging probability inference* [18, 19], which will be introduced in Sect. 6.5.

PU learning is a good learning setting for many research problems in social networks, like link prediction, network alignment, and recommendations. By labeling the known friendship links, anchor links, and product purchase actions as the positive instances while the remaining unknown ones as the unlabeled instances, these tasks aforementioned can all be formulated as the PU learning problems. In Sect. 6.5, we will introduce more information about the network alignment model based on PU learning.

6.3 Semi-supervised Network Alignment

In the real-world online social networks involving millions even billions of users, labeling a large number of known anchor links is almost an infeasible task. In a real-world setting, we can usually have a small-sized training set (of identified anchor links), and a relatively big unlabeled set of the anchor links. Model building with the small-sized training set can hardly achieve a very good performance. How to involve the unlabeled set in the model building to improve its performance will become necessary. In this part, we will introduce a method to align the online social networks based on the semi-supervised learning setting. A new linear model similar to S3VM will be introduced first, and we will introduce how to apply the model to address the network alignment problem [20].

6.3.1 Loss Function for Labeled and Unlabeled Instances

Here, we denote all the set of potential anchor links between networks $G^{(1)}$ and $G^{(2)}$ as set $\mathcal{L} = \mathcal{U}^{(1)} \times \mathcal{U}^{(2)}$. Meanwhile the set of positively labeled anchor links (i.e., the existing anchor links) can be represented as set $\mathcal{A}$, and the remaining unlabeled anchor links can be denoted as set $\mathcal{U} = \mathcal{L} \setminus \mathcal{A}$ for simplicity.

For all the links in set $\mathcal{L}$ (involving links in both $\mathcal{A}$ and $\mathcal{U}$), a set of features will be extracted. For instance, we can represent the feature vector extracted for link $l \in \mathcal{L}$ as vector $\mathbf{x}_l \in \mathbb{R}^d$. Meanwhile, we can denote the label of link $l \in \mathcal{L}$ as $y_l \in \mathcal{Y} = \{0, +1\}$ (here, we use 0 to denote the negative class label and $+1$ to denote the positive class label). All the links in set $\mathcal{A}$ will be assigned with known positive labels $+1$, while the labels of links in set $\mathcal{U}$ are unknown. Therefore, based on these features and labels, all the links in set $\mathcal{L}$ can be represented as a tuple set $\{(\mathbf{x}_l, y_l)\}_{l \in \mathcal{L}}$.

Depending on the separability of the anchor links in set $\mathcal{L}$, different kinds of models can be applied. Here, if we use a linear model to fit the link instances, the model $f : \mathbb{R}^d \rightarrow \{+1, 0\}$ to be learned can be represented as a linear combination of the features parameterized with weight $\mathbf{w}$. For instance, with model $f_{\mathbf{w}}(\cdot)$, we can represent the inferred label of link instance l as $f_{\mathbf{w}}(\mathbf{x}_l) = \mathbf{w}^\top \mathbf{x}_l + w_0$, where w_0 is a bias term. By adding a dummy feature 1 for all the link instances, we can also incorporate w_0 into the variable vector $\mathbf{w}$. Therefore, we will use vector $\mathbf{w}$ to represent the weights for the features as well as the bias term when referring to the model variables, and simply use $f_{\mathbf{w}}(\mathbf{x}_l) = \mathbf{w}^\top \mathbf{x}_l$ to denote the model mathematical representation. Based on the known links in set $\mathcal{A}$, we can represent the training loss term as

$$L(f_{\mathbf{w}}, \mathcal{A}) = \sum_{l \in \mathcal{A}} \max(1 - f_{\mathbf{w}}(\mathbf{x}_l) \cdot y_l, 0) = \sum_{l \in \mathcal{A}} \max\left(1 - (\mathbf{w}^\top \mathbf{x}_l) \cdot y_l, 0\right). \quad (6.17)$$

Meanwhile, for the unlabeled links, we have no idea about their true labels in the training process. By following the intuition introduced for the S3VM model, we can represent the loss introduced by the unlabeled links as

$$L(f_{\mathbf{w}}, \mathcal{U}) = \sum_{l \in \mathcal{U}} \max\left(1 - |\mathbf{w}^\top \mathbf{x}_l|, 0\right). \quad (6.18)$$

By combining the loss function defined for the labeled and unlabeled links, we can represent the combined joint optimization function as follows:

$$\min_{\mathbf{w}, \{y_l\}_{l \in \mathcal{U}}} \frac{c_1}{2} L(f_{\mathbf{w}}, \mathcal{A}) + \frac{1}{2} \|\mathbf{w}\|_2^2 + \frac{c_2}{2} L(f_{\mathbf{w}}, \mathcal{U})$$

$$s.t. \ \ y_l \in \{+1, 0\}, \forall l \in \mathcal{U}. \quad (6.19)$$

Here, $\|\mathbf{w}\|_2^2$ is a regularization term on the model variable $\mathbf{w}$, and c_1, c_2 represent the weights of loss terms of labeled and unlabeled links, respectively.

In the above objective function, the variables to be learned include the weight variable $\mathbf{w}$, as well as the labels of links in the unlabeled set $\mathcal{U}$, i.e., $\{y_l\}_{l \in \mathcal{U}}$. Some approximation methods have been introduced to solve the problem. For example, by assuming that the labels of the link instances in set $\mathcal{U}$ can be correctly inferred by the built model, i.e., $y_l = sign(\mathbf{w}^\top \mathbf{x}_l), l \in \mathcal{U}$. Depending on the value of y_l, we can rewrite the loss introduced by link l as follows:

$$L(f_{\mathbf{w}}, l) = \begin{cases} \max\left(1 - \mathbf{w}^\top \mathbf{x}_l, 0\right), & \text{if } \mathbf{w}^\top \mathbf{x}_l > 0, \\ \max\left(1 + \mathbf{w}^\top \mathbf{x}_l, 0\right), & \text{if } \mathbf{w}^\top \mathbf{x}_l < 0. \end{cases} \quad (6.20)$$

Here, we will preserve the general representation for the loss term introduced by the *unlabeled anchor links* as follows:

$$L(f_{\mathbf{w}}, \mathcal{U}) = \sum_{l \in \mathcal{U}} \max\left(1 - y_l \cdot (\mathbf{w}^\top \mathbf{x}_l), 0\right), \quad (6.21)$$

where labels y_l of these unlabeled links are the variables to be inferred in the model as well.

6.3.2   Cardinality Constraint on Anchor Links

As introduced before, the anchor links in the networks are subject to the *one-to-one* cardinality constraint [11]. Such a constraint will control the maximum number of links incident to the nodes across the networks. Subject to the link cardinality constraints, the prediction tasks of anchor links between the network are no longer independent. For instance, for the links subject to the *one-to-one* constraint, if we can know/infer that the link (u, v) is a positive link (i.e., an existing anchor link), then all the remaining links incident to u or v in the unlabeled set $\mathcal{U}$ will be negative by default. Viewed in such a perspective, the cardinality constraint on links should be incorporated into the problem definition and the result can be improved significantly with such a constraint. In this part, we will introduce the link cardinality constraint and use it to define a set of mathematical constraints on node degrees.

The anchor links studied in this book are assumed to be bi-directional and the node in and out degrees denote the number of links going into/out from them. To represent the node–link incidence relationships, we introduce the node–link in and out matrices $\mathbf{A}_i, \mathbf{A}_o \in \{0, 1\}^{|\mathcal{V}| \times |\mathcal{L}|}$. Entry $A_i(i, j) = 1$ iff the directed link $l_j \in \mathcal{L}$ ends with node n_i, while entry $A_o(i, j) = 1$ iff the directed link $l_j \in \mathcal{L}$ starts with node n_i.

According to the analysis provided before, we can represent the labels of links in $\mathcal{L}$ as vector $\mathbf{y} \in \{+1, 0\}^{|\mathcal{L}| \times 1}$, where entry $y(i)$ represents the label of link $l_i \in \mathcal{L}$. Depending on which group l_i belongs to, its value has different representations

$$
y(i) = \begin{cases} +1, & \text{if } l_i \in \mathcal{A}, \\ \text{variable to be inferred}, & \text{if } l_i \in \mathcal{U} \setminus \mathcal{U}_q. \end{cases} \tag{6.22}
$$

Furthermore, based on the known and inferred labels of links in $\mathbf{y}$, we can represent the node degrees according to the following theorem.

Theorem 6.1 *The in and out degrees of node u_i in either network $G^{(1)}$ or $G^{(2)}$ can be represented as* $\mathbf{A}_i(i, :)\mathbf{y}$ *and* $\mathbf{A}_o(i, :)\mathbf{y}$, *respectively, where* $\mathbf{A}_i(i, :)$ *and* $\mathbf{A}_o(i, :)$ *denote the rows corresponding to* u_i.

Proof As introduced before, for the node u_i, we can get the set of links going out from u_i from the i_{th} row of matrix $\mathbf{A}_o$, i.e., $\mathbf{A}_o(i, :)$. For the entries with value 1 in $\mathbf{A}_o(i, :)$, u_i will have a potential link from u_i to the corresponding node. Therefore, the product $\mathbf{A}_o(i, :)\mathbf{y}$ will remove the remaining links, and sum all the labels of links starting from node u_i. Considering that the labels have value either $+1$ or 0, $\mathbf{A}_o(i, :)\mathbf{y}$ will actually denote the degree of node u_i. In a similar way, we can also obtain that the node in degree d_i equals to $\mathbf{A}_i(i, :)\mathbf{y}$.

Let $\mathbf{0}$ and $\mathbf{1}$ denote the vectors with all 0s and 1s of length $|\mathcal{L}|$, respectively. According to the previous analysis, the link cardinality constraint can be applied to define the degree constraint of nodes in the network, which can be represented as follows:

$$
\mathbf{0} \preccurlyeq \mathbf{A}_i \mathbf{y} \preccurlyeq \mathbf{1}, \tag{6.23}
$$

$$
\mathbf{0} \preccurlyeq \mathbf{A}_o \mathbf{y} \preccurlyeq \mathbf{1}. \tag{6.24}
$$

6.3.3   Joint Objective Function for Semi-supervised Network Alignment

By adding the node degree constraint to the objective function introduced before, we can represent the joint optimization objective function as

$$\min_{\mathbf{w},\mathbf{y}} \frac{c_1}{2} L(f_{\mathbf{w}}, \mathcal{A}) + \frac{1}{2} \|\mathbf{w}\|_2^2 + \frac{c_2}{2} L(f_{\mathbf{w}}, \mathcal{U})$$

$$s.t. \ \ y_l \in \{+1, 0\}, \forall l \in \mathcal{L}, \ y_l = +1, \forall l \in \mathcal{A},$$

$$\mathbf{0} \preccurlyeq \mathbf{A}_i \mathbf{y} \preccurlyeq \mathbf{1}, \ \mathbf{0} \preccurlyeq \mathbf{A}_o \mathbf{y} \preccurlyeq \mathbf{1}. \tag{6.25}$$

The objective function involves two variables, and it is easy to see that it is not jointly convex in terms of these two variables. To solve the function, techniques like alternative updating can be applied here. By fixing one variable, we can keep updating the other variable. Such an alternative updating process will continue until convergence.

Step 1: By fixing $\mathbf{y}$, the objective function will be reduced to the objective function of traditional SVM model involving variable $\mathbf{w}$ only:

$$\min_{\mathbf{w}} \frac{c_1}{2} \sum_{l \in \mathcal{A}} \max\left(1 - y_l \cdot (\mathbf{w}^\top \mathbf{x}), 0\right) + \frac{1}{2} \|\mathbf{w}\|_2^2 + \frac{c_2}{2} \sum_{l \in \mathcal{U}} \max\left(1 - y_l \cdot (\mathbf{w}^\top \mathbf{x}), 0\right). \tag{6.26}$$

The learning methods for the SVM model introduced in Sect. 2.3.3 can be applied to learn the optimal model variable, which will not be introduced here again.

Step 2: By fixing $\mathbf{w}$, the objective function will be reduced to the link selection problem we introduced before in Sect. 4.5. Let $\hat{y}_l = \mathbf{w}^\top \mathbf{x}_l$ denote the inferred label of link l, and the objective function will be reduced to

$$\min_{\mathbf{y}} \frac{c_1}{2} \sum_{l \in \mathcal{A}} \max\left(1 - \hat{y}_l \cdot y_l, 0\right) + \frac{1}{2} \|\mathbf{w}\|_2^2 + \frac{c_2}{2} \sum_{l \in \mathcal{A}} \max\left(1 - \hat{y}_l \cdot y_l, 0\right)$$

$$s.t. \ \ y_l \in \{+1, 0\}, \forall l \in \mathcal{L}, \ y_l = +1, \forall l \in \mathcal{A},$$

$$\mathbf{0} \preccurlyeq \mathbf{A}_i \mathbf{y} \preccurlyeq \mathbf{1}, \ \mathbf{0} \preccurlyeq \mathbf{A}_o \mathbf{y} \preccurlyeq \mathbf{1}. \tag{6.27}$$

Some algorithms like greedy link selection introduced in Sect. 4.5 can be applied to determine the label vector $\mathbf{y}$. Here, we will not talk about that algorithm again, and more information about the selection algorithm is provided in Algorithm 3 in Sect. 4.5.

6.4   Active Network Alignment

In this section, we will introduce an algorithm to address the network alignment problem based on active learning [16]. Different from the traditional active learning problems, due to the one-to-one constraint on anchor links, if an unlabeled anchor link $a = (u, v)$ is identified as positive (i.e., existing), all the other unlabeled anchor links incident to u or v will be negative (i.e., non-existing) automatically. Viewed in such a perspective, querying for the labels of potential positive anchor links in the unlabeled set will be much more rewarding in the active network alignment problem, since the identification of one positive anchor link will help identify a bunch of negative anchor links simultaneously. Various novel anchor link information gain measures will be defined in this section, based on which several active network alignment methods will be introduced.

Active learning aims at minimizing the labeling cost of the training set by asking the model to choose which examples to query for the labels. We can represent the set of labeled anchor links as set $\mathcal{A} = \{(\mathbf{x}_1, y_1), (\mathbf{x}_2, y_2), \ldots, (\mathbf{x}_l, y_l)\}$, which involves the positively labeled anchor links existing between the networks. The active learning algorithm will train an anchor link prediction model M with the training set $\mathcal{A}$. During the training process, what the active learner needs to do is to select a query pool of unlabeled anchor links from the unlabeled anchor link set $\mathcal{U} = \mathcal{U}^{(1)} \times \mathcal{U}^{(2)} \setminus \mathcal{A}$. The selection strategy is to pick the most valuable anchor link(s) according to the values computed by applying M on $\mathcal{U}$, which can be represented as set $\mathcal{U}_Q \subset \mathcal{U}$. The data instances in $\mathcal{U}_Q$ together with their labels will be added to $\mathcal{U}$ to update the model M. The training process and query process will be repeated until the limit of query cost has been reached.

In this section, we will first introduce several *anchor link query strategy* [25] for the network alignment problem first, based on which we will introduce the objective function of *active network alignment* [14] afterwards and provide the solutions.

6.4.1 Anchor Link Label Query Strategy

The main challenge in the active network alignment problem will be the query process of the unlabeled data instances. In each round of query process, traditional active learning methods usually just add the newly queried samples to the training set. However, via the one-to-one constraint, the constrained active learning methods will be able to infer the labels of some unlabeled anchor links after identifying one positive anchor link, and thus the samples to be added to the training set can be more than the queried samples.

Example 6.2 As shown in Fig. 6.2, there are 4 unlabeled anchor links in the query pool, i.e., $\{(u_1^{(1)}, u_1^{(2)}), (u_1^{(1)}, u_2^{(2)}), (u_2^{(1)}, u_1^{(2)}), (u_2^{(1)}, u_2^{(2)})\}$. Let's assume after querying an oracle, we get the label for link $(u_1^{(1)}, u_1^{(2)})$ to be $+1$ (i.e., $u_1^{(1)}$ and $u_1^{(2)}$ are the same user). Traditional active learners will just add $(u_1^{(1)}, u_1^{(2)})$ to the positive training set. However, a constrained active learner will firstly infer that $(u_1^{(1)}, u_2^{(2)})$ and $(u_2^{(1)}, u_1^{(2)})$ to be "negative" according to the one-to-one constraint, and then add $(u_1^{(1)}, u_1^{(2)})$ to the positive training set, as well as $(u_1^{(1)}, u_2^{(2)})$ and $(u_2^{(1)}, u_1^{(2)})$ to the negative training set. In this way, the constrained active learning methods can incorporate two more negatively labeled data instances (i.e., anchor links) than the traditional active learning methods under the same

Fig. 6.2 An example of active anchor link query

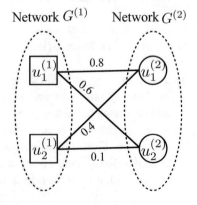

query limitation (where the query limitation means the cost of achieving anchor link labels within one round of query).

6.4.1.1 Regular Active Network Alignment

Many active learning methods usually involve the evaluation of the informativeness of unlabeled instances. However, due to the challenges created by the *one-to-one* cardinality constraint, many query methods used in active learning cannot be applied to the anchor link prediction task. Among the existing query methods [16], the simplest and most commonly used query strategy is uncertainty sampling, where the learner will query the labels of instances that it is the least certain about. There exist several commonly used sampling strategies in uncertainty sampling, including least confidence sampling, the margin sampling, and the entropy based sampling. Compared with the former two sampling strategies, the entropy based sampling generalizes more easily to complex structured instances. It is because by computing the entropy, we can compare the amount of information contained in different multi-structured samples in a uniform metric. The active network alignment method to be introduced here is based on the entropy theory, and aims to calculate the potential entropy $H(l)$ for each unlabeled link $l \in \mathcal{U}$. Here we define $H(l)$ as the evaluated amount of information that the active network alignment model can gain by identifying the label of anchor link l.

Here, we use notation $\Gamma(l)$ to represent the related anchor link set of a given anchor link l, i.e., the set of all anchor links in $\mathcal{U}$ that are incident to the nodes forming l. The major idea of the regular active learning method is to calculate $H(l)$ for each of the unlabeled anchor link $l \in \mathcal{U}$, and select the anchor link with the highest score to query for its label. If the label for the link is "negative," the link will be added to the training set. Meanwhile, if the label of the link is "positive," besides this link, we will also extract the remaining incident anchor links, i.e., those in $\Gamma(l)$, and add them as "negative" instances into the training set.

Formally, the information entropy of the anchor link $l \in \mathcal{A}$ can be represented as

$$H(l) = - \sum_{y \in \mathcal{Y}} P_M(y|\mathbf{x}_l) log P_M(y|\mathbf{x}_l), \tag{6.28}$$

where the term $P_M(y|\mathbf{x}_l)$ denotes the posterior probability of anchor link $l \in \mathcal{A}$ inferred by an anchor link inference model M. Literally, for the anchor link that can introduce a larger entropy, the learned model M is less certain about its label. Querying for the label of such links can actually introduce the maximum information gain.

6.4.1.2 Biased Constrained Active Network Alignment

For the network alignment task, generally identifying the potentially positive instances can lead to more information, since the identification of one positive instance can lead to a bunch of identified negative anchor links at the same time due to the *one-to-one* constraint. As we discussed before, because of the sparsity of anchor links, acquiring enough informative positive anchor links under a limited cost is very important. However, in the regular active network alignment method introduced in the previous subsection, there may not be enough mechanism to increase the probability of each identified link to be positive. So if we can explore such a mechanism, and integrate it into the active network alignment model, we will be able to achieve better results. In this part, we will present the biased constrained active network alignment method, which prefers the potential positive links over the negative ones in the query process.

According to [11], under different circumstances, when predicting the existing anchor links, by incorporating the *one-to-one* cardinality constraint into the learning model, it will bring about a much higher accuracy. So in the biased constrained network alignment approach to be introduced here, the learner should firstly apply the existing non-active network alignment models (e.g., MNA [11] as introduced in Sect. 4.3) to predict the potentially positive anchor links in $\mathcal{U}$, which can be recognized as set $\mathcal{U}_+ \subset \mathcal{U}$. In each round, links from set $\mathcal{U}_+$ will be selected to query for their labels, where those which can introduce the maximum information gain can be the optimum here. Different strategies can be applied to rank the anchor links in set $\mathcal{U}_+$. Meanwhile, considering that the anchor links to be inferred are correlated, ranking of the candidates in set $\mathcal{U}_+$ depends on not only these links themselves but also the other links incident to them.

Here, we would like to introduce two new strategies, i.e., *biased likelihood* and *biased entropy*, for ranking the links in $\mathcal{U}_+$. Given two links $l, l' \in \mathcal{L}$, we can use notation $l \cap l'$ to denote the set of shared nodes by l and l'. Given a link $l \in \mathcal{U}_+$, we can represent the links incident to l (i.e., sharing a common node) as set $\Gamma(l) = \{l' | l' \in \mathcal{L}, l \cap l' \neq \emptyset\}$.

Biased Likelihood: By labeling link l to be positive, we can know that links in set $\Gamma(l)$ will be negative by default. The likelihood of such a case can be represented as

$$P(l, \Gamma(l)) = P(y_l = +1 | \mathbf{x}_l) \cdot \prod_{l' \in \Gamma(l)} P(y_{l'} = -1 | \mathbf{x}_{l'}). \tag{6.29}$$

Links in set $\mathcal{U}_+$ can be sorted according to the probability $P(l, \Gamma(l))$ and those with higher probability can be queried for the labels.

Biased Entropy: Besides the likelihood, a similar measure like the entropy can be defined based on the intuition as well, which considers not only the positive link labels but also the uncertainty. For instance, after querying for the label of link l, we can have two scenarios:

- l *is positive*: If link l is positive, links in $\Gamma(l)$ will be negative for sure.
- l *is negative*: If link l is negative, links in $\Gamma(l)$ can be either positive or negative.

Therefore, the uncertainty about the labels of link l and its incident set $\Gamma(l)$ can be represented as

$$H(l, \Gamma(l))$$
$$= P(y_l = +1 | \mathbf{x}_l) \cdot H(l, \Gamma(l) | y_l = +1) + P(y_l = -1 | \mathbf{x}_l) \cdot H(l, \Gamma(l) | y_l = -1). \tag{6.30}$$

In the above equation, we have $H(l, \Gamma(l) | y_l = +1)$ and $H(l, \Gamma(l) | y_l = +1)$ denotes the conditional entropy as follows:

$$H(l, \Gamma(l) | y_l = +1) = - P_M(y_l = +1 | \mathbf{x}_l) log P_M(y_l = +1 | \mathbf{x}_l)$$
$$- \sum_{l' \in \Gamma(l)} P_M(y_{l'} = -1 | \mathbf{x}_{l'}) log P_M(y_{l'} = -1 | \mathbf{x}_{l'}) \tag{6.31}$$

and

$$H(l, \Gamma(l) | y_l = -1) = - P_M(y_l = -1 | \mathbf{x}_l) log P_M(y_l = -1 | \mathbf{x}_l)$$
$$- \sum_{l' \in \Gamma(l)} \sum_{y \in \mathcal{Y}} P_M(y_{l'} = y | \mathbf{x}_{l'}) log P_M(y_{l'} = y | \mathbf{x}_{l'}). \tag{6.32}$$

For the link l with a larger *biased entropy*, we will be less sure about the results of l and its incident neighbor set $\Gamma(l)$. All the potentially positive anchor links in set $\mathcal{U}_+$ can be sorted according to their entropy scores, and those with larger *biased entropy* scores can be picked for labeling.

6.4.2 Active Network Alignment Objective Function

Here, we will use the loss function for labeled anchor links and cardinality constraints introduced in Sect. 6.3, but for the unlabeled anchor links, we propose to further query for the labels of a subset of the data instances. Given the unlabeled anchor link set $\mathcal{U}$, we can denote the subset of anchor links to be selected for querying the labels as $\mathcal{U}_Q$. The true label of link $l \in \mathcal{U}_Q$ after query can be represented as $\tilde{y}_l \in \{+1, -1\}$. The remaining links in set $\mathcal{U}$ can be represented as $\mathcal{U} \setminus \mathcal{U}_Q$, whose labels are still unknown. Based on the loss functions introduced before, depending on whether the labels of links are queried or not, we can further specify the loss function for set $\mathcal{U}$ as

$$
\begin{aligned}
L(f_{\mathbf{w}}, \mathcal{U}) &= L(f_{\mathbf{w}}, \mathcal{U}_Q) + L(f_{\mathbf{w}}, \mathcal{U} \setminus \mathcal{U}_Q) \\
&= \sum_{l \in \mathcal{U}_Q} (\mathbf{w}^\top \mathbf{x}_l - \tilde{y}_l)^2 + \sum_{l \in \mathcal{U} \setminus \mathcal{U}_Q} (\mathbf{w}^\top \mathbf{x}_l - y_l)^2.
\end{aligned}
\tag{6.33}
$$

Here, we need to add more remarks that notation $\tilde{y}_l$ denotes the queried label of link $l \in \mathcal{U}_Q$ which will be a known value, while y_l will be a variable to be inferred in the model for all the links $\mathcal{U} \setminus \mathcal{U}_Q$.

By combining the loss functions for links in different subsets together with the anchor link cardinality constraint, we can represent the objective function for *active network alignment* to be

$$
\min_{\mathbf{w}, \mathbf{y}, \mathcal{U}_Q} \frac{c_1}{2} L(f_{\mathbf{w}}, \mathcal{A}) + \frac{1}{2} \|\mathbf{w}\|_2^2 + \frac{c_2}{2} L(f_{\mathbf{w}}, \mathcal{U}_Q) + \frac{c_3}{2} L(f_{\mathbf{w}}, \mathcal{U} \setminus \mathcal{U}_Q)
$$

$$
s.t. \ |\mathcal{U}_Q| \le b,
$$

$$
y_l \in \{+1, -1\}, \forall l \in \mathcal{L}, \ y_l = +1, \forall l \in \mathcal{A}, \ y_l = \tilde{y}_l, \forall l \in \mathcal{U}_Q,
$$

$$
\mathbf{0} \preccurlyeq \mathbf{A}_o \mathbf{y} \preccurlyeq \mathbf{1}, \ \mathbf{0} \preccurlyeq \mathbf{A}_i \mathbf{y} \preccurlyeq \mathbf{1}, \tag{6.34}
$$

where c_1, c_2, and c_3 denote the weights of the loss terms and b represents the available query budget in the learning process.

As shown in the above objective function, besides the variable $\mathbf{w}$ of the model and the link labels $\mathbf{y}$ to be inferred, we also need to select the optimal node set $\mathcal{U}_Q$ to query for the labels in active learning. The selection of different node subsets can affect the link prediction result greatly, and the selection of the optimal query node set renders the problem to be much more challenging. In the above objective function, the optimal query node selection is actually a combinatorial problem, which is NP-hard with a search space involving $\binom{|\mathcal{U}|}{|\mathcal{U}_Q|}$ different options.

For simplicity, we assume the weights c_1, c_2, c_3 all to be c, i.e., all the links in the networks are assumed to be of similar importance in training. And the new loss term of all the links in $\mathcal{A}, \mathcal{U}_Q$ and $\mathcal{U} \setminus \mathcal{U}_Q$ can be simplified as

$$
\begin{aligned}
&\frac{c_1}{2} L(f_{\mathbf{w}}, \mathcal{A}) + \frac{c_2}{2} L(f_{\mathbf{w}}, \mathcal{U}_Q) + \frac{c_3}{2} L(f_{\mathbf{w}}, \mathcal{U} \setminus \mathcal{U}_Q) \\
&= \frac{c}{2} L(f_{\mathbf{w}}, \mathcal{L}) \\
&= \frac{c}{2} \|\mathbf{w}\mathbf{X} - \mathbf{y}\|_2^2,
\end{aligned}
\tag{6.35}
$$

where matrix $\mathbf{X} = [\mathbf{x}_{l_1}^\top, \mathbf{x}_{l_2}^\top, \ldots, \mathbf{x}_{l_{|\mathcal{L}|}}^\top]^T$ denotes the feature matrix about all these anchor links in the potential anchor link set $\mathcal{L}$.

Here, we can see that the objective function involves multiple variables, like $\mathbf{w}$, $\mathbf{y}$, and the query set $\mathcal{U}_Q$, and the objective is not jointly convex with regarding these variables. What's more, the inference of the label variable $\mathbf{y}$ and the query set $\mathcal{U}_Q$ are both combinatorial problems. In this section, we propose to update the variables alternatively, while fixing the remaining ones, and design a hierarchical alternative variable updating process for solving the problem instead:

1. fix $\mathcal{U}_Q$, and update $\mathbf{y}$ and $\mathbf{w}$,
   (1–1) with fixed $\mathcal{U}_Q$, fix $\mathbf{y}$, update $\mathbf{w}$,
   (1–2) with fixed $\mathcal{U}_Q$, fix $\mathbf{w}$, update $\mathbf{y}$,
2. fix $\mathbf{y}$ and $\mathbf{w}$, and update $\mathcal{U}_Q$.

A remark to be added here: we can see that variable $\mathcal{U}_Q$ is different from the remaining two, which involves the label query process with the oracle subject to the specified budget. To differentiate these two iterations, we call the iterations (1) and (2) as the *external iteration*, while we call (1–1) and (1–2) as the *internal iteration*. Next, we will illustrate the detailed alternative learning algorithm as follows.

- **External Iteration Step (1)**: Fix $\mathcal{U}_Q$, update $\mathbf{y}$, $\mathbf{w}$.
  - **Internal Iteration Step (1–1)**: Fix $\mathcal{U}_Q$, $\mathbf{y}$, update $\mathbf{w}$.
    With $\mathbf{y}, \mathcal{U}_Q$ fixed, we can represent the objective function involving variable $\mathbf{w}$ as

    $$\min_{\mathbf{w}} \frac{c}{2} \|\mathbf{X}\mathbf{w} - \mathbf{y}\|_2^2 + \frac{1}{2} \|\mathbf{w}\|_2^2 . \tag{6.36}$$

    The objective function is a quadratic convex function, and its optimal solution can be represented as

    $$\mathbf{w} = \mathbf{H}\mathbf{y} = c(\mathbf{I} + c\mathbf{X}^\top\mathbf{X})^{-1}\mathbf{X}^\top\mathbf{y}, \tag{6.37}$$

    where $\mathbf{H} = c(\mathbf{I} + c\mathbf{X}^\top\mathbf{X})^{-1}\mathbf{X}^\top$ is a constant matrix. Therefore, the weight vector $\mathbf{w}$ depends only on the $\mathbf{y}$ variable.
  - **Internal Iteration Step (1–2)**: Fix $\mathcal{U}_Q$, $\mathbf{w}$, update $\mathbf{y}$.
    With $\mathcal{U}_Q$, $\mathbf{w}$ fixed, together with the constraint, we know that terms $L(f_\mathbf{w}, \mathcal{A})$, $L(f_\mathbf{w}, \mathcal{U}_Q)$, and $\|\mathbf{w}\|_2^2$ are all constant. And the objective function will be

    $$\min_{\mathbf{y}} \|\mathbf{X}\mathbf{w} - \mathbf{y}\|_2^2$$

    $$s.t.\ y_l \in \{+1, 0\}, \forall l \in \mathcal{U} \setminus \mathcal{U}_Q,$$

    $$y_l = \tilde{y}_l, \forall l \in \mathcal{U}_Q \text{ and } y_l = +1, \forall l \in \mathcal{A},$$

    $$\mathbf{0} \preccurlyeq \mathbf{A}_i\mathbf{y} \preccurlyeq \mathbf{1}, \text{ and } \mathbf{0} \preccurlyeq \mathbf{A}_o\mathbf{y} \preccurlyeq \mathbf{1}. \tag{6.38}$$

    It is an integer programming problem, which has been shown to be NP-hard and no efficient algorithm exists that leads to the optimal solution. Here, we will introduce the greedy link selection algorithm proposed in [20] based on values $\hat{\mathbf{y}} = \mathbf{X}\mathbf{w}$, which has been proven to achieve $\frac{1}{2}$-approximation of the optimal solution.

- **External Iteration Step (2)**: Fix $\mathbf{w}$, $\mathbf{y}$, update $\mathcal{U}_Q$.

  Selecting the optimal set $\mathcal{U}_Q$ at one time involves the search of all the potential b link instance combinations from the unlabeled set $\mathcal{U}$, whose search space is $\binom{|\mathcal{U}|}{b}$, and there is no known efficient approach for solving the problem in polynomial time. Therefore, instead of selecting them all at one time, we propose to choose several link instances greedily in each iteration. Due to the one-to-one constraint, the unlabeled anchor links no longer bear equal information, and querying for labels of potential positive anchor links will be more "informative" compared with negative anchor links. Formally, the strategies introduced in Sect. 6.4.1 can all be applied to rank the links either based on their *biased likelihood* or their *biased entropy*, and we will not introduce them again here.

6.5 Positive and Unlabeled (PU) Network Alignment

In the previous sections, the anchor links to be inferred are subject to the *one-to-one* cardinality constraint, where each user is assumed to have at most one account within one social network. However, in some scenarios, as mentioned in Sect. 4.5, users may create multiple accounts in the same social networks, where each account will be for different purposes, e.g., personal socialization vs professional socialization, or family socialization vs external socialization. In such a case, each user can be connected with multiple anchor links across networks, and the cardinality constraint on the anchor links will become *many-to-one* or *many-to-many* instead. The identification of positive anchor links can no longer help to infer the other potential negative anchor links.

In this section, we will introduce a network alignment to address the aforementioned problem based on the PU learning settings [12]. Before we talk about the detailed information about the PU network alignment model [18, 19], we will first introduce the formulation and the preliminary used in the studied problem.

6.5.1 PU Network Alignment Problem Formulation and Preliminary

Formally, given the partially aligned online social networks $\mathcal{G} = ((G^{(2)}, G^{(1)}), (\mathcal{A}))$ with the set of anchor links $\mathcal{A}$ connecting the shared users across network $G^{(1)}$ and $G^{(2)}$, we can represent the set of existing and non-existing anchor links between these two networks as $\mathcal{A}$ and $\mathcal{U} = \mathcal{U}^{(1)} \times \mathcal{U}^{(2)} \setminus \mathcal{A}$. If these existing and non-existing anchor links are treated as the "*positive*" and "*unlabeled*" anchor links, the task of building a model to infer the existence of anchor links across networks will be formulated as a PU learning problem.

As introduced in [19], across these two networks, we can extract the set of both existing and unidentified anchor links. To differentiate these links, a term named "*connection state*": $z \in \{-1, +1\}$ was introduced in [19]. If a certain link (u, v) is an existing anchor link across the networks, then $z(u, v) = +1$; if (u, v) is an unidentified anchor link, then $z(u, v) = -1$. Meanwhile, besides the "*connection state*," all the anchor links can also have their own *labels*, i.e., $y \in \{-1, +1\}$. In this section, if an anchor link (u, v) is/will be identified to be existing, then $y(u, v) = +1$; if (u, v) is not an anchor link, then $y(u, v) = -1$. As shown in Fig. 6.3, for all the existing anchor links across the networks, their connection states z and labels y are all $+1$, while the connection states z of all initially unidentified anchor links are -1 but the labels y of these anchor links can be either $+1$ or -1, as these unidentified anchor links include both anchor links that should either exist or not exist. These unidentified anchor links are referred to as the unlabeled anchor links in the PU network alignment problem.

Fig. 6.3 Example of
connection states and
labels of links in PU link
prediction

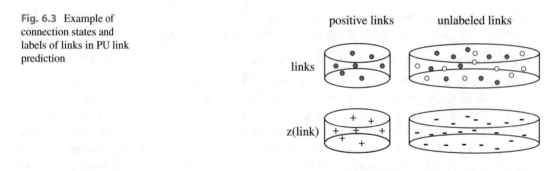

Based on the problem formulation and preliminary concepts introduced above, we will introduce the PU network alignment model in the following part, which will illustrate the relationships of the anchor link *connection state* and their *labels*.

6.5.2   PU Network Alignment Model

For each anchor link, a set of features can be extracted from the networks. For instance, the feature vector extracted for certain anchor/social link (u, v) can be represented as $\mathbf{x}(u, v)$. As a result, each anchor link (u, v) across the networks can be denoted as a tuple $\langle \mathbf{x}(u, v), y(u, v), z(u, v) \rangle$. Let $p(\mathbf{x}, y, z)$ be the joint distribution of $\mathbf{x}$, y, and z. As shown in Fig. 6.3, all the existing links ($z = 1$) are positive links ($y = 1$). In other words, we have

$$p(y = 1|\mathbf{x}, z = 1) = p(y = 1|z = 1) = 1.0. \tag{6.39}$$

A basic assumption in the PU network alignment model is that *the existing positive links are randomly sampled from the whole positive link set* [10], which means that for two arbitrary positive links (u_1, v_1) and (u_2, v_2) we have

$$p(z(u_1, v_1) = 1|\mathbf{x}(u_1, v_1), y(u_1, v_1) = 1)$$
$$= p(z(u_2, v_2) = 1|\mathbf{x}(u_2, v_2), y(u_2, v_2) = 1). \tag{6.40}$$

Based on such an assumption, the conditional distribution $p(z = 1|\mathbf{x}, y = 1)$ is independent of variable $\mathbf{x}$, i.e.,

$$p(z = 1|y = 1) = \sum_{link \in \mathcal{L}} p(z = 1|\mathbf{x}(link), y = 1) \cdot p(\mathbf{x}(link)|y = 1)$$

$$= p(z = 1|\mathbf{x}, y = 1) \cdot \sum_{link \in \mathcal{L}} p(\mathbf{x}(link)|y = 1)$$

$$= p(z = 1|\mathbf{x}, y = 1), \tag{6.41}$$

where $\mathcal{L} = \mathcal{A} \cup \mathcal{U}$ denotes all the potential anchor links across networks.

Meanwhile, the probabilities that anchor link l is predicted to be "existing" ($z = +1$) and "positively labeled" ($y = +1$) can be defined as the *existence probability* (i.e., $p(z = 1|\mathbf{x})$) and "*positively labeled probability*" (i.e., $p(y = 1|\mathbf{x})$), respectively [19]. The relationship between links' "*existence probability*" and "*positively labeled probability*" can be formally represented as follows:

$$p(z = 1|\mathbf{x}) = p(z = 1|\mathbf{x}) \cdot p(y = 1|\mathbf{x}, z = 1) = p(y = 1, z = 1|\mathbf{x})$$

$$= p(y = 1|\mathbf{x}) \cdot p(z = 1|\mathbf{x}, y = 1)$$

$$= p(y = 1|\mathbf{x}) \cdot p(z = 1|y = 1). \tag{6.42}$$

Based on the above equation, the links' *positively labeled probabilities* can be inferred from their *existence probabilities* if we can know $p(z = 1|y = 1)$ in advance, where $p(z = 1|y = 1)$ is called the *bridging probability* formally in the PU network alignment model.

Definition 6.2 (Bridging Probability) The term $p(z = 1|y = 1)$ is formally defined as the *bridging probability* between the existence probability and the positively labeled probability.

The bridging probability can actually be inferred with the binary classification models built with the existing ($z = +1$) and unconnected ($z = -1$) links [10]. To achieve such a goal, we can split all the existing and unconnected links into "training set" and "validation set" via cross validation. Classification models built based on the training set can be applied to the validation set. Let *Pos* be the subset of links that are positive in the validation set. We have

Bridging Probability Inference Equation

$$p(z = 1|y = 1) = \frac{1}{|\mathrm{Pos}|} \sum_{link \in \mathrm{Pos}} p(z = 1|y = 1)$$

$$= \frac{1}{|\mathrm{Pos}|} \sum_{link \in \mathrm{Pos}} p(z = 1|\mathbf{x}, y = 1), \tag{6.43}$$

where $p(z = 1|y = 1) = p(z = 1|\mathbf{x}, y = 1)$ can hold according to proof in the previous part. For links in *Pos*, we have $p(y = 1|\mathbf{x}) = 1$, $p(z = 1|\mathbf{x}, y = -1) = 0$ and $p(y = -1|\mathbf{x}) = 0$. Therefore, we have

$$p(z = 1|y = 1) = \frac{1}{|\mathrm{Pos}|} \sum_{link \in \mathrm{Pos}} (p(z = 1|\mathbf{x}, y = 1)p(y = 1|\mathbf{x})$$

$$+ p(z = 1|\mathbf{x}, y = -1)p(y = -1|\mathbf{x}))$$

$$= \frac{1}{|\mathrm{Pos}|} \sum_{link \in \mathrm{Pos}} p(z = 1|\mathbf{x}). \tag{6.44}$$

As a result, the average existence possibility of links in *Pos* works as an estimator of the bridging probability, which clearly clarifies the correlation between link's existence probability and positively labeled probability. Based on the inferred bridging probability $p(z = 1|y = 1)$, we can predict the positively labeled probabilities of anchor and social links based on their existence probabilities. Meanwhile, inferring the anchor links' *existence probability* (i.e., $p(z = 1|\mathbf{x})$) is an easy task, which can be addressed with the supervised models for anchor link prediction introduced in Sect. 4 perfectly. Based on the links' inferred *existence probability* together with the above *bridging probability*, we will be able to compute the anchor links' *positively labeled probability* as the final output.

6.6   Summary

In this chapter, we introduced a new type of learning tasks using both labeled and unlabeled data instances for model building, which are formally named as the semi-supervised learning tasks. Based on three different semi-supervised learning settings, we talked about the network alignment problem and introduced several different network alignment approaches to address the problem.

To provide readers with the background knowledge about semi-supervised learning, we used one section to introduce the semi-supervised learning problems and algorithms. Based on different assumptions, e.g., smoothness assumption, cluster assumption, and manifold assumption, existing semi-supervised learning models adopt different ways to use the unlabeled data instances. Three different semi-supervised learning models, i.e., S3VM, graph based models, and generative models, were introduced in this section. In addition to regular semi-supervised learning, active learning and PU learning were introduced as two special cases of the semi-supervised learning problem.

We introduced a model similar to S3VM to resolve the network alignment task with both labeled and unlabeled anchor link instances. However, slightly different from the regular semi-supervised learning tasks, the anchor links studied in the network alignment problem are not independent, which are strongly correlated with each other due to the one-to-one cardinality constraint. By modeling the anchor link cardinality constraint as a mathematical constraint on node degrees, the semi-supervised network alignment problem was formulated as an optimization problem instead.

We also introduced a network alignment model based on active learning in this section, which can help address the lack of training data problem. Instead of requiring a large number of training data initially, the introduced active network alignment model adopted an active query strategy to get the labels of unlabeled anchor links from an oracle subject to a pre-specified query budget. However, due to the one-to-one cardinality constraint on anchor links, the information that can be provided by the positive and negative anchor link is no longer balanced. Two query strategies, i.e., biased likelihood and biased entropy, were introduced for anchor link selection.

At the end of this chapter, we introduced to model the network alignment problem as a positive and unlabeled (PU) learning problem. In the PU network alignment tasks, the anchor links will be represented as tuples, involving the feature representation, connection state, and label, respectively. Based on the correlation between the existence probability and positively labeled probability, the introduced model is able to infer the anchor links positively labeled probabilities from their existence probabilities effectively.

6.7   Bibliography Notes

For the readers who are interested in the semi-supervised learning, the textbook [8] is highly recommended, which provides a comprehensive introduction about semi-supervised learning problems and algorithms. The readers can also take a look at the survey article [23], which covers the literature review of the semi-supervised learning algorithms and potential applications. A survey about active learning is available in [16], and the PU learning papers include [10, 12].

The PU network alignment problem was initially proposed in [18], which proposed a semi-supervised method to infer both anchor links and social links simultaneously across aligned social networks. The proposed method analyzes the correlation between links existence probability and formation probability (i.e., positively labeled probability introduced in Sect. 6.5). Active network alignment is studied in [14, 25] for the first time. Via a step-wise selection of anchor links for label query, active network alignment models are capable to achieve very good performance with a small amount of labeling efforts. The semi-supervised network alignment problem formulation was originally introduced in [14, 20, 21], and the readers may also refer to these academic papers for more information about relating works.

6.8   Exercises

1. (Easy) In the S3VM model, we use Eq. (6.1) to denote the loss for the unlabeled data instances. Please briefly explain why it can work well to utilize information of unlabeled data instances in learning the models.
2. (Easy) In the *local and global consistency model*, please briefly talk about the physical meaning of the regularization term $reg = \mathbf{f}^\top \mathbf{L}_n \mathbf{f}$ (as indicated in Eq. (6.10)), and explain why it can work well in regularizing the model.
3. (Easy) Please briefly introduce what is active learning, and provide the advantages of active learning compared against other learning tasks.
4. (Easy) According to Sect. 6.2.4, please briefly explain the ideas of PU learning and the circumstances where it can work well.
5. (Medium) Please read article [9], and briefly introduce how to learn the semi-supervised generative model based on the EM algorithm.
6. (Medium) When introducing the active network alignment model in Sect. 6.4, we mention that "positive and negative anchor links may have different amounts of information." Please explain what does that sentence mean, and briefly talk about why the introduced biased query strategies we introduced in Sect. 6.4.1.2 can work better than regular query strategies, e.g., the *entropy* based strategy introduced in Sect. 6.4.1.1.
7. (Medium) Please briefly introduce the *bridging probability inference equation* used in Sect. 6.5, and explain why it can work in inferring the potential *bridging probability*.
8. (Hard) Please use your preferred programming language to implement the semi-supervised learning algorithm introduced in Sect. 6.3, and compare its advantages over the regular supervised network alignment approach that we introduced in Sect. 4.3 with experiments on some toy data sets.
9. (Hard) Please try to implement the active network alignment algorithm introduced in Sect. 6.4 with one of your preferred programming language, and test its effectiveness on a toy network data set.
10. (Hard) Please try to implement the PU network alignment algorithm introduced in Sect. 6.5 with one of your preferred programming language, and test its effectiveness on a toy network data set.

References

1. E. Allgower, K. Georg, *Numerical Continuation Methods: An Introduction* (Springer, Berlin, 1990)
2. M. Belkin, I. Matveeva, P. Niyogi, Regularization and semi-supervised learning on large graphs, in *Learning Theory*, ed. by J. Shawe-Taylor, Y. Singer (2004)
3. M. Belkin, P. Niyogi, V. Sindhwani, Manifold regularization: a geometric framework for learning from labeled and unlabeled examples. J. Mach. Learn. Res. **7**, 2399–2434 (2006)

4. K. Bennett, A. Demiriz, Semi-supervised support vector machines, in *Proceedings of the 1998 Conference on Advances in Neural Information Processing Systems II* (1999)
5. S. Berchtold, C. Bohm, H. Kriegel, The pyramid-technique: towards breaking the curse of dimensionality, in *Proceedings ACM SIGMOD International Conference on Management of Data* (1998)
6. A. Blum, S. Chawla, Learning from labeled and unlabeled data using graph Mincuts, in *Proceedings of the Eighteenth International Conference on Machine Learning (ICML '01)* (2001)
7. A. Blum, T. Mitchell, Combining labeled and unlabeled data with co-training, in *Proceedings of the Eleventh Annual Conference on Computational Learning Theory (COLT' 98)* (1998)
8. O. Chapelle, B. Schlkopf, A. Zien, *Semi-Supervised Learning*, 1st edn. (The MIT Press, Cambridge, 2010)
9. A. Dempster, N. Laird, D. Rubin, Maximum likelihood from incomplete data via the EM algorithm. J. R. Stat. Soc. Ser. B **39**(1), 1–38 (1977)
10. C. Elkan, K. Noto, Learning classifiers from only positive and unlabeled data, in *Proceedings of the 14th ACM SIGKDD International Conference on Knowledge Discovery and Data Mining (KDD '08)* (2008)
11. X. Kong, J. Zhang, P. Yu, Inferring anchor links across multiple heterogeneous social networks, in *Proceedings of the 22nd ACM International Conference on Information & Knowledge Management (CIKM '13)* (2013)
12. B. Liu, Y. Dai, X. Li, W. Lee, P. Yu, Building text classifiers using positive and unlabeled examples, in *Third IEEE International Conference on Data Mining* (2003)
13. L. Manevitz, M. Yousef, One-class SVMs for document classification. J. Mach. Learn. Res. **2**, 139–154 (2002)
14. Y. Ren, C.C. Aggarwal, J. Zhang, Meta diagram based active social networks alignment. CoRR (2019). http://arxiv.org/abs/1902.04220
15. K. Rose, Deterministic annealing for clustering, compression, classification, regression, and related optimization problems. Proc. IEEE **86**(11), 2210–2239 (1998)
16. B. Settles, Active learning literature survey. Computer sciences technical report, University of Wisconsin–Madison (2009)
17. A. Yuille, A. Rangarajan, The concave-convex procedure (CCCP), in *Proceedings of the 14th International Conference on Neural Information Processing Systems: Natural and Synthetic (NIPS'01)* (2002)
18. J. Zhang, P. Yu, Integrated anchor and social link predictions across partially aligned social networks, in *Proceedings of the 24th International Conference on Artificial Intelligence (IJCAI'15)* (2015)
19. J. Zhang, P. Yu, Z. Zhou, Meta-path based multi-network collective link prediction, in *Proceedings of the 20th ACM SIGKDD International Conference on Knowledge Discovery and Data Mining (KDD '14)* (2014)
20. J. Zhang, J. Chen, J. Zhu, Y. Chang, P. Yu, Link prediction with cardinality constraints, in *Proceedings of the Tenth ACM International Conference on Web Search and Data Mining (WSDM '17)* (2017)
21. J. Zhang, J. Chen, S. Zhi, Y. Chang, P. Yu, J. Han, Link prediction across aligned networks with sparse low rank matrix estimation, in *2017 IEEE 33rd International Conference on Data Engineering (ICDE)* (2017)
22. D. Zhou, O. Bousquet, T. Lal, J. Weston, B. Schölkopf, Learning with local and global consistency, in *Proceedings of the 16th International Conference on Neural Information Processing Systems (NIPS'03)* (2003)
23. X. Zhu, Semi-supervised learning literature survey. Technical Report 1530, Computer Sciences, University of Wisconsin-Madison (2005)
24. X. Zhu, Semi-supervised Learning with Graphs. PhD thesis, Pittsburgh, PA, USA, 2005. AAI3179046
25. J. Zhu, J. Zhang, Q. Wu, Y. Jia, B. Zhou, X. Wei, P. Yu, Constrained active learning for anchor link prediction across multiple heterogeneous social networks. Sensors **17**(8), 1786 (2017)

Part III

**Broad Learning: Knowledge Discovery
Across Aligned Networks**

Link Prediction

7

7.1 Overview

Given a screenshot of the online social networks, the problem of inferring the missing links or the links to be formed in the networks in the future is called the *link prediction* problem. Link prediction problem has concrete applications in the real world, and many concrete services can be cast to the link prediction problem. For instance, the friend recommendations problem [15] in online social networks can be modeled as the friendship link prediction problem among users. Users' trajectory prediction problem [4] can be formulated as the prediction task of potential check-in links between users and offline POIs (points of interest). The user identifier resolution problem [19, 26] across networks (i.e., the network alignment problem introduced in the previous chapters) can be modeled as the anchor link prediction problem of user accounts across different online social networks.

In this chapter, we will take the friendship link as an example to introduce the general link prediction problem in online social networks. Formally, given the training set $\mathcal{T}_{train}$ involving links belong to different classes ($\mathcal{Y} = \{+1, -1\}$, where $+1$ denotes the positive class and -1 denotes the negative class; sometimes we also use 0 to denote the negative class) and the test set $\mathcal{T}_{test}$ (with unknown labels for the links), the link prediction problem aims at building a mapping $f : \mathcal{T}_{train} \cup \mathcal{T}_{test} \rightarrow \mathcal{Y}$ to project these links to their potential labels in $\mathcal{Y}$.

Depending on the scenarios where we study the link prediction problem, the existing link prediction works can be divided into several different categories. Traditional link prediction problems are mainly focused on inferring the links in one single homogeneous network, like inferring the friendship links [62] among users in online social networks or co-author links [39] in bibliographic networks. As the network structures are becoming more and more complicated, many complex network structures can be modeled as the heterogeneous networks involving different types of nodes and complex connections among them. The heterogeneity of the networks leads to many new link prediction problems, like predicting the links between nodes belonging to different categories [58] and the concurrent inference of multiple types of links in the heterogeneous networks [54, 58]. In recent years, many online social networks have appeared, and lots of new research opportunities exist for researchers and practitioners to study the link prediction problems from the cross-network perspective [57, 58, 61].

Meanwhile, depending on the learning settings used in the link prediction problem formulation and models, the existing link prediction works can be categorized into different groups according to the supervision information involved in the model building. For some of the link prediction models, they

© Springer Nature Switzerland AG 2019
J. Zhang, P. S. Yu, *Broad Learning Through Fusions*,
https://doi.org/10.1007/978-3-030-12528-8_7

calculate the user-pair closeness [53] as the link prediction result without any training data, which are referred to as the *unsupervised link prediction models*. For some other models, they will assign the links with different labels, and use them as the training set to learn a supervised classification models as the base model instead. These models are called the *supervised link prediction models* [14]. Usually, manual labeling of the links is very expensive and tedious. In recent years, many of the works have proposed to apply *semi-supervised learning* techniques [54, 59, 61] in the link prediction problem to utilize the links without labels.

In this chapter, we will introduce the social link prediction problems in online social networks. In Sect. 7.2, we will introduce the social link prediction works in one single homogeneous social network, and the models involve the unsupervised models [23], classification based supervised model [14, 25], and the matrix factorization based link prediction model [1]. The link prediction works in the heterogeneous networks will be introduced in Sect. 7.3, including both the supervised link prediction model [14, 25] and the prediction task of multiple types of links [54, 58]. In the following sections, we will talk about the link prediction problem across multiple heterogeneous social networks. In Sect. 7.4, we introduce a novel cross-network social link prediction model to predict social links for new users specifically [57]. In Sect. 7.5, we will focus on introducing the social link prediction model with positive and unlabeled (PU) learning models [54, 59]. Finally, to overcome the domain difference problem, in Sect. 7.6, we will introduce a matrix estimation based social link prediction model with both positive and unlabeled links [61].

7.2 Traditional Single Homogeneous Network Link Prediction

Traditional link prediction problems are mainly studied based on one single homogeneous network, involving one single type of nodes and links. In this section, we will first briefly introduce how to use the social closeness measures [23, 53] introduced in Sect. 3.3.3 for the link prediction tasks. To integrate different social closeness measures together for the link prediction, we will introduce the supervised link prediction model [14]. Some models formulate the link prediction task as a recommendation problem, and propose to apply the matrix factorization method [1] to address the problem. In this section, we will introduce these three types of link prediction models for the traditional one single homogeneous network.

7.2.1 Unsupervised Link Prediction

Given a screenshot of a homogeneous network $G = (\mathcal{V}, \mathcal{E})$, the unsupervised link prediction models [23, 53] introduced here aim at inferring the potential links that will be formed in the future. Usually, the unsupervised link prediction models will calculate some metrics for the links, which will be used as the predicted confidence scores for these links. Depending on the specific scenario and the link formation assumptions applied, different metrics have been proposed for the link prediction tasks already.

Many of the link prediction metrics are based on the assumption that "*close users are more likely to be friends*," and use the social closeness measure as the link prediction confidence score. The social closeness measures introduced in Sect. 3.3.3, like the *local closeness measures* (e.g., common neighbor, Jaccard's coefficient, Adamic/Adar), the *global path based closeness measures* (e.g., shortest path, Katz), and the *random walk based closeness measures* (e.g., hitting time, commute time, and cosine similarity), can all be used to infer potential connections among users.

In this part, we will not talk about these measures again, and the readers may refer to the previous section for more information. Next, we will introduce the general learning settings and evaluation metrics for the link prediction problem with the unsupervised learning models.

7.2.1.1 Unsupervised Link Prediction Problem Setting

Suppose in the network $G = (\mathcal{V}, \mathcal{E})$, each link is associated with a timestamp. For instance, for link $e \in \mathcal{E}$, we can denote the formation timestamp of link e as $t(e)$. Given three time points $t_p < t_c < t_f$ denoting a past time point t_p, the current time point t_c, and a future time point t_f, we can retrieve the network structure formed in the time range $[t_p, t_c]$ as the current network G_{t_p, t_c} and the network structure to be formed in the future time range $(t_c, t_f]$ as the future network G_{t_c, t_f} respectively. The current network can serve as the input of a link prediction algorithm, i.e., G_{t_p, t_c}, which can infer the new connections to be formed in the future time range $(t_c, t_f]$, i.e., the future network structure G_{t_c, t_f}.

As the network structure evolves, new links will be formed and new nodes will be formed as well. However, for a new node to join in the network in the time range $(t_c, t_f]$, we have no historical knowledge about it and can hardly predict links incident to it, which is also referred to as the *cold start* problem [57,63]. Generally, in the link prediction problems, we have a subset of the nodes as the *core set* $\mathcal{V}^c \subset \mathcal{V}$, and we will be focused on studying the links incident to nodes in the *core set* only. For all the new links to be formed in time range $(t_c, t_f]$, we will sample a subset of the links incident to these nodes in the *core set* only to study the link prediction problem. For the cold start link prediction problem regarding the new users, we will address it in Sect. 7.4 specifically.

Given the current network structure G_{t_p, t_c} and the *core set* $\mathcal{V}^c$, we can represent the formed links among the *core set* users as set $\mathcal{E}^c_{t_p, t_c} = \mathcal{E}_{t_p, t_c} \cap \mathcal{V}^c \times \mathcal{V}^c$. Meanwhile, the remaining links among the *core set* users can be represented as set $\mathcal{E}^r_{t_p, t_c} = \mathcal{V}^c \times \mathcal{V}^c \setminus (\{(u, u)\}_{u \in \mathcal{V}^c} \cup \mathcal{E}^c_{t_p, t_c})$. The link prediction model aims inferring: "among all the links in $\mathcal{E}^r_{t_p, t_c}$, which will be formed in the time range $(t_c, t_f]$ and appear in the future network structure G_{t_c, t_f} (i.e., in set $\mathcal{E}^c_{t_c, t_f} = \mathcal{E}_{t_c, t_f} \cap \mathcal{V}^c \times \mathcal{V}^c$)".

7.2.1.2 Unsupervised Link Prediction Models

In the unsupervised link prediction model, we use the *social closeness* as the prediction confidence measure based on the assumption that "*close users tend to be friends*". In Table 7.1, we summarize some closeness measures we have introduced before. For instance, if we use "*Common Neighbor*" as the social closeness measure, we can represent the closeness score of all the user pairs in set $\mathcal{E}^r_{t_p, t_c}$ as $\{C(e)\}_{e \in \mathcal{E}^r_{t_p, t_c}}$.

The scores of all these remaining links, i.e., $\{C(e)\}_{e \in \mathcal{E}^r_{t_p, t_c}}$, can be outputted as the result. Meanwhile, when determining the links to be recommended for each user in the *core set*, we can pick either the *top-k* links with the highest predicted scores or set a threshold to select the links with scores greater than the threshold as the ones to be formed. In other words, these selected links will be assigned with positive labels, while the remaining unselected links will be assigned with negative labels instead.

Table 7.1 Features extracted from user pair u and v from the homogeneous network

Features	Closeness measures
Common Neighbor (CN)	$\Gamma(u) \cap \Gamma(v)$
Jaccard's Coefficient (JC)	$\frac{|\Gamma(u) \cap \Gamma(v)|}{|\Gamma(u) \cup \Gamma(v)|}$
Adamic/Adar (AA)	$\sum_{w \in (\Gamma(u) \cap \Gamma(v))} \frac{1}{\log|\Gamma(w)|}$
Preferential Attachment (PA)	$|\Gamma(u)| \cdot |\Gamma(v)|$
Shortest Path (SP)	$\min\{|p|\}_{p \in \mathcal{P}_{u,v}}$
Katz	$\sum_{l=1}^{l_{max}} \beta^l |\mathcal{P}^l_{u,v}|$

7.2.1.3 Unsupervised Link Prediction Result Evaluation

Given all the remaining links in set $\mathcal{E}^r_{t_p,t_c}$ and the newly formed links in the future network G_{t_c,t_f}, we can represent their ground truth labels as vector $\mathbf{y} \in \{-1, +1\}^{|\mathcal{E}^r_{t_p,t_c}| \times 1}$. For all the links formed in the future network G_{t_c,t_f}, we can assign them with label $+1$, while for the remaining links, they will be assigned with label -1 instead.

Given the calculated scores, e.g., $\{C(e)\}_{e \in \mathcal{E}^r_{t_p,t_c}}$, and the ground truth label vector, we can evaluate the performance of the link prediction model by calculating the AUC score (i.e., the area under ROC curve). Among all the links, the top k links can be picked, the prediction result can also be evaluated with metrics like $nDCG@k$ [18].

Meanwhile, if the top k links are selected to assign with labels, the output of the link prediction model will be the prediction label of these links. By comparing them with the ground truth label vector, metrics like Precision, Recall, F1, and Accuracy (at top k) can be calculated as the performance evaluation results.

7.2.2   Supervised Link Prediction

In some cases, links in the networks are explicitly categorized into different groups, like links denoting friends vs those representing enemies, friends (with connections) vs strangers (no connections). Given a set of labeled links, e.g., set $\mathcal{E}$, containing links belonging to different classes, the *supervised link prediction* problem [14] aims at building a supervised learning model to address the link prediction problem. The learned model will be applied to determine the labels of links in the test set. In this part, we still take the link formation problem as an example to illustrate the supervised link prediction model.

7.2.2.1 Supervised Link Prediction Problem Setting

Given the network structure $G = (\mathcal{V}, \mathcal{E})$ with the formed links in set $\mathcal{E}$, we can represent all the potential links among users in network G as set $\mathcal{L} = \mathcal{V} \times \mathcal{V} \setminus \{(u, u)\}_{u \in \mathcal{V}} \setminus \mathcal{E}$. These existing links can be labeled as the positive training set, while a subset of links in $\mathcal{L}$ are identified as the links will never be formed, which can be denoted as $\mathcal{L}_n \subset \mathcal{L}$ and labeled as the negative training set. These positively and negatively labeled links can be treated as the training set, i.e., $\mathcal{L}_{train} = \mathcal{E} \cup \mathcal{L}_n$, and the remaining links with unknown labels can be used as the testing set $\mathcal{L}_{test} = \mathcal{L} \setminus \mathcal{L}_n$. In the supervised link prediction problem, we aim at building a supervised classification/regression model with the training set $\mathcal{L}_{train}$ and apply the learned model to infer the label of links in the testing set $\mathcal{L}_{test}$.

7.2.2.2 Supervised Link Prediction Feature Extraction

To represent each of the social links, like link $l = (u, v) \in \mathcal{E}$ between nodes u and v, a set of features representing the characteristics of the link l as well as nodes u, v will be extracted in the model building. Normally, the features can be extracted for links in the prediction task can be divided into two categories:

- *Features of Nodes*: The characteristics of the nodes can be denoted by the measures introduced in Chap. 3.3, like these various node centrality measures. For instance, for the link (u, v), based on the known links in the training set, we can compute the centrality measures based on degree, normalized degree, eigen-vector, Katz, PageRank, Betweenness of nodes u and v as part of the features for link (u, v), which can be denoted as vectors $\mathbf{x}_u$ and $\mathbf{x}_v$ respectively.

- *Features of Links*: The characteristics of the links in the networks can be calculated by computing the closeness between the nodes composing the nodes. For instance, for link (u, v), based on the known links in the training set, we can compute the closeness measures based on common neighbor, Jaccard's coefficient, Adamic/Adar, shortest path, Katz, hitting time, commute time, etc. between nodes u and v as the features for link (u, v), which can be denoted as vector $\mathbf{x}_{u,v}$ formally.

We can append the features for nodes u, v and those for link (u, v) together and represent the extracted feature vector for link $l = (u, v)$ as vector $\mathbf{x}_l = [\mathbf{x}_u^\top, \mathbf{x}_v^\top, \mathbf{x}_{u,v}^\top]^\top \in \mathbb{R}^k$ of length k.

7.2.2.3 Supervised Link Prediction Model

With the training set $\mathcal{L}_{train}$, we can represent the feature vectors and labels for the links in $\mathcal{L}_{train}$ as the training data $\{(\mathbf{x}_l, y_l)\}_{l \in \mathcal{L}_{train}}$. Meanwhile, with the testing set $\mathcal{L}_{test}$, we can represent the features extracted for the links in it as $\{\mathbf{x}_l\}_{l \in \mathcal{L}_{train}}$. Different classification models can be used as the base model for the link prediction task, like the decision tree model, artificial neural network model, and support vector machine (SVM) model introduced in Sect. 2.3. These models can be trained with the training data, and the labels of links in the testing set can be determined by applying models to the testing data instances.

Depending on the specific models being applied, the output of the link prediction result can include (1) the predicted labels of the links in $\mathcal{L}_{test}$, and (2) the prediction confidence scores/probability scores of links in $\mathcal{L}_{test}$.

7.2.2.4 Supervised Link Prediction Result Evaluation

Different evaluation metrics can be used for measuring the performance of the link prediction models. For the models producing the prediction labels of the test set, evaluation metrics like precision, recall, F1, and accuracy can be used in performance evaluation. Meanwhile, for the models producing the confidence score list as the output, evaluation metrics like AUC, and Precision@k, nDCG@k can be used in performance evaluation. For these metrics aforementioned, higher evaluation scores will correspond to better link prediction performance.

7.2.3 Matrix Factorization Based Link Prediction

Besides unsupervised and supervised link prediction models, many other methods based on matrix factorization can also be applied to solve the link prediction task in homogeneous networks [1, 10, 42].

7.2.3.1 Matrix Factorization Based Link Prediction Problem Setting

Given a homogeneous social network $G = (\mathcal{V}, \mathcal{E})$ and the existing social links among users in set $\mathcal{E}$, the remaining potential links among users can be represented as $\mathcal{L} = \mathcal{V} \times \mathcal{V} \setminus \{(u, u)\}_{u \in \mathcal{V}} \setminus \mathcal{E}$. The links in set $\mathcal{E}$ are the formed links and can be labeled as the positive instances, while those in set $\mathcal{L}$ contain both the links to be formed and those will never be formed (i.e., involve both positive and negative links) and should be unlabeled.

The training set available involves both the positively labeled links in set $\mathcal{E}$ and the unlabeled links in set $\mathcal{L}$. The testing set is the unlabeled set $\mathcal{L}$, and we aim at inferring the labels of these potential links with a matrix factorization based approach.

7.2.3.2 Matrix Factorization Based Link Prediction Model

Formally, given the homogeneous social network $G = (\mathcal{V}, \mathcal{E})$ and the existing social links among users in set $\mathcal{E}$, we can organize these links into the social adjacency matrix $\mathbf{A} \in \{0, 1\}^{|\mathcal{V}| \times |\mathcal{V}|}$. Given the adjacency matrix $\mathbf{A}$ of network G, a low-rank compact representation matrix, $\mathbf{U} \in \mathbb{R}^{|\mathcal{V}| \times d}, d < |\mathcal{V}|$, can be used to store the social information for each user in the network. Matrix $\mathbf{U}$ can be obtained by solving the following optimization objective function:

$$\min_{\mathbf{U},\mathbf{V}} \left\| \mathbf{A} - \mathbf{UVU}^\top \right\|_F^2 , \tag{7.1}$$

where $\mathbf{U}$ is the low rank matrix and matrix $\mathbf{V}$ contains the correlation among the rows of $\mathbf{U}$, $\|\cdot\|_F$ denotes the Frobenius norm of the matrix.

To avoid overfitting, regularization terms $\|\mathbf{U}\|_F^2$ and $\|\mathbf{V}\|_F^2$ are added to the object function as follows [42]:

$$\min_{\mathbf{U},\mathbf{V}} \left\| \mathbf{A} - \mathbf{UVU}^\top \right\|_F^2 + \alpha \cdot \|\mathbf{U}\|_F^2 + \beta \cdot \|\mathbf{V}\|_F^2 ,$$

$$s.t. \mathbf{U} \geq \mathbf{0}, \mathbf{V} \geq \mathbf{0}, \tag{7.2}$$

where α and β are the weights of terms $\|\mathbf{U}\|_F^2$, $\|\mathbf{V}\|_F^2$ respectively.

This object function is very hard to achieve the global optimal result for both $\mathbf{U}$ and $\mathbf{V}$. A alternative optimization schema can be used here, which can update $\mathbf{U}$ and $\mathbf{V}$ alternatively. The Lagrangian function of the object equation should be:

$$\mathcal{F} = Tr(\mathbf{AA}^\top) - Tr(\mathbf{AUV}^\top \mathbf{U}^\top)$$

$$- Tr(\mathbf{UVU}^\top \mathbf{A}^\top) + Tr(\mathbf{UVU}^\top \mathbf{UV}^\top \mathbf{U}^\top)$$

$$+ \alpha Tr(\mathbf{UU}^\top) + \beta Tr(\mathbf{VV}^\top) - Tr(\Theta \mathbf{U}) - Tr(\Omega \mathbf{V}) \tag{7.3}$$

where Θ and Ω are the multipliers for the constraints on $\mathbf{U}$ and $\mathbf{V}$ respectively.

By taking derivatives of $\mathcal{F}$ with regard to $\mathbf{U}$ and $\mathbf{V}$ respectively, the partial derivatives of $\mathcal{F}$ will be

$$\frac{\partial \mathcal{F}}{\partial \mathbf{U}} = -2\mathbf{A}^\top \mathbf{UV} - 2\mathbf{AUV}^\top + 2\mathbf{UV}^\top \mathbf{U}^\top \mathbf{UV}^\top$$

$$+ 2\mathbf{UVU}^\top \mathbf{UV}^\top + 2\alpha \mathbf{U} - \Theta^\top \tag{7.4}$$

$$\frac{\partial \mathcal{F}}{\partial \mathbf{V}} = -2\mathbf{U}^\top \mathbf{AU} + 2\mathbf{U}^\top \mathbf{UVU}^\top \mathbf{U} + 2\beta \mathbf{V} - \Omega^\top \tag{7.5}$$

By making $\frac{\partial \mathcal{F}}{\partial \mathbf{U}} = \mathbf{0}$ and $\frac{\partial \mathcal{F}}{\partial \mathbf{V}} = \mathbf{0}$ and using the KKT complementary condition, we can get:

$$\mathbf{U}(i, j) \leftarrow \mathbf{U}(i, j) \sqrt{\frac{\left(\mathbf{A}^\top \mathbf{UV} + \mathbf{AUV}^\top\right)(i, j)}{\left(\mathbf{UV}^\top \mathbf{U}^\top \mathbf{UV} + \mathbf{UVU}^\top \mathbf{UV}^\top + \alpha \mathbf{U}\right)(i, j)}}, \tag{7.6}$$

$$\mathbf{V}(i, j) \leftarrow \mathbf{V}(i, j) \sqrt{\frac{\left(\mathbf{U}^\top \mathbf{AU}\right)(i, j)}{\left(\mathbf{U}^\top \mathbf{UVU}^\top \mathbf{U} + \beta \mathbf{V}\right)(i, j)}}. \tag{7.7}$$

The low-rank matrix $\mathbf{U}$ captures the information of each user from the adjacency matrix. The matrix $\mathbf{U}$ can be used in different ways. For instance, each row of $\mathbf{U}$ represents the *latent feature vectors* of users in the network, which can be used in many link prediction models, e.g., supervised link prediction models. Meanwhile, based on the matrix $\mathbf{V}$ learnt from the model, we can also represent the predicted score of link (u, v) as $\mathbf{U}_u \mathbf{V} \mathbf{U}_v^\top$, where notations $\mathbf{U}_u$ and $\mathbf{U}_v$ represent the rows in matrix $\mathbf{U}$ corresponding to users u and v, respectively.

7.2.3.3 Matrix Factorization Based Link Prediction Result Evaluation

Given the ground-truth labels of links in the unlabeled set $\mathcal{L}$ and their corresponding inferred scores based on the matrices $\mathbf{U}$ and $\mathbf{V}$ learnt from the model, we can evaluate the performance of the model with metrics like AUC and Precision@k as well as nDCG@k. In the exercise at the end of this chapter, we will ask the readers to try to implement the above link prediction algorithms with a preferred programming language, and compare their performance in inferring the social links within a homogeneous network.

7.3 Heterogeneous Network Collective Link Prediction

Homogeneous networks with one single type of nodes/links is a very simple network representation. In the real-world online social networks, there usually exist many different kinds of nodes, like *users, offline POIs, posts*. Users can also perform various kinds of actions, like *follow other users* and *check-in at some places*, which will create very complex connections among these nodes. Formally, for the online social networks with such a complex structure, they are called the heterogeneous information networks. There exist very diverse online social networks in the real world. In this section, we will be mainly focused on the online social networks providing the geographic services, which are called the location based social networks (LBSNs) [7], and study the *collective link prediction* task based on the LBSNs [58].

7.3.1 Introduction to LBSNs

Location-based social networks (LBSNs) are one kind of online social networks that can provide geographic services, e.g., location check-ins and posting reviews, and have been attracting much attention in recent years [7, 33, 45, 48, 49]. LBSNs usually have very complex structures, including multiple kinds of nodes (e.g., users, locations, etc.) and different types of links among these nodes (e.g., social links among users and location links between users and locations). For example, Foursquare[1] is a mainstream LBSN. It involves millions of users and locations. Foursquare users can add friends, check-in at different locations with their mobile phones, write reviews, and share the locations with their friends.

Many important services offered by LBSNs can be cast as the link prediction problems. For example, friend recommendation involves predicting social links among users; location recommendation aims at predicting location links between users and locations. LBSNs can benefit a lot from the high-quality social link and location link prediction results. The reason is that well-established social ties can improve user's engagement in social networks [21]. Meanwhile, in location-based social networks, high-quality predicted location links can enhance the value of the location services in the networks.

Conventional link prediction researches on LBSNs mostly focus on predicting either social links [33, 45] or location links [7, 48] and usually assume that the prediction tasks of different types of links to be independent. However, in many real-world LBSNs, the link prediction tasks for social links

---

[1]https://foursquare.com.

and location links are strongly correlated and mutually influential to each other [58]. For example, if two users are friends with each other, they are more likely to check-in at similar locations. Thus the performance of location recommendation can be significantly improved if we could make accurate friendship predictions. Similarly, if two users often check-in at similar locations, they are more likely to know each other and make friends in the real life. Viewed in this perspective, the performance of friend recommendation can be greatly improved if we could make accurate location-link predictions.

7.3.2 Collective Link Prediction

In this section, we study the collective link prediction problem for LBSNs as introduced in [58] and the links to be predicted include both social links and location links. The problem is very challenging to solve due to the fact that social links and location links in LBSNs are correlated instead of being independent. The prediction tasks on social links and location links should be considered at the same time. Many existing works mainly focus on predicting one single type of links in LBSNs [7,33,45,48], which fail to consider the correlations between different link prediction tasks.

In the following part, we will introduce a supervised collective linkage transferring method, TRAIL (TRAnsfer heterogeneous lInks across LBSNs), proposed in [58] to address the above challenges. TRAIL can accumulate auxiliary information for locations from online posts which have check-ins at them and can extract heterogeneous features for both social links and location links. TRAIL can predict social links and location links simultaneously.

Let $G = (\mathcal{V}, \mathcal{E})$ be the networks studied in this section, where $\mathcal{V} = \bigcup_i \mathcal{V}_i$ is the union of different types of nodes and $\mathcal{V}_i, i \in \{1, 2, \ldots, \}$ is the set of nodes of the i_{th} type. $\mathcal{E} = \bigcup_j \mathcal{E}_j$ is the union of link sets among nodes in $\mathcal{V}$ and $\mathcal{E}_j, j \in \{1, 2, \ldots\}$ is the set of links of the j_{th} type. Specially, for a LBSN, node set $\mathcal{V} = \mathcal{U} \cup \mathcal{L} \cup \mathcal{T} \cup \mathcal{W}$ is the union of node sets of users, locations, time, and words. The link set $\mathcal{E} = \mathcal{E}_s \cup \mathcal{E}_l \cup \mathcal{E}_t \cup \mathcal{E}_w$ is the union of link sets consisting of social friendships links, and the links between users with location check-ins, active time, and published words, respectively.

Given an LBSN $G = (\mathcal{V}, \mathcal{E})$ with the existing social links $\mathcal{E}_s$ and location links $\mathcal{E}_l$, what we want to predict in the studied network are a subset of potential social links among users in G: $\mathcal{L}_s \subset (\mathcal{U} \times \mathcal{U} \setminus \mathcal{E}_s)$ and a subset of potential location check-in links in G: $\mathcal{L}_l \subset (\mathcal{U} \times \mathcal{L} \setminus \mathcal{E}_l)$. In other words, we want to build a mapping: $f : \{\mathcal{L}_s, \mathcal{L}_l\} \to \{-1, 1\}$ to decide whether potential links in $\{\mathcal{L}_s, \mathcal{L}_l\}$ exist or not and a confidence score function $P : \{\mathcal{L}_s, \mathcal{L}_l\} \to [0, 1]$ denoting their existence probabilities. In the following parts, we will introduce the supervised collective link transferring method, TRAIL, in detail to address the problem.

7.3.3 Information Accumulation and Feature Extraction

TRAIL is based on a supervised learning setting and, as a result, we need to extract features for both social links and location links using the heterogeneous information in the network. Slightly different from users, the locations in online social networks cannot generate information on their own. Before introducing the extracted features, we will introduce a method to accumulate information for locations at first.

7.3.3.1 Information Accumulation for Locations
Locations are represented as $(latitude, longitude)$ pairs in the studied problem, which possess no auxiliary information except location links with users in the network. As a result, we will confront problems of lacking auxiliary information when extracting heterogeneous features for location links

Fig. 7.1 Example of information accumulation for locations from online posts (both Lincoln Park and Scarlet Bar are located in Chicago, US)

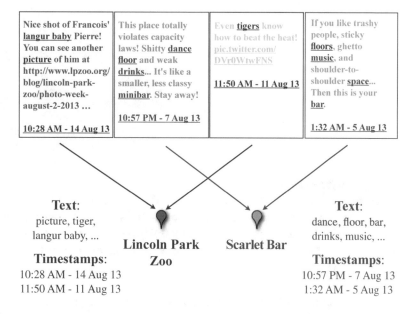

between users and locations. Actually, we notice that users can publish online posts at the locations, and the textual contents and timestamps information of the online posts checked in at a certain location can be accumulated as the auxiliary information possessed by that location.

From a statistical point of view, information from posts published at a certain location, including both timestamps and text contents, can reveal some properties of the location. For example, the timestamps of most posts published at nightlife sites are after 6:00 PM. While those of posts published at restaurants serving brunch are during the daytime. Posts published at national parks can contain some phrases depicting the scenes, while posts published at basketball court may be mostly talking about games, teams, and players. So, we can know more about the locations from the information accumulated from online posts.

Example 7.1 For example, in Fig. 7.1, we have two totally different locations: the Lincoln Park Zoo[2] and Scarlet Bar.[3] The Lincoln Park Zoo is the largest free zoo in Chicago and is open during 10:00 AM–5:00 PM. The Scarlet Bar is one of the most famous bars in Chicago, where people can drink with friends, dance to enjoy their night life, and it is open during 8:00 PM–2:00 AM.

We also have 4 online posts published by people at these two places in either Foursquare or Twitter. From the contents of these posts, we find that people usually publish words about animals, pictures, and the scene at the Lincoln Park Zoo. However, people who visit the Scarlet Bar mainly talk about the atmosphere in the bar, the drinks, the dance floor, and the music there. So, users who frequently talk about animals in daily life can be interested in the Lincoln Park Zoo, while those who usually post words about the drinks may like the Scarlet Bar more. Meanwhile, we can also accumulate the timestamps of posts published at these two places. The timestamps of posts published at the Lincoln Park Zoo are mostly during the daytime, while those of posts published at the Scarlet Bar are at night. So, users who are usually active in the daytime can be more likely to visit the Lincoln Park Zoo, while people who are active during the night may prefer the Bar.

---

[2]http://www.lpzoo.org.

[3]http://www.scarletbarchicago.com.

Table 7.2 Features
extracted from vector x
and y

Features	Descriptions
Extended Degree Count (EDC)	$\|x\|_1, \|y\|_1$
Extended Degree Ratio (EDR)	$\|x\|_1/\|y\|_1$
Extended Common Neighbor (ECN)	$x \cdot y$
Extended Jaccard's Coefficient (EJC)	$\frac{x \cdot y}{\|x\|_1 \cdot \|y\|_1}$
Extended Preferential Attachment (EPA)	$\|x\|_1 \cdot \|y\|_1$
Euclidean Distance (ED)	$(\sum_k (x_k - y_k)^2)^{1/2}$
Cosine Similarity (CS)	$\frac{x \cdot y}{\|x\|_2 + \|y\|_2}$

7.3.3.2 Heterogeneous Features

Based on the heterogeneous information in the networks, we will extract 4 different categories
of features for both social links and location links from the heterogeneous information in the
network, which include *social features*, *spatial distribution features*, *text usage features*, and *temporal
distribution features*. A summary of frequently used features is available in Table 7.2, where $\|x\|_p = (\sum_{i=1}^{|x|} |x_i|^p)^{1/p}$ denotes the L_p-norm of vector x.

- **Features of Social Links:** For a certain social link (u_i, u_j), we can get their neighbors from the
  network, which can be represented as sets $\Gamma(u_i)$ and $\Gamma(u_j)$, respectively. Based on $\Gamma(u_i)$, we can
  construct the social link weight vector $\tilde{s}(u_i)$ for u_i, where $\tilde{s}(u_i) = (p_{1,i}, p_{2,i}, \ldots, p_{k,i}, \ldots, p_{n,i})^\top$
  and $n = |\mathcal{U}|$ is the size of user set and $p_{k,i}$ is the weight of social link $(u_k, u_i), \forall u_k \in \mathcal{U}$: if
  $u_k \in (\mathcal{U} \setminus \Gamma(u_i))$, $p_{k,i} = 0.0$; if $u_k \in \Gamma(u_i)$ and link (u_k, u_i) exists originally, then $p_{k,i} = 1.0$;
  otherwise, $p_{k,i}$ is the existence probability of link (u_k, u_i). Similarly, we can construct vector $\tilde{s}(u_j)$
  for user u_j, which is of the same length as $\tilde{s}(u_i)$. From $\tilde{s}(u_i)$ and $\tilde{s}(u_j)$, 7 different *social features*
  are extracted for social link (u_i, u_j), which are summarized in Table 7.2.

  In a similar way, for a certain social link (u_i, u_j), we can get the set of locations visited by user
  u_i and u_j as sets $\Phi(u_i)$ and $\Phi(u_j)$, from which we can obtain their location link weight vectors
  as $\tilde{l}(u_i)$ and $\tilde{l}(u_j)$, where the entries denote the times that these users visit the locations. From
  the timestamps of posts published by users, we can obtain the users' active patterns. Each day is
  divided into 24 slots and the ratio of online posts published by user u in each hour is saved in a
  temporal distribution vector $\tilde{t}(u)$, whose length is 24. For social link (u_i, u_j), we can construct the
  temporal distribution vectors: $\tilde{t}(u_i)$ and $\tilde{t}(u_j)$ for u_i and u_j. In addition, we transform the words
  used by two users u_i and u_j into two text usage vectors: $\tilde{w}(u_i)$ and $\tilde{w}(u_j)$ weighted by TF-IDF
  [31], which are of the same length. From these vectors, we can extract the *spatial distribution
  features*, *temporal distribution features*, and *text usage features* similar to the social link features
  summarized in Table 7.2 for social link (u_i, u_j).

- **Features of Location Links:** Similarly, we can obtain the set of users who have visited a location
  and regard them as the "neighbors" of that location. And for a location link (u_i, l_j), we can get
  the sets of neighbors of u_i and l_j as $\Gamma(u_i)$ and $\Psi(l_j)$, from which we can construct the social link
  weight vectors $\tilde{s}(u_i)$ and $\tilde{s}(l_j)$, respectively. From the accumulated text and timestamps information
  of locations and the auxiliary information owned by users, we can also construct the temporal
  distribution vectors $\tilde{t}(u_i)$ and $\tilde{t}(l_j)$ and the text usage vectors $\tilde{w}(u_i)$ and $\tilde{w}(l_j)$ for location link
  (u_i, l_j). From these vectors, we can extract the *social features*, *temporal distribution features*, and
  text usage features for location link (u_i, l_j).

  In addition, according to previous definitions, we can get the locations that user u has visited
  in the past: $\Phi(u)$ and the location link weight vector $\tilde{l}(u)$ of u as well as the neighbors $\Psi(l)$ of a

location l and its social link weight vector: $\tilde{s}(l)$. For a certain location link (u_i, l_j), we extract 3 spatial distribution features for the location links from the network:

(1) average weighted geographic distance between locations in $\Phi(u_i)$ and l_j

$$\frac{\sum_{l_k \in \Phi(u_i)} GeoD(l_k, l_j) \cdot \tilde{l}(u_i)_{l_k}}{||\tilde{l}(u_i)||_1 \cdot |\Phi(u_i)|}, \tag{7.8}$$

where $GeoD(l_k, l_j)$ is the geographic distance (e.g., the manhattan distance [2]) between l_k and l_j and $\tilde{l}(u_i)_{l_k}$ is the weight of location link (u_i, l_k) saved in u_i's location link weight vector.

(2) weighted number of users who have visited both locations in $\Phi(u_i)$ and l_j

$$\sum_{l_k \in \Phi(u_i)} \tilde{s}(l_k) \cdot \tilde{s}(l_j) \cdot \tilde{l}(u_i)_{l_k} \tag{7.9}$$

(3) average weighted number of users who have visited both locations in $\Phi(u_i)$ and l_j

$$\frac{\sum_{l_k \in \Phi(u_i)} \tilde{s}(l_k) \cdot \tilde{s}(l_j) \cdot \tilde{l}(u_i)_{l_k}}{||\tilde{l}(u_i)||_1 \cdot \sum_{l_k \in \Phi(u_i)} ||\tilde{s}(l_k)||_1} \tag{7.10}$$

7.3.4 Collective Link Prediction Model

In this section, we will analyze and formulate the correlations between the social link prediction task and the location link prediction task, and introduce an integrated collective link prediction framework to address both of these two tasks simultaneously.

7.3.4.1 Correlation Between Different Tasks

When predicting a link with the supervised link prediction models introduced before, the classifiers will give a score within range [0, 1] to show its existence probability. Newly predicted social links will update the social link existence probability information in the network, which can affect the prediction of other location links. For example, these updated social link existence probabilities can change the extended common neighbors of a location and a user, and may further change the prediction results. Similarly, the location link prediction task can also influence the social link prediction result.

Example 7.2 For example, in Fig. 7.2, we show an example of different link prediction methods. Figure 7.2a is the input aligned networks, in which there are 4 users and some existing social links (u_3, u_4), (u_1, u_4) and location links (u_2, l_1), (u_3, l_1), (u_1, l_2), (u_1, l_3) as well as many other potential links to be predicted. Based on the information in the network, including social information (e.g., common neighbors), location information (e.g., co-check-ins), and other auxiliary information, traditional link prediction methods can predict social links and locate links independently. Figure 7.2b shows the independent social link prediction result, in which social links (u_2, u_3) and (u_1, u_3) are predicted to be positive (i.e., existing), while the other two social links (u_1, u_2) and (u_2, u_4) are predicted to be negative (i.e., non-existing). Figure 7.2c shows the independent location link prediction result and in the result, location links (u_2, l_2), (u_1, l_1), (u_4, l_3) are predicted to be positive (i.e., existing), while (u_2, l_3) and (u_3, l_3) is predicted to be negative (i.e., non-existing).

From the results in Fig. 7.2b, c, we can find some problematic phenomena. For example, user u_2 and u_1 are predicted that they will visit locations l_1, l_2 and they are also predicted to share a common

Fig. 7.2 An example of
different link prediction
methods. (**a**) The input
network. (**b**), (**c**)
Independent social link and
location link prediction
result. (**d**) The collective
link prediction result

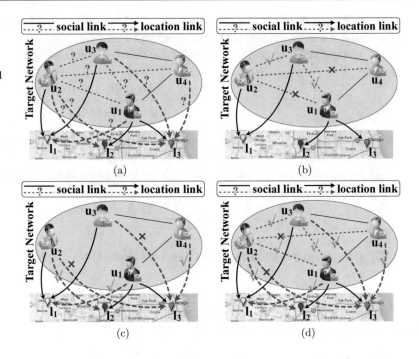

neighbor: u_3. Based on the result, it is highly likely that these two users may know each other, and the potential social link (u_2, u_3) will be predicted to be existing. However, according to the independent prediction result, it is predicted to be non-existing as shown in Fig. 7.2b. Another example is that many neighbors of user u_3, including both the originally existing u_4 and the newly predicted u_1, have visited or are predicted to have visited l_3. Based on such an observation, u_3 is highly likely to be predicted to have visited l_3. However, the location link between u_3 and l_3 is predicted to be non-existing in Fig. 7.2c.

If we consider the correlation between these two link prediction tasks simultaneously, the predicted results of social link (u_1, u_2) and location link (u_3, l_3) are highly likely to be predicted as existing. In Fig. 7.2d, we show a potential result of collective link prediction methods, where the prediction results of social links and location links seem to be much more consistent.

7.3.4.2 Collective Link Prediction

As introduced before, we represent the sets of potential social links and potential location links to be predicted as $\mathcal{L}_s \subset (\mathcal{U} \times \mathcal{U} \setminus \mathcal{E}_s)$ and $\mathcal{L}_l \subset (\mathcal{U} \times \mathcal{L} \setminus \mathcal{E}_l)$, respectively, in the problem formulation section. For links $l_s \in \mathcal{L}_s$ and $l_l \in \mathcal{L}_l$, the supervised models built with the existing information in the network will give them the predicted labels: $y(l_s)$ and $y(l_l)$, as well as the existence probability scores: $P(y(l_s) = 1)$ and $P(y(l_l) = 1)$. Traditional methods predicting social links and location links independently aim at finding the set of labels achieving the maximum likelihood scores for each kind of these links. In other words, let $\hat{\mathcal{Y}}_s \subset \{-1, 1\}^{|\mathbf{L}_s|}, \hat{\mathcal{Y}}_l \subset \{-1, 1\}^{|\mathbf{L}_l|}$ be the sets of optimal labels, the objective functions of the social and location link prediction tasks can be denoted as

$$\hat{\mathcal{Y}}_s = arg \max_{\mathcal{Y}_s} P(y(\mathcal{L}_s) = \mathcal{Y}_s | \mathbf{x}(\mathcal{L}_s)), \tag{7.11}$$

$$\hat{\mathcal{Y}}_l = arg \max_{\mathcal{Y}_l} P(y(\mathcal{L}_l) = \mathcal{Y}_l | \mathbf{x}(\mathcal{L}_l)), \tag{7.12}$$

where $P(y(\mathcal{L}_s) = \mathcal{Y}_s)$ and $P(y(\mathcal{L}_l) = \mathcal{Y}_l)$ denote the probability scores achieved when links in $\mathcal{L}_s$ and $\mathcal{L}_l$ are assigned with labels in $\mathcal{Y}_s$ and $\mathcal{Y}_l$.

However, considering connections between these two link prediction tasks, the inferred social link or location link information should be incorporated into the same framework. The jointly optimal label sets $\hat{\mathcal{Y}}_s$ and $\hat{\mathcal{Y}}_l$ will be

$$\hat{\mathcal{Y}}_s, \hat{\mathcal{Y}}_l = arg \max_{\mathcal{Y}_s, \mathcal{Y}_l} P(y(\mathcal{L}_s) = \mathcal{Y}_s | y(\mathcal{L}_l) = \mathcal{Y}_l, \mathbf{x}(\mathcal{L}_s))$$

$$\times P(y(\mathcal{L}_l) = \mathcal{Y}_l | y(\mathcal{L}_s) = \mathcal{Y}_s, \mathbf{x}(\mathcal{L}_l)) \tag{7.13}$$

For the given optimization equation, there are many different solutions. In this part, we will give an iterative method, TRAIL, to address it, which can predict the social links and location links iteratively until convergence. Let τ be the τ_{th} iteration and the optimal label sets of social links and location links achieved in the τ_{th} iteration be $\hat{\mathcal{Y}}_s^{(\tau)}$ and $\hat{\mathcal{Y}}_l^{(\tau)}$, then we have

$$\hat{\mathcal{Y}}_s^{(\tau)} = arg \max_{\mathcal{Y}_s} P(y(\mathcal{L}_s) = \mathcal{Y}_s | G, y(\mathcal{L}_s) = \hat{\mathcal{Y}}_s^{(\tau-1)}, y(\mathcal{L}_l) = \hat{\mathcal{Y}}_l^{(\tau-1)}) \tag{7.14}$$

$$\hat{\mathcal{Y}}_l^{(\tau)} = arg \max_{\mathcal{Y}_l} P(y(\mathcal{L}_l) = \mathcal{Y}_s | G, y(\mathcal{L}_s) = \hat{\mathcal{Y}}_s^{(\tau)}, y(\mathcal{L}_l) = \hat{\mathcal{Y}}_l^{(\tau-1)}). \tag{7.15}$$

The pseudo code of TRAIL is available in Algorithm 1. Here, we mainly focus on providing the overall framework of TRAIL and haven't specified the classifier models to be used. Actually, any classification algorithms (e.g., SVM, Neural Networks) we have introduced before can all be adopted as the base classifier in the framework.

Algorithm 1 TRAIL

Require: heterogeneous LBSN, G.
        existing social links and location links: E_s, E_l
        potential social links and location links: L_s, L_l
Ensure: the inferred labels and existence probabilities of links in L_s and L_l: $\hat{\mathcal{Y}}_s, \hat{\mathcal{P}}_s, \hat{\mathcal{Y}}_l, \hat{\mathcal{P}}_l$
 1: construct training sets, test sets with E_s, E_l, L_s and L_l.
 2: $converge \leftarrow False$
 3: **while** $converge$ is $False$ **do**
 4:    extract features $\mathbf{x}(E_s)$ and $\mathbf{x}(L_s)$ for social links in E_s and L_s from G.
 5:    $C_s \leftarrow$ **train**$([\mathbf{x}(E_s)^T, \mathbf{x}^s(E_s)^T, y^s(E_s)]^T, y(E_s))$
 6:    $\hat{\mathcal{Y}}_s, \hat{\mathcal{P}}_s \leftarrow C_s.$**classify**$([\mathbf{x}(L_s)^T, \mathbf{x}^s(L_s)^T, y^s(L_s)]^T)$
 7:    update G with $\hat{\mathcal{Y}}_s, \hat{\mathcal{P}}_s$
 8:    Accumulate information for locations
 9:    extract features $\mathbf{x}(E_l)$ and $\mathbf{x}(L_l)$ for location links in E_l and L_l from G.
10:    $C_l \leftarrow$ **train**$([\mathbf{x}(E_l)^T, \mathbf{x}^s(E_l)^T, y^s(E_l)]^T, y(E_l))$
11:    $\hat{\mathcal{Y}}_l, \hat{\mathcal{P}}_l \leftarrow C_l.$**classify**$([\mathbf{x}(L_l)^T, \mathbf{x}^s(L_l)^T, y^s(L_l)]^T)$
12:    update G with $\hat{\mathcal{Y}}_l, \hat{\mathcal{P}}_l$
13:    **if** $\hat{\mathcal{Y}}_s, \hat{\mathcal{P}}_s, \hat{\mathcal{Y}}_l, \hat{\mathcal{P}}_l$ all converge **then**
14:        $converge \leftarrow True$
15:    **end if**
16: **end while**
17: Return $\hat{\mathcal{Y}}_s, \hat{\mathcal{P}}_s, \hat{\mathcal{Y}}_l, \hat{\mathcal{P}}_l$

7.4   Cold Start Link Prediction for New Users

In this section, we study the problem of predicting social links for new users, who have created their accounts for just a short period of time. Generally, new users who have just created the accounts, they are more likely to accept the recommendations to establish their social communities. However, the limited information available for these new users can pose a great challenge on high quality recommendations of friends. Meanwhile, for the users who are new in one network, they may have been involved in other online social networks for a long time. Information can be transferred from these mature source networks for these users to the target network that we are focused on to resolve the lack of information problem.

7.4.1   New User Link Prediction Problem Description

Many of previous works on link prediction [1, 13, 14, 23, 53] focus on predicting potential links that will appear among all the users, based upon a snapshot of the social network. These works treat all users equally and try to predict social links for all users in the network. However, in real-world social networks, many new users are joining in the online social networks every day. It has been shown in previous works that there is a negative correlation between the age of nodes in the network and their link attachment rates. Predicting social links for these new users are more important than for those existing active users in the network as it will leave a good first impression on the new users. First impression often has a lasting impact on a new user and may decide whether he/she will become an active user. A bad first impression can turn a new user away. So it is important to make meaningful recommendations to new users to create a good first impression and attract them to participate more. For simplicity, we refer users that have been actively using the network for a long time as "old users".

The link prediction problem for new users is different from traditional link prediction problems. Conventional supervised link prediction methods implicitly or explicitly assume that the information are identically distributed over all the nodes in the network without considering the joining time of the users. The models trained over one part of the network can be directly used to predict links in other parts of the network. However, in real-world social networks, the information distributions of the new users could be very different from that of old users. New users may have only a few activities or even no activities (i.e., no social links or other auxiliary information) in the network; while old users usually have abundant activities and auxiliary information in the network. In Figs. 7.3 and 7.4, we show the degree distributions of the new users who registered their accounts within 3 months and the old users who registered more than 3 months before in Twitter and Foursquare, respectively. In

Fig. 7.3 Degree distributions of users in Foursquare network

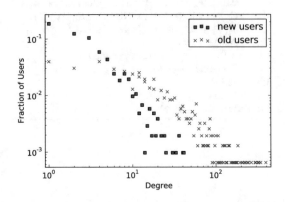

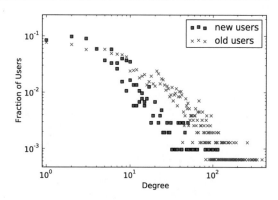

Fig. 7.4 Degree distributions of users in Twitter network

the plots, the x axis denotes the node degrees and the y axis denotes the fraction of users with certain degrees. We observe that the social link distributions of new users and old users are totally different from each other in both Foursquare and Twitter. As a result, conventional supervised link prediction models trained over old users based upon structural features, such as *common neighbors*, may not work well on the new users.

Another challenging problem in link prediction for new users is that information owned by new users can be very rare or even totally missing. Conventional methods based upon one single network will not work well due to the lack of historical data about the new users. In order to solve this problem, we need to transfer additional information about the new users from other sources. Nowadays, people are usually involved in multiple social networks to enjoy more services. For example, people will join Foursquare to search for nearby restaurants to have dinner with their family. Meanwhile, they tend to use Face book to socialize with their friends and involve in Twitter to post comments about recent news. The accounts of the same user in different networks can be linked through account alignments. For example, when users register their Foursquare accounts, they can use their Face book or Twitter accounts to sign in the Foursquare network. Such links among accounts of the same user are named as "anchor links" [19, 57–59] according to the description in Sect. 3.4.3, which could help align users' accounts across multiple social networks. For example, in Fig. 7.5, there are many users in two networks, respectively. We find that the accounts in these two networks are actually owned by 6 different users in reality and we add an *anchor link* between each pair of user accounts corresponding to the same user. Via the *anchor links*, we could locate users' corresponding accounts in the other networks.

New users in one social network (i.e., *target* network) might have been using other social networks (i.e., *source* networks) for a long time. These user accounts in the source networks can provide additional information about the new users in the source network. This additional information is crucial for link prediction about these new users, especially when the new users have little activities or no activities in the target network (i.e., cold start problem).

Example 7.3 For instance, in Fig. 7.5, we have two social networks, i.e., the target network and the source network, with aligned user accounts. In the target network, there are many old users with abundant social links and auxiliary information, such as posts, spatial and temporal activities. In addition, there are also some new users, i.e., user u_1^t and u_2^t, in the target network. These two new users have just created their accounts in the target network and have not yet created many social links or auxiliary information. However, we can see that there is abundant information about these two new users in the source network, based on their "anchor linked" user accounts u_1^s and u_2^s in the source

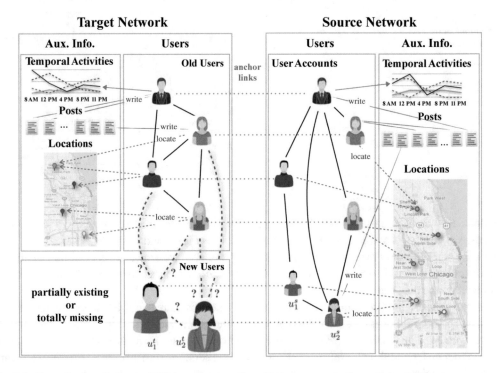

Fig. 7.5 Example of predicting social links across two aligned heterogeneous online social networks

network. The new users' information in source networks can be exploited to help improve the link prediction performances in the target network.

In order to solve these problems, in this section, we will introduce a novel supervised cross aligned networks link recommendation method, SCAN, proposed in [57]. Different from previous works, SCAN extracts heterogeneous features from other aligned networks to improve link prediction results for new users in the target network. SCAN analyzes the problem about the differences in information distributions between new users and old users in great detail and proposes a within-network personalized sampling method to accommodate that difference. What's more, SCAN can also solve the cold start social link prediction problem assisted by other aligned source networks. Intra- and inter-network information transfer can be conducted simultaneously in SCAN to make a full use of the information contained in these aligned networks to improve the prediction results.

7.4.2　Cold Start Link Prediction Problem Formulation

The problem studied in this section is social link prediction for new users. We will introduce a supervised method based on aligned heterogeneous networks. Let $\mathcal{G} = ((G^t, G^s), (\mathcal{A}^{t,s}))$ be two aligned heterogeneous social networks, where G^t is the target network and G^s is an aligned source network. $\mathcal{A}^{t,s}$ denotes the set of anchor links between G^t and G^s. We want to predict social links for the new users in the target network. Let $\mathcal{U}^t = \mathcal{U}^t_{new} \cup \mathcal{U}^t_{old}$ be the user set in G^t, where $\mathcal{U}^t_{new}$ and $\mathcal{U}^t_{old}$ are the sets of new users and old users, respectively, and $\mathcal{U}^t_{new} \cap \mathcal{U}^t_{old} = \emptyset$. What we want to predict is a subset of potential social links between the new users and all other users: $\mathcal{L} \subseteq \mathcal{U}^t_{new} \times \mathcal{U}^t$. In other words, we want to build a function $f : \mathcal{L} \rightarrow \{0, 1\}$, which could decide whether certain links related to new users exist in the target network or not.

7.4.3 Link Prediction Within Target Network

Based on the heterogeneous information available in the online social networks, a set of features can be extracted for the social links as introduced in Sect. 7.3.3.2. Next we will introduce how to use these features to build supervised methods to predict links for new users in the target network. Before doing that, we notice that the new users' information distribution can be totally different from that of the old users in the target network. However, information of both new users and old users is so important that should be utilized. In this section, we will analyze the differences in information distributions of new users and old users in the target network and propose a personalized within-network sampling method to process old users' information to accommodate the differences. Then, we will extend the traditional supervised link prediction method by using the old users' sampled information in the target network to improve the prediction results.

7.4.3.1 Sampling Old Users' Information

A natural challenge inherent in the usage of the target network to predict social links for new users is the differences in information distributions of new users and old users as mentioned before. To address this problem, the SCAN model proposes to accommodate old users' and new users' sub-networks by using a within-network personalized sampling method to process old users' information. Totally different from the link prediction with sampling problem studied in [3], SCAN conducts personalized sampling within the target network, which contains heterogeneous information, rather across multiple non-aligned homogeneous networks. And the link prediction target are the new users in the target network. By sampling the old users' sub-network, we want to achieve the following objectives:

- *Maximizing Relevance*: We aim at maximizing the relevance of the old users' sub-network and the new users' sub-network to accommodate differences in information distributions of new users and old users in the heterogeneous target network.
- *Information Diversity*: Diversity of old users' information after sampling is still of great significance and should be preserved.
- *Structure Maintenance*: Some old users possessing sparse social links should have higher probability to survive after sampling to maintain their links so as to maintain the network structure.

Let the heterogeneous target network be $G^t = \{\mathcal{V}^t, \mathcal{E}^t\}$, and $\mathcal{U}^t = \mathcal{U}^t_{old} \cup \mathcal{U}^t_{new} \subset \mathcal{V}^t$ is the set of user nodes (i.e., set of old users and new users) in the target network. Personalized sampling is conducted on the old users' part: $G^t_{old} = \{\mathcal{V}^t_{old}, \mathcal{E}^t_{old}\}$, in which each node is sampled independently with the sampling rate distribution vector $\boldsymbol{\delta} = (\delta_1, \delta_2, \ldots, \delta_n)$, where $n = |\mathcal{U}^t_{old}|$, $\sum_{i=1}^{n} \delta_i = 1$ and $\delta_i \geq 0$. Old users' heterogeneous sub-network after sampling is denoted as $\bar{G}^t_{old} = \{\bar{\mathcal{V}}^t_{old}, \bar{\mathcal{E}}^t_{old}\}$.

The main objective of the old users' information sampling is to make the old users' sub-network as relevant to new users' as possible. To measure the similarity score of a user u_i and a heterogeneous network G, we define a relevance function as follows:

$$R(u_i, G) = \frac{1}{|\mathcal{U}|} \sum_{u_j \in \mathcal{U}} S(u_i, u_j) \tag{7.16}$$

where $\mathcal{U}$ is the user set of network G and $S(u_i, u_j)$ measures the similarity between user u_i and u_j in the network. Each user has social relationships as well as other auxiliary information and $S(u_i, u_j)$ is defined as the average of similarity scores of these two parts:

$$S(u_i, u_j) = \frac{1}{2}(S_{aux}(u_i, u_j) + S_{social}(u_i, u_j)) \tag{7.17}$$

In our problem settings, the auxiliary information of each users could also be divided into 3 categories: *location*, *temporal*, and *text*. So, $S_{aux}(u_i, u_j)$ is defined as the mean of these three aspects.

$$S_{aux}(u_i, u_j) = \frac{1}{3}(S_{text}(u_i, u_j) + S_{loc}(u_i, u_j) + S_{temp}(u_i, u_j)) \tag{7.18}$$

There are many different methods measuring the similarities of these auxiliary information in different aspects, e.g. cosine similarity [16, 53]. As to the social similarity, Jaccard's coefficient [17] can be used to depict how similar two users are in their social relationships. We will not talk about these measures in this part.

The relevance between the sampled old users' network and the new users' network could be defined as the expectation value of function $R(\bar{u}_{old}^t, G_{new}^t)$:

$$
\begin{aligned}
R(\bar{G}_{old}^t, G_{new}^t) &= \mathbb{E}(R(\bar{u}_{old}^t, G_{new}^t)) \\
&= \frac{1}{|\mathcal{U}_{new}^t|} \sum_{j=1}^{|\mathcal{U}_{new}^t|} \mathbb{E}(S(\bar{u}_{old}^t, u_{new,j}^t)) \\
&= \frac{1}{|\mathcal{U}_{new}^t|} \sum_{j=1}^{|\mathcal{U}_{new}^t|} \sum_{i=1}^{|\mathcal{U}_{old}^t|} \delta_i \cdot S(\bar{u}_{old,i}^t, u_{new,j}^t) \\
&= \delta^\top \mathbf{s}
\end{aligned}
\tag{7.19}
$$

where vector $\mathbf{s}$ equals:

$$\frac{1}{|\mathcal{U}_{new}^t|}[\sum_{j=1}^{|\mathcal{U}_{new}^t|} S(\bar{u}_{old,1}^t, u_{new,j}^t), \ldots, \sum_{j=1}^{|\mathcal{U}_{new}^t|} S(\bar{u}_{old,n}^t, u_{new,j}^t)]^\top \tag{7.20}$$

and $|\mathcal{U}_{old}^t| = n$. Besides the relevance, we also need to ensure that the diversity of information in the sampled old users' sub-network could be preserved. Similarly, it also includes diversities of the auxiliary information and social relationships. The diversity of auxiliary information is determined by the sampling rate δ_i, which could be defined with the averaged *Simpson Index* [36] over the old users' sub-network.

$$D_{aux}(\bar{G}_{old}^t) = \frac{1}{|\mathcal{U}_{old}^t|} \cdot \sum_{i=1}^{|\mathcal{U}_{old}^t|} \delta_i^2 \tag{7.21}$$

As to the diversity in the social relationship, we could get the existence probability of a certain social link (u_i, u_j) after sampling to be proportional to $\delta_i \cdot \delta_j$. So, the diversity of social links in the sampled network could be defined as average existence probabilities of all the links in the old users' sub-network.

$$D_{social}(\bar{G}^t_{old}) = \frac{1}{\left|\mathcal{E}^t_{old,s}\right|} \cdot \sum_{i=1}^{\left|\mathcal{U}^t_{old}\right|} \sum_{j=1}^{\left|\mathcal{U}^t_{old}\right|} \delta_i \cdot \delta_j \times \mathbb{I}(u_i, u_j) \qquad (7.22)$$

where $\left|\mathcal{E}^t_{old,s}\right|$ is the size of social link set of old users' sub-network and $\mathbb{I}(u_i, u_j)$ is an indicator function $\mathbb{I} : (u_i, u_j) \to \{0, 1\}$ to show whether a certain social link exists or not originally before sampling. For example, if link (u_i, u_j) is a social link in the target network originally before sampling, then $\mathbb{I}(u_i, u_j) = 1$, otherwise it will be equal to 0.

By considering these two terms simultaneously, we could have the diversity of information in the sampled old users' sub-network to be the average diversities of these two parts:

$$D(\bar{G}^t_{old}) = \frac{1}{2}(D_{social}(\bar{G}^t_{old}) + D_{aux}(\bar{G}^t_{old}))$$

$$= \frac{1}{2}(\sum_{i=1}^{\left|\mathcal{U}^t_{old}\right|} \sum_{j=1}^{\left|\mathcal{U}^t_{old}\right|} \frac{1}{\left|\mathcal{E}^t_{old,s}\right|} \cdot \delta_i \cdot \delta_j \times \mathbb{I}(u_i, u_j) + \sum_{i=1}^{\left|\mathcal{U}^t_{old}\right|} \frac{1}{\left|\mathcal{U}^t_{old}\right|} \cdot \delta_i^2)$$

$$= \delta^\top \cdot (\frac{1}{2\left|\mathcal{E}^t_{old,s}\right|} \cdot \mathbf{A}^t_{old} + \frac{1}{2\left|\mathcal{U}^t_{old}\right|} \cdot \mathbf{I}_{\left|\mathcal{U}^t_{old}\right|}) \cdot \delta \qquad (7.23)$$

where matrix $\mathbf{I}_{\left|\mathcal{U}^t_{old}\right|}$ is the diagonal identity matrix of dimensions $\left|\mathcal{U}^t_{old}\right| \times \left|\mathcal{U}^t_{old}\right|$ and $\mathbf{A}^t_{old}$ is the adjacency matrix of old users' sub-network.

To ensure that the structure of the original old users' subnetwork is not destroyed, we need to ensure that users with few links could also preserve their links. So, we could add a regularization term to increase the sampling rate for these users as well as their neighbors by maximizing the following terms:

$$Reg(\bar{G}^t_{old}) = \min\{|\Gamma(u_i)|, \min_{u_j \in \Gamma(u_i)}\{|\Gamma(u_j)|\}\} \times \delta_i^2 = \delta^\top \cdot \mathbf{M} \cdot \delta \qquad (7.24)$$

where matrix $\mathbf{M}$ is a diagonal matrix with $M_{i,i} = \min\{|\Gamma(u_i)|, \min_{u_j \in \Gamma(u_i)}\{|\Gamma(u_j)|\}\}$ on its diagonal, where $\Gamma(u_i)$ denotes the neighbor set of user u_j. So, if a user or his/her neighbors have few links, then this user as well as his/her neighbors should have higher sampling rates so as to preserve the links between them.

Example 7.4 For example, in Fig. 7.6, we have 6 users. To decide the sampling rate of user u_1^t, we need to consider his/her social structure. We find that since u_1^t's neighbor u_2^t has no other neighbors except u_1^t. To preserve the social link between u_1^t and u_2^t we need to increase the sampling rate of u_2^t. However, the existence probability of link (u_1^t, u_2^t) is also decided by the sampling rate of user u_1^t, which also needs to be increased too.

Fig. 7.6 Personalized
sampling preserving
network structures

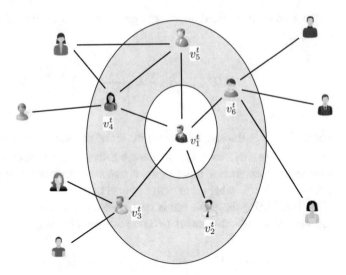

Combining the diversity term and the structure preservation term, we could define the regularized
diversity of information after sampling to be

$$D_{Reg}(\bar{G}_{old}^t) = D(\bar{G}_{old}^t) + Reg(\bar{G}_{old}^t) = \boldsymbol{\delta}^\top \cdot \mathbf{N} \cdot \boldsymbol{\delta} \tag{7.25}$$

where $\mathbf{N} = \frac{1}{2|\mathcal{U}_{old}^t|} \cdot \mathbf{I}_{|\mathcal{U}_{old}^t|} + \frac{1}{2|\mathcal{E}_{old,s}^t|} \cdot \mathbf{A}_{old}^t + \mathbf{M}$.

The optimal value of $\boldsymbol{\delta}$ should be able to maximize the relevance of new users' sub-network and
old users' as well as the regularized diversity of old users' information in the target network

$$\boldsymbol{\delta} = \underset{\boldsymbol{\delta}}{\arg\max}\ R(\bar{G}_{old}^t, G_{new}^t) + \theta \cdot D_{Reg}(\bar{G}_{old}^t)$$

$$= \underset{\boldsymbol{\delta}}{\arg\max}\ \boldsymbol{\delta}^\top \mathbf{s} + \theta \cdot \boldsymbol{\delta}^\top \cdot \mathbf{N} \cdot \boldsymbol{\delta}$$

$$s.t.\ \sum_{i=1}^{|\mathcal{U}_{old}^t|} \delta_i = 1\ and\ \delta_i \geq 0, \tag{7.26}$$

where parameter θ denotes the weight of the regularization term on information diversity.

7.4.3.2 TRAD

A traditional supervised link prediction method TRAD (Traditional Link Prediction) can be applied
for our task by using the existing links in the target network to train a classifier and applying it to
classify the potential social links for new users. In method TRAD, only the target network is used,
which consists of new users and unsampled old users. To overcome the differences in information
distribution between new users and old users in the target network, we revise it a little bit and get
method: TRAD-PS (Traditional Link Prediction with Personalized Sampling). TRAD-PS consists of
two steps: (1) personalized sampling of the old users' sub-network with the previous method; (2) usage
of similar techniques as TRAD to predict links based on the sampled network. Theoretically, TRAD
and TRAD-PS could work well by using information in the target network. However, considering the
fact that it is impossible for new users to possess a large amount of information actually, TRAD and

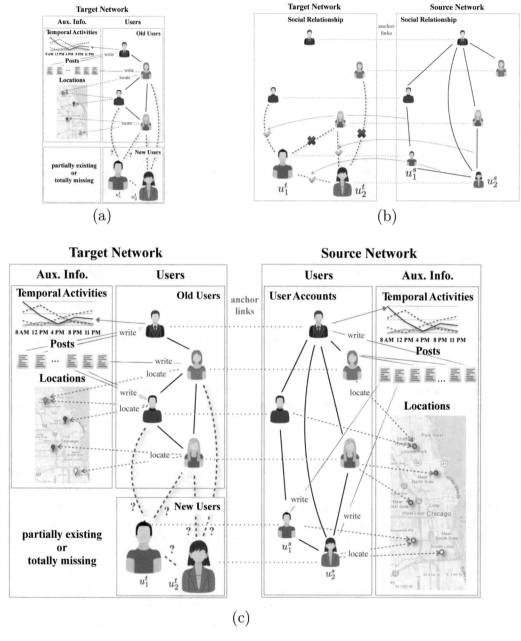

Fig. 7.7 Different methods to predict social link for new users. (**a**) TRAD method. (**b**) NAIVE method. (**c**) SCAN method

TRAD-PS would suffer from the long-standing cold start problem caused by the lack of historical information indicating these new users' preferences. This problem will be even worse when dealing with brand-new users, who have no information at all in the target network.

Example 7.5 For example, in Fig. 7.7a, user u_1^t and u_2^t are two new users in the target network, who possess very few social links with other users and little auxiliary information. We cannot get any information about these two new users and the information we could use is that possessed by other old users. As a result, the links that TRAD and TRAD-PS predicted could hardly be of a high quality.

In order to deal with such a problem, we will introduce a method to use aligned networks simultaneously in the next section.

7.4.4 Cold-Start Link Prediction

In the current problem settings, we have two aligned social networks and the methods proposed in the previous section using the target network may suffer from the cold start problems when processing brand-new users. In this section, we will introduce two methods to utilize the aligned source network to help solve the problem and improve the prediction results.

7.4.4.1 NAIVE
Suppose we have a new user u_i^t in the target network, a naive way to use the aligned source network to recommend social links for user u_i^t is to recommend all the corresponding social links related to this user's aligned account u_i^s in the aligned source network to him/her. Based on this intuition, a cold start link prediction method NAIVE (Naive Link Prediction) as proposed in [57] can be applied. To clarify how NAIVE works in the reality, we will give an example next. And before that, we will introduce a new term *pseudo label* [57] to denote the existence of corresponding links in the aligned source network.

Definition 7.1 (Pseudo Label) The pseudo label of a link (u_i^t, u_j^t) in the target denotes the existence of its corresponding link (u_i^s, u_j^s) in the aligned source network and it is 1 if (u_i^s, u_j^s) exists and 0 otherwise.

Example 7.6 For instance, in Fig. 7.7b, to decide whether to recommend u_1^t to u_2^t in the target network or not, we could find their aligned accounts: u_1^s and u_2^s, and their social link: (u_1^s, u_2^s) in the aligned source network with the help of *anchor links*. We find that u_1^s and u_2^s are friends in the aligned source network and link (u_1^s, u_2^s) exists in the aligned source network. As a result, the pseudo label of link (u_1^t, u_2^t) is 1 and in the target network, we could recommend u_2^t to u_1^t. And that is the reason why the social link between u_1^t and u_2^t is predicted to be existing by method NAIVE. Other links in Fig. 7.7b can be predicted in a similar way.

Method NAIVE is very simple and could work well in addressing the cold start link prediction task even when these new users are brand new, which means that we could overcome the cold start problem by using this method. However, it may still suffer from some disadvantages: (1) the social structures of different networks are not always identical which will degrade the performance of NAIVE a lot; (2) NAIVE only utilizes these new users' social linkage information in the source network and ignores all other information.

7.4.4.2 SCAN
To overcome all these disadvantages mentioned above, a new method SCAN (Supervised Cross Aligned Networks Link Prediction with Personalized Sampling) is proposed in [57]. As shown in Fig. 7.7c, it could use heterogeneous information existing in both the target network and the aligned source and it is built across two aligned social networks. By taking the advantages of the anchor links, we could locate the users' aligned accounts and their information in the aligned source network

exactly. If two aligned networks are used simultaneously, different categories of features can be extracted from aligned networks.

To use multiple networks, these feature vectors extracted for the corresponding links in aligned networks are merged into an expanded feature vector. The expanded feature vector together with the labels from the target network are used to build a cross-network classifier to decide the existence of social links related to these new users in the target network. This is how method SCAN works. SCAN is quite stable and could overcome the cold start problem for the reason that the information about all these users in the aligned source network doesn't change much with the variation of the target network and we get the information showing of these new users' preferences from the information he/she leaves in the aligned source network. As the old users' information inside the target network is also used in SCAN, personalized sampling is also conducted to preprocess the old users' information in the target network.

In addition to features mentioned before, SCAN also utilizes the information used by NAIVE, i.e., the *pseudo label* defined before, by treating it as an extra feature.

- **An Extra Feature**: SCAN uses the social link *pseudo label* as an extract feature to denote the existence of the corresponding links in the aligned source network.

Compared with SCAN with NAIVE, SCAN has many advantages: (1) SCAN utilizes multiple categories of information; (2) SCAN can make use of the information hidden in the old users' network by incorporating them into the training set; and (3) SCAN doesn't rely on the assumption that the social relationships in different networks are identical, which is very risky actually.

Compared with TRAD and TRAD-PS, SCAN can solve the cold start problem as it could have access to information owned by these new users in other aligned source networks. Similar to TRAD and TRAD-PS, these new users' information is used if they are not very new and other old users' information in the target is also preprocessed by using the within-network personalized sampling method before the intra-network knowledge transfer.

7.5   Spy Technique Based Inter-Network PU Link Prediction

Besides the link prediction problems in one single target network, some research works have been done on simultaneous link prediction in multiple aligned online social networks concurrently. In the supervised link prediction model introduced before, among all the non-existing social links, a subset of the links can be identified and labeled as the negative instances. However, in the real world, labeling the links which will never be formed can be extremely hard and almost impossible. In this section, we will study the cross-network concurrent link prediction problem with PU learning, and introduce a spy technique based link prediction method MLI proposed in [59].

7.5.1   Cross-Network Concurrent Link Prediction Problem

Traditional link prediction problems which aim at predicting one single kind of links in one network [7,33,45,48] have been studied for many years. Dozens of different link prediction methods have been proposed so far [5,7,25,33,41,45,48]. Conventional link prediction methods usually assume that there exists sufficient information within the network to compute features (e.g., common neighborhoods [13]) for each pair of nodes. However, as proposed in [19,58], such an assumption can be violated seriously when dealing with social networks containing little information because of the "new network" problems [59].

The *new network problem* can be encountered when online social networks branch into new geographic areas or social groups [58] and information within the new networks can be too sparse to build effective link prediction models. Meanwhile, the recent works [19, 57, 58] notice that users nowadays can participate in multiple online social networks simultaneously. Users who are involved in a new network may use other well-developed networks for a long time, in which they can have plenty of heterogeneous information. To address the new network problem, some papers [57, 58] propose to transfer information from the well-developed networks to overcome the shortage of information problem in the new network. Formally, networks that share some common users are defined as the "*partially aligned networks*" and the common users shared across these *aligned networks* are named as the "*anchor users*" [19,57,58]. Meanwhile, the unshared users are named as the "*non-anchor users*" between the *aligned networks* as introduced in Sect. 3.4.3.

Social networks aligned by the "anchor users" can share common information. Meanwhile, as proposed in [28,50], different online social networks constructed to provide different services usually have distinct characteristics. Moreover, information in various social networks may be of different distributions [28, 50], which is named as the "*network difference problem*" in [59]. The "*network difference problem*" will be an obstacle in link prediction across *multiple partially aligned networks*, as it is likely that information transferred from other aligned networks could deteriorate the prediction performance in a given network.

In this section, we want to predict the formation of social links in multiple *partially aligned networks* simultaneously, which is formally defined as the *multi-network link prediction* problem [59]. As introduced at the beginning of this section, the *multi-network link prediction* problem can have very extensive applications in real-world social networks. As a result, the *multi-network link prediction* problem studied in this section is very important for *multiple partially aligned social networks*.

The *multi-network link prediction* problem studied in this section is also very challenging to solve due to: (1) *lack of features*, (2) *network partial alignment problem*, (3) *network difference problem*, and (4) *simultaneous link prediction in multiple networks*. To solve all these above challenges in the *multi-network link prediction* problem, a novel link prediction framework, MLI proposed in [59], will be introduced in this section. Inspired by Sun's work on meta path [40] as a means to capture similarity of nodes, which are not directly connected in heterogeneous information networks, MLI explores the meta path concept to generate useful features. MLI can generate not only intra-network features via "intra-network meta paths," but also inter-network features via "inter-network meta paths" through the *anchor links*. By judiciously selecting the "inter-network meta paths," MLI can take advantage of the commonality among the *multiple partially aligned networks*, while containing the potential negative transfers from network differences. These derived features can greatly improve the effectiveness of MLI in predicting links for each network. Furthermore, MLI is a general link formation prediction framework that solves the *multi-network link prediction* problem and the *link prediction* tasks in different networks can help each other mutually.

7.5.2   Concurrent Link Prediction Problem Formulation

Let $G^{(1)}, G^{(2)}, \ldots, G^{(n)}$ denote n different *heterogeneous online social network*, where the sets of anchor links among them can be represented as $\mathcal{A}^{(1,2)}, \mathcal{A}^{(1,3)}, \ldots, \mathcal{A}^{(n-1,n)}$. The user set and existing social link set of $G^{(i)}$ can be represented as $\mathcal{U}^{(i)}$ and $\mathcal{E}_{u,u}^{(i)}$, respectively. In network $G^{(i)}$, all the existing links are the formed links and, as a result, the formed links of $G^{(i)}$ can be represented as $\mathcal{P}^{(i)}$, where $\mathcal{P}^{(i)} = \mathcal{E}_{u,u}^{(i)}$. Furthermore, a large set of unconnected user pairs are referred to as the unconnected

Fig. 7.8 Schema of heterogeneous network

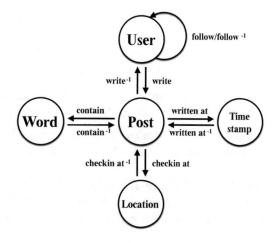

links, $\bar{\mathcal{U}}^{(i)}$, and can be extracted from network $G^{(i)}$: $\bar{\mathcal{U}}^{(i)} = \mathcal{U}^{(i)} \times \mathcal{U}^{(i)} \setminus \mathcal{P}^{(i)}$. However, no information about links that will never be formed can be obtained from the network. With the formed link set $\mathcal{P}^{(i)}$ and unconnected link set $\bar{\mathcal{U}}^{(i)}$, the *link formation prediction* problem can be formulated as a *PU link prediction* problem.

Formally, let $\{\mathcal{P}^{(1)}, \ldots, \mathcal{P}^{(n)}\}$, $\{\bar{\mathcal{U}}^{(1)}, \ldots, \bar{\mathcal{U}}^{(n)}\}$ and $\{\mathcal{L}^{(1)}, \ldots, \mathcal{L}^{(n)}\}$ be the sets of formed links, unconnected links, and links to be predicted of $G^{(1)}$, $G^{(2)}$, $\ldots$, $G^{(n)}$, respectively. With the formed and unconnected links of $G^{(1)}$, $G^{(2)}$, $\ldots$, $G^{(n)}$, we can solve the *multi-network link prediction* problem as the *concurrent PU link prediction* problem.

In the following subsections, we will introduce MLI to solve the *multi-network link prediction* problem. This section includes 3 parts: (1) social meta path based feature extraction and selection; (2) PU link prediction; (3) multi-network concurrent link prediction framework.

7.5.3 Social Meta Path Definition and Selection

Before talking about the link prediction methods, we will introduce the features extracted from the *partially aligned networks* in this subsection at first. The feature extraction in MLI is based on the meta paths as defined in Sect. 3.5. Based on the schema of the network studied in this section, shown in Fig. 7.8, we can define many different kinds of *homogeneous and heterogeneous intra-network social meta paths* for the network, whose physical meanings and notations are listed as follows:

Homogeneous Intra-Network Social Meta Path

- *ID 0. Follow*: User $\xrightarrow{follow}$ User, whose notation is "$U \rightarrow U$" or $\Phi_0(U, U)$.
- *ID 1. Follower of Follower*: User $\xrightarrow{follow}$ User $\xrightarrow{follow}$ User, whose notation is "$U \rightarrow U \rightarrow U$" or $\Phi_1(U, U)$.
- *ID 2. Common Out Neighbor*: User $\xrightarrow{follow}$ User $\xrightarrow{follow^{-1}}$ User, whose notation is "$U \rightarrow U \leftarrow U$" or $\Phi_2(U, U)$.
- *ID 3. Common In Neighbor*: User $\xrightarrow{follow^{-1}}$ User $\xrightarrow{follow}$ User, whose notation is "$U \leftarrow U \rightarrow U$" or $\Phi_3(U, U)$.

Heterogeneous Intra-Network Social Meta Path

- *ID 4. Common Words*: User $\xrightarrow{write}$ Post $\xrightarrow{contain}$ Word $\xrightarrow{contain^{-1}}$ Post $\xrightarrow{write^{-1}}$ User, whose notation is "$U \to P \to W \leftarrow P \leftarrow U$" or $\Phi_4(U, U)$.
- *ID 5. Common Timestamps*: User $\xrightarrow{write}$ Post $\xrightarrow{contain}$ Time $\xrightarrow{contain^{-1}}$ Post $\xrightarrow{write^{-1}}$ User, whose notation is "$U \to P \to T \leftarrow P \leftarrow U$" or $\Phi_5(U, U)$.
- *ID 6. Common Location Check-ins*: User $\xrightarrow{write}$ Post $\xrightarrow{attach}$ Location $\xrightarrow{attach^{-1}}$ Post $\xrightarrow{write^{-1}}$ User, whose notation is "$U \to P \to L \leftarrow P \leftarrow U$" or $\Phi_6(U, U)$.

Social Meta Path based Features: These meta paths can actually cover a large number of path instances connecting users in the network. Formally, we denote that node n (or link l) is an instance of node type T (or link type R) in the network as $n \in T$ (or $l \in R$). Identity function $\mathbb{I}(a, A) = \begin{cases} 1, & \text{if } a \in A \\ 0, & otherwise, \end{cases}$ can check whether node/link a is an instance of node/link type A in the network. To consider the effect of the unconnected links when extracting features for social links in the network, the *Intra-Network Social Meta Path based Features* can be formally defined as follows:

Definition 7.2 (Intra-Network Social Meta Path Based Features) For a given link (u, v), the feature extracted for it based on meta path $\Phi = T_1 \xrightarrow{R_1} T_2 \xrightarrow{R_2} \cdots \xrightarrow{R_{k-1}} T_k$ from the network is defined to be the expected number of formed path instances between u and v in the network:

$$x(u, v) = \mathbb{I}(u, T_1)\mathbb{I}(v, T_k) \sum_{n_1 \in \{u\}, n_2 \in T_2, \dots, n_k \in \{v\}} \prod_{i=1}^{k-1} p(n_i, n_{i+1})\mathbb{I}((n_i, n_{i+1}), R_i), \quad (7.27)$$

where $p(n_i, n_{i+1}) = 1.0$ if $(n_i, n_{i+1}) \in \mathcal{E}_{u,u}$ and otherwise, $p(n_i, n_{i+1})$ denotes the *formation probability* of link (n_i, n_{i+1}) to be introduced in Sect. 7.5.4.

Features extracted by MLI based on $\Phi = \{\Phi_1, \dots, \Phi_6\}$ are named as the *intra-network social meta path* based social features. (Φ_0 will be used in the following subsection only.)

Inter-Network Social Meta Paths: When a network is very new, features extracted based on *intra-network social meta paths* can be very sparse, as there exist few connections in the network.

Example 7.7 Consider, for example, in Fig. 7.9, we want to predict whether social link $(A^{(1)}, B^{(1)})$ in network $G^{(1)}$ will be formed or not. Merely based on the *intra-network social meta paths*, the feature vector of extracted for link $(A^{(1)}, B^{(1)})$ will be **0**. However, we find that $A^{(1)}$ and $B^{(1)}$ can be correlated actually with various inter-network paths, e.g., $B^{(1)} \to B^{(2)} \to A^{(2)} \to A^{(1)}$, $B^{(1)} \to B^{(2)} \to F^{(2)} \to A^{(2)} \to A^{(1)}$ and $B^{(1)} \to B^{(2)} \to G^{(2)} \to A^{(2)} \to A^{(1)}$.

By following this idea, MLI proposes to transfer useful information from aligned networks with the following *anchor meta path* and the *inter-network social meta paths*, whose formal definitions are available in Sect. 3.5. In MLI, we are mainly concerned about *inter-network meta path* starting and ending with users, which are named as the *inter-network social meta path*. Let $\Upsilon(U^{(i)}, U^{(j)})$ denote the *anchor meta path* defined between networks $G^{(i)}$ and $G^{(j)}$. The 4 specific *inter-network social meta paths* used in MLI include:

- **Category 1**: $\Upsilon(U^{(i)}, U^{(j)}) \circ (\Phi(U^{(j)}, U^{(j)}) \cup \Phi_0(U^{(j)}, U^{(j)})) \circ \Upsilon(U^{(j)}, U^{(i)})$, whose notation is $\Psi_1(U^{(i)}, U^{(i)})$;
- **Category 2.**: $(\Phi(U^{(i)}, U^{(i)}) \cup \Phi_0(U^{(i)}, U^{(i)})) \circ \Upsilon(U^{(i)}, U^{(j)}) \circ (\Phi(U^{(j)}, U^{(j)}) \cup \Phi_0(U^{(j)}, U^{(j)})) \circ \Upsilon(U^{(j)}, U^{(i)})$, whose notation is $\Psi_2(U^{(i)}, U^{(i)})$;
- **Category 3.**: $\Upsilon(U^{(i)}, U^{(j)}) \circ (\Phi(U^{(j)}, U^{(j)}) \cup \Phi_0(U^{(j)}, U^{(j)})) \circ \Upsilon(U^{(j)}, U^{(i)}) \circ (\Phi(U^{(i)}, U^{(i)}) \cup \Phi_0(U^{(i)}, U^{(i)}))$, whose notation is $\Psi_3(U^{(i)}, U^{(i)})$;
- **Category 4.**: $(\Phi(U^{(i)}, U^{(i)}) \cup \Phi_0(U^{(i)}, U^{(i)})) \circ \Upsilon(U^{(i)}, U^{(j)}) \circ (\Phi(U^{(j)}, U^{(j)}) \cup \Phi_0(U^{(j)}, U^{(j)})) \circ \Upsilon(U^{(j)}, U^{(i)}) \circ (\Phi(U^{(i)}, U^{(i)}) \cup \Phi_0(U^{(i)}, U^{(i)}))$, whose notation is $\Psi_4(U^{(i)}, U^{(i)})$;

where $\Phi(U^{(i)}, U^{(i)}) \cup \Phi_0(U^{(i)}, U^{(i)}) = \{\Phi_0(U^{(i)}, U^{(i)}), \ldots, \Phi_6(U^{(i)}, U^{(i)})\}$ denote the 7 *intra-network social meta paths* in network $G^{(i)}$ introduced before.

Let $\Psi = \{\Psi_1, \Psi_2, \Psi_3, \Psi_4\}$. Ψ is a comprehensive *inter-network social meta path* set and features extracted based on Ψ can transfer information for both anchor users and non-anchor users from other aligned networks.

Example 7.8 For example, in Fig. 7.9, by following path "$B^{(1)} \to B^{(2)} \to A^{(2)} \to A^{(1)}$," we can go from an *anchor user* $B^{(1)}$ to another *anchor user* $A^{(1)}$ and such path is an instance of $\Psi_1(U^{(1)}, U^{(1)})$; by following path $C^{(1)} \to A^{(1)} \to A^{(2)} \to D^{(2)} \to D^{(1)}$, we can go from a *non-anchor user* $C^{(1)}$ to an *anchor user* $D^{(1)}$, which is an instance of $\Psi_2(U^{(1)}, U^{(1)})$; in addition, by following path $C^{(1)} \to A^{(1)} \to A^{(2)} \to B^{(2)} \to B^{(1)} \to E^{(1)}$, we can go from a *non-anchor user* $C^{(1)}$ to another *non-anchor user* $E^{(1)}$, which is an instance of $\Psi_4(U^{(1)}, U^{(1)})$.

Social Meta Path Selection: As introduced in Sect. 7.5.1, information transferred from aligned networks is helpful for improving link prediction performance in a given network but can be misleading as well, which is called the *network difference problem*. To solve the *network difference*

Fig. 7.9 Meta path across aligned networks

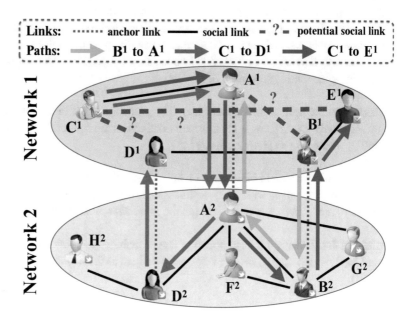

problem, MLI proposes to rank and select the top K features from the feature vector extracted based on the *intra-network* and *inter-network social meta paths*, $[\mathbf{x}_\Phi^\top, \mathbf{x}_\Psi^\top]^\top$, from the multiple *partially aligned heterogeneous networks*.

Let variable $X_i \in [\mathbf{x}_\Phi^\top, \mathbf{x}_\Psi^\top]^\top$ be a feature extracted based on a meta path in $\{\Phi, \Psi\}$ and variable Y be the *label*. $P(Y = y)$ denotes the *prior probability* that links in the training set having label y and $P(X_i = x)$ represents the *frequency* that feature X_i has value x. Information theory related measure *mutual information* (mi) [43] is used as the ranking criteria:

$$mi(X_i) = \sum_x \sum_y P(X_i = x, Y = y) \log \frac{P(X_i = x, Y = y)}{P(X_i = x)P(Y = y)} \tag{7.28}$$

Let $[\bar{\mathbf{x}}_\Phi^\top, \bar{\mathbf{x}}_\Psi^\top]^\top$ be the features of the top K *mi* score selected from $[\mathbf{x}_\Phi^\top, \mathbf{x}_\Psi^\top]^\top$. In the next subsection, we will use the selected feature vector $[\bar{\mathbf{x}}_\Phi^\top, \bar{\mathbf{x}}_\Psi^\top]^\top$ to build a novel PU link prediction model.

7.5.4 Spy Technique Based PU Link Prediction

In this subsection, we will first introduce a method to solve the *PU link prediction* problem in one single network. As introduced in Sect. 7.5.2, from a given network, e.g., G, we can get two disjoint sets of links: connected (i.e., formed) links $\mathcal{P}$ and unconnected links $\mathcal{U}$. To differentiate these links, we define a new concept "*connection state*," z, to show whether a link is connected (i.e., formed) or unconnected in network G. For a given link l, if l is connected in the network, then $z(l) = +1$; otherwise, $z(l) = -1$. As a result, we can have the "*connection states*" of links in $\mathcal{P}$ and $\mathcal{U}$ to be: $z(\mathcal{P}) = +\mathbf{1}$ and $z(\mathcal{U}) = -\mathbf{1}$.

Besides the "*connection state*," links in the network can also have their own "*labels*," y, which can represent whether a link is to be formed or will never be formed in the network. For a given link l, if l has been formed or to be formed, then $y(l) = +1$; otherwise, $y(l) = -1$. Similarly, we can have the "*labels*" of links in $\mathcal{P}$ and $\mathcal{U}$ to be: $y(\mathcal{P}) = +\mathbf{1}$ but $y(\mathcal{U})$ can be either $+1$ or -1, as $\mathcal{U}$ can contain both links to be formed and links that will never be formed.

By using $\mathcal{P}$ and $\mathcal{U}$ as the positive and negative training sets, we can build a *link connection prediction model* $\mathcal{M}_c$, which can be applied to predict whether a link exists in the original network, i.e., the *connection state* of a link. Let l be a link to be predicted, by applying $\mathcal{M}_c$ to classify l, we can get the *connection probability* of l to be:

Definition 7.3 (Connection Probability) The probability that link l's *connection states* is predicted to be *connected* (i.e., $z(l) = +1$) is formally defined as the *connection probability* of link l: $p(z(l) = +1|\mathbf{x}(l))$, where $\mathbf{x}(l) = [\bar{\mathbf{x}}_\Phi(l)^\top, \bar{\mathbf{x}}_\Psi(l)^\top]^\top$.

Meanwhile, if we can obtain a set of links that "will never be formed," i.e., "-1" links, from the network, which together with $\mathcal{P}$ ("+1" links) can be used to build a *link formation prediction model* $\mathcal{M}_f$. Here, model $\mathcal{M}_f$ can be used to get the *formation probability* of l to be:

Definition 7.4 (Formation Probability) The probability that link l's *label* is predicted to be *formed or will be formed* (i.e., $y(l) = +1$) is formally defined as the *formation probability* of link l: $p(y(l) = +1|\mathbf{x}(l))$.

Fig. 7.10 PU link
prediction

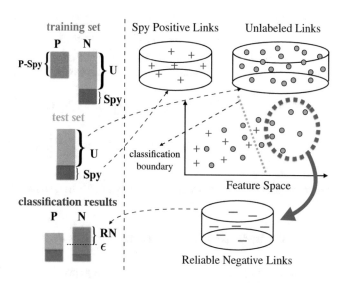

However, from the network, we have no information about "links that will never be formed" (i.e., "-1" links). As a result, the *formation probabilities* of potential links that we aim to obtain as proposed in Sect. 7.5.2 can be very challenging to calculate. Meanwhile, the correlation between link l's *connection probability* and *formation probability* has been proved in existing works [11] to be:

$$p(y(l) = +1|\mathbf{x}(l)) \propto p(z(l) = +1|\mathbf{x}(l)). \tag{7.29}$$

In other words, for links whose *connection probabilities* are low, their *formation probabilities* will be relatively low as well. This rule can be utilized to extract links which can be more likely to be the reliable "-1" links from the network. The *link connection prediction model* $\mathcal{M}_c$ built with $\mathcal{P}$ and $\bar{\mathcal{U}}$ can be applied to classify links in $\bar{\mathcal{U}}$ to extract the *reliable negative link set*.

Definition 7.5 (Reliable Negative Link Set) The *reliable negative links* in the *unconnected link* set $\bar{\mathcal{U}}$ are those whose *connection probabilities* predicted by the *link connection prediction model*, $\mathcal{M}_c$, are lower than threshold $\epsilon \in [0, 1]$:

$$\mathcal{RN} = \{l | l \in \bar{\mathcal{U}}, p(z(l) = +1|\mathbf{x}(l)) < \epsilon\}. \tag{7.30}$$

Some heuristic based methods have been proposed to set the optimal threshold ϵ, e.g., the *spy technique* proposed in [24]. As shown in Fig. 7.10, we randomly selected a subset of links in $\mathcal{P}$ as the spy, $\mathcal{SP}$, whose proportion is controlled by $s\%$ ($s\% = 15\%$ is used as the default sample rate as introduced in [59]). Sets $(\mathcal{P} \setminus \mathcal{SP})$ and $(\bar{\mathcal{U}} \cup \mathcal{SP})$ are used as positive and negative training sets to the *spy prediction* model, $\mathcal{M}_s$. By applying $\mathcal{M}_s$ to classify links in $(\bar{\mathcal{U}} \cup \mathcal{SP})$, we can get their *connection probabilities* to be:

$$p(z(l) = +1|\mathbf{x}(l)), l \in (\bar{\mathcal{U}} \cup \mathcal{SP}), \tag{7.31}$$

and parameter ϵ is set as the minimal *connection probability* of spy links in $\mathcal{SP}$:

$$\epsilon = \min_{l \in \mathcal{SP}} p(z(l) = +1|\mathbf{x}(l)). \tag{7.32}$$

With the extracted *reliable negative link set* $\mathcal{RN}$, we can solve the *PU link prediction* problem with *classification based link prediction methods*, where $\mathcal{P}$ and $\mathcal{RN}$ are used as the positive and negative training sets, respectively. Meanwhile, when applying the built model to predict links in $\mathcal{L}^{(i)}$, their optimal labels, i.e., $\hat{\mathcal{Y}}^{(i)}$, should be those which can maximize the following *formation probabilities*:

$$
\begin{aligned}
\hat{\mathcal{Y}}^{(i)} &= \arg\max_{\mathcal{Y}^{(i)}} p(y(\mathcal{L}^{(i)}) = \mathcal{Y}^{(i)} | G^{(1)}, G^{(2)}, \dots, G^{(k)}) \\
&= \arg\max_{\mathcal{Y}^{(i)}} p(y(\mathcal{L}^{(i)}) = \mathcal{Y}^{(i)} | \left[\bar{\mathbf{x}}_{\Phi}(\mathcal{L}^{(i)})^{\top}, \bar{\mathbf{x}}_{\Psi}(\mathcal{L}^{(i)})^{\top} \right]^{\top})
\end{aligned}
\tag{7.33}
$$

where $y(\mathcal{L}^{(i)}) = \mathcal{Y}^{(i)}$ represents that links in $\mathcal{L}^{(i)}$ have labels $\mathcal{Y}^{(i)}$.

7.5.5 Multi-Network Concurrent PU Link Prediction Framework

Method MLI to be introduced in this part is a general link prediction framework and can be applied to predict social links in n *partially aligned networks* simultaneously. When it comes to n partially aligned network formulated in Sect. 7.5.2, the optimal labels of potential links $\{\mathcal{L}^{(1)}, \mathcal{L}^{(2)}, \dots, \mathcal{L}^{(n)}\}$ of networks $G^{(1)}, G^{(2)}, \dots, G^{(n)}$ will be:

$$
\begin{aligned}
\hat{\mathcal{Y}}^{(1)}, \hat{\mathcal{Y}}^{(2)}, \dots, \hat{\mathcal{Y}}^{(n)} = \arg\max_{\mathcal{Y}^{(1)}, \mathcal{Y}^{(2)}, \dots, \mathcal{Y}^{(n)}} p(y(\mathcal{L}^{(1)}) = \mathcal{Y}^{(1)}, y(\mathcal{L}^{(2)}) = \mathcal{Y}^{(2)}, \\
\dots, y(\mathcal{L}^{(n)}) = \mathcal{Y}^{(n)} | G^{(1)}, G^{(2)}, \dots, G^{(n)})
\end{aligned}
\tag{7.34}
$$

The above target function is very complex to solve and, in [59], MLI obtains the solution by updating one variable, e.g., $\mathcal{Y}^{(1)}$, and fix other variables, e.g., $\mathcal{Y}^{(2)}, \dots, \mathcal{Y}^{(n)}$, alternatively with the following equation:

$$
\begin{cases}
(\hat{\mathcal{Y}}^{(1)})^{(\tau)} &= \arg\max_{\mathcal{Y}^{(1)}} p(y(\mathcal{L}^{(1)}) = \mathcal{Y}^{(1)} | G^{(1)}, G^{(2)}, \dots, G^{(n)}, \\
& \qquad (\hat{\mathcal{Y}}^{(2)})^{(\tau-1)}, (\hat{\mathcal{Y}}^{(3)})^{(\tau-1)}, \dots, (\hat{\mathcal{Y}}^{(n)})^{(\tau-1)}) \\
(\hat{\mathcal{Y}}^{(2)})^{(\tau)} &= \arg\max_{\mathcal{Y}^{(2)}} p(y(\mathcal{L}^{(2)}) = \mathcal{Y}^{(2)} | G^{(1)}, G^{(2)}, \dots, G^{(n)}, \\
& \qquad (\hat{\mathcal{Y}}^{(1)})^{(\tau)}, (\hat{\mathcal{Y}}^{(3)})^{(\tau-1)}, \dots, (\hat{\mathcal{Y}}^{(n)})^{(\tau-1)}) \\
\quad \dots \dots \\
(\hat{\mathcal{Y}}^{(n)})^{(\tau)} &= \arg\max_{\mathcal{Y}^{(n)}} p(y(\mathcal{L}^{(n)}) = \mathcal{Y}^{(n)} | G^{(1)}, G^{(2)}, \dots, G^{(n)}, \\
& \qquad (\hat{\mathcal{Y}}^{(1)})^{(\tau)}, (\hat{\mathcal{Y}}^{(2)})^{(\tau)}, \dots, (\hat{\mathcal{Y}}^{(n-1)})^{(\tau)})
\end{cases}
\tag{7.35}
$$

The architecture of framework MLI is shown in Fig. 7.11. When predicting social links in network $G^{(i)}$, MLI can extract features based on the *intra-network social meta path*, $\mathbf{x}_{\Phi}$, extracted from $G^{(i)}$ and those extracted based on the *inter-network social meta path*, $\mathbf{x}_{\Psi}$, across $G^{(1)}, G^{(2)}, \dots, G^{(i-1)}$, $G^{(i+1)}, \dots, G^{(n)}$ for links in $\mathcal{P}^{(i)}, \bar{\mathcal{U}}^{(i)}$ and $\mathcal{L}^{(i)}$. Feature vectors $\mathbf{x}_{\Phi}(\mathcal{P}), \mathbf{x}_{\Phi}(\bar{\mathcal{U}})$ and $\mathbf{x}_{\Psi}(\mathcal{P}), \mathbf{x}_{\Psi}(\bar{\mathcal{U}})$ as well as the labels, $y(\mathcal{P}), y(\bar{\mathcal{U}})$, of links in $\mathcal{P}$ and $\bar{\mathcal{U}}$ are passed to the PU link prediction model $\mathcal{M}^{(i)}$ and the meta path selection model $\mathcal{MS}^{(i)}$. The formation probabilities of links in $\mathcal{L}^{(i)}$ predicted by model $\mathcal{M}^{(i)}$ will be used to update the network by replacing the weights of $\mathcal{L}^{(i)}$ with the newly predicted formation probabilities. The initial weights of these potential links in $\mathcal{L}^{(i)}$ are set as 0 (i.e., the *formation probability* of links mentioned in Definition 7.2). After finishing these steps on $G^{(i)}$, we

Fig. 7.11 Multi-PU link
prediction framework

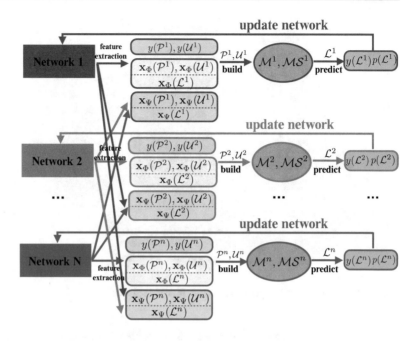

will move to conduct similar operations on $G^{(i+1)}$. We predict links in $G^{(1)}$ to $G^{(n)}$ alternatively in a sequence until the results in all of these networks converge.

7.6 Sparse and Low Rank Matrix Estimation Based PU Link Prediction

Different online social networks usually have different functions, and information in them follows totally different distributions. When predicting the links across multiple aligned online social networks, the link prediction models aforementioned, which merely append the feature vectors from different sources, can hardly address the domain difference problem at all. In this section, we will introduce a new cross-network link prediction model proposed in [61], which embeds the feature vectors of links from aligned networks into a shared feature space. The knowledge from the source networks is transferred to the target network in the shared feature space.

7.6.1 Problem Description

In this section, we will study the link prediction problem for the target network, which is aligned with multiple source networks concurrently. Formally, the problem is named as the "*Social Link Transfer*" (SLT) problem. Formally, given the *multiple aligned online social networks* $\mathcal{G} = (\{G^t, G^{(1)}, G^{(2)}, \ldots, G^{(K)}\}, \{\mathcal{A}^{(t,1)}, \mathcal{A}^{(t,2)}, \ldots, \mathcal{A}^{(K-1,K)}\})$, the SLT problem to be studied in this section aims at inferring the potential social connections among users in the target network G^t with information across all these networks. Formally, based on information available in $\mathcal{G}$, the objective of SLT is to build a social link prediction function $S : \mathcal{U}^t \times \mathcal{U}^t \setminus \mathcal{E}_u^t \to [0, 1]$ to infer the confidence scores of all the potential social connections among the users in the target network G^t, where $\mathcal{U}^t$ and $\mathcal{E}_u^t$ represent the existing users and social links in G^t, respectively.

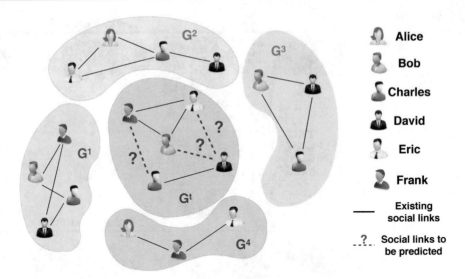

Fig. 7.12 An example of link prediction across aligned networks

Example 7.9 An example to illustrate the SLT problem is provided in Fig. 7.12. In Fig. 7.12, network G^t is the target network, and $G^{(1)}, \ldots, G^{(4)}$ are the other aligned source networks, which share a number of common users with G^t. With the information across networks $\{G^t, G^{(1)}, \ldots, G^{(4)}\}$, the objective of SLT is to infer potential social links (i.e., the red dashed lines) to be formed in the target network G^t.

The SLT problem studied in this section is based on the same setting as those in [57, 59], but we will introduce a new model to address the problem. We summarize the differences of this work from these existing works as follows. Firstly, the link prediction model proposed in this section is based on the matrix estimation, which is totally different from the classification based models proposed in [57, 59] and will not suffer from the class imbalance problem. Secondly, considering the connections among users in the networks are usually very sparse and users tend to form densely connected local communities, a sparse regularizer and a low-rank regularizer are incorporated in the objective function. Thirdly, these existing works [57, 59] transfer information across different networks without considering the domain differences. Meanwhile, based on the known anchor and social link information, our model overcomes the domain difference problem by mapping the feature vectors extracted for links from the aligned networks to a shared lower-dimensional latent feature space instead.

The SLT problem studied in this section is very hard to solve mainly due to the following challenges caused by (1) the *heterogeneity of networks*, (2) the *multiple aligned networks* setting, (3) the *sparse and low-rank property* of the target network, and (4) the *objective function* is hard to solve. To overcome these challenges, a novel link prediction model named SLAMPRED (*Sparse Low-rAnk Matrix* estimation based *Pred*iction) [61] will be introduced in this part. SLAMPRED formulates the link prediction problem as a sparse and low-rank matrix estimation problem. Heterogeneous information is used to calculate the similarity among users, and similar users tend be linked. With the existing anchor and social link information, SLAMPRED proposes to map the feature vectors of the social links extracted from the target and other aligned source networks to a common low-dimensional latent feature space. Two regularizers are introduced in the objective function of SLAMPRED to preserve the sparse and low-rank properties. Furthermore, SLAMPRED solves the objective function with the iterative CCCP (convex concave procedure), and in each iteration the involved non-differentiable sparsity and low-rank regularizers are effectively handled by the proximal operators.

In the following parts, we will first introduce the link prediction model built with the observed network connection information and other heterogeneous attribute information available in the target network. After that, we will talk about the target network link prediction problem with information across multiple aligned networks, where the features extracted from different networks are projected to a lower-dimensional feature space to accommodate the domain differences. Finally, we will introduce the joint optimization objective function, which can be resolved by the proximal operator based iterative CCCP algorithm effectively.

7.6.2 Intra-Network Link Prediction

Users' diverse online social activities may generate heterogeneous information in the online social networks, which include both the network structure information and the different categories of attribute information about the users. In this subsection, we will introduce the link prediction method with the heterogeneous information available in the target network.

7.6.2.1 Intra-Network Link Prediction with Link Information

Given the target network G^t involving users $\mathcal{U}^t$, we can represent the observed social connection among the users with the binary social adjacency matrix $\mathbf{A}^t \in \{0, 1\}^{|\mathcal{U}^t| \times |\mathcal{U}^t|}$, where entry $A^t(i, j) = 1$ iff the corresponding social link (u_i^t, u_j^t) exists between users u_i^t and u_j^t in G^t. In the SLT problem, our objective is to infer the potential unobserved social links for the target network, which can be achieved by finding a sparse and low-rank predictor matrix $\mathbf{S} \in \mathcal{S}$ from some convex admissible set $\mathcal{S} \subset \mathbb{R}^{|\mathcal{U}^t| \times |\mathcal{U}^t|}$. Meanwhile, the inconsistency between the inferred matrix $\mathbf{S}$ and the observed social adjacency matrix $\mathbf{A}^t$ can be represented as the loss function $l(\mathbf{S}, \mathbf{A}^t)$. The optimal social link predictor for the target network can be achieved by minimizing the loss term, i.e.,

$$\arg \min_{\mathbf{S} \in \mathcal{S}} l(\mathbf{S}, \mathbf{A}^t). \tag{7.36}$$

The loss function $l(\mathbf{S}, \mathbf{A}^t)$ can be defined in many different ways, and the *loss function* can be approximated by counting the loss introduced by the existing social links in $\mathcal{E}_u^t$, i.e.,

$$l(\mathbf{S}, \mathbf{A}^t) = \frac{1}{|\mathcal{E}_u^t|} \sum_{(u_i^t, u_j^t) \in \mathcal{E}_u^t} \mathbb{1}\left(\left(A^t(i, j) - \frac{1}{2}\right) \cdot S(i, j) \leq 0\right). \tag{7.37}$$

7.6.2.2 Intra-Network Link Prediction with Heterogeneous Attribute Information

Besides the connection information, there also exists a large amount of attribute information available in the target network, e.g., *location check-in records*, *online social activity temporal patterns*, and *text usage patterns*, etc. Based on the attribute information, a set of features can be extracted for all the potential user pairs to denote their closeness, which are called the *intimacy features* formally. For instance, given a user pair (u_i^t, u_j^t) in the target network, we can represent its *intimacy features* as vector $\mathbf{x}_{i,j}^t \in \mathbb{R}^{d^t}$ (d^t denotes the extracted intimacy feature number). According to the existing works [14, 57], different intimacy features can be extracted from the attribute information.

More generally, we can represent the feature vectors extracted for user pairs as a 3-way tensor $\mathbf{X}^t \in \mathbb{R}^{d^t \times |\mathcal{U}^t| \times |\mathcal{U}^t|}$, where slice $\mathbf{X}^t(k, :, :)$ denotes all the k_{th} intimacy features among all the user pairs. In online social networks, *homophily* principle [27] has been observed to widely structure the users' online social connections, and users who are close to each other are more likely to be friends. Based on such an intuition, we can infer the potential social connection matrix $\mathbf{S}$ by maximizing the overall intimacy scores of the inferred new social connections, i.e.,

$$\arg \max_{\mathbf{S} \in \mathcal{S}} int(\mathbf{S}, \mathbf{X}^t). \tag{7.38}$$

SLAMPRED proposes to define the intimacy score term $int(\mathbf{S}, \mathbf{X}^t)$ by enumerating and summing the *intimacy scores* of the inferred social connections, i.e.,

$$int(\mathbf{S}, \mathbf{X}^t) = \sum_{k=1}^{d^t} \left\| \mathbf{S} \circ \mathbf{X}^t(k, :, :) \right\|_1, \tag{7.39}$$

where operator $\circ$ denotes the Hadamard product (i.e., entrywise product) of two matrices.

7.6.2.3 Joint Optimization Function for Intra-Network Link Prediction

By considering the link and attribute information in the target network at the same time, we can represent the joint optimization for link prediction in the target network to be

$$\arg \min_{\mathbf{S} \in \mathcal{S}} l(\mathbf{S}, \mathbf{A}^t) - \alpha^t \cdot int(\mathbf{S}, \mathbf{X}^t) + \gamma \cdot \|\mathbf{S}\|_1 + \tau \cdot \|\mathbf{S}\|_*. \tag{7.40}$$

Considering that the social connections in online social networks are usually very sparse and of low-rank, the regularizers $\|\mathbf{S}\|_1$ and $\|\mathbf{S}\|_*$ are added to preserve the *sparse* and *low rank* properties of the inferred predictor matrix $\mathbf{S}$. Parameters α^t, γ, τ denote the importance scalars of different terms in the objective function.

7.6.3   Inter-Network Link Prediction

Besides the information available in the target network, a large amount of information about the users' social activities is available in other external source networks as well, which can be transferred to the target network to help improve the link prediction results, especially when the target network suffers from information sparsity problem. To be general, we can represent the *intimacy* features extracted for user pairs in source network $G^{(i)}$ ($i \in \{1, 2, \ldots, K\}$) as a 3-way tensor $\mathbf{X}^{(i)} \in \mathbb{R}^{d^{(i)} \times |\mathcal{U}^{(i)}| \times |\mathcal{U}^{(i)}|}$, where $\mathcal{U}^{(i)}$ denotes the user set in $G^{(i)}$ and $d^{(i)}$ is the extracted feature number.

Meanwhile, different online social networks are constructed for different purposes, information from which may follow totally different distributions actually. To adopt the information domains of these different aligned networks, SLAMPRED proposes to project the extracted feature vectors from different networks (both G^t and aligned source networks $G^{(1)}, \ldots, G^{(K)}$) to a common lower-dimensional feature space instead. Given the $K + 1$ partially aligned social networks, we formulate the information domain adaption problem as a mapping function inference problem instead. Our objective is to construct $K + 1$ mapping functions, $f^t : \mathbb{R}^{d^t} \to \mathbb{R}^c, \ldots, f^{(K)} : \mathbb{R}^{d^{(K)}} \to \mathbb{R}^c$ to map the $K + 1$ input features to a new c-dimensional latent space, where certain properties about the networks are still preserved.

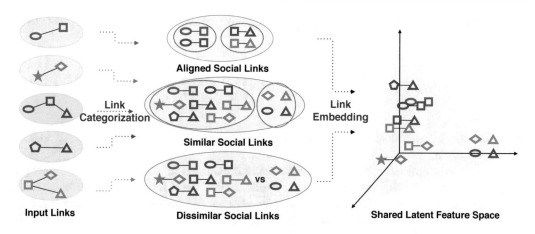

Fig. 7.13 An example of method SLAMPRED on social link embedding

SLAMPRED achieves the objective by utilizing the existing anchor links and social links across the networks. As shown in Fig. 7.13, the links in different social networks are first categorized into different sets: (1) social links aligned by anchor links (i.e., the *aligned social links* to be introduced later), (2) *similar social links* (i.e., connected user pairs or unconnected user pairs), and (3) *dissimilar social links* (i.e., the connected user pairs vs. the unconnected ones). Based on the categorization information about the links, in the link embedding process, we aim at placing aligned social links and similar social links closely in the common latent feature space, while placing the dissimilar ones far away from each other in the feature space. More information about these concepts and the embedding process will be introduced in the following parts in great detail.

7.6.3.1 Anchor Link Based Feature Space Projection

Before introducing the anchor link based feature space projection method, we first introduce the concept of *aligned social link* as follows:

Definition 7.6 (Aligned Social Link) Given two social links (u_i^t, u_j^t) and $(u_m^{(k)}, u_n^{(k)})$ in networks G^t and $G^{(k)}$, respectively, if $u_i^t, u_m^{(k)}$ and $u_j^t, u_n^{(k)}$ are both aligned by the anchor links (i.e., $(u_i^t, u_m^{(k)}) \in \mathcal{A}^{(t,k)}$ and $(u_j^t, u_n^{(k)}) \in \mathcal{A}^{(t,k)}$), then (u_i^t, u_j^t) and $(u_m^{(k)}, u_n^{(k)})$ are called the *aligned social links*.

Let sets $\mathcal{L}^t$ and $\mathcal{L}^{(k)}$ denote all the potential social links in networks G^t and $G^{(k)}$, respectively, where $\mathcal{L}^t = \mathcal{U}^t \times \mathcal{U}^t \setminus \{(u, u)\}_{u \in \mathcal{U}^t}$ and $\mathcal{L}^{(k)} = \mathcal{U}^{(k)} \times \mathcal{U}^{(k)} \setminus \{(u, u)\}_{u \in \mathcal{U}^{(k)}}$. Based on the anchor links between networks G^t and $G^{(k)}$ (i.e., $\mathcal{A}^{(t,k)}$), we can denote all the aligned social links with the *aligned social link indicator matrix* $\mathbf{W}_A^{(t,k)} \in \{0, 1\}^{|\mathcal{L}^t| \times |\mathcal{L}^{(k)}|}$, where entry $W_A^{(t,k)}(i, j) = 1$ iff the corresponding social links $l_i^t \in \mathcal{L}^t$ and $l_j^{(k)} \in \mathcal{L}^{(k)}$ are *aligned social links*.

Generally, the *aligned social links* are actually connecting the accounts of the same users, and the feature vectors extracted for them from different networks should be mapped to close areas in a low-dimensional latent feature space. Based on such an intuition, we can define the inconsistency introduced in projecting the features for aligned social links between networks G^t and other external source networks as term $Cost_A$:

$$Cost_A = \mu \sum_{m=t}^{K} \sum_{n=t}^{K} \sum_{i=1}^{|\mathcal{L}^{(m)}|} \sum_{j=1}^{|\mathcal{L}^{(n)}|} \left\| f^{(m)}(\mathbf{x}_{l_i^m}^m) - f^{(n)}(\mathbf{x}_{l_j^n}^n) \right\|^2 W_A^{(m,n)}(i,j), \qquad (7.41)$$

where notation $\sum_{m=t}^{K}$ denotes the enumeration of all the networks in the set $\{G^t, G^{(1)}, \ldots, G^{(K)}\}$, and μ is the scalar.

Minimizing the cost term will encourage the features extracted for social links corresponding to the aligned social links being mapped to similar locations in the latent feature space. Furthermore, for all the pairwise networks, we can group all the *aligned social link indicator matrices* together as the big *joint aligned social link indicator matrix* $\mathbf{W}_A \in \{0,1\}^{|\mathcal{L}| \times |\mathcal{L}|}$, where $\mathcal{L} = \mathcal{L}^t \cup \mathcal{L}^{(1)} \cup \cdots \cup \mathcal{L}^{(K)}$. Formally, matrix $\mathbf{W}_A$ can be represented as

$$\mathbf{W}_A = \begin{bmatrix} \mathbf{W}_A^{(t,t)} & \mathbf{W}_A^{(t,1)} & \cdots & \mathbf{W}_A^{(t,K)} \\ \mathbf{W}_A^{(1,t)} & \mathbf{W}_A^{(1,1)} & \cdots & \mathbf{W}_A^{(1,K)} \\ \vdots & \vdots & \ddots & \vdots \\ \mathbf{W}_A^{(K,t)} & \mathbf{W}_A^{(K,1)} & \cdots & \mathbf{W}_A^{(K,K)} \end{bmatrix}. \qquad (7.42)$$

In addition, we can represent its Laplacian matrix as $\mathbf{L}_A = \mathbf{D}_A - \mathbf{W}_A$, where matrix $\mathbf{D}_A$ denotes the diagonal row sum matrix of $\mathbf{W}_A$ with entry $D_A(i,i) = \sum_j W_A(i,j)$ on the diagonal. Matrix $\mathbf{L}_A$ will be used in the projection function inference to be introduced in the following parts.

7.6.3.2 Existing Social Link Based Feature Space Projection

Besides the anchor link information, we also propose to utilize the existing social connections among the users to help infer the *feature mapping functions*. Before introducing the detailed method, we will define the concept of *link existence label* $y(\cdot)$ first as follows:

Definition 7.7 (Link Existence Label) Given a link $l_i^{(k)} \in \mathcal{L}^{(k)}$ in network $G^{(k)}, k \in \{t, 1, 2, \ldots, K\}$, if link $l_i^{(k)}$ exists in the network then its corresponding *link existence label* $y(l_i^{(k)}) = 1$, otherwise $y(l_i^{(k)}) = 0$.

Since our ultimate goal is to infer the potential feature vector mappings to the latent feature space to transfer information for the link prediction tasks, the social link existence information will plan a very important role in identifying the potential feature space mappings. Based on the known social connections in a pair of aligned networks G^t and $G^{(k)}$ ($k \in \{1, 2, \ldots, K\}$), we can construct the *similar link existence label indicator matrix* $\mathbf{W}_S^{(t,k)} \in \{0,1\}^{|\mathcal{L}^t| \times |\mathcal{L}^{(k)}|}$ and *dissimilar link existence label indicator matrix* $\mathbf{W}_D^{(t,k)} \in \{0,1\}^{|\mathcal{L}^t| \times |\mathcal{L}^{(k)}|}$ between networks G^t and $G^{(k)}$. For any link instances $l_i^t \in \mathcal{L}^t$ and $l_j^{(k)} \in \mathcal{L}^{(k)}$, if l_i^t and $l_j^{(k)}$ share the same *link existence label*, we will assign the corresponding entry in $\mathbf{W}_S^{(t,k)}$ with value 1 (and the corresponding entry in $\mathbf{W}_D^{(t,k)}$ with value 0); otherwise, we will assign the corresponding entry in $\mathbf{W}_S^{(t,k)}$ with value 0 (and the corresponding entry in $\mathbf{W}_D^{(t,k)}$ with value 1). Therefore, matrices $\mathbf{W}_S^{(t,k)}$ and $\mathbf{W}_D^{(t,k)}$ store all the link existence information in the networks G^t and $G^{(k)}$.

As pointed out in [44], the instances which share common labels tend to be projected together in the latent feature space, while those having different labels will be projected to be apart from each other instead. Based on such an intuition, terms $Cost_S$ and $Cost_D$ can be defined to denote the mapping costs introduced by the *link existence label* information (for the links having *similar* and *different* labels), respectively:

$$Cost_S = \sum_{m=t}^{K} \sum_{n=t}^{K} \sum_{i=1}^{|\mathcal{L}^{(m)}|} \sum_{j=1}^{|\mathcal{L}^{(n)}|} \left\| f^{(m)}(\mathbf{x}_{l_i^m}^m) - f^{(n)}(\mathbf{x}_{l_j^n}^n) \right\|^2 W_S^{(m,n)}(i,j), \tag{7.43}$$

$$Cost_D = \sum_{m=t}^{K} \sum_{n=t}^{K} \sum_{i=1}^{|\mathcal{L}^{(m)}|} \sum_{j=1}^{|\mathcal{L}^{(n)}|} \left\| f^{(m)}(\mathbf{x}_{l_i^m}^m) - f^{(n)}(\mathbf{x}_{l_j^n}^n) \right\|^2 W_D^{(m,n)}(i,j). \tag{7.44}$$

If link instances l_i^t and $l_j^{(k)}$ in networks G^t and $G^{(k)}$ share the same *link existence label* (i.e., $W_S^{(t,k)}(i,j) = 1$), but their embeddings are far away from each other, then $Cost_S$ will be larger. Meanwhile, if link instances l_i^t and $l_j^{(k)}$ have different *link existence labels* (i.e., $W_D^{(t,k)}(i,j) = 1$), and their embeddings are close to each other, the introduced $Cost_D$ will be small. Therefore, minimizing $Cost_S$ and maximizing $Cost_D$ simultaneously will encourage the link instances of the same label to be projected to similar areas, while those of different labels to be projected separately instead.

What's more, in a similar way, we can also group all the network pairwise *similar link existence label indicator matrices* and *dissimilar link existence label indicator matrices* together in the same order as matrix $\mathbf{W}_A$, which can be represented as $\mathbf{W}_S$ and $\mathbf{W}_D$. Their corresponding Laplacian matrices can be denoted as $\mathbf{L}_S$ and $\mathbf{L}_D$, respectively.

7.6.3.3 Joint Mapping Function Inference

We may want to ensure the mapping functions can achieve the above three objectives at the same time, which can be achieved by minimizing the overall cost function

$$\min Cost(f^t, f^{(1)}, f^{(2)}, \ldots, f^{(K)}) = \frac{Cost_A + Cost_S}{Cost_D}. \tag{7.45}$$

The projection mappings can be of different forms, and we will take the linear mapping as an example here. In other words, the mappings $f^t, f^{(1)}, f^{(2)}, \ldots, f^{(K)}$ can be represented as $K + 1$ matrices $\mathbf{F}^t \in \mathbb{R}^{d^t \times c}$, $\mathbf{F}^{(1)} \in \mathbb{R}^{d^{(1)} \times c}$, ..., $\mathbf{F}^{(K)} \in \mathbb{R}^{d^{(K)} \times c}$, respectively, where $d^t, d^{(1)}, \ldots, d^{(K)}$ denote the length of features from networks $G^t, G^{(1)}, \ldots, G^{(K)}$ and c is the dimension of the projected feature space.

Formally, given all the feature vectors extracted for potential user pairs in the networks $G^t, G^{(1)}, \ldots, G^{(K)}$, we can group them together and represent it as matrix

$$\mathbf{Z} = \begin{bmatrix} \mathbf{Z}^t & \mathbf{0} & \cdots & \mathbf{0} \\ \mathbf{0} & \mathbf{Z}^{(1)} & \cdots & \mathbf{0} \\ \vdots & \vdots & \vdots & \vdots \\ \mathbf{0} & \cdots & \mathbf{0} & \mathbf{Z}^{(K)} \end{bmatrix}, \tag{7.46}$$

where submatrix $\mathbf{Z}^{(k)} = (\mathbf{z}_1^{(k)}, \mathbf{z}_2^{(k)}, \ldots, \mathbf{z}_{|\mathcal{L}^{(k)}| \times |\mathcal{L}^{(k)}|}^{(k)})$ and vector $\mathbf{z}_i^{(k)} \in \mathbb{R}^{d^{(k)} \times 1}$ represents the feature vector extracted for the i_{th} social link in network $G^{(k)}$. Furthermore, we can group all the projection function together and represent it as a $(d^t + d^{(1)} + \cdots + d^{(K)}) \times c$ dimensional matrix

$$\mathbf{F} = \left((\mathbf{F}^t)^\top, (\mathbf{F}^{(1)})^\top, \ldots, (\mathbf{F}^{(K)})^\top \right)^\top, \tag{7.47}$$

which can be effectively inferred with the following theorem.

Theorem 7.1 *The projection functions that minimize the overall cost function are given by the eigenvectors corresponding to the smallest non-zero eigenvalues of the generalized eigenvalue decomposition*

$$\mathbf{Z}(\mu \mathbf{L}_A + \mathbf{L}_S)\mathbf{Z}^\top \mathbf{x} = \lambda \mathbf{Z} \mathbf{L}_D \mathbf{Z}^\top \mathbf{x}. \tag{7.48}$$

Proof Depending on the specific value of c, the theorem can be proven by considering two cases:

Case 1 if $c > 1$, with the above defined matrices, we can rewrite the introduced cost terms $Cost_A$, $Cost_S$, and $Cost_D$ in the linear algebra representation:

$$Cost_A = \mathrm{Tr}(\mathbf{F}^\top \mathbf{Z} \mu \mathbf{L}_A \mathbf{Z}^\top \mathbf{F}), \tag{7.49}$$

$$Cost_S = \mathrm{Tr}(\mathbf{F}^\top \mathbf{Z} \mathbf{L}_S \mathbf{Z}^\top \mathbf{F}), \tag{7.50}$$

$$Cost_D = \mathrm{Tr}(\mathbf{F}^\top \mathbf{Z} \mathbf{L}_D \mathbf{Z}^\top \mathbf{F}). \tag{7.51}$$

Furthermore, the objective function can be represented as

$$\arg\min_{\mathbf{F}} \frac{\mathrm{Tr}(\mathbf{F}^\top \mathbf{Z}(\mu \mathbf{L}_A + \mathbf{L}_S)\mathbf{Z}^\top \mathbf{F})}{\mathrm{Tr}(\mathbf{F}^\top \mathbf{Z} \mathbf{L}_D \mathbf{Z}^\top \mathbf{F})}. \tag{7.52}$$

According to [44, 46], the matrix $\mathbf{F}$ which can minimize the objective function are actually the c eigenvectors corresponding to the c smallest non-zero eigenvalues of the following generalized eigenvalue decomposition function:

$$\mathbf{Z}(\mu \mathbf{L}_A + \mathbf{L}_S)\mathbf{Z}^\top \mathbf{x} = \lambda \mathbf{Z} \mathbf{L}_D \mathbf{Z}^\top \mathbf{x}. \tag{7.53}$$

Case 2 if $c = 1$, then matrix $\mathbf{F}$ to be inferred is actually a vector and the cost terms can be simply represented as

$$Cost_A = \mathbf{F}^\top \mathbf{Z} \mu \mathbf{L}_A \mathbf{Z}^\top \mathbf{F}, \tag{7.54}$$

$$Cost_S = \mathbf{F}^\top \mathbf{Z} \mathbf{L}_S \mathbf{Z}^\top \mathbf{F}, \tag{7.55}$$

$$Cost_D = \mathbf{F}^\top \mathbf{Z} \mathbf{L}_D \mathbf{Z}^\top \mathbf{F}. \tag{7.56}$$

The optimization objective function can be rewritten with the new cost representations as

$$\arg\min_{\mathbf{F}} \frac{\mathbf{F}^\top \mathbf{Z}(\mu \mathbf{L}_A + \mathbf{L}_S)\mathbf{Z}^\top \mathbf{F}}{\mathbf{F}^\top \mathbf{Z} \mathbf{L}_D \mathbf{Z}^\top \mathbf{F}}, \tag{7.57}$$

which is actually the *Rayleigh quotient* of $(\mu \mathbf{L}_A + \mathbf{L}_S)$ relative to $\mathbf{L}_D$. According to the existing books on linear algebra and related works [34, 38], the optimal solution to the objective function can be represented as the eigenvectors corresponding to the c small non-zero eigenvalues of the generalized eigenvalue problem:

$$\mathbf{Z}(\mu \mathbf{L}_A + \mathbf{L}_S)\mathbf{Z}^\top \mathbf{x} = \lambda \mathbf{Z} \mathbf{L}_D \mathbf{Z}^\top \mathbf{x}. \tag{7.58}$$

Therefore, we can formally represent the feature tensors of network G^k (including both the target and aligned source networks) after the domain adaption as $\hat{\mathbf{X}}^k \in \mathbb{R}^{|\mathcal{U}^k| \times |\mathcal{U}^k| \times c}$ ($\forall k \in \{t, 1, 2, \dots, K\}$), where feature vector

$$\hat{\mathbf{X}}^k(i, j, :) = (\mathbf{F}^k)^\top \mathbf{X}^k(i, j, :). \tag{7.59}$$

7.6.3.4 Inter-Network Link Prediction Objective Function

With the information from the external source networks, we can obtain more knowledge about the users and their social patterns. Based on the adapted feature tensors $\hat{\mathbf{X}}^{(1)}, \dots, \hat{\mathbf{X}}^{(K)}$, we can represent the intimacy scores of the potential social links as

$$int(\mathbf{S}, \hat{\mathbf{X}}^{(1)}, \dots, \hat{\mathbf{X}}^{(K)}) = \sum_{k=1}^{K} \alpha^{(k)} \cdot int(\mathbf{S}, \hat{\mathbf{X}}^{(k)}) \tag{7.60}$$

where term $int(\mathbf{S}, \hat{\mathbf{X}}^{(k)}) = \left\| \mathbf{S} \circ \hat{\mathbf{X}}^{(k)} \right\|_1$, and users in $\hat{\mathbf{X}}^{(k)}$ are organized in the same order as $\mathbf{X}^t$. Parameters $\alpha^{(i)}$ denote the importance of the information transferred from the source network $G^{(i)}$. Furthermore, by adding the intimacy terms about the source networks into the objective function, we can rewrite it as follows:

$$\arg\min_{\mathbf{S} \in \mathcal{S}} \; l(\mathbf{S}, \mathbf{A}^t) - \alpha^t \cdot int(\mathbf{S}, \hat{\mathbf{X}}^t) - \sum_{k=1}^{K} \alpha^{(i)} \cdot int(\mathbf{S}, \hat{\mathbf{X}}^{(k)})) + \gamma \|\mathbf{S}\|_1 + \tau \|\mathbf{S}\|_* \tag{7.61}$$

7.6.4 Proximal Operator Based CCCP Algorithm

By studying the objective function, we observe that the intimacy terms are convex while the empirical loss term $l(\mathbf{S}, \mathbf{A}^t)$ is non-convex, which can be approximated with other classical loss functions (e.g., the hinge loss and the Frobenius norm), and the convex squared Frobenius norm loss function is used (i.e., $l(\mathbf{S}, \mathbf{A}^t) = \left\| \mathbf{S} - \mathbf{A}^t \right\|_F^2$). Therefore, the above objective function can be represented as a convex loss term minus another convex term together with two convex non-differentiable regularizers, which actually renders the objective function non-trivial. According to the existing works [37,51], this kind of objective function can be addressed with the concave-convex procedure (CCCP). CCCP is a majorization-minimization algorithm that solves the difference of convex functions problems as a sequence of convex problems. Meanwhile, the regularization terms can be effectively handled with the proximal operators [29] in each iteration of the CCCP process.

7.6.4.1 CCCP Algorithm

Formally, we can decompose the objective function into two convex functions:

$$u(\mathbf{S}) = l(\mathbf{S}, \mathbf{A}^t) + \gamma \cdot \|\mathbf{S}\|_1 + \tau \cdot \|\mathbf{S}\|_*, \tag{7.62}$$

$$v(\mathbf{S}) = \alpha^t \cdot int(\mathbf{S}, \hat{\mathbf{X}}^t) + \sum_{k=1}^{K} \alpha^{(k)} \cdot int(\mathbf{S}, \hat{\mathbf{X}}^{(k)}), \tag{7.63}$$

With $u(\mathbf{S})$ and $v(\mathbf{S})$, we can rewrite the objective function to be

$$\arg\min_{\mathbf{S} \in \mathcal{S}} u(\mathbf{S}) - v(\mathbf{S}). \tag{7.64}$$

The CCCP algorithm can address the objective function with an iterative procedure that solves the following sequence of convex problems:

$$\mathbf{S}^{(h+1)} = \arg\min_{\mathbf{S} \in \mathcal{S}} u(\mathbf{S}) - \mathbf{S}^\top \nabla v(\mathbf{S}^{(h)}) \tag{7.65}$$

It is easy to show that function $v(\mathbf{S})$ differentiable, and the derivative of function $v(\mathbf{S})$ is actually a constant term

$$\nabla v(\mathbf{S}) = \sum_{k=t}^{K} \alpha^{(i)} \sum_{i=1}^{c} \hat{\mathbf{X}}^{(k)}(i,:,:). \tag{7.66}$$

By relying on the Zangwill's global convergence theory [52] of iterative algorithms, it is theoretically proven in [37] that as such a procedure continues, the generated sequence of the variables $\{\mathbf{S}^{(h)}\}_{h=0}^{\infty}$ will converge to some stationary points $\mathbf{S}_*$ in the inference space $\mathcal{S}$.

7.6.4.2 Proximal Operators
Meanwhile, in each iteration of the CCCP updating process, objective function is not easy to address due to the non-differentiable regularizers. Some works have been done to deal with the objective function involving non-smooth functions. The Forward-Backward splitting method proposed in [8] can handle such a kind of optimization function with one single non-smooth regularizer based on the introduced proximal operators. More specifically, as introduced in [8], we can represent the proximal operators for the trace norm and L_1 norm as follows:

$$\text{prox}_{\tau\|\cdot\|_*}(\mathbf{S}) = \mathbf{U}\text{diag}((\sigma_i - \tau)_+)_i \mathbf{V}^\top, \tag{7.67}$$

$$\text{prox}_{\gamma\|\cdot\|_1}(\mathbf{S}) = \text{sgn}(\mathbf{S}) \circ (|\mathbf{S}| - \gamma)_+, \tag{7.68}$$

where $\mathbf{S} = \mathbf{U}\text{diag}(\sigma_i)_i \mathbf{V}^\top$ denotes the singular decomposition of matrix $\mathbf{S}$, and $\text{diag}(\sigma_i)_i$ represents the diagonal matrix with values σ_i on the diagonal.

Recently, some works have proposed the generalized Forward-Backward algorithm to tackle the case with $q(q \geq 2)$ non-differentiable convex regularizers [30]. These methods alternate the gradient step and the proximal steps to update the variables. For instance, given the above objective function in iteration h of the CCCP, we can represent the alternative updating equations in step k to address the objective function as follows:

$$\begin{cases} \mathbf{S}^{(k)} = \mathbf{S}^{(k-1)} - \theta \cdot \nabla_{\mathbf{S}}\left(l(\mathbf{S}, \mathbf{A}) - \mathbf{S}^\top \nabla v(\mathbf{S}^{(h)})\right), \\ \mathbf{S}^{(k)} = \text{prox}_{\theta\tau\|\cdot\|_*}(\mathbf{S}^{(k)}), \\ \mathbf{S}^{(k)} = \text{prox}_{\theta\gamma\|\cdot\|_1}(\mathbf{S}^{(k)}), \end{cases} \tag{7.69}$$

Algorithm 2 Proximal operator based CCCP algorithm

Require: social adjacency matrix $\mathbf{A}$

projected feature tensors $\hat{\mathbf{X}}^t, \hat{\mathbf{X}}^1, \ldots, \hat{\mathbf{X}}^K$

Ensure: link predictor matrix $\mathbf{S}$

1: Initialize matrix $\mathbf{S}_{cccp} = \mathbf{A}$

2: Initialize CCCP convergence CCCP-tag = False

3: **while** CCCP-tag == False **do**

4:     Initialize Proximal convergence Proximal-tag = False

5:     Solve optimization function $\min_{\mathbf{S} \in \mathcal{S}} u(\mathbf{S}) - \mathbf{S}^\top \nabla v(\mathbf{S}_{cccp})$

6:     Initialize $\mathbf{S}_{po} = \mathbf{S}_{cccp}$

7:     **while** Proximal-tag == False **do**

8:         $\mathbf{S}_{po} = \mathbf{S}_{po} - \theta \nabla_{\mathbf{S}} \left(l(\mathbf{S}_{po}, \mathbf{A}) - \mathbf{S}_{po}^\top \nabla v(\mathbf{S}_{cccp}) \right)$

9:         $\mathbf{S}_{po} = \mathrm{prox}_{\theta \tau \|\cdot\|_*}(\mathbf{S}_{po})$

10:        $\mathbf{S}_{po} = \mathrm{prox}_{\theta \gamma \|\cdot\|_1}(\mathbf{S}_{po})$

11:        **if** $\mathbf{S}_{po}$ converges **then**

12:            Proximal-tag = True

13:            $\mathbf{S}_{cccp} = \mathbf{S}_{po}$

14:        **end if**

15:    **end while**

16:    **if** $\mathbf{S}_{cccp}$ converges **then**

17:        CCCP-tag = True

18:    **end if**

19: **end while**

20: Return $\mathbf{S}_{cccp}$

---

where the parameter θ denotes the learning rate and it is assigned with a very small value to ensure the converge of the above functions [32]. We will also give the convergence analysis about the model in the experiment section.

The pseudo-code of the Proximal Operators based CCCP algorithm is available in Algorithm 2.

7.7 Summary

In this chapter, we introduced the link prediction problem in social networks, where various social network services can all be cast as the link prediction problem for simplicity. To address the problem, we introduced the traditional link prediction models for one single homogeneous networks, including the unsupervised link prediction models, supervised link prediction models, and the matrix factorization based link prediction models.

We also introduced the collective link prediction model for heterogeneous social networks, where we took the location-based social networks as an example to describe the problem setting and the proposed model. In the studied problem, we aimed at inferring multiple types of links in the location-based social networks simultaneously, including both the social links and location links. We provided a brief introduction of an integrated link prediction framework, which integrates these sub-problems into one unified framework.

To address the cold start problem in predicting potential links of new users, we introduced the cold start link prediction model, which can be built by utilizing the information about the "old users" within and across the networks. To accommodate the information distribution difference problem about the new users and old users, we introduced a method to sample the old users' subnetwork. Features extracted from multiple aligned heterogeneous can promisingly resolve the cold start problem in the proposed model.

Instead of modeling the non-existing links as the negative instances, we introduced an approach to address the link prediction problem as a PU learning problem, where those non-existing anchor links are treated as unlabeled instead. To identify a subset of the unlabeled anchor links which are highly likely to be negative (i.e., the reliable negative instances), we introduced to apply the spy techniques in the introduced model, which can work well to infer the social links in multiple networks concurrently.

To overcome the domain difference problem, at the end of this chapter, we introduced an approach to address the link prediction as a PU learning problem, where the link representations from different networks are projected into a shared low-dimensional feature space. Considering that the social network structure formed by the users are usually very sparse and users tend to form some small groups inside the social networks, the adjacency matrix of the social networks can have both the sparsity and low-rank properties. The introduced model resolves the problem as an optimization problem, where CCCP and proximal operators are adopted to learn the potential social links among the users.

7.8   Bibliography Notes

Link prediction problems is a traditional research problem studied in various areas, which aims at inferring the connections among nodes in the graph. To this context so far, dozens of link prediction works have been published already [3, 6, 9, 25, 47]. Depending on the learning setting utilized, the existing link prediction models for information networks can be divided into several categories. Initially, researchers study the link prediction problem based on an unsupervised learning setting [22], which predicts links by calculating the similarity scores among nodes with the assumption that close nodes are more likely to be connected. Afterwards, to utilize the supervision information and incorporate multiple closeness measures altogether, researchers introduce the supervised classification based link prediction models [14], where the existing and non-existing links are labeled as the positive and negative instances, respectively. Recently, researchers point that labeling the non-existing as negative instances is not reasonable, since some of the links will be formed, which should be unlabeled actually [54, 59]. Based on such an intuition, link prediction framework based on PU (Positive and Unlabeled) learning setting is introduced in [54, 59].

Most existing works solve link prediction problem with a single source of information. Nowadays, the researchers have pushed the problem boundary further forward by proposing the link prediction across multiple domains. Tang et al. [41] focus on inferring the particular type of links over multiple heterogeneous networks and develop a framework for classifying the type of social ties. To deal with the differences in information distributions of multiple networks, Qi et al. [3] propose to use biased cross-network sampling to do link prediction across networks. Meanwhile, some works have also been done on predicting multiple kinds of links simultaneously. Konstas et al. [20] propose to recommend multiple kinds of links with collaborative filtering methods. Fouss et al. [12] propose to use a traditional model, random walk, to predict multiple kinds of links simultaneously in networks.

Since Zhang et al. [19, 57] propose the concept of "*aligned social networks*," "*anchor links*," "*anchor users*," the social network studies across *multiple aligned social networks* have become a hot research area in recent years. Dozens of papers have been published around various problems about the multiple aligned networks, including *network alignment* [19, 55, 56] and *link prediction* [54, 57–60]. The link prediction models introduced in [54, 57–59] propose to combine the information from different sites by simply merging the extracted feature vectors together without considerations about the domain differences at all, which are totally different from the model introduced in this chapter. The recent paper [60] aims at unifying the link prediction problems subject to different cardinality constraints, like *one-to-one*, *one-to-many* and *many-to-many*, and introduce a general scalable link prediction framework to solve the problem.

To gain a more comprehensive knowledge about existing link prediction works, please refer to the survey paper [13, 35, 53] for more information.

7.9 Exercises

1. (Easy) Please implement the various unsupervised link predictors introduced in Sect. 7.2.1.2 and compare their effectiveness in inferring the friendship links within an online social network.
2. (Easy) Please explain the advantages of the supervised link prediction model over the unsupervised link predictors based on the closeness measures, e.g., *common neighbor* and *Jaccard's coefficient*.
3. (Easy) Please explain why the SCAN model introduced in Sect. 7.4 can resolve the *cold start problem* in predicting links for new users.
4. (Easy) Please define several of *inter-network meta paths* across aligned networks, and explain their physical meanings.
5. (Medium) Please try to implement the supervised link prediction model, and evaluate its performance with a synthetic network dataset.
6. (Medium) Please explain why the spy technique can help identify a set of *reliable negative* instances from the unlabeled set.
7. (Medium) Please explain why the L_1-norm and trace-norm introduced in Sect. 7.6 can maintain the sparse and low-rank properties of the matrix to be estimated.
8. (Hard) Please implement the matrix factorization based link prediction model introduced in Sect. 7.2.3, and evaluate its performance on a synthetic network dataset.
9. (Hard) Please implement the spy technique based PU learning algorithm introduced in Sect. 7.5.4, and use in the link prediction task.
10. (Hard) Please try to implement the sparse and low-rank matrix estimation based link prediction model introduced in Algorithm 2 with a preferred programming language.

References

1. K. Aditya, A. Menon, C. Elkan, Link prediction via matrix factorization, in *Proceedings of the 2011 European Conference on Machine Learning and Knowledge Discovery in Databases - Volume Part II (ECML PKDD'11)* (2011)
2. C. Aggarwal, A. Hinneburg, D. Keim, On the surprising behavior of distance metrics in high dimensional spaces, in *Proceedings of the 8th International Conference on Database Theory (ICDT '01)* (2001)
3. C. Aggarwal G. Qi, T. Huang, Link prediction across networks by biased cross-network sampling, in *2013 IEEE 29th International Conference on Data Engineering (ICDE)* (2013)
4. A. Alahi, K. Goel, V. Ramanathan, A. Robicquet, F. Li, S. Savarese, Social LSTM: human trajectory prediction in crowded spaces, in *2016 IEEE Conference on Computer Vision and Pattern Recognition (CVPR)* (2016)
5. M. Bilgic, G. Namata, L. Getoor, Combining collective classification and link prediction, in *Seventh IEEE International Conference on Data Mining Workshops (ICDMW 2007)* (2007)
6. B. Cao, N. Liu, Q. Yang, Transfer learning for collective link prediction in multiple heterogeneous domains, in *Proceedings of the 27th International Conference on International Conference on Machine Learning (ICML'10)* (2010)
7. E. Cho, S. Myers, J. Leskovec, Friendship and mobility: user movement in location-based social networks, in *Proceedings of the 17th ACM SIGKDD International Conference on Knowledge Discovery and Data Mining (KDD '11)* (2011)
8. P.L. Combettes, V. Wajs, Signal recovery by proximal forward-backward splitting. Multiscale Model. Simul. **4**(4), 1168–1200 (2005)

9. Y. Dong, J. Tang, S. Wu, J. Tian, N. Chawla, J. Rao, H. Cao, Link prediction and recommendation across heterogeneous social networks, in *Proceedings of the 2012 IEEE 12th International Conference on Data Mining (ICDM '12)* (2012)
10. D. Dunlavy, T. Kolda, E. Acar, Temporal link prediction using matrix and tensor factorizations, in *ACM Transactions on Knowledge Discovery from Data (TKDD)* (2011)
11. C. Elkan, K. Noto, Learning classifiers from only positive and unlabeled data, in *Proceedings of the 14th ACM SIGKDD International Conference on Knowledge Discovery and Data Mining (KDD '08)* (2008)
12. F. Fouss, A. Pirotte, J. Renders, M. Saerens, Random-walk computation of similarities between nodes of a graph with application to collaborative recommendation. IEEE Trans. Knowl. Data Eng. **19**, 355–369 (2007)
13. M. Hasan, M.J. Zaki, A survey of link prediction in social networks, in *Social Network Data Analytics*, ed. by C.C. Aggarwal (2011)
14. M. Hasan, V. Chaoji, S. Salem, M. Zaki, Link prediction using supervised learning, in *Proceedings of SDM 06 Workshop on Link Analysis, Counterterrorism and Security* (2006)
15. W. Hsu, A. King, M. Paradesi, T. Pydimarri, T. Weninger, Collaborative and structural recommendation of friends using weblog-based social network analysis, in *AAAI Spring Symposium: Computational Approaches to Analyzing Weblogs* (2006)
16. A. Huang, Similarity measures for text document clustering, in *Proceedings of the Sixth New Zealand Computer Science Research Student Conference (NZCSRSC2008), Christchurch, New Zealand* (2008)
17. P. Jaccard, Étude comparative de la distribution florale dans une portion des alpes et des jura. Bulletin del la Société Vaudoise des Sciences Naturelles **37**, 547–579 (1901)
18. K. Järvelin, J. Kekäläinen, Cumulated gain-based evaluation of IR techniques. ACM Trans. Inf. Syst. **20**(4) (2002)
19. X. Kong, J. Zhang, P. Yu, Inferring anchor links across multiple heterogeneous social networks, in *Proceedings of the 22nd ACM International Conference on Information & Knowledge Management (CIKM '13)* (2013)
20. I. Konstas, V. Stathopoulos, J.M. Jose, On social networks and collaborative recommendation, in *Proceedings of the 32nd International ACM SIGIR Conference on Research and Development in Information Retrieval (SIGIR '09)* (2009)
21. H. Kwak, C. Lee, H. Park, S. Moon, What is twitter, a social network or a news media? in *Proceedings of the 19th International Conference on World Wide Web (WWW '10)* (2010)
22. D. Liben-Nowell, J. Kleinberg, The link prediction problem for social networks, in *Proceedings of the Twelfth International Conference on Information and Knowledge Management (CIKM '03)* (2003)
23. D. Liben-Nowell, J. Kleinberg, The link-prediction problem for social networks. J. Am. Soc. Inf. Sci. Technol. **58**(7) (2007)
24. B. Liu, Y. Dai, X. Li, W. Lee, P. Yu, Building text classifiers using positive and unlabeled examples, in *Proceedings of the Third IEEE International Conference on Data Mining (ICDM '03)* (2003)
25. Z. Lu, B. Savas, W. Tang, I. Dhillon, Supervised link prediction using multiple sources, in *2010 IEEE International Conference on Data Mining* (2010)
26. C. Lynch, Identifiers and their role in networked information applications. Bull. Am. Soc. Inf. Sci. Technol. **24**(2), 17–20 (2005)
27. M. McPherson, L. Smith-Lovin, J. Cook, Birds of a feather: homophily in social networks. Annu. Rev. Sociol. **27**, 415–444 (2001)
28. S. Pan, Q. Yang, A survey on transfer learning. IEEE Trans. Knowl. Data Eng. **22**, 1345–1359 (2010)
29. N. Parikh, S. Boyd, Proximal algorithms. Found. Trends Optim. **1**(3), (2014)
30. H. Raguet, J. Fadili, G. Peyré, A generalized forward-backward splitting. SIAM J. Imag. Sci. (2013)
31. J. Ramos, Using TF-IDF to determine word relevance in document queries, in *Proceedings of the First Instructional Conference on Machine Learning* (1999), pp. 133–142
32. E. Richard, P. Savalle, N. Vayatis, Estimation of simultaneously sparse and low rank matrices, in *Proceedings of the 29th International Coference on Machine Learning (ICML'12)* (2012)
33. S. Scellato, A. Noulas, C. Mascolo, Exploiting place features in link prediction on location-based social networks, in *Proceedings of the 17th ACM SIGKDD International Conference on Knowledge Discovery and Data Mining (KDD '11)* (2011)
34. J. Shi, J. Malik, Normalized cuts and image segmentation. IEEE Trans. Pattern Anal. Mach. Intell. **22**(8), 888–905 (2000)
35. C. Shi, Y. Li, J. Zhang, Y. Sun, P. Yu, A survey of heterogeneous information network analysis. *CoRR*, abs/1511.04854 (2015)
36. E. Simpson, Measurement of diversity, in *Nature* (1949)
37. B. Sriperumbudur, G. Lanckriet, On the convergence of concave-convex procedure, in *Proceedings of the 22nd International Conference on Neural Information Processing Systems (NIPS'09)* (2009)
38. Y. Sun, A. Damle, Matrix analysis notes. http://web.stanford.edu/~damle/refresher/notes/matrixanalysis.pdf

39. Y. Sun, R. Barber, M. Gupta, C.C. Aggarwal, J. Han, Co-author relationship prediction in heterogeneous bibliographic networks, in *2011 International Conference on Advances in Social Networks Analysis and Mining* (2011)

40. Y. Sun, J. Han, X. Yan, P. Yu, T. Wu, Pathsim: meta path-based top-k similarity search in heterogeneous information networks. *Proceedings of 2011 International Conference on Very Large Data Bases (VLDB'11)* (2011)

41. J. Tang, T. Lou, J. Kleinberg, Inferring social ties across heterogeneous networks, in *Proceedings of the Fifth ACM International Conference on Web Search and Data Mining (WSDM '12)* (2012)

42. J. Tang, H. Gao, X. Hu, H. Liu, Exploiting homophily effect for trust prediction, in *Proceedings of the Sixth ACM International Conference on Web Search and Data Mining (WSDM '13)* (2013)

43. K. Torkkola, Feature extraction by non parametric mutual information maximization. J. Mach. Learn. Res. **3** (2003)

44. C. Wang, S. Mahadevan, Heterogeneous domain adaptation using manifold alignment, in *Proceedings of the Twenty-Second International Joint Conference on Artificial Intelligence - Volume Two (IJCAI'11)* (2011)

45. D. Wang, D. Pedreschi, C. Song, F. Giannotti, A. Barabasi, Human mobility, social ties, and link prediction, in *Proceedings of the 17th ACM SIGKDD International Conference on Knowledge Discovery and Data Mining (KDD '11)* (2011)

46. S. Wilks, *Mathematical Statistics* (Wiley, Hoboken, 1963)

47. W. Xi, B. Zhang, Z. Chen, Y. Lu, S. Yan, W. Ma, E. Fox, Link fusion: a unified link analysis framework for multi-type interrelated data objects, in *13th International Conference on World Wide Web* (2004), pp. 319–327

48. M. Ye, P. Yin, W. Lee, Location recommendation for location-based social networks, in *Proceedings of the 18th SIGSPATIAL International Conference on Advances in Geographic Information Systems (GIS '10)* (2010)

49. M. Ye, D. Shou, W. Lee, P. Yin, K. Janowicz, On the semantic annotation of places in location-based social networks, in *Proceedings of the 17th ACM SIGKDD International Conference on Knowledge Discovery and Data Mining (KDD '11)* (2011)

50. J. Ye, H. Cheng, Z. Zhu, M. Chen, Predicting positive and negative links in signed social networks by transfer learning, in *Proceedings of the 22nd International Conference on World Wide Web (WWW '13)* (2013)

51. A. Yuille, A. Rangarajan, The concave-convex procedure. Neural Comput. **15**(4), 915–936 (2003)

52. W. Zangwill, *Nonlinear Programming* (Prentice-Hall, Englewood Cliffs, 1969)

53. J. Zhang, P. Yu, Link prediction across heterogeneous social networks: a survey (2014)

54. J. Zhang, P. Yu, Integrated anchor and social link predictions across partially aligned social networks, in *Twenty-Fourth International Joint Conference on Artificial Intelligence* (2015)

55. J. Zhang, P. Yu, Multiple anonymized social networks alignment, in *2015 IEEE International Conference on Data Mining* (2015)

56. J. Zhang, P. Yu, PCT: partial co-alignment of social networks, in *Proceedings of the 25th International Conference on World Wide Web (WWW '16)* (2016)

57. J. Zhang, X. Kong, P. Yu, Predicting social links for new users across aligned heterogeneous social networks, in *2013 IEEE 13th International Conference on Data Mining* (2013)

58. J. Zhang, X. Kong, P. Yu, Transferring heterogeneous links across location-based social networks, in *Proceedings of the 7th ACM International Conference on Web Search and Data Mining (WSDM '14)* (2014)

59. J. Zhang, P. Yu, Z. Zhou, Meta-path based multi-network collective link prediction, in *Proceedings of the 20th ACM SIGKDD International Conference on Knowledge Discovery and Data Mining (KDD '14)* (2014)

60. J. Zhang, J. Chen, J. Zhu, Y. Chang, P. Yu, Link prediction with cardinality constraints, in *Proceedings of the Tenth ACM International Conference on Web Search and Data Mining (WSDM '17)* (2017)

61. J. Zhang, J. Chen, S. Zhi, Y. Chang, P. Yu, J. Han, Link prediction across aligned networks with sparse low rank matrix estimation, in *2017 IEEE 33rd International Conference on Data Engineering (ICDE)* (2017)

62. E. Zheleva, L. Getoor, J. Golbeck, U. Kuter, Using friendship ties and family circles for link prediction, in *Advances in Social Network Mining and Analysis*, ed. by L. Giles, M. Smith, J. Yen, H. Zhang (2010)

63. K. Zhou, S. Yang, H. Zha, Functional matrix factorizations for cold-start recommendation, in *Proceedings of the 34th International ACM SIGIR Conference on Research and Development in Information (SIGIR '11)* (2011)

Community Detection

8

Birds of a feather flock together.

8.1 Overview

In the real-world online social networks, users also tend to form different social groups [2]. Users belonging to the same groups usually have more frequent interactions with each other, while those in different groups will have less interactions on the other hand [61]. Formally, such social groups form by users in online social networks are called the online social communities [52]. Online social communities will partition the network into a number of components, where the intra-community social connections are usually far more dense compared with the inter-community social connections [52]. Meanwhile, from the mathematical representation perspective, due to these online social communities, the social network adjacency matrix tend to be not only sparse but also low-rank [58].

Identifying the social communities formed by users in online social networks is formally defined as the *community detection* problem [16, 52, 53]. Community detection is a very important problem for online social network studies, as it can be a crucial prerequisite for numerous concrete social network services: (1) a better organization of users' friends in online social networks (e.g., Facebook and Twitter), which can be achieved by applying community detection techniques to partition users' friends into different categories, e.g., schoolmates, family, celebrities, etc. [10]; (2) a better recommender systems for users with common shopping preference in e-commerce social sites (e.g., Amazon and Epinions), which can be addressed by grouping users with similar purchase records into the same clusters prior to recommender system building [36]; and (3) a better identification of influential users [44] for advertising campaigns in online social networks, which can be attained by selecting the most influential users in each community as the seed users in the viral marketing [35].

In this chapter, we will focus on introducing the *social community detection* problem in online social networks. Given a heterogeneous network G with node set $\mathcal{V}$, we can represent the involved user nodes in network G as set $\mathcal{U} \subset \mathcal{V}$. Based on both the social structures among users and the diverse attribute information from the network G, the *social community detection* problem aims at partitioning the user set $\mathcal{U}$ into several subsets $\mathcal{C} = \{\mathcal{U}_1, \mathcal{U}_2, \ldots, \mathcal{U}_k\}$, where each subset $\mathcal{U}_i, i \in \{1, 2, \ldots, k\}$ is called a social community. Term k formally denotes the total number of partitioned communities, which is usually provided as a hyper-parameter in the problem.

© Springer Nature Switzerland AG 2019
J. Zhang, P. S. Yu, *Broad Learning Through Fusions*,
https://doi.org/10.1007/978-3-030-12528-8_8

Depending on whether the users are allowed to be partitioned into multiple communities simultaneously or not, the *social community detection* problem can actually be categorized into two different types:

- *Hard Social Community Detection*: In the *hard social community detection* problem, each user will be partitioned into one single community, and all the social communities are disjoint without any overlap. In other words, given the communities $C = \{U_1, U_2, \ldots, U_k\}$ detected from network G, we have $U = \bigcup_i U_i$ and $U_i \cap U_j = \emptyset, \forall i, j \in \{1, 2, \ldots, k\} \wedge i \neq j$.
- *Soft Social Community Detection*: In the *soft social community detection* problem, users can belong to multiple social communities simultaneously. For instance, if we apply the *Mixture-of-Gaussian Soft Clustering* algorithm as the base community detection model (introduced in Sect. 2.5.4), each user can belong to multiple communities with certain probabilities. In the *soft social community detection* result, the communities are no longer disjoint and will share some common users with other communities.

Meanwhile, depending on the network connection structures, the *community detection* problem can be categorized as *directed network community detection* [28] and *undirected network community detection* [61]. Based on the heterogeneity of the network information, the *community detection* problem can be divided into the *homogeneous network community detection* [46] and *heterogeneous network community detection* [37, 40, 59, 60]. Furthermore, according to the number of networks involved, the *community detection* problem involves *single network community detection* [22] and *multiple network community detection* [16, 52, 53, 59, 60]. In this chapter, we will take the *hard community detection problem* as an example to introduce both the existing models proposed for conventional (one single) *homogeneous social network*, and especially the recent broad learning based (multiple aligned) *heterogeneous social networks* [20, 54–56], respectively.

This chapter is organized as follows. At the beginning, in Sect. 8.2, we will introduce the community detection problem and the existing methods proposed for traditional one single homogeneous networks. After that, we will talk about the latest research works on social community detection across multiple aligned heterogeneous networks. The cold start community detection [53] is introduced in Sect. 8.3, in which we propose a new information transfer algorithm to propagate information from other developed source networks to the emerging target network. In Sect. 8.4, we will be focused on the concurrent mutual community detection [52] across multiple aligned heterogeneous networks simultaneously, where information from other aligned networks will be applied to refine their community detection results mutually. Finally, in Sect. 8.5, we talk about the synergistic community detection across multiple large-scale networks based on the distributed computing platform [16].

8.2   Traditional Homogeneous Network Community Detection

Social community detection problem has been studied for a long time, and many community detection models have been proposed based on different types of techniques. In this section, we will talk about the social community detection problem for one single homogeneous network G, whose objective is to partition the user set U in network G into k disjoint subsets $C = \{U_1, U_2, \ldots, U_k\}$, where $U = \bigcup_i U_i$ and $U_i \cap U_j = \emptyset, \forall i, j \in \{1, 2, \ldots, k\}$. Several different community detection methods will be introduced, which include the *node proximity based community detection*, *modularity maximization based community detection*, and *spectral clustering based community detection*.

8.2.1 Node Proximity Based Community Detection

The *node proximity based community detection* method assumes that "close nodes tend to be in the same communities, while the nodes far away from each other will belong to different communities." Therefore, the *node proximity based community detection* model partitions the nodes into different clusters based on the node proximity measures [24]. Various node proximity measures can be used here, including the node *structural equivalence* to be introduced as follows, as well as various node closeness measures as introduced in Sect. 3.3.3.

In a homogeneous network G, the proximity of nodes, like u and v, can be calculated based on their positions and connections in the network structure.

Example 8.1 For instance, in Fig. 8.1, we show an example of a homogeneous network G involving 9 nodes and 14 links among them. For the nodes 1 and 3, they have equivalent positions in the network structure. According to the connections around 1 and 3, we can observe the neighbors of node 1 are $\Gamma(1) = \{2, 3, 4\}$, while the neighbors of node 3 include $\Gamma(3) = \{1, 2, 4\}$. They share two common neighbors $\{1, 2\}$, and also connect with each other. If we switch their positions, the network structure will still be the same as the original one and the neighbors of nodes 1 and 3 will both remain the same.

Definition 8.1 (Structural Equivalence) Given a network $G = (\mathcal{V}, \mathcal{E})$, two nodes $u, v \in \mathcal{V}$ are said to be *structural equivalent* iff

1. Nodes u and v are not connected and u and v share the same set of neighbors (i.e., $(u, v) \notin \mathcal{E} \wedge \Gamma(u) = \Gamma(v)$),

2. Or u and v are connected and excluding themselves, u and v share the same set of neighbors (i.e., $(u, v) \in \mathcal{E} \wedge \Gamma(u) \setminus \{v\} = \Gamma(v) \setminus \{u\}$).

As mentioned before, for the nodes which are *structural equivalent*, they are *substitutable* and switching their positions will not change the overall network structure. The *structural equivalence* concept can be applied to partition the nodes into different communities. For the nodes which are *structural equivalent*, they can be grouped into the same communities, while for the nodes which are not equivalent in their positions, they will be partitioned into different groups. However, the *structural equivalence* can be too restricted for practical use in detecting the communities in real-world social networks. Computing the *structural equivalence* relationships among all the node pairs in the network can lead to very high time cost. What's more, the *structural equivalence* relationship will partition the social network structure into lots of small-sized fragments, since the users will have different social patterns in making friends online and few users will have identical neighbors actually.

To avoid the weakness mentioned above, some other measures are proposed to measure the proximity among nodes in the networks. For instance, as introduced in Sect. 3.3.3, the node closeness

Fig. 8.1 Example of homogeneous network (nodes 1 and 3 are structural equivalent)

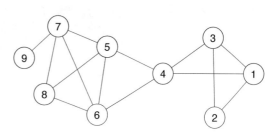

measures based on the social connections can all be applied here to compute the node proximity, e.g., "common neighbor," "Jaccard's coefficient." Here, if we use "common neighbor" as the proximity measure, by applying the "common neighbor" measure to the network G, we can transform the network G into a set of instances $\mathcal{V}$ with mutual closeness scores $\{c(u, v)\}_{u,v \in \mathcal{V}}$. Some existing similarity/distance based clustering algorithms, like k-Medoids (a variant of k-Means as introduced in Sect. 2.5.2), can be applied to partition the users into different communities.

8.2.2   Modularity Maximization Based Community Detection

Besides the pairwise proximity of nodes in the network, the connection strength of a community is also very important in the community detection process. Different measures have been proposed to compute the strength of a community like the *modularity* measure [29] to be introduced in this part.

The *modularity* measure takes into account of the node degree distribution. For instance, given the network G, the expected number of links existing between nodes u and v with degrees $D(u)$ and $D(v)$ can be represented as $\frac{D(u) \cdot D(v)}{2|\mathcal{E}|}$. Meanwhile, in the network, the real number of links existing between u and v can be denoted as entry $A[u, v]$ in the social adjacency matrix $\mathbf{A}$. For the user pair (u, v) with a low expected connection confidence score, if they are connected in the real world, it indicates that u and v have a relatively strong relationship with each other. Meanwhile, if the community detection algorithm can partition such user pairs into the same group, it will be able to identify very strong social communities from the network.

Based on such an intuition, the strength of a community, e.g., $\mathcal{U}_i \in \mathcal{C}$, can be defined as

$$\sum_{u,v \in \mathcal{U}_i} \left(A[u, v] - \frac{D(u) \cdot D(v)}{2|\mathcal{E}|} \right). \tag{8.1}$$

Example 8.2 For instance, let's take network shown in Fig. 8.1 as an example. We assume the network nodes are partitioned into two groups, i.e., $\mathcal{C} = \{\mathcal{U}_1, \mathcal{U}_2\}$, where $\mathcal{U}_1 = \{1, 2, 3, 4\}$ and $\mathcal{U}_2 = \{5, 6, 7, 8, 9\}$. According to the network structure, we can compute the expected number of links between user pairs within community $\mathcal{U}_1$ in Table 8.1.

According to the above equation, we can compute the strength of community $\mathcal{U}_1 = \{1, 2, 3, 4\}$ as

$$\sum_{u,v \in \mathcal{U}_1} \left(A[u, v] - \frac{D(u) \cdot D(v)}{2|\mathcal{E}|} \right) = 4.857. \tag{8.2}$$

Furthermore, the strength of the overall community detection result $\mathcal{C} = \{\mathcal{U}_1, \mathcal{U}_2, \ldots, \mathcal{U}_k\}$ can be defined as the *modularity* of the communities as follows.

Table 8.1   Numerical analysis of community $\mathcal{U}_1 = \{1, 2, 3, 4\}$

(u,v)	(1, 1)	(1, 2)	(1,3)	(1,4)	(2, 1)	(2, 2)	(2, 3)	(2, 4)	(3, 1)	(3, 2)	(3, 3)	(3, 4)	(4, 4)	...
D(u), D(v)	3, 3	3, 2	3, 3	3, 4	2, 3	2, 2	2, 3	2, 4	3, 3	3, 2	3, 3	3, 4	4, 3	...
A[u,v]	0	1	1	1	1	0	1	0	1	1	0	1	1	...
$\frac{D(u) \cdot D(v)}{2|\mathcal{E}|}$	$\frac{9}{28}$	$\frac{6}{28}$	$\frac{9}{28}$	$\frac{12}{28}$	$\frac{9}{28}$	$\frac{4}{28}$	$\frac{6}{28}$	$\frac{8}{28}$	$\frac{9}{28}$	$\frac{6}{28}$	$\frac{9}{28}$	$\frac{12}{28}$	$\frac{12}{28}$	...
$A[u, v] - \frac{D(u) \cdot D(v)}{2|\mathcal{E}|}$	$\frac{-9}{28}$	$\frac{22}{28}$	$\frac{19}{28}$	$\frac{16}{28}$	$\frac{19}{28}$	$\frac{-4}{28}$	$\frac{22}{28}$	$\frac{-8}{28}$	$\frac{19}{28}$	$\frac{22}{28}$	$\frac{-9}{28}$	$\frac{16}{28}$	$\frac{16}{28}$	...

Definition 8.2 (Modularity) Given the community detection result $\mathcal{C} = \{\mathcal{U}_1, \mathcal{U}_2, \ldots, \mathcal{U}_k\}$, the modularity of the community structure is defined as

$$Q(\mathcal{C}) = \frac{1}{2|\mathcal{E}|} \sum_{\mathcal{U}_i \in \mathcal{C}} \sum_{u,v \in \mathcal{U}_i} \left(A[u,v] - \frac{D(u) \cdot D(v)}{2|\mathcal{E}|} \right). \tag{8.3}$$

The *modularity* concept effectively measures the strength of the detected community structure. Generally, for a community structure with a larger *modularity* score, it indicates a good community detection result.

Example 8.3 By following the analysis provided in Example 8.2, we can also compute the strength of community $\mathcal{U}_2$ as

$$\sum_{u,v \in \mathcal{U}_2} \left(A[u,v] - \frac{D(u) \cdot D(v)}{2|\mathcal{E}|} \right) = 4.857. \tag{8.4}$$

Therefore, we can compute the *modularity* of community detection results $\mathcal{C} = \{\mathcal{U}_1, \mathcal{U}_2\}$ as

$$Q(\mathcal{C}) = \frac{1}{2|\mathcal{E}|} \sum_{\mathcal{U}_i \in \mathcal{C}} \sum_{u,v \in \mathcal{U}_i} \left(A[u,v] - \frac{D(u) \cdot D(v)}{2|\mathcal{E}|} \right) = 0.347. \tag{8.5}$$

Another way to explain the *modularity* is from the number of links within and across communities. By rewriting the above *modularity* equation, we can have

$$Q(\mathcal{C}) = \frac{1}{2|\mathcal{E}|} \sum_{\mathcal{U}_i \in \mathcal{C}} \sum_{u,v \in \mathcal{U}_i} \left(A[u,v] - \frac{D(u) \cdot D(v)}{2|\mathcal{E}|} \right)$$

$$= \frac{1}{2|\mathcal{E}|} \left(\sum_{\mathcal{U}_i \in \mathcal{C}} \sum_{u,v \in \mathcal{U}_i} A[u,v] - \sum_{\mathcal{U}_i \in \mathcal{C}} \sum_{u,v \in \mathcal{U}_i} \frac{D(u) \cdot D(v)}{2|\mathcal{E}|} \right)$$

$$= \frac{1}{2|\mathcal{E}|} \left(\sum_{\mathcal{U}_i \in \mathcal{C}} \sum_{u,v \in \mathcal{U}_i} A[u,v] - \frac{1}{2|\mathcal{E}|} \sum_{\mathcal{U}_i \in \mathcal{C}} \sum_{u \in \mathcal{U}_i} D(u) \cdot \sum_{u \in \mathcal{U}_i} D(v) \right)$$

$$= \frac{1}{2|\mathcal{E}|} \left(\sum_{\mathcal{U}_i \in \mathcal{C}} \sum_{u,v \in \mathcal{U}_i} A[u,v] - \frac{1}{2|\mathcal{E}|} \sum_{\mathcal{U}_i \in \mathcal{C}} \left(\sum_{u \in \mathcal{U}_i} D(u) \right)^2 \right). \tag{8.6}$$

In the above equation, term $\sum_{u,v \in \mathcal{U}_i} A[u,v]$ denotes the number of links connecting users within the community $\mathcal{U}_i$ (which will be 2 times the intra-community links for undirected networks, as each link will be counted twice). Term $\sum_{u \in \mathcal{U}_i} D(u)$ denotes the sum of node degrees in community $\mathcal{U}_i$, which equals to the number of intra-community and inter-community links connected to nodes in community $\mathcal{U}_i$. If there exist lots of inter-community links, then the *modularity* measure will have a smaller value. On the other hand, if the inter-community links are very rare, the *modularity* measure will have a larger value. Therefore, maximizing the community *modularity* measure is equivalent to minimizing the inter-community link numbers.

The *modularity* measure can also be represented with linear algebra equations. Let matrix $\mathbf{A}$ denote the adjacency matrix of the network, and vector $\mathbf{d} \in \mathbb{R}^{|\mathcal{V}| \times 1}$ denote the degrees of nodes in the network. We can define the *modularity matrix* as

$$\mathbf{B} = \mathbf{A} - \frac{\mathbf{d}\mathbf{d}^\top}{2|\mathcal{E}|}. \tag{8.7}$$

Let matrix $\mathbf{H} \in \{0, 1\}^{|\mathcal{V}| \times k}$ denote the communities that users in $\mathcal{V}$ belong to. In real application, such a binary constraint can be relaxed to allow real value solutions for matrix $\mathbf{H}$. The optimal community detection result which can maximize the *modularity* can be obtained by solving the following objective function

$$\max \frac{1}{2|\mathcal{E}|}\mathrm{Tr}(\mathbf{H}^\top \mathbf{B}\mathbf{H})$$

$$s.t. \mathbf{H}^\top \mathbf{H} = \mathbf{I}, \tag{8.8}$$

where constraint $\mathbf{H}^\top \mathbf{H} = \mathbf{I}$ ensures there is no overlap in the community detection result.

The above objective function looks very similar to the objective function of *spectral clustering* to be introduced in the next section. After obtaining the optimal $\mathbf{H}$, the communities can be obtained by applying the K-Means algorithm to $\mathbf{H}$ to determine the cluster labels of each node in the network.

8.2.3   Spectral Clustering Based Community Detection

In the community detection process, besides maximizing the proximity of nodes belonging to the same communities (as introduced in Sect. 8.2.1), minimizing the connections among nodes in different clusters is also an important factor. Different from the previous proximity based community detection algorithms, another way to address the community detection problem is from the cost perspective. Partitioning the nodes into different clusters will cut the links among the clusters. To ensure the nodes partitioned into different clusters have less connections with each other, the number of links to be cut in the community detection process should be as small as possible [45].

8.2.3.1 Cut
Formally, given the community structure $\mathcal{C} = \{\mathcal{U}_1, \mathcal{U}_2, \ldots, \mathcal{U}_k\}$ detected from network G. The number of links cut [38] between communities $\mathcal{U}_i, \mathcal{U}_j \in \mathcal{C}$ can be represented as

$$cut(\mathcal{U}_i, \mathcal{U}_j) = \sum_{u \in \mathcal{U}_i} \sum_{v \in \mathcal{U}_j} \mathbb{I}(u, v), \tag{8.9}$$

where function $\mathbb{I}(u, v) = 1$ if $(u, v) \in \mathcal{E}$; otherwise, it will be 0.

The total number of links cut in the partition process can be represented as

$$cut(\mathcal{C}) = \sum_{\mathcal{U}_i \in \mathcal{C}} cut(\mathcal{U}_i, \bar{\mathcal{U}}_i), \tag{8.10}$$

where set $\bar{\mathcal{U}}_i = \mathcal{C} \setminus \mathcal{U}_i$ denotes the remaining communities except $\mathcal{U}_i$.

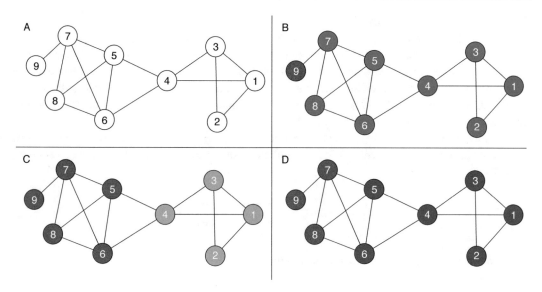

Fig. 8.2 Comparison of *cut*, *ratio-cut*, and *normalized-cut* measures in social network community detection ((**a**) input network; (**b**)–(**d**) three different ways to partition the network into two communities)

By minimizing the cut cost introduced in the partition process, we can obtain the optimal community detection result with the minimum number of cross-community links.

Example 8.4 For instance, in Fig. 8.2, we show 3 different community detection results (i.e., plots (b)–(d)) of the input network as illustrated in plot (a). For the 9 nodes in the network, plot (b) partitions the node into 2 communities: $\mathcal{C} = \{\mathcal{U}_1, \mathcal{U}_2\}$, where $\mathcal{U}_1 = \{9\}$ and $\mathcal{U}_2 = \{1, 2, 3, 4, 5, 6, 7, 8\}$. Link $(7, 9)$ is the only link between different communities in the network. According to the above definition, the introduced cut can be represented as

$$cut(\mathcal{C}) = cut(\mathcal{U}_1, \bar{\mathcal{U}}_1) + cut(\mathcal{U}_2, \bar{\mathcal{U}}_2) = 2, \qquad (8.11)$$

where $\bar{\mathcal{U}}_1 = \mathcal{U}_2, \bar{\mathcal{U}}_2 = \mathcal{U}_1$ and $cut(\mathcal{U}_1, \mathcal{U}_2) = |\{(7, 9)\}| = 1$.

Meanwhile, for the community detection results in plot (c), its introduced cut will be 2×2, since two edges $\{(4, 5), (4, 6)\}$ are between the two communities, i.e., $\mathcal{U}_1 = \{5, 6, 7, 8, 9\}$ and $\mathcal{U}_2 = \{1, 2, 3, 4\}$. Community detection results in plot (d) introduce a cut of 8, and 4 edges $\{(5, 7), (5, 8), (6, 7), (6, 8)\}$ connect those two detected communities.

Considering that we don't allow empty communities, plot (b) actually identifies the optimal community structure of the input network data, where the cut cost is minimized. However, we can also observe that the achieved community structure is also extremely imbalanced, where community $\mathcal{U}_1 = \{9\}$ contains a singleton node, while $\mathcal{U}_2 = \{1, 2, 3, 4, 5, 6, 7, 8\}$ contains 8 nodes. Such a problem will be much more severe when it comes to the real-world social network data. In the following part of this section, we will introduce two other cost measures that can help achieve more balanced community detection results.

8.2.3.2 *Ratio-Cut* and *Normalized-Cut*

As shown in the example, the minimum cut cost treats all the links in the network equally, and can usually achieve very imbalanced partition results (e.g., a singleton node as a cluster) when applied in the real-world community detection problem. To overcome such a disadvantage, some models have been proposed to take the community size into consideration. The community size can be calculated by counting the number of nodes or links in each community, which will lead to two new cost measures: *ratio-cut* [45] and *normalized-cut* [45].

Formally, given the community detection result $C = \{\mathcal{U}_1, \mathcal{U}_2, \ldots, \mathcal{U}_k\}$ in network G, the *ratio-cut* and *normalized-cut* costs introduced in the community detection result can be defined as follows, respectively.

$$ratio - cut(C) = \frac{1}{k} \sum_{\mathcal{U}_i \in C} \frac{cut(\mathcal{U}_i, \bar{\mathcal{U}}_i)}{|\mathcal{U}_i|}, \tag{8.12}$$

where $|\mathcal{U}_i|$ denotes the number of nodes in community $\mathcal{U}_i$.

$$ncut(C) = \frac{1}{k} \sum_{\mathcal{U}_i \in C} \frac{cut(\mathcal{U}_i, \bar{\mathcal{U}}_i)}{vol(\mathcal{U}_i)}, \tag{8.13}$$

where $vol(\mathcal{U}_i)$ denotes the degree sum of nodes in community $\mathcal{U}_i$.

Example 8.5 For instance, by following the example as illustrated in Example 8.4 and Fig. 8.2, we have already computed the cut cost introduced by the community detection results in plots (b), (c), (d), which are 2, 4 and 8, respectively. Here, if we also consider the node number of node degree volume of each community, we can get the *ratio-cut* and *ncut* of these community detection results as follows:

- Plot (b): $\mathcal{U}_1 = \{9\}$ and $\mathcal{U}_2 = \{1, 2, 3, 4, 5, 6, 7, 8\}$. We have $cut(\mathcal{U}_1, \bar{\mathcal{U}}_1) = cut(\mathcal{U}_2, \bar{\mathcal{U}}_2) = 1$. The sizes and volumes of these communities are $|\mathcal{U}_1| = 1$, $|\mathcal{U}_2| = 8$, and $vol(\mathcal{U}_1) = 1$, $vol(\mathcal{U}_2) = 27$. Therefore, we have

$$ratio\text{-}cut(C) = \frac{1}{2} \left(\frac{cut(\mathcal{U}_1, \bar{\mathcal{U}}_1)}{|\mathcal{U}_1|} + \frac{cut(\mathcal{U}_2, \bar{\mathcal{U}}_2)}{|\mathcal{U}_2|} \right) = \frac{1}{2} \left(\frac{1}{1} + \frac{1}{8} \right) = \frac{9}{16}, \tag{8.14}$$

$$ncut(C) = \frac{1}{2} \left(\frac{cut(\mathcal{U}_1, \bar{\mathcal{U}}_1)}{vol(\mathcal{U}_1)} + \frac{cut(\mathcal{U}_2, \bar{\mathcal{U}}_2)}{vol(\mathcal{U}_1)} \right) = \frac{1}{2} \left(\frac{1}{1} + \frac{1}{27} \right) = \frac{14}{27}. \tag{8.15}$$

- Plot (c): $\mathcal{U}_1 = \{5, 6, 7, 8, 9\}$ and $\mathcal{U}_2 = \{1, 2, 3, 4\}$. We have $cut(\mathcal{U}_1, \bar{\mathcal{U}}_1) = cut(\mathcal{U}_2, \bar{\mathcal{U}}_2) = 2$. The sizes and volumes of these communities are $|\mathcal{U}_1| = 5$, $|\mathcal{U}_2| = 4$, and $vol(\mathcal{U}_1) = 16$, $vol(\mathcal{U}_2) = 12$. Therefore, we have

$$ratio\text{-}cut(C) = \frac{1}{2} \left(\frac{cut(\mathcal{U}_1, \bar{\mathcal{U}}_1)}{|\mathcal{U}_1|} + \frac{cut(\mathcal{U}_2, \bar{\mathcal{U}}_2)}{|\mathcal{U}_2|} \right) = \frac{1}{2} \left(\frac{1}{5} + \frac{1}{4} \right) = \frac{9}{40}, \tag{8.16}$$

$$ncut(C) = \frac{1}{2} \left(\frac{cut(\mathcal{U}_1, \bar{\mathcal{U}}_1)}{vol(\mathcal{U}_1)} + \frac{cut(\mathcal{U}_2, \bar{\mathcal{U}}_2)}{vol(\mathcal{U}_1)} \right) = \frac{1}{2} \left(\frac{1}{16} + \frac{1}{12} \right) = \frac{7}{96}. \tag{8.17}$$

- Plot (d): $\mathcal{U}_1 = \{7, 8, 9\}$ and $\mathcal{U}_2 = \{1, 2, 3, 4, 5, 6\}$. We have $cut(\mathcal{U}_1, \bar{\mathcal{U}}_1) = cut(\mathcal{U}_2, \bar{\mathcal{U}}_2) = 4$. The sizes and volumes of these communities are $|\mathcal{U}_1| = 3$, $|\mathcal{U}_2| = 6$, and $vol(\mathcal{U}_1) = 20$, $vol(\mathcal{U}_2) = 8$. Therefore, we have

$$ratio\text{-}cut(\mathcal{C}) = \frac{1}{2}\left(\frac{cut(\mathcal{U}_1, \bar{\mathcal{U}}_1)}{|\mathcal{U}_1|} + \frac{cut(\mathcal{U}_2, \bar{\mathcal{U}}_2)}{|\mathcal{U}_2|}\right) = \frac{1}{2}\left(\frac{1}{3} + \frac{1}{6}\right) = \frac{1}{4}, \tag{8.18}$$

$$ncut(\mathcal{C}) = \frac{1}{2}\left(\frac{cut(\mathcal{U}_1, \bar{\mathcal{U}}_1)}{vol(\mathcal{U}_1)} + \frac{cut(\mathcal{U}_2, \bar{\mathcal{U}}_2)}{vol(\mathcal{U}_1)}\right) = \frac{1}{2}\left(\frac{1}{20} + \frac{1}{8}\right) = \frac{7}{80}. \tag{8.19}$$

As shown in the above example, from the computed costs, we find that the community detected in plot (c) achieves much lower *ratio-cut* and *ncut* costs compared with those in plots (b) and (d). Compared against the regular *cut* cost, both *ratio-cut* and *normalized-cut* prefer a balanced partition of the social network.

8.2.3.3 Spectral Clustering

Actually the objective function of both *ratio-cut* and *normalized-cut* can be unified as the following linear algebra equation

$$\min_{\mathbf{H}\in\{0,1\}^{|\mathcal{V}|\times k}} \mathrm{Tr}(\mathbf{H}^\top \bar{\mathbf{L}} \mathbf{H}), \tag{8.20}$$

where matrix $\mathbf{H} \in \{0, 1\}^{|\mathcal{V}|\times k}$ denotes the communities that users in $\mathcal{V}$ belong to.

Let $\mathbf{A} \in \{0, 1\}^{|\mathcal{V}|\times|\mathcal{V}|}$ denote the social adjacency matrix of the network, and we can represent the corresponding diagonal matrix of $\mathbf{A}$ as matrix $\mathbf{D}$, where $\mathbf{D}$ has value $D(i, i) = \sum_j A(i, j)$ on its diagonal. The Laplacian matrix of the network adjacency matrix $\mathbf{A}$ can be represented as $\mathbf{L} = \mathbf{D} - \mathbf{A}$. Depending on the specific measures applied, matrix $\bar{\mathbf{L}}$ can be represented as

$$\bar{\mathbf{L}} = \begin{cases} \mathbf{L}, & \text{for } ratio\text{-}cut \text{ measure,} \\ \mathbf{D}^{\frac{-1}{2}}\mathbf{L}\mathbf{D}^{\frac{-1}{2}}, & \text{for } normalized\text{-}cut \text{ measure.} \end{cases} \tag{8.21}$$

The binary constraint on the variable $\mathbf{H}$ renders the problem a non-linear integer programming problem, which is very hard to solve. One common practice to learn the variable $\mathbf{H}$ is to apply spectral relaxation to replace the binary constraint with the orthogonality constraint.

$$\min \mathrm{Tr}(\mathbf{H}^\top \bar{\mathbf{L}} \mathbf{H}),$$

$$s.t. \mathbf{H}^\top \mathbf{H} = \mathbf{I}. \tag{8.22}$$

As proposed in [38], the optimal solution $\mathbf{H}^*$ to the above objective function equals to the eigenvectors corresponding to the k smallest eigenvalues of matrix $\bar{\mathbf{L}}$.

8.3   Emerging Network Community Detection

The community detection algorithms introduced in the previous section are mostly proposed for one single homogeneous network. However, in the real world, most of the online social networks are actually heterogeneous containing very complex information. In recent years, lots of new online social networks have emerged and start to provide services, the information available for the users in these emerging networks is usually very limited. Meanwhile, many of the users are also involved in multiple online social networks simultaneously. For users who are using these emerging networks, they may also be involved in other developed social networks for a long time [50,51]. The abundant information available in these mature networks can actually be useful for the community detection in the emerging networks. In this section, we will introduce the cross-network community detection for emerging networks with information transferred from other mature social networks [53].

8.3.1 Background Knowledge

Witnessing the incredible success of popular online social networks, e.g., Facebook and Twitter, a large number of new social networks offering specific services also spring up overnight to compete for the market share. Generally, emerging networks are the networks containing very sparse information and can be (1) the social networks which are newly constructed and start to provide social services for a very short period of time; or (2) even more mature ones that start to branch into new geographic areas or social groups [55]. These emerging networks can be of a wide variety, which include (1) location-based social networks, e.g., Foursquare and Jiepang; (2) photo organizing and sharing sites, e.g., Pinterest and Instagram; and (3) educational social sites, e.g., Stage 32.

Community detection in emerging networks is a new problem and conventional community detection methods for well-developed networks cannot be applied directly. Compared with well-developed networks, information in emerging networks can be too sparse to support traditional community detection methods to calculate effective closeness scores and achieve good results. According to the market report from DRM,[1] by the end of 2013, the total number of registered users in Foursquare has reached 45 million but these Foursquare users have only post 40 million tips. In other words, each user has posted less than one tip in Foursquare on average. Meanwhile, the 1 billion registered Twitter users have published more than 300 billion tweets by the end of 2013 and each Twitter user has written more than 300 tweets. We also provide a statistics investigation on a crawled dataset, which include both Foursquare and Twitter, and the information distribution results are given in Fig. 8.3. As shown in Fig. 8.3a–c, users in Twitter have far more social connections, posts, and location check-ins than users in Foursquare. The shortage of information encountered in community detection problems for emerging networks can be a serious obstacle for traditional community detection methods to achieve good performance and is urgent to be solved.

In this section, we will introduce the social community detection for *emerging networks* with information propagated across multiple *partially aligned social networks*, which is formally defined as the "*emerging network community detection*" problem. Especially, when the network is brand new, the problem will be the "*cold start community detection*" problem. *Cold start problem* is mostly prevalent in *recommender systems* [54], where the system cannot draw any inferences for users or items, for which it has not yet gathered sufficient information, but few works have been done on studying the *cold start problem* in clustering/community detection problems. The "*emerging network community detection*" problem and "*cold start community detection*" problem studied in this section are both novel problems and very different from other existing works on community detection.

---

[1] http://expandedramblings.com.

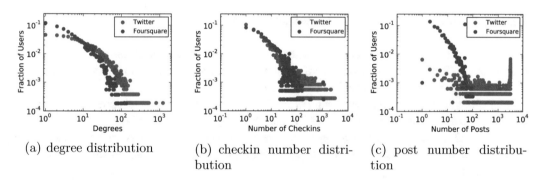

(a) degree distribution   (b) checkin number distri-   (c) post number distribu-
                            bution   tion

Fig. 8.3 Information and anchor user distributions in Foursquare and Twitter. (**a**) social degree distribution, (**b**) number of check-ins distribution, (**c**) number of posts distribution

8.3.2 Problem Formulation

Networks studied in this section can be formulated as two partially aligned attribute augmented heterogeneous networks: $\mathcal{G} = ((G^t, G^s), A^{t,s})$, where G^t and G^s are the emerging target network and well-developed source network, respectively and $A^{t,s}$ is the set of anchor links between G^t and G^s. Both G^t and G^s can be formulated as the attribute augmented heterogeneous social network, e.g., $G^t = (\mathcal{V}^t, \mathcal{E}^t, \mathcal{A}^t)$ (where sets $\mathcal{V}^t$, $\mathcal{E}^t$, and $\mathcal{A}^t$ denote the user nodes, social links, and diverse attributes in the network). With information propagated across $\mathcal{G}$, we can calculate the *intimacy matrix*, **H**, among users in $\mathcal{V}^t$. The *emerging network community detection* problem aims at partitioning user set $\mathcal{V}^t$ of the emerging network G^t into K disjoint clusters, $\mathcal{C} = \{C_1, C_2, \ldots, C_K\}$, based on the *intimacy matrix*, **H**, where $\bigcup_i^K C_i = \mathcal{V}^t$ and $C_i \cap C_j = \emptyset, \forall i, j \in \{1, 2, \ldots, K\}, i \neq j$. When the target network G^t is brand new, i.e., $\mathcal{E}^t = \emptyset$ and $\mathcal{A}^t = \emptyset$, the problem will be the *cold start community detection* problem. The "*emerging network community detection*" studied in this section is also very challenging to solve due to the following reasons:

- *network heterogeneity problem*: Proper definition of closeness measure among users with link and attribute information in the heterogeneous social networks is very important for community detection problems.
- *shortage of information*: Community detection for emerging networks can suffer from the shortage of information problem, i.e., the "*cold start problem*" [54, 55].
- *network difference problem*: Different networks can have different properties. Some information propagated from other well-developed networks can be useful for solving the *emerging network community detection* problem but some can be misleading on the other hand.
- *high memory space cost*: Community detection across multiple aligned networks can involve too many nodes and connections, which will lead to high space cost.

To solve all the above challenges, a novel community detection method, CAD [53], will be introduced in great detail in this section: (1) CAD introduces a new concept, *intimacy*, to measure the closeness relationships among users with both link and attribute information in online social networks; (2) CAD can propagate useful information from aligned well-developed networks to the emerging network to solve the shortage of information problem; (3) CAD addresses the network heterogeneity and difference problems with both *micro-level* and *macro-level* control of the link and attribute information proportions, whose parameters can be adjusted by CAD automatically; (4) effective and efficient cross-network information propagation models are introduced in this section to solve the high space cost problem.

8.3.3 Intimacy Matrix of Homogeneous Network

The CAD model is built based on the closeness scores among users, which is formally called the *intimacy scores* in this section. Here, we will introduce the *intimacy scores* and *intimacy matrix* used in CAD from an information propagation perspective.

For a given homogeneous network, e.g., $G = (\mathcal{V}, \mathcal{E})$, where $\mathcal{V}$ is the set of users and $\mathcal{E}$ is the set of social links among users in $\mathcal{V}$, we can define the adjacency matrix of G to be $\mathbf{A} \in \mathbb{R}^{|\mathcal{V}| \times |\mathcal{V}|}$, where $A(i, j) = 1$, iff $(u_i, u_j) \in \mathcal{E}$. Meanwhile, via the social links in $\mathcal{E}$, information can propagate among the users within the network, whose propagation paths can reflect the closeness among users [33]. Formally, term

$$p_{ji} = \frac{A(j, i)}{\sqrt{\sum_m A(j, m) \sum_n A(n, i)}} \qquad (8.23)$$

is called the information *transition probability* from u_j to u_i, which equals to the proportion of information propagated from u_j to u_i in one step.

We can use an example to illustrate how information propagates within the network more clearly. Let's assume that user $u_i \in \mathcal{V}$ injects a stimulation into network G initially and the information will be propagated to other users in G via the social interactions afterwards. During the propagation process, users receive stimulation from their neighbors and the amount is proportional to the difference of the amount of information reaching the user and his neighbors. Let vector $\boldsymbol{f}^{(\tau)} \in \mathbb{R}^{|\mathcal{V}|}$ denote the states of all users in $\mathcal{V}$ at time τ, i.e., the proportion of stimulation at users in $\mathcal{V}$ at τ. The change of stimulation at u_i at time $\tau + \Delta t$ is defined as follows:

$$\frac{f^{(\tau + \Delta t)}(i) - f^{(\tau)}(i)}{\Delta t} = \alpha \sum_{u_j \in \mathcal{V}} p_{ji}(f^{(\tau)}(j) - f^{(\tau)}(i)), \qquad (8.24)$$

where coefficient α can be set as 1 as proposed in [62]. The *transition probabilities* $p_{ij}, i, j \in \{1, 2, \ldots, |\mathcal{V}|\}$ can be represented with the *transition matrix*

$$\mathbf{X} = (\mathbf{D}^{-\frac{1}{2}} \mathbf{A} \mathbf{D}^{-\frac{1}{2}}) \qquad (8.25)$$

of network G, where $\mathbf{X} \in \mathbb{R}^{|\mathcal{V}| \times |\mathcal{V}|}$, $X(i, j) = p_{ij}$ and diagonal matrix $\mathbf{D} \in \mathbb{R}^{|\mathcal{V}| \times |\mathcal{V}|}$ has value $D(i, i) = \sum_{j=1}^{|\mathcal{V}|} A(i, j)$ on its diagonal.

Definition 8.3 (Social Transition Probability Matrix) The *social transition probability matrix* of network G can be represented as $\mathbf{Q} = \mathbf{X} - \mathbf{D_X}$, where $\mathbf{X}$ is the *transition matrix* defined above and diagonal matrix $D_{\mathbf{X}}$ has value $D_{\mathbf{X}}(i, i) = \sum_{j=1}^{|\mathcal{V}|} X(i, j)$ on its diagonal.

Furthermore, by setting $\Delta t = 1$, denoting that stimulation propagates step by step in a discrete time through network, we can rewrite the propagation updating equation as:

$$\boldsymbol{f}^{(\tau)} = \boldsymbol{f}^{(\tau-1)} + \alpha(\mathbf{X} - \mathbf{D_X})\boldsymbol{f}^{(\tau-1)} = (\mathbf{I} + \alpha\mathbf{Q})\boldsymbol{f}^{(\tau-1)}$$

$$= (\mathbf{I} + \alpha\mathbf{Q})^{\tau} \boldsymbol{f}^{(0)}. \qquad (8.26)$$

Such a propagation process will stop when $f^{(\tau)} = f^{(\tau-1)}$, i.e.,

$$(\mathbf{I} + \alpha \mathbf{Q})^{(\tau)} = (\mathbf{I} + \alpha \mathbf{Q})^{(\tau-1)}. \tag{8.27}$$

The smallest τ that can stop the propagation is defined as the *stop step*. To obtain the *stop step* τ, CAD needs to keep checking the powers of $(\mathbf{I} + \alpha \mathbf{Q})$ until it doesn't change as τ increases, i.e., the *stop criteria*.

Definition 8.4 (Intimacy Matrix) Matrix

$$\mathbf{H} = (\mathbf{I} + \alpha \mathbf{Q})^{\tau} \in \mathbb{R}^{|\mathcal{V}| \times |\mathcal{V}|} \tag{8.28}$$

is defined as the *intimacy matrix* of users in $\mathcal{V}$, where τ is the *stop step* and $H(i, j)$ denotes the *intimacy score* between u_i and $u_j \in \mathcal{V}$ in the network.

8.3.4 Intimacy Matrix of Attributed Heterogeneous Network

Real-world social networks can usually contain various kinds of information, e.g., links and attributes, and can be formulated as $G = (\mathcal{V}, \mathcal{E}, \mathcal{A})$ as introduced in Sect. 8.3.2. Attribute set $\mathcal{A} = \{a_1, a_2, \ldots, a_m\}$, $a_i = \{a_{i1}, a_{i2}, \ldots, a_{in_i}\}$ can have n_i different values for $i \in \{1, 2, \ldots, m\}$. An example of attribute augmented heterogeneous network is given in Fig. 8.4, where Fig. 8.4a is the input *attribute augmented heterogeneous network*. Figure 8.4b–d shows the attribute information in the network, which include timestamps, text, and location check-ins. Including the attributes as a special type of nodes in the graph definition provides a conceptual framework to handle social links and node attributes in a unified framework. The effect on increasing the dimensionality of the network will be handled as in Lemma 8.1 in a lower dimensional space.

Definition 8.5 (Attribute Transition Probability Matrix) The connections between users and attributes, e.g., a_i, can be represented as the *attribute adjacency matrix* $\mathbf{A}_{a_i} \in \mathbb{R}^{|\mathcal{V}| \times n_i}$. Based on $\mathbf{A}_{a_i}$, CAD formally defines the *attribute transition probability matrix* from users to attribute a_i to be $\mathbf{R}_i \in \mathbb{R}^{|\mathcal{V}| \times n_i}$, where

$$\mathbf{R}_i(i, j) = \frac{1}{\sqrt{\left(\sum_{m=1}^{n_i} \mathbf{A}_{a_i}(i, m)\right)\left(\sum_{n=1}^{|\mathcal{V}|} \mathbf{A}_{a_i}(n, j)\right)}} \mathbf{A}_{a_i}(i, j). \tag{8.29}$$

Similarly, CAD defines the *attribute transition probability matrix* from attribute a_i to users in $\mathcal{V}$ as $\mathbf{S}_i = \mathbf{R}_i^{\top}$.

The importance of different information types in calculating the closeness measure among users can be different. To handle the *network heterogeneity problem*, the CAD model proposes to apply the *micro-level* control by giving different information sources distinct weights to denote their differences: $\boldsymbol{\omega} = [\omega_0, \omega_1, \ldots, \omega_m]^{\top}$, where $\sum_{i=0}^{m} \omega_i = 1.0$, ω_0 is the weight of link information and ω_i is the weight of attribute a_i, for $i \in \{1, 2, \ldots, m\}$.

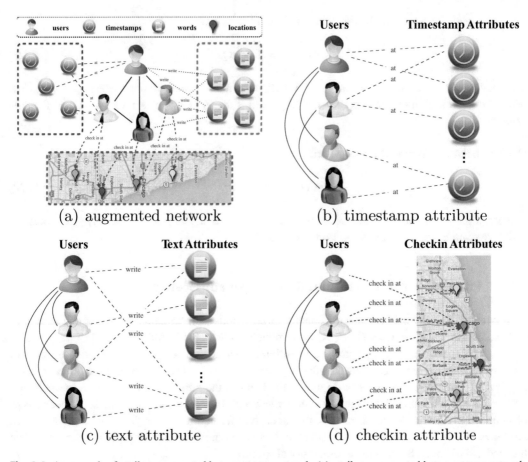

Fig. 8.4 An example of attribute augmented heterogeneous network. (**a**) attribute augmented heterogeneous network, (**b**) timestamp attribute, (**c**) text attribute, (**d**) location check-in attribute

Definition 8.6 (Weighted Attribute Transition Probability Matrix) With weights ω, CAD can define matrices

$$\tilde{\mathbf{R}} = [\omega_1\mathbf{R}_1, \ldots, \omega_n\mathbf{R}_n],$$
(8.30)

$$\tilde{\mathbf{S}} = [\omega_1\mathbf{S}_1, \ldots, \omega_n\mathbf{S}_n]^\top$$
(8.31)

to be the *weighted attribute transition probability matrices* between users and all attributes, where $\tilde{\mathbf{R}} \in \mathbb{R}^{|\mathcal{V}| \times (n_{aug} - |\mathcal{V}|)}$, $\tilde{\mathbf{S}} \in \mathbb{R}^{(n_{aug} - |\mathcal{V}|) \times |\mathcal{V}|}$, $n_{aug} = (|\mathcal{V}| + \sum_{i=1}^{m} n_i)$ is the number of all user and attribute nodes in the augmented network.

Definition 8.7 (Network Transition Probability Matrix) Furthermore, the *transition probability matrix* of the whole attribute augmented heterogeneous network G is defined as

$$\tilde{\mathbf{Q}}_{aug} = \begin{bmatrix} \tilde{\mathbf{Q}} & \tilde{\mathbf{R}} \\ \tilde{\mathbf{S}} & \mathbf{0} \end{bmatrix},$$
(8.32)

where $\tilde{\mathbf{Q}}_{aug} \in \mathbb{R}^{n_{aug} \times n_{aug}}$ and block matrix $\tilde{\mathbf{Q}} = \omega_0 \mathbf{Q}$ is the *weighted social transition probability matrix* of social links in $\mathcal{E}$.

In the real world, heterogeneous social networks can contain large amounts of attributes, i.e., n_{aug} can be extremely large. The *weighted transition probability matrix*, i.e., $\tilde{\mathbf{Q}}_{aug}$, can be of extremely high dimensions and can hardly fit in the memory. As a result, it will be impossible to update the matrix until the *stop criteria* meets to obtain the *stop step* and the *intimacy matrix*. To solve such problem, CAD proposes to obtain the *stop step* and the *intimacy matrix* by applying partitioned block matrix operations with the following Lemma 8.1.

Lemma 8.1 $(\tilde{\mathbf{Q}}_{aug})^k = \begin{bmatrix} \tilde{\mathbf{Q}}_k & \tilde{\mathbf{Q}}_{k-1}\tilde{\mathbf{R}} \\ \tilde{\mathbf{S}}\tilde{\mathbf{Q}}_{k-1} & \tilde{\mathbf{S}}\tilde{\mathbf{Q}}_{k-2}\tilde{\mathbf{R}} \end{bmatrix}$, $k \geq 2$, *where*

$$\tilde{\mathbf{Q}}_k = \begin{cases} \mathbf{I}, & if\ k = 0, \\ \tilde{\mathbf{Q}}, & if\ k = 1, \\ \tilde{\mathbf{Q}}\tilde{\mathbf{Q}}_{k-1} + \tilde{\mathbf{R}}\tilde{\mathbf{S}}\tilde{\mathbf{Q}}_{k-2}, & if\ k \geq 2 \end{cases} \tag{8.33}$$

and the intimacy matrix among users in $\mathcal{V}$ can be represented as

$$\begin{aligned} \tilde{\mathbf{H}}_{aug} &= \left(\mathbf{I} + \alpha\tilde{\mathbf{Q}}_{aug}\right)^{\tau} (1 : |\mathcal{V}|, 1 : |\mathcal{V}|) \\ &= \left(\sum_{t=0}^{\tau} \binom{\tau}{t} \alpha^t (\tilde{\mathbf{Q}}_{aug})^t\right) (1 : |\mathcal{V}|, 1 : |\mathcal{V}|) \\ &= \left(\sum_{t=0}^{\tau} \binom{\tau}{t} \alpha^t \left((\tilde{\mathbf{Q}}_{aug})^t (1 : |\mathcal{V}|, 1 : |\mathcal{V}|)\right)\right) \\ &= \left(\sum_{t=0}^{\tau} \binom{\tau}{t} \alpha^t \tilde{\mathbf{Q}}_t\right), \end{aligned} \tag{8.34}$$

where $\mathbf{X}(1 : |\mathcal{V}|, 1 : |\mathcal{V}|)$ is a sub-matrix of $\mathbf{X}$ with indexes in range $[1, |\mathcal{V}|]$. Notation τ is the stop step, achieved when $\tilde{\mathbf{Q}}_\tau = \tilde{\mathbf{Q}}_{\tau-1}$, i.e., the stop criteria, and $\tilde{\mathbf{Q}}_\tau$ is called the stationary matrix of the attributed augmented heterogeneous network.

Proof The lemma can be proved by induction on k [63], which will be left as an exercise for the readers. Considering that $(\tilde{\mathbf{R}}\tilde{\mathbf{S}}) \in \mathbb{R}^{|\mathcal{V}| \times |\mathcal{V}|}$ can be precomputed in advance, the space cost of Lemma 8.1 is $O(|\mathcal{V}|^2)$, where $|\mathcal{V}| \ll n_{aug}$.

Since we are only interested in the *intimacy* and *transition matrices* among user nodes instead of those between the augmented items and users for the community detection task, CAD creates a reduced dimensional representation only involving users for $\tilde{\mathbf{Q}}_k$ and $\tilde{\mathbf{H}}$ such that CAD can capture the effect of "user-attribute" and "attribute-user" transition on "user-user" transition. $\tilde{\mathbf{Q}}_k$ is a reduced dimension representation of $\tilde{\mathbf{Q}}_{aug}^k$, while eliminating the augmented items, it can still capture the "user-user" transitions effectively.

8.3.5   Intimacy Matrix Across Aligned Heterogeneous Networks

When G^t is new, the *intimacy matrix* $\tilde{\mathbf{H}}$ among users calculated based on the information in G^t can be very sparse. To solve this problem, CAD proposes to propagate useful information from other well-developed aligned networks to the emerging network. Information propagated from other aligned well-developed networks can help solve the shortage of information problem in the emerging network [54, 55]. However, as proposed in [32], different networks can have different properties and information propagated from other well-developed aligned networks can be very different from that of the emerging network as well.

To handle this problem, CAD model proposes to apply the *macro-level control* technique by using weights, $\rho^{s,t}$, $\rho^{t,s} \in [0, 1]$, to control the proportion of information propagated between the developed network G^s and the emerging network G^t. If information from G^s is helpful for improving the community detection results in G^t, CAD can set a higher $\rho^{s,t}$ to propagate more information from G^s. Otherwise, CAD can set a lower $\rho^{s,t}$ instead. The weights $\rho^{s,t}$ and $\rho^{t,s}$ can be adjusted automatically with a method to be introduced in Sect. 8.3.7.

Definition 8.8 (Anchor Transition Matrix) To propagate information across networks, CAD introduces the *anchor transition matrices* between G^t and G^s to be $\mathbf{T}^{t,s} \in \mathbb{R}^{|\mathcal{V}^t| \times |\mathcal{V}^s|}$ and $\mathbf{T}^{s,t} \in \mathbb{R}^{|\mathcal{V}^s| \times |\mathcal{V}^t|}$, where $\mathbf{T}^{t,s}(i, j) = \mathbf{T}^{s,t}(j, i) = 1$, iff $(u_i^t, u_j^s) \in A^{t,s}$, $u_i^t \in \mathcal{V}^t$, $u_j^s \in \mathcal{V}^s$.

Meanwhile, with weights $\rho^{s,t}$ and $\rho^{t,s}$, the *weighted network transition probability matrix* of G^t and G^s are represented as

$$\bar{\mathbf{Q}}_{aug}^t = (1 - \rho^{t,s}) \begin{bmatrix} \tilde{\mathbf{Q}}^t & \tilde{\mathbf{R}}^t \\ \tilde{\mathbf{S}}^t & \mathbf{0} \end{bmatrix} \tag{8.35}$$

and

$$\bar{\mathbf{Q}}_{aug}^s = (1 - \rho^{s,t}) \begin{bmatrix} \tilde{\mathbf{Q}}^s & \tilde{\mathbf{R}}^s \\ \tilde{\mathbf{S}}^s & \mathbf{0} \end{bmatrix}, \tag{8.36}$$

where $\bar{\mathbf{Q}}_{aug}^t \in \mathbb{R}^{n_{aug}^t \times n_{aug}^t}$ and $\bar{\mathbf{Q}}_{aug}^s \in \mathbb{R}^{n_{aug}^s \times n_{aug}^s}$, n_{aug}^t and n_{aug}^s are the numbers of all nodes in G^t and G^s, respectively.

Furthermore, to accommodate the dimensions, CAD introduces the *weighted anchor transition matrices* between G^s and G^t to be

$$\bar{\mathbf{T}}^{t,s} = (\rho^{t,s}) \begin{bmatrix} \mathbf{T}^{t,s} & \mathbf{0} \\ \mathbf{0} & \mathbf{0} \end{bmatrix}, \tag{8.37}$$

$$\bar{\mathbf{T}}^{s,t} = (\rho^{s,t}) \begin{bmatrix} \mathbf{T}^{s,t} & \mathbf{0} \\ \mathbf{0} & \mathbf{0} \end{bmatrix}, \tag{8.38}$$

where $\bar{\mathbf{T}}^{t,s} \in \mathbb{R}^{n_{aug}^t \times n_{aug}^s}$ and $\bar{\mathbf{T}}^{s,t} \in \mathbb{R}^{n_{aug}^s \times n_{aug}^t}$. Nodes corresponding to entries in $\bar{\mathbf{T}}^{t,s}$ and $\bar{\mathbf{T}}^{s,t}$ are of the same order as those in $\bar{\mathbf{Q}}_{aug}^t$ and $\bar{\mathbf{Q}}_{aug}^s$, respectively.

By combining the weighted intra-network transition probability matrices together with the weighted anchor transition matrices, CAD defines the *transition probability matrix* across *aligned networks* as

$$\bar{\mathbf{Q}}_{align} = \begin{bmatrix} \bar{\mathbf{Q}}_{aug}^{t} & \bar{\mathbf{T}}^{t,s} \\ \bar{\mathbf{T}}^{s,t} & \bar{\mathbf{Q}}_{aug}^{s} \end{bmatrix} \tag{8.39}$$

where $\bar{\mathbf{Q}}_{align} \in \mathbb{R}^{n_{align} \times n_{align}}$, and $n_{align} = n_{aug}^{t} + n_{aug}^{s}$ is the number of all nodes across the aligned networks.

Definition 8.9 (Aligned Network Intimacy Matrix) Based on the previous remarks, with $\bar{\mathbf{Q}}_{align}$, CAD can obtain the *intimacy matrix*, $\bar{\mathbf{H}}_{align}$, of users in G^t to be

$$\bar{\mathbf{H}}_{align} = (\mathbf{I} + \alpha \bar{\mathbf{Q}}_{align})^{\tau} (1 : |\mathcal{V}^t|, 1 : |\mathcal{V}^t|), \tag{8.40}$$

where $\bar{\mathbf{H}}_{align} \in \mathbb{R}^{|\mathcal{V}^t| \times |\mathcal{V}^t|}$, τ is the *stop step*.

Meanwhile, the structure of $(\mathbf{I} + \alpha \bar{\mathbf{Q}}_{align})$ cannot meet the requirements of Lemma 8.1 as it doesn't have a zero square matrix at the bottom right corner. As a result, methods introduced in Lemma 8.1 cannot be applied. To obtain the *stop step*, there is no other choice but to keep calculating powers of $(\mathbf{I} + \alpha \bar{\mathbf{Q}}_{align})$ until the *stop criteria* can meet, which can be very time consuming. In this part, we will introduce with the following Lemma 8.2 adopted by CAD model for efficient computation of the high-order powers of matrix $(\mathbf{I} + \alpha \bar{\mathbf{Q}}_{align})$.

Lemma 8.2 *For the given matrix* $(\mathbf{I} + \alpha \bar{\mathbf{Q}}_{align})$, *its kth power meets*

$$(\mathbf{I} + \alpha \bar{\mathbf{Q}}_{align})^{k} \mathbf{P} = \mathbf{P} \boldsymbol{\Lambda}^{k}, k \geq 1, \tag{8.41}$$

matrices $\mathbf{P}$ *and* $\boldsymbol{\Lambda}$ *contain the eigenvector and eigenvalues of* $(\mathbf{I} + \alpha \bar{\mathbf{Q}}_{align})$. *The ith column of matrix* $\mathbf{P}$ *is the eigenvector of* $(\mathbf{I} + \alpha \bar{\mathbf{Q}}_{align})$ *corresponding to its ith eigenvalue* λ_i *and diagonal matrix* $\boldsymbol{\Lambda}$ *has value* $\Lambda(i, i) = \lambda_i$ *on its diagonal.*

Proof The Lemma can be proved by induction on k [34] as follows:

Base Case When $k = 1$, let $\mathbf{p}_i$ and λ_i be the *ith* eigenvector and eigenvalue of matrix $\mathbf{Q}$, respectively, where

$$\mathbf{Q}\mathbf{p}_i = \lambda_i \mathbf{p}_i. \tag{8.42}$$

Organizing all the eigenvectors and eigenvalues of $\mathbf{Q}$ in matrix $\mathbf{P}$ and $\boldsymbol{\Lambda}$, we can have

$$\mathbf{Q}^1 \mathbf{P} = \mathbf{P} \boldsymbol{\Lambda}^1. \tag{8.43}$$

Inductive Assumption When $k = m, m \geq 1$, let's assume the lemma holds when $k = m, m \geq 1$. In other words, the following equation holds:

$$\mathbf{Q}^m \mathbf{P} = \mathbf{P} \boldsymbol{\Lambda}^m. \tag{8.44}$$

Induction When $k = m + 1, m \geq 1$,

$$\mathbf{Q}^{(m+1)}\mathbf{P} = \mathbf{Q}\mathbf{Q}^m\mathbf{P} = \mathbf{Q}\mathbf{P}\boldsymbol{\Lambda}^m = \mathbf{P}\boldsymbol{\Lambda}\boldsymbol{\Lambda}^m = \mathbf{P}\boldsymbol{\Lambda}^{(m+1)}. \tag{8.45}$$

In sum, the lemma holds for $k \geq 1$.

The time cost of calculating $\boldsymbol{\Lambda}^k$ is $O(n_{align})$, which is far less than that required to calculate $(\mathbf{I} + \alpha\bar{\mathbf{Q}}_{align})^k$.

Definition 8.10 (Eigen-Decomposition Based Aligned Network Intimacy Matrix) In addition, if $\mathbf{P}$ is invertible, we can have

$$(\mathbf{I} + \alpha\bar{\mathbf{Q}}_{align})^k = \mathbf{P}\boldsymbol{\Lambda}^k\mathbf{P}^{-1}, \tag{8.46}$$

where $\boldsymbol{\Lambda}^k$ has $\Lambda(i,i)^k$ on its diagonal. And the intimacy calculated based on eigenvalue decomposition will be

$$\bar{\mathbf{H}}_{align} = \left(\mathbf{P}\boldsymbol{\Lambda}^\tau\mathbf{P}^{-1}\right)(1 : |\mathcal{V}^t|, 1 : |\mathcal{V}^t|). \tag{8.47}$$

where the *stop step* τ can be obtained when $\mathbf{P}\boldsymbol{\Lambda}^\tau\mathbf{P}^{-1} = \mathbf{P}\boldsymbol{\Lambda}^{\tau-1}\mathbf{P}^{-1}$, i.e., *stop criteria*.

8.3.6 Approximated Intimacy to Reduce Dimension

Eigendecomposition based method proposed in Lemma 8.2 enables CAD to calculate the powers of $(\mathbf{I} + \alpha\mathbf{Q}_{align})$ very efficiently. However, when applying Lemma 8.2 to calculate the *intimacy matrix* of real-world partially aligned networks, it can suffer from many serious problems. The reason is that the dimension of $(\mathbf{I} + \alpha\mathbf{Q}_{align})$, i.e., $n_{align} \times n_{align}$, is so high that matrix $(\mathbf{I} + \alpha\mathbf{Q}_{align})$ can hardly fit in the memory. To solve that problem, CAD proposes to calculate the approximated *intimacy matrix* $\bar{\mathbf{H}}_{align}^{approx}$ with less space and time costs instead.

Let's define the transition probability matrices of G^t and G^s to be $\tilde{\mathbf{Q}}_{aug}^t$ and $\tilde{\mathbf{Q}}_{aug}^s$, respectively. By applying Lemma 8.1, we can get their *stop step* and the *stationary matrices* to be τ^t, τ^s, $\tilde{\mathbf{Q}}_{\tau^t}^t$ and $\tilde{\mathbf{Q}}_{\tau^s}^t$, respectively. *Stationary matrices* $\tilde{\mathbf{Q}}_{\tau^t}^t$, $\tilde{\mathbf{Q}}_{\tau^s}^s$ together with the *anchor transition matrix*, $\mathbf{T}^{t,s}$ and $\mathbf{T}^{t,s}$, can be used to define a low-dimensional *reduced aligned network transition probability matrix*, which only involves users explicitly, while the effect of "attribute-user" or "user-attribute" transition is implicitly absorbed into $\tilde{\mathbf{Q}}_{\tau^t}^t$ and $\tilde{\mathbf{Q}}_{\tau^s}^s$:

$$\bar{\mathbf{Q}}_{align}^{user} = \begin{bmatrix} (1 - \rho^{t,s})\tilde{\mathbf{Q}}_{\tau^t}^t & (\rho^{t,s})\mathbf{T}^{t,s} \\ (\rho^{s,t})\mathbf{T}^{s,t} & (1 - \rho^{s,t})\tilde{\mathbf{Q}}_{\tau^s}^s \end{bmatrix}, \tag{8.48}$$

where $\bar{\mathbf{Q}}_{align}^{user} \in \mathbb{R}^{(|\mathcal{V}|^t + |\mathcal{V}^s|)^2}$ and $(|\mathcal{V}|^t + |\mathcal{V}^s|) \ll n_{align}$.

Definition 8.11 (Approximated Aligned Network Intimacy Matrix) Furthermore, with Lemma 8.2, we can get *intimacy matrix* of users in G^t based on $\bar{\mathbf{Q}}_{align}^{user}$ to be:

$$\bar{\mathbf{H}}_{align}^{approx} = \left(\mathbf{P}^*(\boldsymbol{\Lambda}^*)^{\tau}(\mathbf{P}^*)^{-1}\right)(1:|\mathcal{V}^t|, 1:|\mathcal{V}^t|), \tag{8.49}$$

where $\left(\mathbf{I} + \alpha\bar{\mathbf{Q}}_{align}^{user}\right) = \mathbf{P}^*\boldsymbol{\Lambda}^*(\mathbf{P}^*)^{-1}$ and τ is the *stop step*.

The approximated intimacy matrix computation method introduced above can greatly reduce the time and space costs. Let $|\mathcal{V}^t| = n^t$, the size of intimacy matrix $\bar{\mathbf{H}}_{align}$ will be $(n^t)^2$. However, to obtain $\bar{\mathbf{H}}_{align}$, we need to calculate the transition probability matrix $\bar{\mathbf{Q}}_{align}$ in advance, whose size is $(n_{align})^2$.

Space Cost: In eigendecomposition based method, we have to calculate and store matrices $\bar{\mathbf{Q}}_{align}^{eigen}$, $\mathbf{P}$, $\mathbf{P}^{-1}$, $\boldsymbol{\Lambda} \in \mathbb{R}^{n_{align} \times n_{align}}$, whose space costs are $O\left(4n_{align}^2\right)$. However, in the approximation based method, we just need to store matrices $\tilde{\mathbf{Q}}^x \in \mathbb{R}^{n^x \times n^x}$, $\tilde{\mathbf{R}}^x \in \mathbb{R}^{n^x \times (\sum_i n_i^x)}$, $\tilde{\mathbf{S}}^x \in \mathbb{R}^{(\sum_i n_i^x) \times n^x}$, $x \in \{s, t\}$, as well as $\bar{\mathbf{Q}}_{align}^{approx} \in \mathbb{R}^{(n^t+n^s) \times (n^t+n^s)}$, whose space cost will be $O(\max\{(n^t + n^s)^2, n^t(\sum_i n_i^t), n^s(\sum_i n_i^s)\}) < O\left(4n_{align}^2\right)$.

Time Cost: In eigendecomposition based method, the matrix eigendecomposition of $\bar{\mathbf{Q}}_{align}^{eigen}$, inversion $\mathbf{P}^{-1}$, and multiplication of $\mathbf{P}\boldsymbol{\Lambda}^k\mathbf{P}^{-1}$ are all time-consuming operations, whose time costs are $O\left(kn_{align}^2\right)$ [47], $O\left(n_{align}^2 \log(n_{align})\right)$ [11] and $O\left(2n_{align}^3\right)$, respectively. As a result, the time cost of eigendecomposition based method is about $O\left(2n_{align}^3\right)$. However, in approximation based methods, we need to apply Lemma 2 to get $\bar{\mathbf{H}}^t$ and $\bar{\mathbf{H}}^t$, whose time cost is

$$O\left(\max\left\{\tau\left((n^t)^3 + (n^t)^2\left(\sum_i a_i^t\right)\right), \tau\left((n^s)^3 + (n^s)^2\left(\sum_i a_i^s\right)\right)\right\}\right), \tag{8.50}$$

which is much smaller than that of eigendecomposition based methods.

8.3.7 Clustering and Weight Self-adjustment

Intimacy matrix $\bar{\mathbf{H}}_{align}$ (or $\bar{\mathbf{H}}_{align}^{approx}$) stores the intimacy scores among users in $\mathcal{V}^t$ and can be used to detect the communities in the network. CAD will use the low-rank matrix factorization method used proposed in [43] to get the latent feature vectors, $\mathbf{U}$, for each user. To avoid overfitting, CAD introduces two regularization terms to the object function as follows:

$$\min_{\mathbf{U},\mathbf{V}} \left\|\bar{\mathbf{H}}_{align} - \mathbf{U}\mathbf{V}\mathbf{U}^{\top}\right\|_F^2 + \theta \cdot \|\mathbf{U}\|_F^2 + \beta \cdot \|\mathbf{V}\|_F^2,$$

$$s.t. \mathbf{U} \geq \mathbf{0}, \mathbf{V} \geq \mathbf{0}, \tag{8.51}$$

where $\mathbf{U}$ is the latent feature vectors, $\mathbf{V}$ stores the correlation among rows of $\mathbf{V}$, θ and β are the weights of $\|\mathbf{U}\|_F^2$, $\|\mathbf{V}\|_F^2$, respectively.

This object function is hard to solve and obtaining the global optimal result for both $\mathbf{U}$ and $\mathbf{V}$ simultaneously can be very challenging. CAD proposes to solve the objective function by fixing one variable, e.g., $\mathbf{U}$, and update another variable, e.g.,$\mathbf{V}$, alternatively. The Lagrangian function of the object equation can be represented as:

$$\mathcal{F} = Tr\left(\bar{\mathbf{H}}_{align}\bar{\mathbf{H}}_{align}^{\top}\right) - Tr\left(\bar{\mathbf{H}}_{align}\mathbf{U}\mathbf{V}^{\top}\mathbf{U}^{\top}\right)$$
$$- Tr\left(\mathbf{U}\mathbf{V}\mathbf{U}^{\top}\bar{\mathbf{H}}_{align}^{\top}\right) + Tr\left(\mathbf{U}\mathbf{V}\mathbf{U}^{\top}\mathbf{U}\mathbf{V}^{\top}\mathbf{U}^{\top}\right)$$
$$+ \theta Tr\left(\mathbf{U}\mathbf{U}^{\top}\right) + \beta Tr\left(\mathbf{V}\mathbf{V}^{\top}\right) - Tr(\boldsymbol{\Theta}\mathbf{U}) - Tr(\boldsymbol{\Omega}\mathbf{V}) \qquad (8.52)$$

where $\boldsymbol{\Theta}$ and $\boldsymbol{\Omega}$ are the multipliers for the constraint of $\mathbf{U}$ and $\mathbf{V}$, respectively. By taking derivatives of $\mathcal{F}$ with regard to $\mathbf{U}$ and $\mathbf{V}$, we can get

$$\frac{\partial \mathcal{F}}{\partial \mathbf{U}} = -2\left(\bar{\mathbf{H}}_{align}^{\top}\mathbf{U}\mathbf{V} + \bar{\mathbf{H}}_{align}\mathbf{U}\mathbf{V}^{\top} - \mathbf{U}\mathbf{V}^{\top}\mathbf{U}^{\top}\mathbf{U}\mathbf{V}^{\top} - \mathbf{U}\mathbf{V}\mathbf{U}^{\top}\mathbf{U}\mathbf{V}^{\top} - \theta\mathbf{U}\right) - \boldsymbol{\Theta}^{\top} \qquad (8.53)$$

$$\frac{\partial \mathcal{F}}{\partial \mathbf{V}} = -2\left(\mathbf{U}^{\top}\bar{\mathbf{H}}_{align}\mathbf{U} - \mathbf{U}^{\top}\mathbf{U}\mathbf{V}\mathbf{U}^{\top}\mathbf{U} - \beta\mathbf{V}\right) - \boldsymbol{\Omega}^{\top} \qquad (8.54)$$

Let $\frac{\partial \mathcal{F}}{\partial \mathbf{U}} = \mathbf{0}$ and $\frac{\partial \mathcal{F}}{\partial \mathbf{V}} = \mathbf{0}$ and use the KKT complementary condition, we can get

$$\mathbf{U}(i, j) \leftarrow \mathbf{U}(i, j)\sqrt{\frac{\left(\bar{\mathbf{H}}_{align}^{\top}\mathbf{U}\mathbf{V} + \bar{\mathbf{H}}_{align}\mathbf{U}\mathbf{V}^{\top}\right)(i, j)}{\left(\mathbf{U}\mathbf{V}^{\top}\mathbf{U}^{\top}\mathbf{U}\mathbf{V} + \mathbf{U}\mathbf{V}\mathbf{U}^{\top}\mathbf{U}\mathbf{V}^{\top} + \theta\mathbf{U}\right)(i, j)}}, \qquad (8.55)$$

$$\mathbf{V}(i, j) \leftarrow \mathbf{V}(i, j)\sqrt{\frac{\left(\mathbf{U}^{\top}\bar{\mathbf{H}}_{align}\mathbf{U}\right)(i, j)}{\left(\mathbf{U}^{\top}\mathbf{U}\mathbf{V}\mathbf{U}^{\top}\mathbf{U} + \beta\mathbf{V}\right)(i, j)}}. \qquad (8.56)$$

The low-rank matrix $\mathbf{U}$ captures the information of each users from the intimacy matrix and can be used as latent numerical feature vectors to cluster users in G^t with traditional clustering methods, e.g., Kmeans [15].

Meanwhile, to handle the *information heterogeneity problem* in each network and the *network difference problem* across networks, CAD uses weights, ω^t, ω^s, $\rho^{t,s}$, and $\rho^{s,t}$ to denote the importance of information in G^t, G^s and that propagated from G^t and G^s, respectively. For simplicity, CAD sets $\omega^t = \omega^s = \omega = [\omega_0, \omega_1, \ldots, \omega_m]$ and $\rho^{t,s} = \rho^{s,t} = \rho$ in CAD. Let $\mathcal{C}$ be the community detection result achieved by CAD in G^t. The optimal choices of parameters ω and ρ, evaluated by some metrics, e.g., *entropy* [62], can be achieved with the following equation:

$$\omega, \rho = \min_{\omega, \rho} E(\mathcal{C}). \qquad (8.57)$$

The optimization problem is very difficult to solve. CAD proposes a method to adjust ω and ρ automatically to enable CAD to achieve better results.

The weight adjustment method used to deal with ω can work as follows: for example, in network G^t, we have relational information and attribute information $\mathcal{E}$ and $\mathcal{A} = \{A_1, A_2, \ldots, A_m\}$, whose weights are initialized to be $\omega = \{\omega_0, \omega_1, \ldots, \omega_m\}$. For $\omega_i \in \omega, i \in \{0, 1, \ldots, m\}$, CAD keeps checking if increasing ω_i by a ratio of γ, i.e., $(1 + \gamma)\omega_i$, can improve the performance or not. If so, $(1 + \gamma)\omega_i$ after re-normalization is used as the new value of ω_i; otherwise, CAD restores the old ω_i before increase and study ω_{i+1}. In the experiment, γ is set as 0.05. Similarly, for the weight of different networks, i.e., ρ, CAD can adjust them with the same methods to find the optimal ρ. The pseudo code of CAD is available in Algorithm 1.

Algorithm 1 CAD with Parameter Self-Adjustment

Require: aligned network: $\mathcal{G} = \{\{G^t, G^s\}, \{A^{t,s}, A^{s,t}\}\}$
parameters: $\omega, \rho, \gamma, \alpha, \beta$ and method type M
Ensure: community detection results of G^t: $\mathcal{C}$
1: $\omega_{old} = \omega, \rho_{old} = \rho, E_{old} = \infty$
2: **for** parameter $\delta \in \omega \cup \{\rho\}$ **do**
3:    **while** $True$ **do**
4:       $\delta = (1 + \gamma)\delta$ and renormalize ω if $\delta \in \omega$ to get ω_{new}, ρ_{new}
5:       construct transition probability matrix $\bar{Q}_{align}$
6:       **if** $M = approximation$ **then**
7:          construct $\bar{Q}_{align}^{user}$ with $\tilde{Q}_{\tau^t}^t, \tilde{Q}_{\tau^s}^s$ calculated according to Lemma 1
8:          calculate $\bar{H}_{align}^{approx}$ with $\bar{Q}_{align}^{user}$ according to Lemma 2
9:          $\bar{H}_{align} = \bar{H}_{align}^{approx}$
10:      **else**
11:         calculate $\bar{H}_{align}$ with $\bar{Q}_{align}$ according to Lemma 2
12:      **end if**
13:      get lower-dimensional latent feature vectors $\mathbf{U}$
14:      $\mathcal{C} = Kmeans(\mathbf{U})$
15:      $E_{new} = -\sum_{i=1}^{K} P(i) \log P(i), P(i) = \frac{|U_i|}{\sum_{i=1}^{K} |U_i|}, U_i \in \mathcal{C}$
16:      **if** $E_{new} < E_{old}$ **then**
17:         $\omega_{old} = \omega, \rho_{old} = \rho, E_{old} = E_{new}$
18:      **else**
19:         $\omega = \omega_{old}, \rho = \rho_{old}$
20:         break
21:      **end if**
22:   **end while**
23: **end for**

8.4 Mutual Community Detection

Besides the knowledge transfer from developed networks to the emerging networks to overcome the cold start problem, information in developed networks can also be transferred mutually to help refine the detected community structure detected from each of them. In this section, we will introduce the mutual community detection problem across multiple aligned heterogeneous networks and talk about a new cross-network mutual community detection model MCD [52]. To refine the community structures, a new concept named *discrepancy* is introduced to help preserve the consensus of the community detection result of the shared anchor users.

8.4.1 Background Knowledge

In this section, we will focus on the simultaneous community detection of each network across multiple *partially aligned social networks* simultaneously, which is formally defined as the *Mutual Community Detection* problem [52]. The goal is to distill relevant information from other aligned social network to complement knowledge directly derivable from each network to improve the clustering or community detection, while preserving the distinct characteristics of each individual network. The *Mutual Community Detection* problem is very important for online social networks and can be the prerequisite for many concrete social network applications: (1) *network partition*: detected communities can usually represent small-sized subgraphs of the network, and (2) *comprehensive understanding of user social behaviors*: community structures of the shared users in multiple aligned networks can provide a complementary understanding of their social interactions in the online social world.

Besides its importance, the *Mutual Community Detection* problem is a novel problem and different from existing clustering problems, including: (1) *consensus clustering* [12, 23, 26, 27, 31], which aims at achieving a consensus result of several input clustering results about the same data; (2) *multi-view clustering* [4, 6], whose target is to partition objects into clusters based on their different representations, e.g., clustering webpages with text information and hyperlinks; (3) *multi-relational clustering* [3, 49], which focuses on clustering objects in one relation (called target relation) using information in multiple inter-linked relations; and (4) *co-regularized multi-domain graph clustering* [7], which relaxes the *one-to-one* constraints on node correspondence relationships between different views in multi-view clustering to "*uncertain*" mappings. Unlike these existing clustering problems, the *Mutual Community Detection* problem aims at detecting the communities for multiple networks involving both anchor and non-anchor users simultaneously and each network contains heterogeneous information about users' social activities.

8.4.2 Problem Formulation

For the given multiple aligned heterogeneous networks, i.e., those in $\mathcal{G} = ((G^{(1)}, G^{(2)}, \ldots, G^{(n)}),$ $(\mathcal{A}^{(1,2)}, \mathcal{A}^{(1,3)}, \ldots, \mathcal{A}^{(n-1,n)}))$, the *Mutual Community Detection* problem aims to obtain the optimal communities $\{\mathcal{C}^{(1)}, \mathcal{C}^{(2)}, \ldots, \mathcal{C}^{(n)}\}$ for $\{G^{(1)}, G^{(2)}, \ldots, G^{(n)}\}$ simultaneously, where $\mathcal{C}^{(i)} = \left\{U_1^{(i)}, U_2^{(i)}, \ldots, U_{k^{(i)}}^{(i)}\right\}$ is a partition of the users set $\mathcal{U}^{(i)}$ in $G^{(i)}$, $k^{(i)} = \left|\mathcal{C}^{(i)}\right|$, $U_l^{(i)} \cap U_m^{(i)} = \emptyset$, $\forall \, l, m \in \{1, 2, \ldots, k^{(i)}\}$ and $\bigcup_{j=1}^{k^{(i)}} U_j^{(i)} = \mathcal{U}^{(i)}$. Users in each detected social community are more densely connected with each other than with users in other communities. In this section, we focus on studying the hard (i.e., non-overlapping) community detection of users in online social networks.

The *Mutual Community Detection* problem studied in this section is very challenging to solve due to:

- *Closeness Measure*: Users in heterogeneous social networks can be connected with each other by various direct and indirect connections. A general closeness measure among users with such connection information is the prerequisite for addressing the *Mutual Community Detection* problem.
- *Network Characteristics*: Social networks usually have their own characteristics, which can be reflected in the community structures formed by users. Preservation of each network's characteristics (i.e., some unique structures in each network's detected communities) is very important in the *Mutual Community Detection* problem.
- *Mutual Community Detection*: Information in different networks can provide us with a more comprehensive understanding about the anchor users' social structures. For anchor users whose community structures are not clear based on information in one network, utilizing the heterogeneous information in aligned networks can help refine and disambiguate the community structures about the anchor users. However, how to achieve such a goal is still an open problem.

To solve all these challenges, a novel cross-network community detection method, MCD (Mutual Community Detector), is proposed in this section. MCD maps the complex relationships in the social network into a heterogeneous information network [41] and introduces a novel meta-path based closeness measure, *HNMP-Sim*, to utilize both direct and indirect connections among users in closeness scores calculation. With full considerations of the network characteristics, MCD exploits the information in aligned networks to refine and disambiguate the community structures of the multiple networks concurrently. More detailed information about the MCD model will be introduced as follows.

8.4.3 Meta Path Based Social Proximity Measure

Many existing similarity measures, e.g., "common neighbor" [13], "Jaccard's coefficient" [13], defined for homogeneous networks cannot capture all the connections among users in heterogeneous networks. To use both direct and indirect connections among users in calculating the similarity score among users in the heterogeneous information network, MCD introduces meta path based similarity measure HNMP-Sim, whose information will be introduced as follows.

In heterogeneous networks, pairs of nodes can be connected by different paths, which are sequences of links in the network. Meta paths [40, 41] in heterogeneous networks, i.e., *heterogeneous network meta paths* (HNMPs), can capture both direct and indirect connections among nodes in a network. The length of a meta path is defined as the number of links that constitute it. Meta paths in networks can start and end with various node types. However, in this section, we are mainly concerned about those starting and ending with users, which are formally defined as the *social HNMPs*. A formal definition of *social HNMPs* is available in [52, 56, 57]. The notation, definition, and semantics of 7 different *social HNMPs* used in MCD are listed in Table 8.2. To extract the social meta paths, prior domain knowledge about the network structure is required.

These 7 different social HNMPs in Table 8.2 can cover lots of connections among users in networks. Some meta path based similarity measures have been proposed so far, e.g., the *PathSim* proposed in [41], which is defined for undirected networks and considers different meta paths to be of the same importance. To measure the social closeness among users in directed heterogeneous information networks, we extend *PathSim* to propose a new closeness measure as follows.

Definition 8.12 (HNMP-Sim) Let $\mathcal{P}_i(x \rightsquigarrow y)$ and $\mathcal{P}_i(x \rightsquigarrow \cdot)$ be the sets of path instances of the ith HNMP going from x to y and those going from x to other nodes in the network. The HNMP-Sim (HNMP based Similarity) of node pair (x, y) is defined as

$$\text{HNMP-Sim}(x, y) = \sum_i \omega_i \left(\frac{|\mathcal{P}_i(x \rightsquigarrow y)| + |\mathcal{P}_i(y \rightsquigarrow x)|}{|\mathcal{P}_i(x \rightsquigarrow \cdot)| + |\mathcal{P}_i(y \rightsquigarrow \cdot)|} \right), \tag{8.58}$$

Table 8.2 Summary of HNMPs

ID	Notation	Heterogeneous network meta path	Semantics
1	$U \rightarrow U$	User $\xrightarrow{follow}$ User	Follow
2	$U \rightarrow U \rightarrow U$	User $\xrightarrow{follow}$ User $\xrightarrow{follow}$ User	Follower of follower
3	$U \rightarrow U \leftarrow U$	User $\xrightarrow{follow}$ User $\xrightarrow{follow^{-1}}$ User	Common out neighbor
4	$U \leftarrow U \rightarrow U$	User $\xrightarrow{follow^{-1}}$ User $\xrightarrow{follow}$ User	Common in neighbor
5	$U \rightarrow P \rightarrow W \leftarrow P \leftarrow U$	User $\xrightarrow{write}$ Post $\xrightarrow{contain}$ Word	Posts containing
		$\xrightarrow{contain^{-1}}$ Post $\xrightarrow{write^{-1}}$ User	Common words
6	$U \rightarrow P \rightarrow T \leftarrow P \leftarrow U$	User $\xrightarrow{write}$ Post $\xrightarrow{contain}$ Time	Posts containing
		$\xrightarrow{contain^{-1}}$ Post $\xrightarrow{write^{-1}}$ User	Common timestamps
7	$U \rightarrow P \rightarrow L \leftarrow P \leftarrow U$	User $\xrightarrow{write}$ Post $\xrightarrow{attach}$ Location	Posts attaching
		$\xrightarrow{attach^{-1}}$ Post $\xrightarrow{write^{-1}}$ User	Common check-ins

where ω_i is the weight of the ith HNMP and $\sum_i \omega_i = 1$. In MCD, the weights of different HNMPs can be automatically adjusted by applying a similar greedy search technique as introduced in Sect. 8.3.7.

Let $\mathbf{A}_i$ be the *adjacency matrix* corresponding to the ith HNMP among users in the network and $A_i(m, n) = k$ iff there exist k different path instances of the ith HNMP from user m to n in the network. Furthermore, the similarity score matrix among users of HNMP $\# i$ can be represented as $\mathbf{S}_i = \left(\mathbf{D}_i + \bar{\mathbf{D}}_i\right)^{-1} \left(\mathbf{A}_i + \mathbf{A}_i^{\top}\right)$, where $\mathbf{A}_i^{\top}$ denotes the transpose of $\mathbf{A}_i$, diagonal matrices $\mathbf{D}_i$ and $\bar{\mathbf{D}}_i$ have values $\mathbf{D}_i(l, l) = \sum_m \mathbf{A}_i(l, m)$ and $\bar{\mathbf{D}}_i(l, l) = \sum_m (\mathbf{A}_i^{\top})(l, m)$ on their diagonals, respectively. The HNMP-Sim matrix of the network which can capture all possible connections among users is represented as follows:

$$\mathbf{S} = \sum_i \omega_i \mathbf{S}_i = \sum_i \omega_i \left(\left(\mathbf{D}_i + \bar{\mathbf{D}}_i\right)^{-1} \left(\mathbf{A}_i + \mathbf{A}_i^{\top}\right) \right). \tag{8.59}$$

8.4.4 Network Characteristic Preservation Clustering

Clustering each network independently can preserve each networks characteristics effectively as no information from external networks will interfere with the clustering results. Partitioning users of a certain network into several clusters will cut connections in the network and lead to some costs inevitably. Optimal clustering results can be achieved by minimizing the clustering costs.

For a given network G, let $\mathcal{C} = \{U_1, U_2, \ldots, U_k\}$ be the community structures detected from G. Term $\overline{U}_i = \mathcal{U} - U_i$ is defined to be the complement of set U_i in G. Various cost measures of partition $\mathcal{C}$ can be used, e.g., *cut* and *normalized cut*.

$$cut(\mathcal{C}) = \frac{1}{k} \sum_{i=1}^{k} S(U_i, \overline{U}_i) = \frac{1}{k} \sum_{i=1}^{k} \sum_{u \in U_i, v \in \overline{U}_i} S(u, v), \tag{8.60}$$

$$ncut(\mathcal{C}) = \frac{1}{k} \sum_{i=1}^{k} \frac{S(U_i, \overline{U}_i)}{S(U_i, \cdot)} = \frac{1}{k} \sum_{i=1}^{k} \frac{cut(U_i, \overline{U}_i)}{S(U_i, \cdot)}, \tag{8.61}$$

where $S(u, v)$ denotes the HNMP-Sim between u, v and $S(U_i, \cdot) = S(U_i, \mathcal{U}) = S(U_i, U_i) + S(U_i, \overline{U}_i)$.

For all users in $\mathcal{U}$, their clustering result can be represented in the *result confidence matrix* $\mathbf{H}$, where $\mathbf{H} = [\mathbf{h}_1, \mathbf{h}_2, \ldots, \mathbf{h}_n]^{\top}$, $n = |\mathcal{U}|$, $\mathbf{h}_i = (h_{i,1}, h_{i,2}, \ldots, h_{i,k})$ and $h_{i,j}$ denotes the confidence that $u_i \in \mathcal{U}$ is in cluster $U_j \in \mathcal{C}$. The optimal $\mathbf{H}$ that can minimize the normalized-cut cost can be obtained by solving the following objective function [45]:

$$\min_{\mathbf{H}} \ \text{Tr}(\mathbf{H}^{\top} \mathbf{L} \mathbf{H}),$$

$$s.t. \ \mathbf{H}^{\top} \mathbf{D} \mathbf{H} = \mathbf{I}. \tag{8.62}$$

where $\mathbf{L} = \mathbf{D} - \mathbf{S}$, diagonal matrix $\mathbf{D}$ has $D(i, i) = \sum_j S(i, j)$ on its diagonal, and $\mathbf{I}$ is an identity matrix.

8.4.5 Discrepancy Based Clustering of Multiple Networks

Besides the shared information due to common network construction purposes and similar network features [53], anchor users can also have unique information (e.g., social structures) across aligned networks, which can provide us with a more comprehensive knowledge about the community structures formed by these users. Meanwhile, by maximizing the consensus (i.e., minimizing the "*discrepancy*") of the clustering results about the anchor users in multiple partially aligned networks, model MCD will be able to refine the clustering results of the anchor users with information in other aligned networks mutually. We can represent the clustering results achieved in $G^{(1)}$ and $G^{(2)}$ as $\mathcal{C}^{(1)} = \{U_1^{(1)}, U_2^{(1)}, \ldots, U_{k^{(1)}}^{(1)}\}$ and $\mathcal{C}^{(2)} = \{U_1^{(2)}, U_2^{(2)}, \ldots, U_{k^{(2)}}^{(2)}\}$, respectively.

Let u_i and u_j be two anchor users in the network, whose accounts in $G^{(1)}$ and $G^{(2)}$ are $u_i^{(1)}, u_i^{(2)}, u_j^{(1)}$ and $u_j^{(2)}$, respectively. If users $u_i^{(1)}$ and $u_j^{(1)}$ are partitioned into the same cluster in $G^{(1)}$ but their corresponding accounts $u_i^{(2)}$ and $u_j^{(2)}$ are partitioned into different clusters in $G^{(2)}$, then it will lead to a *discrepancy* [37, 52] between the clustering results of $u_i^{(1)}, u_i^{(2)}, u_j^{(1)}$ and $u_j^{(2)}$ in aligned networks $G^{(1)}$ and $G^{(2)}$.

Definition 8.13 (Discrepancy) The discrepancy between the clustering results of u_i and u_j across aligned networks $G^{(1)}$ and $G^{(2)}$ is defined as the difference of confidence scores of u_i and u_j being partitioned in the same cluster across aligned networks. Considering that in the clustering results, the confidence scores of $u_i^{(1)}$ and $u_j^{(1)}$ ($u_i^{(2)}$ and $u_j^{(2)}$) being partitioned into $k^{(1)}$ ($k^{(2)}$) clusters can be represented as vectors $\mathbf{h}_i^{(1)}$ and $\mathbf{h}_j^{(1)}$ ($\mathbf{h}_i^{(2)}$ and $\mathbf{h}_j^{(2)}$), respectively, while the confidence that u_i and u_j are in the same cluster in $G^{(1)}$ and $G^{(2)}$ can be denoted as $\mathbf{h}_i^{(1)}(\mathbf{h}_j^{(1)})^\top$ and $\mathbf{h}_i^{(2)}(\mathbf{h}_j^{(2)})^\top$. Formally, the discrepancy of the clustering results about u_i and u_j is defined to be $d_{ij}(\mathcal{C}^{(1)}, \mathcal{C}^{(2)}) = \left(\mathbf{h}_i^{(1)}(\mathbf{h}_j^{(1)})^\top - \mathbf{h}_i^{(2)}(\mathbf{h}_j^{(2)})^\top\right)^2$ if u_i, u_j are both anchor users; and $d_{ij}(\mathcal{C}^{(1)}, \mathcal{C}^{(2)}) = 0$ otherwise. Furthermore, the discrepancy of $\mathcal{C}^{(1)}$ and $\mathcal{C}^{(2)}$ will be:

$$d(\mathcal{C}^{(1)}, \mathcal{C}^{(2)}) = \sum_i^{n^{(1)}} \sum_j^{n^{(2)}} d_{ij}(\mathcal{C}^{(1)}, \mathcal{C}^{(2)}), \tag{8.63}$$

where $n^{(1)} = |\mathcal{U}^{(1)}|$ and $n^{(2)} = |\mathcal{U}^{(2)}|$. In the definition, non-anchor users are not involved in the discrepancy calculation.

However, considering that $d(\mathcal{C}^{(1)}, \mathcal{C}^{(2)})$ is highly dependent on the number of anchor users and anchor links between $G^{(1)}$ and $G^{(2)}$, minimizing $d(\mathcal{C}^{(1)}, \mathcal{C}^{(2)})$ can favor highly consented clustering results when the anchor users are abundant but have no significant effects when the anchor users are very rare. To solve this problem, we propose to minimize the *normalized discrepancy* instead.

Definition 8.14 (Normalized Discrepancy) The normalized discrepancy measure computes the differences of clustering results in two aligned networks as a fraction of the discrepancy with regard to the number of anchor users across partially aligned networks:

$$nd(\mathcal{C}^{(1)}, \mathcal{C}^{(2)}) = \frac{d(\mathcal{C}^{(1)}, \mathcal{C}^{(2)})}{\left(|A^{(1,2)}|\right)\left(|A^{(1,2)}| - 1\right)}. \tag{8.64}$$

Optimal consensus clustering results of $G^{(1)}$ and $G^{(2)}$ will be $\hat{\mathcal{C}}^{(1)}, \hat{\mathcal{C}}^{(2)}$:

$$\hat{\mathcal{C}}^{(1)}, \hat{\mathcal{C}}^{(2)} = \arg \min_{\mathcal{C}^{(1)}, \mathcal{C}^{(2)}} nd(\mathcal{C}^{(1)}, \mathcal{C}^{(2)}). \tag{8.65}$$

Similarly, the normalized-discrepancy objective function can also be represented with the *clustering results confidence matrices* $\mathbf{H}^{(1)}$ and $\mathbf{H}^{(2)}$ as well. Meanwhile, considering that the networks studied in this section are partially aligned, matrices $\mathbf{H}^{(1)}$ and $\mathbf{H}^{(2)}$ contain the results of both anchor users and non-anchor users, while non-anchor users should not be involved in the discrepancy calculation according to the definition of discrepancy. We propose to prune the results of the non-anchor users with the following *anchor transition matrix* first.

Definition 8.15 (Anchor Transition Matrix) Binary matrix $\mathbf{T}^{(1,2)}$ (or $\mathbf{T}^{(2,1)}$) is defined as the anchor transition matrix from networks $G^{(1)}$ to $G^{(2)}$ (or from $G^{(2)}$ to $G^{(1)}$), where $\mathbf{T}^{(1,2)} = (\mathbf{T}^{(2,1)})^{\top}$, $\mathbf{T}^{(1,2)}(i, j) = 1$ if $(u_i^{(1)}, u_j^{(2)}) \in \mathcal{A}^{(1,2)}$ and 0 otherwise. The row indexes of $\mathbf{T}^{(1,2)}$ (or $\mathbf{T}^{(2,1)}$) are of the same order as those of $\mathbf{H}^{(1)}$ (or $\mathbf{H}^{(2)}$). Considering that the constraint on anchor links is "*one-to-one*" in this section, as a result, each row/column of $\mathbf{T}^{(1,2)}$ and $\mathbf{T}^{(2,1)}$ contains at most one entry filled with 1.

Example 8.6 In Fig. 8.5, we show an example about the clustering discrepancy of two partially aligned networks $G^{(1)}$ and $G^{(2)}$, users in which are grouped into two clusters $\{\{u_1, u_3\}, \{u_2\}\}$ and $\{\{u_A, u_C\}, \{u_B, u_D\}\}$, respectively. Users u_1, u_A and u_3, u_C are identified to be anchor users, based on which we can construct the "anchor transition matrices" $\mathbf{T}^{(1,2)}$ and $\mathbf{T}^{(2,1)}$ as shown in the upper right plot. Furthermore, based on the community structure, we can construct the "*clustering confidence matrices*" as shown in the lower left plot. To obtain the clustering results of anchor users only, the *anchor transition matrix* can be applied to prune the clustering results of non-anchor users from the *clustering confidence matrices*. By multiplying the *anchor transition matrices* $(\mathbf{T}^{(1,2)})^{\top}$ and $(\mathbf{T}^{(2,1)})^{\top}$ with *clustering confidence matrices* $\mathbf{H}^{(1)}$ and $\mathbf{H}^{(2)}$, respectively, we can obtain the "pruned confidence

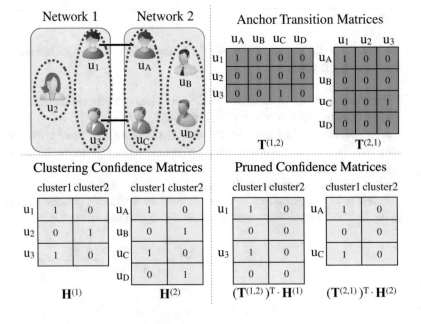

Fig. 8.5 An example to illustrate the clustering discrepancy

matrices" as shown in the lower right plot of Fig. 8.5. Entries corresponding anchor users u_1, u_3, u_A, and u_C are preserved but those corresponding to non-anchor users are all pruned.

In this example, the clustering discrepancy of the partially aligned networks should be 0 according to the above discrepancy definition. Meanwhile, networks $G^{(1)}$ and $G^{(2)}$ are of different sizes and the pruned confidence matrices are of different dimensions, e.g., $(\mathbf{T}^{(1,2)})^\top \mathbf{H}^{(1)} \in \mathbb{R}^{4 \times 2}$ and $(\mathbf{T}^{(2,1)})^\top \mathbf{H}^{(2)} \in \mathbb{R}^{3 \times 2}$. To represent the discrepancy with the clustering confidence matrices, we need to further accommodate the dimensions of different pruned *clustering confidence matrices*. It can be achieved by multiplying one pruned *clustering confidence matrices* with the corresponding *anchor transition matrix* again, which will not prune entries but only adjust the matrix dimensions. Let $\bar{\mathbf{H}}^{(1)} = (\mathbf{T}^{(1,2)})^\top \mathbf{H}^{(1)}$ and $\bar{\mathbf{H}}^{(2)} = (\mathbf{T}^{(1,2)})^\top (\mathbf{T}^{(2,1)})^\top \mathbf{H}^{(2)}$. In the example, we can represent the clustering discrepancy to be

$$\left\| \bar{\mathbf{H}}^{(1)} \left(\bar{\mathbf{H}}^{(1)} \right)^\top - \bar{\mathbf{H}}^{(2)} \left(\bar{\mathbf{H}}^{(2)} \right)^\top \right\|_F^2 = 0, \tag{8.66}$$

where matrix $\bar{\mathbf{H}}\bar{\mathbf{H}}^\top$ indicates whether pairs of anchor users are in the same cluster or not.

Furthermore, the objective function of inferring clustering confidence matrices, which can minimize the normalized discrepancy can be represented as follows:

$$\min_{\mathbf{H}^{(1)}, \mathbf{H}^{(2)}} \frac{\left\| \bar{\mathbf{H}}^{(1)} \left(\bar{\mathbf{H}}^{(1)} \right)^\top - \bar{\mathbf{H}}^{(2)} \left(\bar{\mathbf{H}}^{(2)} \right)^\top \right\|_F^2}{\left\| \mathbf{T}^{(1,2)} \right\|_F^2 \left(\left\| \mathbf{T}^{(1,2)} \right\|_F^2 - 1 \right)},$$

$$s.t. \ (\mathbf{H}^{(1)})^\top \mathbf{D}^{(1)} \mathbf{H}^{(1)} = \mathbf{I}, (\mathbf{H}^{(2)})^\top \mathbf{D}^{(2)} \mathbf{H}^{(2)} = \mathbf{I}. \tag{8.67}$$

where $\mathbf{D}^{(1)}, \mathbf{D}^{(2)}$ are the corresponding diagonal matrices of HNMP-Sim matrices of networks $G^{(1)}$ and $G^{(2)}$, respectively.

8.4.6 Joint Mutual Clustering of Multiple Networks

Normalized-Cut objective function favors clustering results that can preserve the characteristic of each network, however, normalized-discrepancy objective function favors consensus results which are mutually refined with information from other aligned networks. Taking both of these two issues into considerations, the optimal *Mutual Community Detection* results $\hat{\mathcal{C}}^{(1)}$ and $\hat{\mathcal{C}}^{(2)}$ of aligned networks $G^{(1)}$ and $G^{(2)}$ can be achieved as follows:

$$\arg \min_{\mathcal{C}^{(1)}, \mathcal{C}^{(2)}} \alpha \cdot ncut(\mathcal{C}^{(1)}) + \beta \cdot ncut(\mathcal{C}^{(2)}) + \theta \cdot nd(\mathcal{C}^{(1)}, \mathcal{C}^{(2)}) \tag{8.68}$$

where α, β, and θ represent the weights of these terms and, for simplicity, α, β are both set as 1 in MCD.

By replacing $ncut(C^{(1)})$, $cut(C^{(2)})$, $nd(C^{(1)}, C^{(2)})$ with the objective equations derived above, we can rewrite the joint objective function as follows:

$$\min_{\mathbf{H}^{(1)}, \mathbf{H}^{(2)}} \quad \alpha \cdot \mathrm{Tr}((\mathbf{H}^{(1)})^{\top} \mathbf{L}^{(1)} \mathbf{H}^{(1)}) + \beta \cdot \mathrm{Tr}((\mathbf{H}^{(2)})^{\top} \mathbf{L}^{(2)} \mathbf{H}^{(2)})$$

$$+ \theta \cdot \frac{\left\| \bar{\mathbf{H}}^{(1)} \left(\bar{\mathbf{H}}^{(1)}\right)^{\top} - \bar{\mathbf{H}}^{(2)} \left(\bar{\mathbf{H}}^{(2)}\right)^{\top} \right\|_F^2}{\left\| \mathbf{T}^{(1,2)} \right\|_F^2 \left(\left\| \mathbf{T}^{(1,2)} \right\|_F^2 - 1 \right)},$$

$$s.t. \quad (\mathbf{H}^{(1)})^{\top} \mathbf{D}^{(1)} \mathbf{H}^{(1)} = \mathbf{I}, \ (\mathbf{H}^{(2)})^{\top} \mathbf{D}^{(2)} \mathbf{H}^{(2)} = \mathbf{I}, \tag{8.69}$$

where $\mathbf{L}^{(1)} = \mathbf{D}^{(1)} - \mathbf{S}^{(1)}$, $\mathbf{L}^{(2)} = \mathbf{D}^{(2)} - \mathbf{S}^{(2)}$ and matrices $\mathbf{S}^{(1)}$, $\mathbf{S}^{(2)}$ and $\mathbf{D}^{(1)}$, $\mathbf{D}^{(2)}$ are the HNMP-Sim matrices and their corresponding diagonal matrices defined before.

The objective function is a complex optimization problem with orthogonality constraints, which can be very difficult to solve because the constraints are not only non-convex but also numerically expensive to preserve during iterations. Meanwhile, by substituting $(\mathbf{D}^{(1)})^{\frac{1}{2}} \mathbf{H}^{(1)}$ and $(\mathbf{D}^{(2)})^{\frac{1}{2}} \mathbf{H}^{(2)}$ with $\mathbf{X}^{(1)}$, $\mathbf{X}^{(2)}$, we can transform the objective function into a standard form of problems solvable with method proposed in [48]:

$$\min_{\mathbf{X}^{(1)}, \mathbf{X}^{(2)}} \quad \alpha \cdot \left(\mathrm{Tr}((\mathbf{X}^{(1)})^{\top} \tilde{\mathbf{L}}^{(1)} \mathbf{X}^{(1)}) + \beta \cdot \mathrm{Tr}((\mathbf{X}^{(2)})^{\top} \tilde{\mathbf{L}}^{(2)} \mathbf{X}^{(2)}) \right.$$

$$\left. + \theta \cdot \frac{\left\| \tilde{\mathbf{T}}^{(1)} \mathbf{X}^{(1)} \left(\tilde{\mathbf{T}}^{(1)} \mathbf{X}^{(1)}\right)^{\top} - \tilde{\mathbf{T}}^{(2)} \mathbf{X}^{(2)} \left(\tilde{\mathbf{T}}^{(2)} \mathbf{X}^{(2)}\right)^{\top} \right\|_F^2}{\left\| \mathbf{T}^{(1,2)} \right\|_F^2 \left(\left\| \mathbf{T}^{(1,2)} \right\|_F^2 - 1 \right)} \right),$$

$$s.t. \quad (\mathbf{X}^{(1)})^{\top} \mathbf{X}^{(1)} = \mathbf{I}, \ (\mathbf{X}^{(2)})^{\top} \mathbf{X}^{(2)} = \mathbf{I}. \tag{8.70}$$

where $\tilde{\mathbf{L}}^{(1)} = ((\mathbf{D}^{(1)})^{-\frac{1}{2}})^{\top} \mathbf{L}^{(1)} ((\mathbf{D}^{(1)})^{-\frac{1}{2}})$, $\tilde{\mathbf{L}}^{(2)} = ((\mathbf{D}^{(2)})^{-\frac{1}{2}})^{\top} \mathbf{L}^{(2)} ((\mathbf{D}^{(2)})^{-\frac{1}{2}})$ and $\tilde{\mathbf{T}}^{(1)} = (\mathbf{T}^{(1,2)})^{\top} (\mathbf{D}^{(1)})^{-\frac{1}{2}}$, $\tilde{\mathbf{T}}^{(2)} = (\mathbf{T}^{(1,2)})^{\top} (\mathbf{T}^{(2,1)})^{\top} (\mathbf{D}^{(2)})^{-\frac{1}{2}}$.

Wen et al. [48] propose a feasible method to solve the above optimization problems with a constraint-preserving update scheme. They propose to update one variable, e.g., $\mathbf{X}^{(1)}$, while fixing the other variable, e.g., $\mathbf{X}^{(2)}$, alternatively with the curvilinear search with Barzilai-Borwein step method until convergence. For example, when $\mathbf{X}^{(2)}$ is fixed, we can simplify the objective function into

$$\min_{\mathbf{X}} \mathcal{F}(\mathbf{X}), \ s.t. (\mathbf{X})^{\top} \mathbf{X} = \mathbf{I}, \tag{8.71}$$

where $\mathbf{X} = \mathbf{X}^{(1)}$ and $\mathcal{F}(\mathbf{X})$ is the objective function, which can be solved with the curvilinear search with Barzilai-Borwein step method proposed in [48] to update $\mathbf{X}$ until convergence and the variable $\mathbf{X}$ after the $(k+1)th$ iteration will be

$$\mathbf{X}_{k+1} = \mathbf{Y}(\tau_k), \ \mathbf{Y}(\tau_k) = \left(\mathbf{I} + \frac{\tau_k}{2} \mathbf{A} \right)^{-1} \left(\mathbf{I} - \frac{\tau_k}{2} \mathbf{A} \right) \mathbf{X}_k, \tag{8.72}$$

$$\mathbf{A} = \frac{\partial \mathcal{F}(\mathbf{X}_k)}{\partial \mathbf{X}} \mathbf{X}_k^{\top} - \mathbf{X}_k \left(\frac{\partial \mathcal{F}(\mathbf{X}_k)}{\partial \mathbf{X}} \right)^{\top}, \tag{8.73}$$

where let $\hat{\tau} = \left(\dfrac{\mathrm{Tr}((\mathbf{X}_k - \mathbf{X}_{k-1})^{\top}(\mathbf{X}_k - \mathbf{X}_{k-1}))}{\left| \mathrm{Tr}((\mathbf{X}_k - \mathbf{X}_{k-1})^{\top}(\nabla \mathcal{F}(\mathbf{X}_k) - \nabla \mathcal{F}(\mathbf{X}_{k-1}))) \right|} \right)$, $\tau_k = \hat{\tau}\delta^h$, δ is the Barzilai-Borwein step size and h is the smallest integer to make τ_k satisfy

$$\mathcal{F}\left(\mathbf{Y}(\tau_k) \right) \leq C_k + \rho \tau_k \mathcal{F}'_\tau \left(\mathbf{Y}(0) \right). \tag{8.74}$$

Terms C, Q are defined as $C_{k+1} = (\eta Q_k C_k + \mathcal{F}(\mathbf{X}_{k+1})) / Q_{k+1}$ and $Q_{k+1} = \eta Q_k + 1$, $Q_0 = 1$. More detailed derivatives of the curvilinear search method (i.e., Algorithm 2) with Barzilai-Borwein step are available in [48]. Meanwhile, the pseudo-code of method MCD is available in Algorithm 3. Based on the achieved solutions $\mathbf{X}^{(1)}$ and $\mathbf{X}^{(2)}$, we can get $\mathbf{H}^{(1)} = \left(\mathbf{D}^{(1)} \right)^{-\frac{1}{2}} \mathbf{X}^{(1)}$ and $\mathbf{H}^{(2)} = \left(\mathbf{D}^{(2)} \right)^{-\frac{1}{2}} \mathbf{X}^{(2)}$, from which we will be able to achieve the community structures of networks $G^{(1)}$ and $G^{(2)}$ by applying the KMeans algorithms on matrices $\mathbf{H}^{(1)}$ and $\mathbf{H}^{(2)}$.

Algorithm 2 Curvilinear Search Method ($\mathcal{CSM}$)

Require: $\mathbf{X}_k$ C_k, Q_k and function $\mathcal{F}$
   parameters $\epsilon = \{\rho, \eta, \delta, \tau, \tau_m, \tau_M\}$
Ensure: $\mathbf{X}_{k+1}, C_{k+1}, Q_{k+1}$
1: $\mathbf{Y}(\tau) = \left(\mathbf{I} + \frac{\tau}{2}\mathbf{A} \right)^{-1} \left(\mathbf{I} - \frac{\tau}{2}\mathbf{A} \right) \mathbf{X}_k$
2: **while** $\mathcal{F}\left(\mathbf{Y}(\tau) \right) \geq \mathbf{C}_k + \rho \tau \mathcal{F}' \left((\mathbf{Y}(0)) \right)$ **do**
3:    $\tau = \delta\tau$
4:    $\mathbf{Y}(\tau) = \left(\mathbf{I} + \frac{\tau}{2}\mathbf{A} \right)^{-1} \left(\mathbf{I} - \frac{\tau}{2}\mathbf{A} \right) \mathbf{X}_k$
5: **end while**
6: $\mathbf{X}_{k+1} = \mathbf{Y}_k(\tau)$
   $Q_{k+1} = \eta Q_k + 1$
   $C_{k+1} = (\eta Q_k C_k + \mathcal{F}(\mathbf{X}_{k+1})) / Q_{k+1}$
   $\tau = \max\left(\min(\tau, \tau_M), \tau_m \right)$

Algorithm 3 Mutual Community Detector (MCD)

Require: aligned network: $\mathcal{G} = \{\{G^{(1)}, G^{(2)}\}, \{A^{(1,2)}, A^{(2,1)}\}\}$;
   number of clusters in $G^{(1)}$ and $G^{(2)}$: $k^{(1)}$ and $k^{(2)}$;
   HNMP Sim matrices weight: ω;
   parameters: $\epsilon = \{\rho, \eta, \delta, \tau, \tau_m, \tau_M\}$;
   function $\mathcal{F}$ and consensus term weight θ
Ensure: $\mathbf{H}^{(1)}$, $\mathbf{H}^{(2)}$
1: Calculate HNMP Sim matrices, $\mathbf{S}_i^{(1)}$ and $\mathbf{S}_i^{(2)}$
2: $\mathbf{S}^{(1)} = \sum_i \omega_i S_i^{(1)}$, $\mathbf{S}^{(2)} = \sum_i \omega_i S_i^{(2)}$
3: Initialize $\mathbf{X}^{(1)}$ and $\mathbf{X}^{(2)}$ with Kmeans clustering results on $\mathbf{S}^{(1)}$ and $\mathbf{S}^{(2)}$
4: Initialize $C_0^{(1)} = 0$, $Q_0^{(1)} = 1$ and $C_0^{(2)} = 0$, $Q_0^{(2)} = 1$
5: $converge = False$
6: **while** $converge = False$ **do**
7:    /* update $\mathbf{X}^{(1)}$ and $\mathbf{X}^{(2)}$ with $\mathcal{CSM}$ */
      $\mathbf{X}_{k+1}^{(1)}, C_{k+1}^{(1)}, Q_{k+1}^{(1)} = \mathcal{CSM}(\mathbf{X}_k^{(1)}, C_k^{(1)}, Q_k^{(1)}, \mathcal{F}, \epsilon)$
      $\mathbf{X}_{k+1}^{(2)}, C_{k+1}^{(2)}, Q_{k+1}^{(2)} = \mathcal{CSM}(\mathbf{X}_k^{(2)}, C_k^{(2)}, Q_k^{(2)}, \mathcal{F}, \epsilon)$
8:    **if** $\mathbf{X}_{k+1}^{(1)}$ and $\mathbf{X}_{k+1}^{(2)}$ both converge **then**
9:       $converge = True$
10:   **end if**
11: **end while**
12: $\mathbf{H}^{(1)} = \left((\mathbf{D}^{(1)})^{-\frac{1}{2}} \right)^T \mathbf{X}^{(1)}$, $\mathbf{H}^{(2)} = \left((\mathbf{D}^{(2)})^{-\frac{1}{2}} \right)^T \mathbf{X}^{(2)}$

8.5   Large-Scale Network Synergistic Community Detection

The community detection algorithm proposed in the previous section involves very complicated matrix operations, and works well for small-sized network data. However, when being applied to handle real-world online social networks involving millions even billions of users, they will suffer from the time complexity problem a lot. In this section, we will introduce a synergistic community detection algorithm SPMN [16] for multiple large-scale aligned online social networks.

8.5.1   Problem Formulation

The problem to be introduced here follows the same formulation as the one introduced in Sect. 8.4, but the involved networks are of far larger sizes in terms of both node number and the social connection number. Synergistic partitioning across multiple large-scale social networks is very difficult for the following challenges:

- *Social Network*: Distinct from generic data, network structured data usually contain intricate interactions. In addition, the multiple heterogeneous networks formulation also renders the cross-network relationships to be an important part in considerations.
- *Network Scale*: Network size implies it is difficult for stand-alone programs to apply traditional partitioning methods and it is a difficult task to parallelize the existing stand-alone network partitioning algorithms.
- *Distributed Framework*: For distributed algorithms, load balance should be taken into considerations and how to generate balanced partitions is another challenge.

To address the challenges, in this section, we will introduce a network structure based distributed network partitioning framework, namely SPMN [16]. The SPMN model identifies the anchor nodes among the multiple networks, and selects one network as the datum network, then divides it into k balanced partitions and generate ⟨anchor node ID, partition ID⟩ pairs as the main objective. Based on the objective, SPMN coarsens the other aligned networks into smaller ones, which will further divide the smallest networks into k balanced initial partitions, and tries to assign same kinds of anchor nodes into the same initial partition as many as possible. Here, anchor nodes of same kind means that they are divided into the same partition in the datum network. Finally, SPMN projects the initial partitions back to the original networks.

8.5.2   Distributed Multilevel k-Way Partitioning

In this section, we describe the heuristic based framework for synergistic partitioning among multiple large-scale social networks, and we call the framework as SPMN. For large-sized networks, data processing in SPMN can be roughly divided into two stages: datum generation stage and network alignment stage.

When getting the anchor node set $\mathcal{A}^{(1,2)}$ between networks $G^{(1)}$ and $G^{(2)}$, the SPMN framework will apply a distributed multilevel k-way partitioning method onto the datum network to generate k balanced partitions. During this process, the anchor nodes are ignored and all the nodes are treated identically. We call this process as the datum generation stage. When finished, partition result of anchor nodes will be generated, and SPMN stores them in a set-$Map\langle anidx, pidx\rangle$, where $anidx$ is anchor node ID and $pidx$ represents the partition ID the anchor node belongs to. After the datum generation stage, synergistic networks will be partitioned into k partitions according to the $Map\langle anidx, pidx \rangle$ to make the synergistic networks to align to the datum network, and during this process *discrepancy* and *cut* are the objectives to be minimized. We call this process as the network alignment stage.

Algorithms guaranteed to find out near-optimal partitions in a single network have been studied for a long period. But most of the methods are stand-alone, and their performance is limited by the server's capacity. Inspired by the multilevel k-way partitioning (MKP) method proposed by Karypis and Kumar [18, 19] and based on our previous work [1], SPMN uses MapReduce [9] to speed up the MKP method. As the same with other multilevel methods, MapReduce based MKP also includes three phases: coarsening, initial partitioning, and un-coarsening.

Coarsening phase is a multilevel process and a sequence of smaller approximate networks $G_i = (\mathcal{V}_i, \mathcal{E}_i)$ are constructed from the original network $G_0 = (\mathcal{V}, \mathcal{E})$ and so forth, where $|\mathcal{V}_i| < |\mathcal{V}_{i-1}|, i \in \{1, 2, \ldots, n\}$. To construct the coarser networks, node combination and edge collapsing should be performed. The task can be formally defined in terms of matching inside the networks [5]. An intra-network matching can be represented as a set of node pairs $\mathcal{M} = \{(v_i, v_j)\}, i \neq j$ and $(v_i, v_j) \in \mathcal{E}$, in which each node can only appear for no more than once. For a network G_i with a matching $\mathcal{M}_i$, if $(v_j, v_k) \in \mathcal{M}_i$ then v_j and v_k will form a new node $v_q \in \mathcal{V}_{i+1}$ in network G_{i+1} coarsen from G_i. All the links connected to v_j or v_k in G_i will be connected to v_q in G_{i+1}. The total weight of edges and number of nodes will be greatly reduced. Let's define $W(\cdot)$ to be the sum of edge weight in the input set and $N(\cdot)$ to be the number of nodes/components in the input set. In the coarsening process, we have

$$W(\mathcal{E}_{i+1}) = W(\mathcal{E}_i) - W(\mathcal{M}_i), \qquad (8.75)$$

$$N(\mathcal{V}_{i+1}) = N(\mathcal{V}_i) - N(\mathcal{M}_i). \qquad (8.76)$$

Analysis in [17] shows that for the same coarser network, smaller edge-weight corresponds to smaller edge-cut. With the help of MapReduce framework, SPMN uses a local search method to implement an edge-weight based matching (EWM) scheme to collect larger edge weight during the coarsening phase. For the convenience of MapReduce, SPMN designs an emerging network representation format: each line contains essential information about a node and all its neighbors (NN), such as node ID and edge weight (W). The whole network data are distributed in distributed file system, such as HDFS [39], and each data block only contains a part of node set and corresponding connection information. Function $map()$ takes a data block as input and searches locally to find node pairs to match according to the edge weight. Function $reduce()$ is in charge of node combination, renaming and sorting. With the new node IDs and matching, a simple MapReduce job will be able to update the edge information and write the coarser network back onto HDFS. The complexity of EWM is $O(|\mathcal{E}|)$ in each iteration and the pseudo code about EWM is shown in Algorithm 4.

After several iterations, a coarsest weighted network G_s consisting of only hundreds of nodes will be generated. For the network size of G_s, stand-alone algorithms with high computing complexity will be acceptable for initial partitioning. Meanwhile, the weight of edges of coarser networks is set to reflect the weights of the finer network during the coarsening phase, so G_s contains sufficient information to intelligently satisfy the balanced partition and the minimum edge-cut requirements. Plenty of traditional bisection methods are quite qualified for the task. In SPMN, it adopts the KL method with an $O(|\mathcal{E}|^3)$ computing complexity to divide G_s into two partitions and then take recursive invocations of KL method on the partitions to generate balanced k partitions.

Algorithm 4 Edge Weight Based Matching ($\mathcal{EWM}$)

Require: Network G_h
           Maximum weight of a node $maxVW = n/k$
Ensure: A coarser network G_{h+1}
 1: **map**() Function:
 2: **for** node i in current data block **do**
 3:    **if** $match[i] == -1$ **then**
 4:       $maxIdx = -1$
 5:       $sortByEdgeWeight(NN(i))$
 6:       **for** $v_j \in NN(i)$ **do**
 7:          **if** $match[j] == -1$ and $VW(i) + VW(j) < maxVW$ **then**
 8:             maxIdx = j
 9:          **end if**
10:          $match[i] = maxIdx$
11:          $match[maxIdx] = i$
12:       **end for**
13:    **end if**
14: **end for**
15: **reduce**() Function:
16: new $newNodeID[n+1]$
17: new $newVW[n+1]$
18: set $idx = 1$
19: **for** $i \in \{1, 2, \cdots, n\}$ **do**
20:    **if** $i < match[i]$ **then**
21:       set $newNodeID[match[i]] = idx$
22:       set $newNodeID[i] = idx$
23:       set $newVW[i] = newVW[match[i]] = VW(i) + VW(match[i[)$
24:       $idx++$
25:    **end if**
26: **end for**

Un-coarsening phase is the inverse processing of the coarsening phase. With the initial partitions and the matching of the coarsening phase, it is easy to run the un-coarsening process on the MapReduce cluster.

8.5.3   Distributed Synergistic Partitioning Process

In this part, we will talk about the synergistic partitioning process in SPMN based on the synergistic networks with the knowledge of partition results of anchor nodes from datum network. The synergistic partitioning is also an MKP process but quite different from the general MKP methods.

In the coarsening phase, anchor nodes are endowed with a higher priority than non-anchor nodes. When choosing nodes to pair, SPMN assumes that anchor nodes and non-anchor nodes have different tendencies. Let G^d be the datum network. For an anchor node v_i in another aligned networks, at the top of its preference list, it would like to be matched with another anchor node v_i, which has the same partition ID in the datum network, i.e., $pidx(G^d, v_i) = pidx(G^d, v_j)$ (here $pidx(G^d, v_i)$ denotes the community label that v_j belongs to in G^d). Second, if there is no appropriate anchor node, it would try to find a non-anchor node to pair. When planning to find a non-anchor node to pair, the anchor node, assuming to be v_i, would like to find a correct direction, and it would prefer to match with the non-anchor node v_j, which has lots of anchor nodes as neighbors with the same $pidx$ with v_i. When being matched together, the new node will be given the same $pidx$ as the anchor node. To improve

Fig. 8.6 An example of synergistic partition process. In coarsening phase, the networks are stored in two servers, $V_1^i = \{v^i(j)|j \leq |V^i|/2\}$ are stored on a sever and the others are on the other server. Anchor nodes are with colors, and different colors represent different partitions. Node pairs encircled by dotted chains represent the matchings. Numbers on chains mean the order of pairing

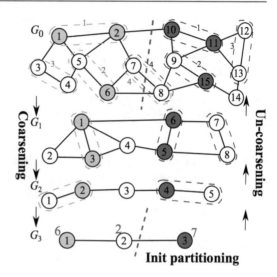

the accuracy of synergistic partitioning among multiple social networks, an anchor node will never try to combine with another anchor node with different $pidx$.

For a non-anchor node, it would prefer to be matched with an anchor node neighbor which belongs to the dominant partition in the non-anchor node's neighbors. Here, dominant partition in a node's neighbors means the number of anchor nodes with this partition ID is the largest. Next, a non-anchor node would choose a general non-anchor node to pair with. At last, a non-anchor node would not like to combine with an anchor node being part of the partitions which are in subordinate status. After being combined together, the new node will be given the same $pidx$ as the anchor node. To ensure the balance among the partitions, about $\frac{1}{3}$ of the nodes in the coarsest network are unlabeled.

Example 8.7 In Fig. 8.6, we show a diagrammatic sketch of the synergistic partitioning process. Take network G_0 for example, nodes $v_1 - v_7$ and the corresponding links information are stored on the same server and the other nodes are stored on another server. Nodes in pairs (v_1, v_2) and (v_{10}, v_{11}) are all with the same pidx, so they should be tackled first. v_6 and v_{15} choose a correct direction to make a pair. Then, v_3 cannot pair with v_1, so it chooses to combine with v_4. Finally, after searching locally, v_7 cannot find a local neighbor to pair, but has to make a pair with its remote neighbor v_8.

In addition to minimizing both the discrepancy and cut loss terms simultaneously, SPMN also tries to balance the size of partitions are the objectives in synergistic partitioning process. However, it is impossible to achieve them simultaneously. So, SPMN tries to make a compromise among them and develop a heuristic method to tackle the problems.

- First, according to the conclusion, smaller edge-weight corresponds to smaller edge-cut and the pairing tendencies, SPMN proposes a modified EWM (MEWM) method to find a matching in the coarsening phase, of which the edge-weight is as large as possible. At the end of the coarsening phase, there is no impurity in any node, meaning that each node contains no more than one type of anchor nodes. Besides, a "*purity*" vector attribute and a $pidx$ attribute are added to each node to represent the percentage of each kind of anchor nodes swallowed up by it and the $pidx$ of the new node, respectively.
- Then, during the initial partitioning phase, SPMN treats the anchor nodes as labeled nodes and uses a modified label propagation algorithm to deal with the non-anchor nodes in the coarsest network.

Algorithm 5 Synergistic Partitioning ($\mathcal{SP}$)

Require: Network G_h
           Anchor Link Map $Map < anidx, pidx >$
           Maximum weight of a node $maxVW = n/k$
Ensure: A coarser network G_{h+1}
 1: Call Synergistic Partitioning-Map Function
 2: Call Synergistic Partitioning-Reduce Function

- At the end of the initial partitioning phase, SPMN will be able to generate balanced k partitions and to maximize the number of same kind of anchor nodes being divided into the same partitions.
- Finally, SPMN projects the coarsest network back to the original network, which is the same as the traditional MKP process.

The pseudo code of the coarsening phase in synergistic partitioning is available in Algorithm 5, which will call the $Map()$ and $Reduce()$ functions in Algorithms 6 and 7, respectively.

8.6   Summary

In this chapter, we focused on the *community detection* problem in online social networks. Community detection has been demonstrated as an important research problem especially for online social networks. We summarized several existing community detection methods for one single homogeneous networks, which are based *node proximity*, *community modularity maximization*, and *spectral clustering*, respectively. These single-homogeneous network community detection methods provide the basis for addressing the problem in more complicated problem settings, e.g., *heterogeneous networks* and *multiple networks*.

We introduced a novel problem setting based on multiple aligned heterogeneous social networks, i.e., the *emerging network community detection*, where the target network lacks sufficient information for detecting effective community structure in it. To address the problem, we introduced an effective and efficient method to compute the intimacy scores among users based on the heterogeneous information available across the social networks. Information from different sources are assigned with different weights, whose values can be adjusted automatically based on a weight selection algorithm.

We also talked about the *mutual community detection* problem of multiple aligned heterogeneous social networks, where information across these networks can be utilized for mutual community structure refinement. To preserve the characteristics of the social networks, we partitioned each social network by minimizing the normalized-cut metric. Meanwhile, to transfer useful knowledge across the networks, we introduced a new metric, i.e., *discrepancy*, minimization of which allows the information from different networks to refine the communities detected from each social network.

Finally, at the end of this chapter, we introduced the *synergistic community detection* of large-scale social networks involving millions even billions of users. We proposed to identify the anchor nodes among the multiple networks, and select one network as the datum network, then divide it into k balanced partitions. By coarsening the other aligned networks into smaller ones, we introduced to further divide the smallest networks into k balanced initial partitions, and try to assign same kinds of anchor nodes into the same initial partition as many as possible. Finally, we proposed to project the initial partitions back to the original networks.

Algorithm 6 Synergistic Partitioning-Map

Require: Network G_h
        Anchor Link Map $Map < anidx, pidx >$
        Maximum weight of a node $maxVW = n/k$
Ensure: A coarser network G_{h+1}
1: **map()** Function:
2: **for** node i in current data block **do**
3:    **if** $match[i] == -1$ **then**
4:      set $flag = false$
5:      $sortByEdgeWeight(NN(i))$
6:      **if** $v_i \in Map < anidx, pidx >$ **then**
7:        **for** $v_j \in NN(i)$ & $match[j] == -1$ **do**
8:          **if** $v_j \in Map < anidx, pidx >$ & $Map.get(v_i) == Map.get(v_j)$ & $VW(i) + VW(j) < maxVW$
            then
9:            $match[i] = j, match[j] = i$
10:            $flag = true$, break
11:          **end if**
12:        **end for**
13:        **if** $flag == false$, no suitable anchor node **then**
14:          **for** $v_j \in NN(i)$ & $match[j] == -1$ & $VW(v_i) + VW(v_j) < maxVW$ **do**
15:            $indirectNeighbor = NN(v_j)$
16:            $sortByEdgeWeight(NN(i))$
17:            **for** $v_k \in indirectNeighbor$ **do**
18:              **if** $v_k \in Map < anidx, pidx >$ & $Map.get(v_i) == Map.get(v_k)$ **then**
19:               $match[i] = j, match[j] = i$
20:               $flag = true$, break
21:              **end if**
22:            **end for**
23:            **if** $flag == true$ **then**
24:              break
25:            **end if**
26:          **end for**
27:        **end if**
28:      **else**
29:        $sortByEdgeWeight(NN(i))$
30:        **for** $v_j \in NN(v_i)$ & $v_j \notin Map < anidx, pidx >$ & $VW(i) + VW(j) < maxVW$ & $match[j] == -1$
         do
31:          $match[i] = j, match[j] = i$, break
32:        **end for**
33:      **end if**
34:    **end if**
35: **end for**

Three novel community detection algorithms CAD, MCD, and SPMN have also been introduced in great detail in this section. These three proposed community detection algorithms learn the social community structures of the multiple aligned networks with the (strong/weak) supervision of anchor links, based on the assumption that anchor users tend to be involved into relatively similar communities in different networks. Meanwhile, they also take considerations of the network properties at the same time, where each social network can maintain their characteristics as well.

Algorithm 7 Synergistic Partitioning-Reduce

Require: Network G_h
      Anchor Link Map $Map < anidx, pidx >$
      Maximum weight of a node $maxVW = n/k$
Ensure: A coarser network G_{h+1}
1: **reduce**() Function:
2: new $newNodeID[n+1]$
3: new $newVW[n+1]$
4: set $idx = 1$
5: **for** $i \in newNodeID[]$ **do**
6:      **if** $i < match[i]$ **then**
7:          set $newNodeID[match[i]] = idx$
8:          set $newNodeID[i] = idx$
9:          set $newVW[i] = newVW[match[i]] = VW(i) + VW(match[i[)$
10:          $idx++$
11:      **end if**
12: **end for**
13: new $newPurity[idx+1]$
14: new $newPidx[idx+1]$
15: **for** $i \in [1, idx]$ **do**
16:      $newPurity[i] = \frac{purity[i]*VW(i)+purity[j]*VW(j)}{VW(i)+VW(j)}$
17:      $newPidx[i] = \max\{pidx[i], pidx[match[i]]\}$
18: **end for**

8.7   Bibliography Notes

Clustering aims at grouping similar objects in the same cluster and many different clustering methods have also been proposed. One type is the hierarchical clustering methods [14], which include agglomerative hierarchical clustering methods [8] and divisive hierarchical clustering methods [8]. Another type is the partition-based clustering methods, which include K-means for instances with numerical attributes [15].

In addition, clustering is also a very broad research area, which include various types of clustering problems, e.g., consensus clustering [26, 27], multi-view clustering [4, 6], multi-relational clustering [49], co-training based clustering [21], and dozens of papers have been published on these topics. Lourenco et al. [27] propose a probabilistic consensus clustering method by using evidence accumulation. Lock et al. propose a Bayesian consensus clustering method in [26]. Meanwhile, Bickel et al. [4] propose to study the multi-view clustering problem, where the attributes of objects are split into two independent subsets. Cai et al. [6] propose to apply multi-view K-Means clustering methods to big data. Yin et al. [49] propose a user-guided multi-relational clustering method, CrossClus, to perform multi-relational clustering under user's guidance. Kumar et al. propose to address the multi-view clustering problem based on a co-training setting in [21].

Clustering based community detection in online social networks is a hot research topic and many different techniques have been proposed to optimize certain measures of the results, e.g., modularity function [30] and normalized cut [38]. Malliaros et al. give a comprehensive survey of correlated techniques used to detect communities in networks in [28] and a detailed tutorial on spectral clustering has been given by Luxburg in [45]. These works are mostly studied based on homogeneous social networks.

In recent years, many community detection works have been done on heterogeneous online social networks. Zhou et al. [63] propose to do graph clustering with relational and attribute information simultaneously. Zhou et al. [62] propose a social influence based clustering method for heterogeneous information networks. Some other works have also been done on clustering with incomplete data. Sun et al. [42] propose to study the clustering problem with complete link information but incomplete attribute information. Lin et al. [25] try to detect the communities in networks with incomplete relational information but complete attribute information.

8.8   Exercises

1. (Easy) Given a network as shown in Fig. 8.7, please find all the nodes which are *structural equivalent*.
2. (Easy) Let $\mathcal{C} = \{\mathcal{U}_1, \mathcal{U}_2\}$ denote the community detected from network in Fig. 8.7, where $\mathcal{U}_1 = \{1, 2, 3\}$ and $\mathcal{U}_2 = \{4, 5, 6, 7, 8, 9\}$. Please compute the community strength of communities $\mathcal{U}_1$ and $\mathcal{U}_2$, respectively, as well as the *modularity* of $\mathcal{C}$.
3. (Easy) Please find all the potential community partition of the network in Fig. 8.7 that can introduce the minimum *cut* costs.
4. (Easy) Given the mutual community detection result in Fig. 8.8, please compute its *normalized discrepancy*.
5. (Easy) Please justify the coarsening process from G_1 to G_2 in Fig. 8.6.
6. (Medium) Given two different partitions (i.e., plots (a) and (b) in Fig. 8.9) of the input social network shown in Fig. 8.7, please justify which partition achieves better community detection results based on *cut*, *ratio-cut*, and *normalized-cut* costs, respectively.
7. (Medium) Given the network in Fig. 8.7, please implement the algorithm introduced in Sect. 8.2.3 to identify the optimal community structure based on *normalized cut* cost, where community number $k = 2$.

Fig. 8.7 An example of homogeneous network

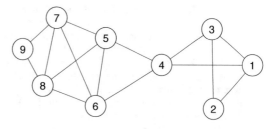

Fig. 8.8 Mutual community detection example

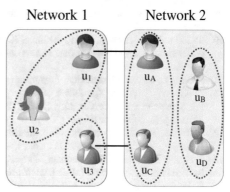

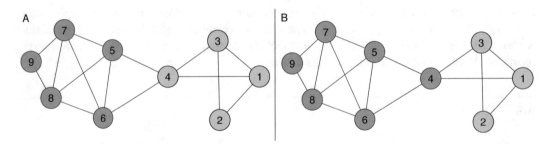

Fig. 8.9 Two potential community detection results

8. (Medium) Please prove Lemma 8.1.
9. (Hard) Please implement the MCD algorithm introduced in Sect. 8.4.
10. (Hard) Please introduce a new aligned heterogeneous network community detection algorithm based on the Gaussian Mixture Model.

References

1. C. Aggarwal, Y. Xie, P. Yu, Gconnect: a connectivity index for massive disk-resident graphs. Proc. VLDB Endow. **2**(1), 862–873 (2009)
2. A. Arenas, L. Danon, A. Díaz-Guilera, P.M. Gleiser, R. Guimerá, Community analysis in social networks. Eur. Phys. J. B **38**(2), 373–380 (2004)
3. I. Bhattacharya, L. Getoor, Relational clustering for multi-type entity resolution, in *Proceedings of the 4th International Workshop on Multi-Relational Mining* (ACM, New York, 2005), pp. 3–12
4. S. Bickel, T. Scheffer, Multi-view clustering, in *Fourth IEEE International Conference on Data Mining (ICDM'04)*, vol 4 (ACM, New York, 2004), pp. 19–26
5. T. Bui, C. Jones, A heuristic for reducing fill-in in sparse matrix factorization, in *Proceedings of the Sixth SIAM Conference on Parallel Processing for Scientific Computing, PPSC 1993, Norfolk, Virginia* (1993), pp. 445–452
6. X. Cai, F. Nie, H. Huang, Multi-view k-means clustering on big data, in *Twenty-Third International Joint Conference on Artificial Intelligence* (2013), pp. 2598–2604
7. W. Cheng, X. Zhang, Z. Guo, Y. Wu, P. Sullivan, W. Wang, Flexible and robust co-regularized multi-domain graph clustering, in *Proceedings of the 19th ACM SIGKDD International Conference on Knowledge Discovery and Data Mining* (ACM, New York, 2013), pp. 320–328
8. P. Cimiano, A. Hotho, S. Staab, Comparing conceptual, divisive and agglomerative clustering for learning taxonomies from text, in *Proceedings of the 16th European Conference on Artificial Intelligence (ECAI'2004)* (2004), pp. 435–439
9. J. Dean, S. Ghemawat, MapReduce: simplified data processing on large clusters. Commun. ACM **51**(1), 107–113 (2008)
10. M. Eslami, A. Aleyasen, R. Moghaddam, K. Karahalios, Friend grouping algorithms for online social networks: preference, bias, and implications, in *International Conference on Social Informatics* (Springer, Cham, 2014), pp. 34–49
11. P. Gács, L. Lovász, Complexity of algorithms. Lect. Notes (1999)
12. A. Goder, V. Filkov, Consensus clustering algorithms: comparison and refinement, in *Proceedings of the 9th Workshop on Algorithm Engineering and Experiments* (Society for Industrial and Applied Mathematics, Philadelphia, 2008), pp. 109–117
13. M. Hasan, M.J. Zaki, A survey of link prediction in social networks, in *Social Network Data Analytics*, ed. by C.C. Aggarwal, (Springer, Boston, 2011), pp. 243–275
14. T. Hastie, R. Tibshirani, J. Friedman, Hierarchical clustering, in *The Elements of Statistical Learning*, 2nd edn., ed. by T. Hastie, R. Tibshirani, J. Friedman (Springer, New York, 2009), pp. 520–528
15. Z. Huang, Extensions to the k-means algorithm for clustering large data sets with categorical values. Data Min. Knowl. Disc. **2**(3), 283–304 (1998)
16. S. Jin, J. Zhang, P. Yu, S. Yang, A. Li, Synergistic partitioning in multiple large scale social networks, in *IEEE BigData* (IEEE, Piscataway, 2014), pp. 281–290

17. G. Karypis, V. Kumar, Analysis of multilevel graph partitioning, in *Supercomputing* (IEEE, Piscataway, 1995), p. 29
18. G. Karypis, V. Kumar, Parallel multilevel k-way partitioning scheme for irregular graphs, in *Proceedings of the 1996 ACM/IEEE Conference on Supercomputing (Supercomputing '96)* (IEEE, Piscataway, 1996)
19. G. Karypis, V. Kumar, Multilevel k-way partitioning scheme for irregular graphs. J. Parallel Distrib. Comput. **20**(1), 359–392 (1998)
20. X. Kong, J. Zhang, P. Yu, Inferring anchor links across multiple heterogeneous social networks, in *Proceedings of the 22nd ACM International Conference on Information & Knowledge Management* (ACM, New York, 2013), pp. 179–188
21. A. Kumar, H. Daumé, A co-training approach for multi-view spectral clustering, in *Proceedings of the 28th International Conference on Machine Learning (ICML-11)* (2011), pp. 393–400
22. J. Leskovec, K. Lang, M. Mahoney, Empirical comparison of algorithms for network community detection, in *Proceedings of the 19th International Conference on World Wide Web* (ACM, New York, 2010), pp. 631–640
23. T. Li, C. Ding, M.I. Jordan, Solving consensus and semi-supervised clustering problems using nonnegative matrix factorization, in *Seventh IEEE International Conference on Data Mining (ICDM 2007)* (IEEE, Piscataway, 2007), pp. 577–582
24. D. Liben-Nowell, J. Kleinberg, The link-prediction problem for social networks. J. Am. Soc. Inf. Sci. Technol. **58**(7), 1019–1031 (2007)
25. W. Lin, X. Kong, P. Yu, Q. Wu, Y. Jia, C. Li, Community detection in incomplete information networks, in *Proceedings of the 21st International Conference on World Wide Web* (ACM, New York, 2012), pp. 341–350
26. E.F. Lock, D.B. Dunson, Bayesian consensus clustering. Bioinformatics **29**(20), 2610–2616 (2013)
27. A. Lourenço, S.R. Bulò, N. Rebagliati, A.L.N. Fred, M.A.T. Figueiredo, M. Pelillo, Probabilistic consensus clustering using evidence accumulation. Mach. Learn. **98**(1–2), 331–357, (2013)
28. F.D. Malliaros, M. Vazirgiannis, Clustering and community detection in directed networks: a survey. CoRR, abs/1308.0971, abs/1308.0971 (2013)
29. M. Newman, Modularity and community structure in networks. Proc. Natl. Acad. Sci. U. S. A. **103**(23), 8577–8582 (2006)
30. M.E.J. Newman, M. Girvan, Finding and evaluating community structure in networks. Phys. Rev. E **69**(2), 026113 (2004)
31. N. Nguyen, R. Caruana, Consensus clusterings, in *Seventh IEEE International Conference on Data Mining (ICDM 2007)* (IEEE, Piscataway, 2007), pp. 607–612
32. S. Pan, Q. Yang, A survey on transfer learning. IEEE Trans. Knowl. Data Eng. **22**(10), 1345–1359 (2010)
33. R. Panigrahy, M. Najork, Y. Xie, How user behavior is related to social affinity, in *Proceedings of the Fifth ACM International Conference on Web Search and Data Mining (WSDM '12)* (ACM, New York, 2012), pp. 713–722
34. P. Petersen, Linear Algebra (Springer, New York, 2012)
35. M. Richardson, P. Domingos, Mining knowledge-sharing sites for viral marketing, in *Proceedings of the Eighth ACM SIGKDD International Conference on Knowledge Discovery and Data Mining (KDD '02)* (ACM, New York, 2002), pp. 61–70
36. R. Roman, Community-based recommendations to improve intranet users' productivity, Master's thesis (2016)
37. W. Shao, J. Zhang, L. He, P. Yu, Multi-source multi-view clustering via discrepancy penalty, in *2016 International Joint Conference on Neural Networks (IJCNN)* (IEEE, Piscataway, 2016), pp. 2714–2721
38. J. Shi, J. Malik, Normalized cuts and image segmentation. IEEE Trans. Pattern Anal. Mach. Intell. **22**(8), 888–905 (2000)
39. K. Shvachko, H. Kuang, S. Radia, R. Chansler, The Hadoop distributed file system, in *2010 IEEE 26th Symposium on Mass Storage Systems and Technologies (MSST)* (IEEE, Piscataway, 2010), pp. 1–10
40. Y. Sun, Y. Yu, J. Han, Ranking-based clustering of heterogeneous information networks with star network schema, in *Proceedings of the 15th ACM SIGKDD International Conference on Knowledge Discovery and Data Mining* (ACM, New York, 2009), pp. 797–806
41. Y. Sun, J. Han, X. Yan, P. Yu, T. Wu, PathSim: meta path-based top-k similarity search in heterogeneous information networks. Proc. VLDB Endow. **4**(11), 992–1003 (2011)
42. Y. Sun, C. Aggarwal, J. Han, Relation strength-aware clustering of heterogeneous information networks with incomplete attributes. Proc. VLDB Endow. **5**(5), 394–405 (2012)
43. J. Tang, H. Gao, X. Hu, H. Liu, Exploiting homophily effect for trust prediction, in *Proceedings of the Sixth ACM International Conference on Web Search and Data Mining* (ACM, New York, 2013), pp. 53–62
44. M. Trusov, A. Bodapati, R. Bucklin, Determining influential users in internet social networks. J. Mark. Res. **47**(4), 643–658 (2010)
45. U. von Luxburg, A tutorial on spectral clustering. Stat. Comput. **17**(4) (2007). arXiv:0711.0189
46. X. Wang, G. Chen, Complex networks: small-world, scale-free and beyond. IEEE Circuits Syst. Mag. **3**(1), 6–20 (2003)

47. S. Wang, Z. Zhang, J. Li, A scalable cur matrix decomposition algorithm: lower time complexity and tighter bound. Mach. Learn. (2012). arXiv:1210.1461

48. Z. Wen, W. Yin, A feasible method for optimization with orthogonality constraints, Technical report, Rice University (2010)

49. X. Yin, J. Han, P. Yu, CrossClus: user-guided multi-relational clustering. Data Min. Knowl. Disc. **15**(3), 321–348 (2007)

50. Q. Zhan, J. Zhang, S. Wang, P. Yu, J. Xie, Influence maximization across partially aligned heterogeneous social networks, in *Pacific-Asia Conference on Knowledge Discovery and Data Mining* (Springer, Cham, 2015), pp. 58–69

51. J. Zhang, P. Yu, Integrated anchor and social link predictions across partially aligned social networks, in *Proceedings of the Twenty-Fourth International Joint Conference on Artificial Intelligence* (2015)

52. J. Zhang, P. Yu, Mcd: mutual clustering across multiple social networks, in *2015 IEEE International Congress on Big Data* (IEEE, Piscataway, 2015). 10.1109/BigDataCongress.2015.127

53. J. Zhang, P. Yu, Community detection for emerging networks, in *Proceedings of the 2015 SIAM International Conference on Data Mining* (Society for Industrial and Applied Mathematics, Philadelphia, 2015), pp. 127–135

54. J. Zhang, X. Kong, P. Yu, Predicting social links for new users across aligned heterogeneous social networks, in *2013 IEEE 13th International Conference on Data Mining* (IEEE, Piscataway, 2013), pp. 1289–1294

55. J. Zhang, X. Kong, P. Yu, Transferring heterogeneous links across location-based social networks, in *Proceedings of the 7th ACM International Conference on Web Search and Data Mining* (ACM, New York, 2014), pp. 303–312

56. J. Zhang, P. Yu, Z. Zhou, Meta-path based multi-network collective link prediction, in *Proceedings of the 20th ACM SIGKDD International Conference on Knowledge Discovery and Data Mining* (ACM, New York, 2014), pp. 1286–1295

57. J. Zhang, P. Yu, Y. Lv, Q. Zhan, Information diffusion at workplace, in *Proceedings of the 25th ACM International on Conference on Information and Knowledge Management* (ACM, Piscataway, 2016), pp. 1673–1682

58. J. Zhang, J. Chen, S. Zhi, Y. Chang, P. Yu, J. Han, Link prediction across aligned networks with sparse low rank matrix estimation, in *2017 IEEE 33rd International Conference on Data Engineering (ICDE)* (IEEE, Piscataway, 2017), pp. 971–982

59. J. Zhang, L. Cui, P. Yu, Y. Lv, BL-ECD: broad learning based enterprise community detection via hierarchical structure fusion, in *Proceedings of the 2017 ACM on Conference on Information and Knowledge Management* (ACM, New York, 2017), pp. 859–868

60. J. Zhang, P. Yu, Y. Lv, Enterprise community detection, in *Proceedings of the 2017 ACM on Conference on Information and Knowledge Management* (ACM, New York, 2017), pp. 859–868

61. Y. Zhao, E. Levina, J. Zhu, Community extraction for social networks. Proc. Natl. Acad. Sci. USA **108**(18), 7321–7326 (2011)

62. Y. Zhou, L. Liu, Social influence based clustering of heterogeneous information networks, in *Proceedings of the 19th ACM SIGKDD International Conference on Knowledge Discovery and Data Mining* (ACM, New York, 2013), pp. 338–346

63. Y. Zhou, H. Cheng, J. Yu, Graph clustering based on structural/attribute similarities. Proc. VLDB Endow. **2**(1), 718–729 (2009)

Information Diffusion

9

9.1 Overview

In the real world, social information can widely spread among people, and information exchange has become one of the most important social activities. The creation of the Internet and online social networks has rapidly facilitated the communication among people. Via the interactions among users in online social networks, information can easily be propagated from one user to other users. For instance, in recent years, online social networks have become the most important social occasion for news acquisition, and many outbreaking social events can get widely spread in the online social networks at a very fast speed. People as the multi-functional "sensors" can detect different kinds of event signals happening in the real world, and write posts to report their discoveries via the online social networks.

In this chapter, we will study the *information diffusion* problem [13, 42] in the online social networks. Formally, *diffusion* denotes the spreading process of certain entities (like information, idea, innovation, even heat in physics and disease in bio-medical science) through certain channels among the target group of objects inside a system. The *entity* to be spread, the propagation *channels*, the target group of *objects*, and the overall *system* can all affect the information diffusion process and lead to different diffusion observations. Therefore, different types of diffusion models have been proposed already, which will be introduced in this chapter.

Depending on the system where the diffusion process is studied, the diffusion models can be divided into (1) information diffusion models in social networks [13], (2) viral spreading models in the bio-medical system [15], and (3) heat diffusion models in physical system [21]. We will focus on the information diffusion in online social networks in this chapter. The channels for information diffusion actually belong to certain sources (or platforms), like the diffusion channels in the *online platform* [13], diffusion channels in the *offline world* [47], as well as the diffusion channels across *multiple online platforms* [38–40, 49]. Meanwhile, depending on the number of diffusion channels as well as the information sources available, the diffusion models include (1) single-channel diffusion model [13], (2) multi-source single-channel diffusion model [41], and (3) multi-source multi-channel diffusion model [38, 47]. On the other hand, based on the number of topics to be spread in the diffusion process, the information diffusion models can be categorized into (1) single topic diffusion models [13, 38, 39], (2) multiple intertwined topics concurrent diffusion models [46, 47].

© Springer Nature Switzerland AG 2019
J. Zhang, P. S. Yu, *Broad Learning Through Fusions*,
https://doi.org/10.1007/978-3-030-12528-8_9

In the following part of this chapter, we will introduce different kinds of diffusion models proposed to depict how information propagates among users in online social networks. We will first talk about the classic diffusion models proposed for the single-network single-channel scenario, including the *threshold based models* [10, 13], *cascades based models* [13], *heat diffusion based models* [21], and *viral diffusion based models* [15]. After that, several diffusion models proposed for much more complicated scenarios will be introduced, including the *intertwined information diffusion model* [46], *signed network diffusion model* [50], *network coupling based diffusion model* [49], *cross-network random walk based diffusion model* [41], and *multi-source multi-channel diffusion model* [47].

9.2   Traditional Information Diffusion Models

The "*diffusion*" phenomenon has been observed in different disciplines, like social science, physics, and bio-medical science. Various diffusion models have been proposed in these areas already. We summarize the traditional information diffusion models in Fig. 9.1. In this section, we will provide a brief introduction to these models, and talk about how to apply or adapt them to describe information diffusion process in online social networks, which covers (1) how the information diffusion process starts, (2) how the information spreads, and (3) how the information diffusion process ends.

Let $G = (\mathcal{V}, \mathcal{E})$ represent the target network structure, based on which we want to study the information diffusion problem. Formally, given a user node $u \in \mathcal{V}$, we can represent the set of neighbors of u as $\Gamma(u) = \{v | v \in \mathcal{V} \wedge (u, v) \in \mathcal{E}\}$. Each user node in the network G will have an indicator denoting whether the user has been activated by certain information or not. We will use notation $s(u) = 1$ to denote that user u has been activated, and $s(u) = 0$ to represent that u is still inactive. Initially, all the users are inactive to a certain information. Information can be propagated from an initial influence seed user set $\mathcal{S} \subset \mathcal{V}$ who are exposed to and activated by the information at the very beginning. At a timestamp in the diffusion process, given user u's neighbor, we can represent

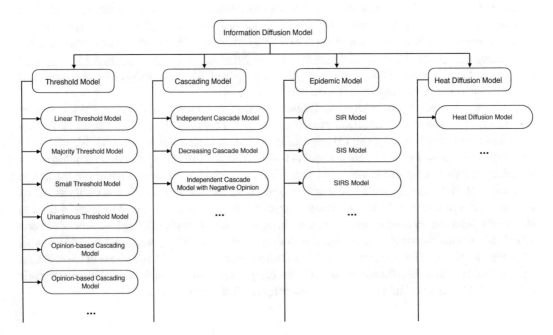

Fig. 9.1 A summary of traditional information diffusion models

the subset of the u's active neighbors as $\Gamma^a(u) = \{v | v \in \Gamma(u), s(v) = 1\} \subseteq \Gamma(u)$. The set of inactive neighbors can be represented as $\Gamma^i(u) = \Gamma(u) \setminus \Gamma^a(u)$. Generally, the information diffusion process will stop if no new activation is possible.

9.2.1   Threshold Based Diffusion Model

In this subsection, we will introduce the *threshold* based models [10], and will use *linear threshold model* [13] as an example to illustrate such a kind of models. Several different variants of the *linear threshold models* will be briefly introduced in this part as well.

Generally, the *threshold models* assume that individuals have a unique threshold indicating the minimum amount of required information for them to be activated by certain information. Information can propagate among the users, and the propagated information amount is determined by the closeness of the users. Close friends can influence each other much more than regular friends and strangers. If the information propagated from other users in the network surpasses the threshold of a certain user, the user will turn to an activated status and also start to influence other users. Therefore, the threshold values can determine both the status and actions of users in the online social networks. Depending on the setting of the thresholds as well as the amount of information propagated among the users, the *threshold models* have different variants.

9.2.1.1 LT Model

In the *linear threshold* (LT) model [13], each user has a unique threshold denoting the minimum required information to active the user. Formally, the threshold of user u can be represented as $\theta_u \in [0, 1]$. In the simulation experiments, the threshold values are normally selected from the uniform distribution $U(0, 1)$. Meanwhile, for each user pair, like $u, v \in \mathcal{V}$, information can be propagated between them. As mentioned before, close friends will have larger influence on each other compared with regular friends and strangers. Formally, the amount of information users u can send to v is denoted as weight $w_{u,v} \in [0, 1]$. Generally, the total amount of information can send out is bounded. For instance, in the LT model, the total amount of information user u can receive is bounded by 1, i.e., $\sum_{v \in \Gamma(u)} w_{v,u} \leq 1$. Different methods have been proposed to define the specific values of the weight $w_{u,v}$. In many of the cases, between the same user pair u and v, weight $w_{u,v}$ can be different from weight $w_{v,u}$, since the information propagation between users is asymmetrical. However, in many simulation experiments, to simplify the setting, for the same user pair, $w_{u,v}$ and $w_{v,u}$ are usually assigned with the same value. For instance, in the LT models, *Jaccard's coefficient* can be used to calculate the closeness between the user pairs, where $w_{u,v}$ will be equal to $w_{v,u}$.

In the LT model, the information sent from the neighbors to user u can be aggregated with linear summation. For instance, the total amount of information user u can receive from his/her neighbors can be denoted as $\sum_{v \in \Gamma(u)} w(v, u) s(v)$ or $\sum_{v \in \Gamma^a(u)} w(v, u)$. To check whether a user can be activated or not, the LT model will confirm whether the following inequality holds or not:

$$\sum_{v \in \Gamma^a(u)} w(v, u) \geq \theta_u. \tag{9.1}$$

The above inequality denotes whether the received information can surpass the activation threshold of user u or not. Here, we also need to notice that inactive neighbors will not send out information, and only the active neighbors can send out information. The information provided so far shows the critical details of the LT model. Next, we will provide the general framework of the LT model to illustrate how it works.

In the LT model, the initial activated seed user set can be represented as $\mathcal{S} \subset \mathcal{V}$, users in which can start the propagation of information to their neighbors. Generally, in the LT model, information propagates among users within the network step by step.

- *Diffusion Starts*: At step 0, only the seed users in $\mathcal{S}$ are active, and all the remaining users have inactive status.
- *Diffusion Spreads*: At step t $(t > 0)$, for each user u, if the information propagated from u's active neighbors is greater than his threshold, i.e., $\sum_{v \in \Gamma^a(u)} w(v, u) \geq \theta_u$, u will be activated with status $s(u) = 1$. All the activated users will remain active in the coming rounds, and can send out information to the neighbors in the next rounds. Active users cannot be activated again.
- *Diffusion Ends*: If no new activation happens in step t, the diffusion process will stop and all the activated users will be returned as the infection result.

Specifically, in the diffusion process, at step t, we don't need to check all the users to see whether they will be activated or not. The reason is that, in the diffusion process, for most of the inactive users, if the status of their neighbors is not changed in the previous step, i.e., step $t - 1$, the influence they can receive in step t will still be the same as in step $t - 1$. And they will remain the same status as they are in the previous step, i.e., "*inactive*." Let $\mathcal{V}^a(t-1)$ denote the set of users who are recently activated in step $t - 1$, we can represent the set of users they can influence as $\bigcup_{u \in \mathcal{V}^a(t-1)} \Gamma^i(u)$. In step t, these recently activated users will make changes to the information their neighbors can receive. Therefore, we only need to check whether the status of inactive users in the set $\bigcup_{u \in \mathcal{V}^a(t-1)} \Gamma^i(u)$ will meet the activation criterion or not.

After the diffusion process stops, a group of users with the active status will indicate the influence these seed users spread to, which can be represented as set $\mathcal{V}^a$. Generally, to denote the impact of certain information, we can introduce a mapping: $\sigma : \mathcal{S} \to |\mathcal{V}^a|$ to project the seed users to their expected number of activated users, which is formally called the influence function. Given the *influence function*, with different seed user sets as the input, the influence they can achieve is usually different. Choosing the optimal seed user who can lead to the maximum influence is named as the *influence maximization* problem [13], which will be introduced in the following Chap. 10 in detail.

9.2.1.2 Other Threshold Models

The LT model assumes the cumulative effects of information propagated from the neighbors, and can illustrate the basic information diffusion process among users in the online social networks. The LT model has been well analyzed, and many other variant models have been proposed as well. Depending on the assignment of the threshold and weight values, many other different diffusion models can all be reduced to a special case of the LT model.

Majority Threshold Model Different from the LT mode, in the *majority threshold model* [27], an inactive user u can be activated if majority of his/her neighbors are activated. The *majority threshold model* can be reduced to the LT model in the case that: (1) the influence weight between any friends (u, v) in the network is assigned with value 1; (2) the threshold of any user u is set as $\frac{1}{2}D(u)$, where $D(u)$ denotes the degree of node u in the network. For the nodes with large degrees, like the central node in the star-structured diagram, their activation will lead to the activation of lots of surrounding nodes in the network.

k-Threshold Model Another diffusion model similar to the LT model is called the *k-threshold diffusion model*, in which users can be activated of at least k if his/her neighbors are active. The *k-threshold model* is equivalent to the LT model with the setting: (1) the influence weight between any friend pairs (u, v) in the network is assigned with value 1; and (2) the activation thresholds of all the users are assigned with a shared value k. For each user u, if k of his/her neighbors have been activated, u will be activated.

Depending on the values of k, the *k-threshold model* will have different performance. When $k = 1$, a user will be activated of at least one of his/her neighbor is active. In such a case, all the users in the same connected components with the initial seed users will be activated finally. When k is a very large value and even greater than the large node degree, e.g., $k > \max_{u \in \mathcal{V}} D(u)$, no nodes can be activated. When k is a medium value, some of the users will be activated as the information propagates, but the other users with less than k neighbors will never be activated.

9.2.2 Cascade Based Diffusion Model

An information cascade [9] occurs when people observe the actions of others and then engage in the same actions. Cascade clearly illustrates the information propagation routes, and the activating actions performed by users on their neighbors. In the cascade model, the information propagation dynamics is carried out in a step-by-step fashion. At each step, users can have trials to activate their neighbors to change their opinions with certain probabilities. If they succeed, the neighbors will change their status to follow the initiators. In the case that multiple users can all have the chance to activate a certain target user, the activation trials will be performed sequentially in an arbitrary order.

Depending on the activation trials and users' reactions to the activation trials, different cascade models have been proposed already. In this section, we will use the *independent cascade* (IC) model [13] as an example of the cascade based models to illustrate the model architecture.

9.2.2.1 IC Model

In the diffusion process, about a certain target user, multiple activation trials can be performed by his/her neighbors. In the *independent cascade* model [13], each activation trial is performed independently regardless of the historical unsuccessful trials. The activation trials are performed step by step. When user u who has been activated in the previous step and tries to activate his/her neighbor v in the current step, the success probability is denoted as $p_{u,v} \in [0, 1]$. Generally, if users u and v are close friends, the activation probability will be larger compared with regular friends and strangers. The specific activation probability value is usually correlated with the social closeness between users u and v, which can also be defined based on the Jaccard's coefficient in the simulation. The activation trials will only happen among the users who are friends. If u succeeds in activating v, then user v will change his/her status to "*active*" and will remain in the status in the following steps. However, if u fails to activate v, u will lose the chance and cannot perform the activation trials any more.

In the IC mode, we can represent the initial seed user as set $\mathcal{S} \subset \mathcal{V}$, who will spread the information to the remaining users. We illustrate the general information propagation procedure in the IC model as follows:

- *Diffusion Starts*: In the initial, the seed users will send out the information and start to activate their neighbors. For the users in set $\mathcal{S}$, the activation trials will start in a random order. For instance, if we pick user $u \in \mathcal{S}$ as the first user, u will activate his/her inactive friends in $\Gamma(u)$ in a random order as well.

- *Diffusion Spreads*: In step t, only the users who have just been activated in the previous step can activate the other users. We can denote the users who have just been activated in the previous as set $\mathcal{V}^a(t-1) \subset \mathcal{V}$. Users in set $\mathcal{V}^a(t-1)$ will start to perform activation trials. For the users who are activated by these users, they will remain active in the following steps and will be added to the set $\mathcal{V}^a(t)$, who will start the activation trials in the next step.
- *Diffusion Ends*: If no new activation happens in a step, the diffusion process will stop and the activated users will be returned as the infection result.

In the IC model, the activation trials are performed by flipping a coin with certain probabilities, whose result is uncertain. Even with the same provided initial seed user set $\mathcal{S}$, the number of users who will be activated by the seed users can be different if we run the IC model twice. Formally, we can represent the set of activated users by the seed users as $\mathcal{V}^a \subset \mathcal{V}$. Therefore, in the experimental simulations, we usually run the diffusion model multiple times and calculate the average number of activated users, i.e., $|\mathcal{V}^a|$, to denote the expected influence achieved by IC on the provided seed user set $\mathcal{S}$.

9.2.2.2 Other Cascade Models

Generally, the independent activation assumption renders the IC model the simplest cascade based diffusion models. In the real world, the diffusion process will be more complicated. For the users, who have failed to be activated by many other users, it probably indicates that the user is not interested in the information. Viewed in such a perspective, the probability for the user to be activated will decrease as more activation trials have been performed. In this part, we will introduce another cascade based diffusion model, *decreasing cascade model* [14].

Decreasing Cascade (DC) Model To illustrate the DC model more clearly and show its difference compared with the IC model, we use notation $P(u \to v|\mathcal{T})$ to represent the probability for user u to activate v given a set of users $\mathcal{T}$ who have performed but failed the activation trials to v in the previous rounds. Let $\mathcal{T}, \mathcal{T}'$ denote two different historical activation user set, where $\mathcal{T} \subseteq \mathcal{T}'$. In the IC model, we have

$$P(u \to v|\mathcal{T}) = P(u \to v|\mathcal{T}'). \tag{9.2}$$

In other words, every activation trial is independent with each other, and user's activation probability will not be changed as more activation trials have been performed.

As introduced at the beginning of this subsection, the fact that users in set $\mathcal{T}$ fail to activate v indicates that v probably is not interested in the information, and the chance for v to be activated afterwards will be lower. Furthermore, as more activation trials are performed by users in $\mathcal{T}'$, the probability for u to active v will be decreased steadily, i.e.,

$$P(u \to v|\mathcal{T}) \geq P(u \to v|\mathcal{T}'). \tag{9.3}$$

Intuitively, this restriction states that a contagious node's probability of activating some v decreases if more nodes have already attempted to activate v, and v is hence more "marketing-saturated." The DC model incorporates the IC model as a special case, and it is a much more general information diffusion model.

9.2.3 Epidemic Diffusion Model

The threshold and cascade based diffusion models introduced in the previous subsections mostly assume that "once a user is activated, he/she will remain in the active status forever." However, in the real world, these activated users can change their minds and they can still have the chance to recover to their originally inactive status. In the bio-medical science, the diffusion problem has also been studied for many years to model the spread of disease, and several *epidemic diffusion models* [15] have been introduced already. In the disease propagation, people who are susceptible to the disease can be get infected by other people. After some time, many of these infected people can get recovered and become immune to the disease, while many other users can recover but may get susceptible to the disease again. Depending on the people's reactions to the disease after recovery, several different *epidemic diffusion models* have been proposed.

In this subsection, we will introduce the *epidemic diffusion models*, and try to use them to model the information diffusion process in online social networks.

9.2.3.1 Susceptible-Infected-Recovered (SIR) Diffusion Model

The SIR model [15] was proposed by W.O. Kermack and A.G. McKendrick in 1927 to model the spread of infectious diseases, which considers a fixed population in three main categories: *susceptible* (S), *infected* (I), and *recovered* (R). As the disease propagates, the individual status can change among $\{S, I, R\}$ with the following flow:

$$S \to I \to R. \tag{9.4}$$

In other words, the individuals who are susceptible to the disease can get infected, while those infected individuals also have the chance to recover from the disease. In this part, we will use the SIR model to describe the information cascading process in online social networks. Let $\mathcal{V}$ denote the set of users in the network. We introduce the following notations to represent the number of users in different categories:

- $S(t)$: the number of users who are *susceptible* to the information at time t, but have not been *infected* yet.
- $I(t)$: the number of users who are currently *infected* by the information, and can spread the information to others in the *susceptible* category.
- $R(t)$: the number of users who have been infected and already recovered from the information infection. After recovery, the users will become immune to the information and cannot be infected again in the future.

Based on the above notations, we have the following equations hold in the SIR model.

$$
\begin{cases}
(1) & S(t) + I(t) + R(t) = |\mathcal{V}|, \\
(2) & \frac{dS(t)}{dt} + \frac{dI(t)}{dt} + \frac{dR(t)}{dt} = 0, \\
(3) & \frac{dS(t)}{dt} = -\beta S(t) I(t), \\
(4) & \frac{dI(t)}{dt} = \beta S(t) I(t) - \gamma I(t), \\
(5) & \frac{dR(t)}{dt} = \gamma I(t).
\end{cases}
\tag{9.5}
$$

In the above equations, the parameter β denotes the infection rate of these *susceptible* users by the *infected* users in a unit time, and γ represents the recovery rate. Generally, all the users in the social network will belong to these three categories, and the total number of users in these three categories will sum to $|\mathcal{V}|$ at any time in the diffusion process. Therefore, we can also get the derivatives of the summation term (i.e., $S(t) + I(t) + R(t)$) with regarding to the time parameter t will be 0. At a unit time, the number of users transit from the *susceptible* status to the *infection* status depends on the available susceptible and infected users at the same time. For each infected user, the number of users he/she can infect is proportional to the available susceptible users, which can be denoted ad $\beta S(t)$. For all the infected users, the total number of users can get infected will be $\beta S(t) I(t)$. For the number of users who are recovered in a unit time, it depends on the number of total infected users $I(t)$ as well as the recovery rate γ, which can be represented as $\gamma I(t)$. Meanwhile, as to the number of infected user changes in a unit time, it is determined by both the number of susceptible users who get infected and the infected users who get recovered.

We have parameters $\beta, \gamma \geq 0$, and the numbers $S(t), I(t), R(t) \geq 0$ to be positive at any time. Therefore, we can know that (1) $\frac{dS(t)}{dt} \leq 0$, and users in the *susceptible* group is non-increasing; (2) $\frac{dR(t)}{dt} \geq 0$, and users in the *recovered* group is non-decreasing; while (3) the sign of term $\frac{dI(t)}{dt}$ can be either positive, zero or negative depending on the parameters β, γ and the users in the *susceptible* and *infected* groups:

- *positive*: if $\beta S(t) > \gamma$;
- *zero*: if $\beta S(t) = \gamma$ or $I(t) = 0$;
- *negative*: if $\beta S(t) < \gamma$.

9.2.3.2 Susceptible-Infected-Susceptible (SIS) Diffusion Model

In some cases, the users cannot get immune to the information and don't exist the *recovery* status actually. For the users, who get infected, they can go to the *susceptible* status and can get *infected* again in the future. To model such a phenomenon, another diffusion model very similar to the SIR model has been proposed, which is called the susceptible-infected-susceptible (SIS) model [12].

In the SIS model, the individual status flow is provided as follows:

$$S \to I \to S. \tag{9.6}$$

Such a status flow will continue, and individuals will switch their status between *susceptible* and *infected* in the information diffusion process. Therefore, the absolute number changes of individuals in these two categories will be the same in unit time.

$$\begin{cases} (1) & S(t) + I(t) + R(t) = |\mathcal{V}|, \\ (2) & \frac{dS(t)}{dt} + \frac{dI(t)}{dt} + \frac{dR(t)}{dt} = 0, \\ (3) & \frac{dS(t)}{dt} = -\beta S(t) I(t) + \gamma I(t), \\ (4) & \frac{dI(t)}{dt} = \beta S(t) I(t) - \gamma I(t). \end{cases} \tag{9.7}$$

9.2.3.3 Susceptible-Infected-Recovered-Susceptible (SIRS) Diffusion Model

The susceptible-infected-recovered-susceptible (SIRS) diffusion model to be introduced in this part is another type of epidemic model, where the individuals in the *recovery* category can lose their immunity and transit to the *susceptible* category. These individuals have the potential to get infected again. Therefore, the individual status flow will be as follows:

$$S \to I \to R \to S. \tag{9.8}$$

We can denote the rate of individuals who lose the immunity as f, and the total number of individuals who may lose the immunity will be $f \cdot R(t)$. Therefore, we can have the derivative of the individual numbers belonging to different categories as follows:

$$\frac{dS(t)}{dt} = -\beta S(t)I(t) + fR(t), \tag{9.9}$$

$$\frac{dI(t)}{dt} = \beta S(t)I(t) - \gamma I(t), \tag{9.10}$$

$$\frac{dR(t)}{dt} = \gamma I(t) - fR(t). \tag{9.11}$$

Besides these epidemic diffusion models introduced in this subsection, there also exist many different versions of the epidemic diffusion models, which consider many other factors in the diffusion process, like the birth/death of individuals. It is also very common in the real-world online social networks, since new users will join in the social network, and existing users will also delete their account and get removed from the social network. Involving such factors will make the diffusion model more complex, and we will not introduce them here due to the limited space. More information about these different epidemic diffusion models is available in [26].

9.2.4 Heat Diffusion Models

Heat diffusion is a well-observed physical phenomenon. Generally, in a medium, heat will always diffuse from regions with a high temperature to the region with a lower temperature. In this subsection, we will talk about the *heat diffusion model* [21] and introduce how to adapt it to model the information diffusion process in online social networks.

9.2.4.1 General Heat Diffusion
Throughout a geometric manifold, let function $f(x, t)$ denote the temperature at region x at time t, and we can represent the initial temperature at different regions as $f_0(x)$. The heat flows with initial conditions can be described by the following differential equation:

$$\begin{cases} \frac{\partial f(x,t)}{\partial t} - \Delta f(x,t) = 0 \\ f(x,0) = f_0(x), \end{cases} \tag{9.12}$$

where $\Delta f(x,t)$ is a *Laplace-Beltrami operator* on function $f(x,t)$.

Many existing works on the heat diffusion studies are mainly focused on the heat kernel matrix. Formally, let $\mathbf{K}_t$ denote the heat kernel matrix at timestamp t, which describes the heat diffusion among different regions in the medium. In the matrix, entry $K_t(x, y)$ denotes the heat diffused from the original region y to region x at time t. However, it is very difficult to represent the medium as a regular geometry with known dimensions. In the next part, we will introduce how to apply the heat diffusion observations to model the information diffusion in the online social networks.

9.2.4.2 Heat Diffusion Model
Given a homogeneous network $G = (\mathcal{V}, \mathcal{E})$, for each node $u \in \mathcal{V}$ in the network, we can represent the information accumulated at u in timestamp t as $f(u, t)$. The initial information available at each of the node can be denoted as $f(u, 0)$. The information can propagate among the nodes in the network

if there exists a pipe (i.e., a link) between them. For instance, with a link $(u, v) \in \mathcal{E}$ in the network, information can propagate between u and v.

Generally, in the diffusion process, the amount of information propagated between different nodes in the network depends on (1) the difference of information available at these two nodes, and (2) the thermal conductivity, i.e., the heat diffusion coefficient α. For instance, at timestamp t, we can represent the amount of information reaching nodes $u, v \in \mathcal{V}$ as $f(u, t)$ and $f(v, t)$. If $f(u, t) > f(v, t)$, information tends to propagate from u to v in the network, and the propagated information amount is denoted as $\alpha \cdot (f(u, t) - f(v, t))$. The propagation direction will be reversed if $f(u, t) < f(v, t)$. The information amount changes at node u at timestamps t and $t + \Delta t$ can be represented as

$$\frac{f(u, t + \Delta t) - f(u, t)}{\Delta t} = - \sum_{v \in \Gamma(u)} \alpha \cdot (f(u, t) - f(v, t)). \tag{9.13}$$

Let's use a vector $\mathbf{f}(t)$ to represent the amount of information available at all the nodes in the network at timestamp t. The above information amount changes can be rewritten with the following equation:

$$\frac{\mathbf{f}(t + \Delta t) - \mathbf{f}(t)}{\Delta t} = \alpha \mathbf{H} \mathbf{f}(t), \tag{9.14}$$

where in the matrix $\mathbf{H} \in \mathbb{R}^{|\mathcal{V}| \times |\mathcal{V}|}$, entry $H(u, v)$ has value

$$H(u, v) = \begin{cases} 1, & \text{if } (u, v) \in \mathcal{E} \vee (v, u) \in \mathcal{E}, \\ -D(u), & \text{if } u = v, \\ 0, & \text{otherwise}, \end{cases} \tag{9.15}$$

where $D(u)$ denotes the degree of node u in the network.

In the limit case $\Delta t \to 0$, we can rewrite the equation as

$$\frac{d\mathbf{f}(t)}{dt} = \alpha \mathbf{H} \mathbf{f}(t). \tag{9.16}$$

By solving the above function, we can represent the amount of information at each node in the network as

$$\mathbf{f}(t) = \exp^{t\alpha\mathbf{H}} \mathbf{f}(0)$$
$$= \left(\mathbf{I} + \alpha t \mathbf{H} + \frac{\alpha^2 t^2}{2!} \mathbf{H}^2 + \frac{\alpha^3 t^3}{3!} \mathbf{H}^3 + \cdots \right) \mathbf{f}(0), \tag{9.17}$$

where term $\exp^{t\alpha\mathbf{H}}$ is also called the diffusion kernel matrix, and it can be expanded as indicated in the above equation according to the Taylor's theorem.

9.3 Intertwined Diffusion Models

For the information diffusion models introduced in the previous section, they are all proposed for modeling the propagation of information in social network with one single diffusion channel and one type of information topic only. However, in the real world, multiple types of information can be propagated within the network simultaneously, relationships among which can be quite intertwined, including *competitive*, *complimentary*, and *independent*. Furthermore, within the networks, even the network structure is homogeneous but the social links among users may be associated with polarities indicating the relationship among the users. For instance, for some of the social links, they denote friendship, while for the other links, they indicate the user pairs are enemies. Formally, the social network structure with polarities associated with the social links are called *signed networks* [18], where the link polarities can affect the information diffusion in them greatly.

In this section, we will introduce two different *intertwined diffusion models* proposed to describe the information propagation process about both (1) the information entities with intertwined relationships [46], and (2) for network structures with links attaching different polarities [50], respectively.

9.3.1 Intertwined Diffusion Models for Multiple Topics

Traditional information diffusion studies mainly focus on one single online social network and have extensive concrete applications in the real world, e.g., product promotion [5, 24] and opinion spread [4]. In the traditional viral marketing setting [7, 13], only one product/idea is to be promoted. However, in the real scenarios, the promotions of multiple products can co-exist in the social networks at the same time, which is referred to as the *intertwined information diffusion problem* [46].

Example 9.1 The relationships among the products to be promoted in the network can be very complicated. For example, in Fig. 9.2, we show 4 different products to be promoted in an online social network and HP printer is our target product. At the product level, the relationships among these products can be:

- *independent*: promotion activities of some products (e.g., HP printer and Pepsi) can be *independent* of each other.
- *competing*: products having common functions will *compete* for the market share [2, 3] (e.g., HP printer and Canon printer). Users who have bought a HP printer are less likely to buy a Canon printer again.
- *complementary*: product cross-sell is also very common in marketing [24]. Users who have bought a certain product (e.g., PC) will be more likely to buy another product (e.g., HP printer) and the promotion of PC is said to be *complementary* to that of HP printer.

Fig. 9.2 Intertwined relationships among products

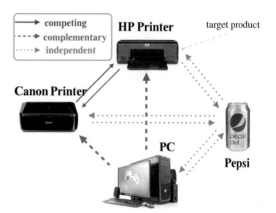

In this section, we will study the information diffusion problem in online social networks, where multiple products are being promoted simultaneously. The relationships among these product can be obtained in advance via effective market research, which can be *independent*, *competitive*, or *complementary*. A novel information diffusion model interTwined Linear Threshold (TLT) proposed in [46] will be introduced in this section. TLT quantifies the *impacts* among products with the *intertwined threshold updating strategy* and can handle the intertwined diffusions of these products at the same time.

Before talking about the detailed diffusion models, we first introduce the definitions of several important terminologies as follows, which will be used in this section.

Definition 9.1 (Social Network) An online *social network* can be represented as $G = (\mathcal{V}, \mathcal{E})$, where $\mathcal{V}$ is the set of users and $\mathcal{E}$ contains the interactions among users in $\mathcal{V}$. The set of n different products to be promoted in network G can be represented as $\mathcal{P} = \{p^1, p^2, \ldots, p^n\}$.

Definition 9.2 (User Status Vector) For a given product $p^j \in \mathcal{P}$, users who are influenced to buy p^j are defined to be "*active*" to p^j, while the remaining users who have not bought p^j are defined to be "*inactive*" to p^j. User u_i's status towards all the products in $\mathcal{P}$ can be represented as "*user status vector*" $\mathbf{s}_i = (s_i^1, s_i^2, \ldots, s_i^n)$, where s_i^j is u_i's status to product p^j. Users can be activated by multiple products at the same time (even competing products), i.e., multiple entries in *status vector* $\mathbf{s}_i$ can be "*active*" concurrently.

Definition 9.3 (Independent, Competing, and Complementary Products) Let $P(s_i^j = 1)$ (or $P(s_i^j)$ for simplicity) denote the probability that u_i is activated by product p^j and $P(s_i^j|s_i^k)$ be the conditional probability given that u_i has been activated by p^k already. For products $p^j, p^k \in \mathcal{P}$, the promotion of p^k is defined to be (1) *independent* to that of p^j if $\forall u_i \in \mathcal{V}$, $P(s_i^j|s_i^k) = P(s_i^j)$, (2) *competing* to that of p^j if $\forall u_i \in \mathcal{V}$, $P(s_i^j|s_i^k) < P(s_i^j)$, and (3) *complementary* to that of p^j if $\forall u_i \in \mathcal{V}$, $P(s_i^j|s_i^k) > P(s_i^j)$.

9.3.1.1 TLT Diffusion Model

To depict the intertwined diffusions of multiple independent, competing or complementary products, we will introduce a new information diffusion model TLT. In the existence of multiple products $\mathcal{P}$, user u_i's influence to his neighbor u_k in promoting product p^j can be represented as $w_{i,k}^j \geq 0$. Similar to the traditional LT model, in TLT, the influence of different products can propagate within the network step by step. User u_i's threshold for product p^j can be represented as θ_i^j and u_i will be activated by his neighbors to buy product p^j if

$$\sum_{u_l \in \Gamma_{out}(u_i)} w_{l,i}^j \geq \theta_i^j. \tag{9.18}$$

Different from traditional LT model, in TLT, users in online social networks can be activated by multiple products at the same time, which can be either *independent*, *competing* or *complementary*. As shown in Fig. 9.2, we observe that users' chance to buy the HP printer will be (1) unchanged given that they have bought Pepsi (i.e., the *independent* product of HP printer), (2) increased if they own PCs (i.e., the *complementary* product of HP printer), and (3) decreased if they already have the Canon printer (i.e., the *competing* product of HP printer).

To model such a phenomenon in TLT, we introduce the following *intertwined threshold updating strategy*, where users' *thresholds* to different products will change *dynamically* as the influence of other products propagates in the network.

Definition 9.4 (Intertwined Threshold Updating Strategy) Assuming that user u_i has been activated by m products $p^{\tau_1}, p^{\tau_2}, \ldots, p^{\tau_m} \in \mathcal{P} \setminus \{p^j\}$ in a sequence, then u_i's *threshold* towards product p^j will be updated as follows:

$$
\left(\theta_i^j\right)^{\tau_1} = \theta_i^j \frac{P\left(s_i^j\right)}{P(s_i^j|s_i^{\tau_1})}, \ \left(\theta_i^j\right)^{\tau_2} = \left(\theta_i^j\right)^{\tau_1} \frac{P\left(s_i^j|s_i^{\tau_1}\right)}{P\left(s_i^j|s_i^{\tau_1}, s_i^{\tau_2}\right)}, \cdots \tag{9.19}
$$

$$
\left(\theta_i^j\right)^{\tau_m} = \left(\theta_i^j\right)^{\tau_{m-1}} \frac{P\left(s_i^j|s_i^{\tau_1}, \ldots, s_i^{\tau_{m-1}}\right)}{P\left(s_i^j|s_i^{\tau_1}, \ldots, s_i^{\tau_{m-1}}, s_i^{\tau_m}\right)}, \tag{9.20}
$$

where $(\theta_i^j)^{\tau_k}$ denotes u_i's threshold to p^j after he has been activated by $p^{\tau_1}, p^{\tau_2}, \ldots, p^{\tau_k}, k \in \{1, 2, \ldots, m\}$.

In this section, we do not focus on the order of products that activate users [4] and to simplify the calculation of the *threshold updating strategy*, we assume only the most recent activation has an effect on updating current thresholds, i.e.,

$$
\frac{P\left(s_i^j|s_i^{\tau_1}, \ldots, s_i^{\tau_{m-1}}\right)}{P\left(s_i^j|s_i^{\tau_1}, \ldots, s_i^{\tau_{m-1}}, s_i^{\tau_m}\right)} \approx \frac{P\left(s_i^j\right)}{P\left(s_i^j|s_i^{\tau_m}\right)} = \phi_i^{\tau_m \to j}. \tag{9.21}
$$

Definition 9.5 (Threshold Updating Coefficient) Term $\phi_i^{l \to j} = \frac{P(s_i^j)}{P(s_i^j|s_i^l)}$ is formally defined as the "*threshold updating coefficient*" of product p^l to product p^j for user u_i, where

$$
\phi_i^{l \to j} \begin{cases} < 1, & \text{if } p^l \text{ is } complementary \text{ to } p^j, \\ = 1, & \text{if } p^l \text{ is } independent \text{ to } p^j, \\ > 1, & \text{if } p^l \text{ is } competing \text{ to } p^j. \end{cases} \tag{9.22}
$$

The *intertwined threshold updating strategy* can be rewritten based on the *threshold updating coefficients* as follows:

$$
(\theta_i^j)^{\tau_m} \approx \theta_i^j \cdot \phi_i^{\tau_1 \to j} \cdot \phi_i^{\tau_2 \to j} \cdots \phi_i^{\tau_m \to j}. \tag{9.23}
$$

Based on the above *threshold updating coefficients* equation together with the detailed information diffusion process described in the LT model, the TLT model introduced in this part can depict the information diffusion process about multiple products with *intertwined relationships*.

9.3.2 Diffusion Models for Signed Networks

In this part, we will introduce another type of *intertwined diffusion model* proposed for the *signed networks* [31, 48] involving polarized links among the nodes specifically. In recent years, signed networks have gained increasing attention because of their ability to represent diverse and contrasting social relationships. Some examples of such contrasting relationships include friends vs enemies [35], trust vs distrust [36], positive attitudes vs negative attitudes [37], and so on. These contrasting relationships can be represented as links of different polarities, which result in signed networks. Signed social networks can provide a meaningful perspective on a wide range of social network studies, like *user sentiment analysis* [34], *social interaction pattern extraction* [18], *trustworthy friend recommendation* [17], and so on.

Information dissemination is common in social networks [29]. Due to the extensive social links among users, information on certain topics, e.g., politics, celebrities and product promotions, can propagate leading to a large number of nodes reporting the same (incorrect) observations rapidly in online social networks. In particular, the links in signed networks are of different polarities and can denote trust and distrust relationships among users [20], which will inevitably have an impact on information propagation inside the networks.

Example 9.2 In Fig. 9.3, an example is provided to help illustrate the information diffusion problem in signed networks more clearly. In the example, users are connected to one another with signed links, depending on their trust and distrust relations. It is noteworthy that the conventions used for the direction of information diffusion in this network are slightly different from traditional influence analysis, because they represent signed links. For instance, if Alice trusts (or follows) Bob, a directed edge exists from Alice to Bob, but the information diffusion direction will be from Bob to Alice. Via the signed links, inactive users in the network can get infected by certain information propagated from their neighbors with either a positive or negative opinion about the information (i.e., the green or red states in the figure). Considering the fact that it is often difficult to directly identify all the user infection states in real settings, we allow for the possibility of some user states in the network to be unknown. Activated users can propagate the information to other users. In general, if a user is activated with a positive or negative opinion about the information, she might activate one or more of her incoming neighbors to trust or distrust the information, depending on the sign of the incoming link.

Fig. 9.3 Example of the information diffusion problem in signed networks

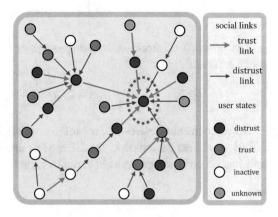

The edges in the network are directed and signed, and they represent trust or distrust relationships. For example, when node i trusts or distrusts node j, we will have a corresponding positive or negative link from node i to node j. In this setting, nodes are associated with states corresponding to a prevailing opinion about the truth of a fact. These states can be drawn from $\{-1, +1, 0, ?\}$, where $+1$ indicates their agreement with a specific fact, -1 indicates their disagreement, 0 indicates the fact that they have no opinion of the fact at hand, and ? indicates their opinion is unknown. The last of these states is necessary to model the fact that the states of many nodes in large-scale networks are often unknown. Note that the use of multiple states of nodes in the network is different from traditional influence analysis. Users are influenced with varying opinions of the fact in question, based on their observation of their neighbors (i.e., states of neighborhood nodes), and their trust or distrust of their neighbor's opinions (i.e., signs of links with them). This model is essentially a signed version of influence propagation models, because the sign of the link plays a critical role in how a specific bit of information is transmitted.

Most existing information diffusion models are designed for unsigned networks. In signed networks, information diffusion is also related to actor-centric trust and distrust, in which notions of node states and the signs on links play an important role. To depict how information propagates in the signed networks, we will introduce the *asyMmetric Flipping Cascade* (MFC) diffusion model proposed in [50] for signed networks specifically.

9.3.2.1 Terminology Definition

Traditional social networks are unsigned in the sense that the links are assumed, by default, to be positive links. Signed social networks are a generalization of this basic concept.

Definition 9.6 (Weighted Signed Social Network) A *weighted signed social network* can be represented as a graph $G = (\mathcal{V}, \mathcal{E}, s, w)$, where $\mathcal{V}$ and $\mathcal{E}$ represent the nodes (users) and directed edges (social links), respectively. In signed networks, each social link has its own polarity (i.e., the sign) and is associated with a weight indicating the intimacy among users, which can be represented with the mappings $s : \mathcal{E} \to \{-1, +1\}$ and $w : \mathcal{E} \to [0, 1]$, respectively.

As discussed before, we interpret the signs from a trust-centric point of view. Information propagated among users is highly associated with the intimacy scores [43] among them: information tends to propagate among close users. To represent the information diffusion process in trust-centric networks, we define the concept of *weighted signed diffusion network* as follows:

Definition 9.7 (Weighted Signed Diffusion Network) Given a *signed social network* G, its corresponding *weighted signed diffusion network* can be represented as $G_D = (\mathcal{V}_D, \mathcal{E}_D, s_D, w_D)$, where $\mathcal{V}_D = \mathcal{V}$ and $\mathcal{E}_D = \{(v, u)\}_{(u,v)\in\mathcal{E}}$. Diffusion links in $\mathcal{E}_D$ share the same sign and weight mappings as those in $\mathcal{E}$, which can be obtained via mappings $s_D : \mathcal{E}_D \to \{-1, +1\}$, $s_D(v, u) = s(u, v)$, $\forall(v, u) \in \mathcal{E}_D$ and $w_D : \mathcal{E}_D \to [0, 1]$, $w_D(v, u) = w(u, v)$, $\forall(v, u) \in \mathcal{E}_D$. For any directed diffusion link $(u, v) \in \mathcal{E}_D$, we can represent its sign and weight to be $s_D(u, v)$ and $w_D(u, v)$, respectively.

Note that we have reversed the direction of the links because of the trust-centric interpretation, in which information diffuses from A to B, when B trusts A. However, in networks with other semantic interpretations, this reversal does not need to be performed. The overall algorithm is agnostic to the specific preprocessing performed in order to fit a particular semantic interpretation of the signed network.

9.3.2.2 MFC Diffusion Model

The signs associated with diffusion links denote the "positive" and "negative" relationships, e.g., trust and distrust, among users. In everyday life, people tend to believe information from people they trust and not believe the information from those they distrust. For example, if someone we trust says that "Hillary Clinton will be the new president," we believe it to be true. However, if someone we distrust says the same thing, we might not believe it. In addition, when receiving contradictory messages, information obtained from the trusted people is usually given higher weights. In other words, the effects of trust and distrust diffusion links are asymmetric in activating users. For instance, when various actors assert that "Hillary Clinton will be the new president," we may tend to follow those we trust, even though the distrusted ones also say it. In addition, if someone we distrust says that "Hillary Clinton will be the new president," we may think it to be false and will not believe it. However, after being activated to distrust it, if we are exposed to contradictory information from a trusted party, we might be willing to change our minds. To model such cases, which are unique to signed and state-centric networks, a number of basic principles are introduced in the MFC model, (1) the effects of positive links in activating users is boosted to give them higher weights in activating users, and (2) users who are activated already will stay active in the subsequential rounds but their activation states can be flipped to follow the people they trust.

In MFC, users have 3 unique known states in the information diffusion process: $\{+1, -1, 0\}$ (i.e., trust, distrust, and inactive, respectively). Users with unknown states are automatically taken into account during the model construction process by assuming states as necessary. For simplicity, we use $s(\cdot)$ to represent both the sign of links and the states of users. If user u trusts the information, then user u is said to have a positive state $s(u) = +1$ towards the information. The initial states of all users in MFC are assigned a value of 0 (i.e., inactive to the information). A set of information seed users $\mathcal{I} \subseteq \mathcal{V}$ activated by the information at the very beginning will have their own attitudes towards the information based on their judgments, which can be represented with $\mathcal{S} = \{+1, -1\}^{|\mathcal{I}|}$. Information seed users in $\mathcal{I}$ spread the information to other users in signed networks step by step. At step τ, user u (activated at $\tau - 1$) is given only one chance to activate (1) inactive neighbor v, as well as (2) active neighbor v but v has different state from u and v trusts u, with the boosted success probability $\overline{w_D}(u, v)$, where $\overline{w_D}(v, u) \in [0, 1]$ can be represented as

$$\overline{w_D}(v, u) = \begin{cases} \min\{\alpha \cdot w_D(v, u), 1\} & \text{if } s_D(v, u) = +1, \\ w_D(v, u), & \text{otherwise.} \end{cases} \tag{9.24}$$

In the above equation, parameter $\alpha > 1$ denotes the boosting of information from u to v and is called the *asymmetric boosting coefficient*.

If u succeeds, v will become active in step $\tau + 1$, whose states can be represented as $s(v) = s(u) \cdot s(u, v)$. For example, if user u thinks the information to be real (i.e., $s(u) = +1$) and v trusts u (i.e., $s(u, v) = +1$), once v get activated by u successfully, the state of v will be $s(v) = +1$ (i.e., believe the information to be true). Otherwise, v will keep its original state (either inactive or activated) and u cannot make any further attempts to activate v in subsequent rounds. All activated users will stay active in the following rounds and the process continues until no more activations are possible.

Example 9.3 MFC can model the information diffusion process in signed social networks much better than traditional diffusion models, such as IC. To illustrate the advantages of MFC, we also give an example in Fig. 9.4, where two different cases: "simultaneous activation" (i.e., the left two plots) and "sequential activation" (i.e., the right two plots) are shown. In the "simultaneous activation" case,

Fig. 9.4 Example of the binary tree transformation

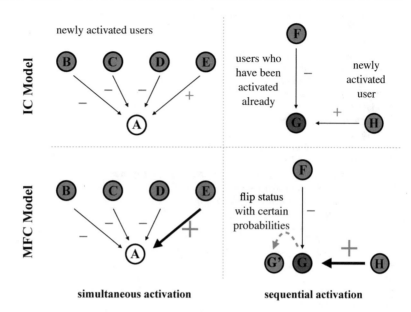

newly activated users

IC Model

MFC Model

users who have been activated already

newly activated user

flip status with certain probabilities

simultaneous activation   **sequential activation**

multiple users (B, C, D, and E) are all just activated at step τ, who all think information to be true and at step $\tau + 1$, B-E will activate their inactive neighbor A. Among these users, A trusts E and distrusts the remaining users. In traditional IC models, signs on links are ignored and B-E are given equal chance to activate A in random order with activation probabilities $w_D(\cdot, A)$, $\cdot \in \{B, C, D, E\}$. However, in the MFC model, signs of links are utilized and the activation probability of positive diffusion (E, A) will be boosted and can be represented as $\min\{\alpha \cdot w_D(E, A), 1\}$. As a result, user A is more likely to be activated by E in MFC. Meanwhile, in the sequential activation case, once a user (e.g., F) succeeds in activating G, G will remain active and other users (e.g., H) cannot reactivate A any longer in traditional IC model. However, in the MFC model, we allow users to flip their activation state by people they trust. For example, if G has been activated by F with state $s(G) = -1$ already, the trusted user H can still have the chance to flip G's state with probability $\min\{\alpha \cdot w_D(H, G), 1\}$.

The pseudo-code of the MFC diffusion model is provided in Algorithm 1.

9.4 Inter-Network Information Diffusion

The information diffusion models introduced in the previous sections are mostly based on one single network, assuming that information will propagate within the network only. However, in the real-world, users are involved in multiple social sites simultaneously, and cross-platform information diffusion is happening all the time. Users, as the bridges, can receive information from one social sites, and share with their friends in another network intentionally. Meanwhile, due to some social network settings, sometimes the activities happening in one social site (e.g., Foursquare) can be reposted to other social sites (e.g., Twitter) automatically, if the users login Foursquare with their Twitter account.

Cross-network information sharing and reposting renders the inter-network information diffusion ubiquitous and very common in the real-world online social networks. By involving in multiple online social networks simultaneously, users can also be exposed to more information from multiple social sites at the same time. In this section, we will study the information diffusion across multiple social platforms, and introduce two inter-network diffusion models across multiple online social

---

Algorithm 1 MFC Information Diffusion Model

Require: input rumor initiators $\mathcal{I}$ with states $\mathcal{S}$
        diffusion network $G_D = (\mathcal{V}_D, \mathcal{E}_D, s_D, w_D)$
Ensure: infected diffusion network G_I
 1: initialize infected user set $\mathcal{U} = \mathcal{I}$, state set $\mathcal{S}_{\mathcal{U}} = \mathcal{S}$
 2: let recently infected user set $\mathcal{R} = \mathcal{I}$
 3: **while** $\mathcal{R} \neq \emptyset$ **do**
 4:    new recently infected user set $\mathcal{N} = \emptyset$
 5:    **for** $u \in \mathcal{R}$ **do**
 6:       let the set of users that u can activate to be $\Gamma(u)$
 7:       **for** $v \in \Gamma(u)$ **do**
 8:          **if** $s(v) = 0$ or $\big(s_D(u, v) = +1$ and $s(u) \neq s(v)\big)$ **then**
 9:             **if** $s_D(u, v) = +1$ **then**
10:                $p = \min\{1.0, \alpha \cdot w_D(u, v)\}$
11:             **else**
12:                $p = w_D(u, v)$
13:             **end if**
14:             **if** u activates v with probability p **then**
15:                $\mathcal{U} = \mathcal{U} \cup \{v\}, \mathcal{S}_{\mathcal{U}} = \mathcal{S}_{\mathcal{U}} \cup \{s(v) = s(u) \cdot s_D(u, v)\}$
16:                $\mathcal{N} = \mathcal{N} \cup \{v\}$
17:             **end if**
18:          **end if**
19:       **end for**
20:    **end for**
21:    $\mathcal{R} = \mathcal{N}$
22: **end while**
23: extract infected diffusion network G_I consisting of infected users $\mathcal{U}$

---

networks [41, 49]. Generally, different social network platforms will create different information diffusion sources, and the interactions available among users in each of the sources can all propagate information among users.

9.4.1 Network Coupling Based Cross-Network Information Diffusion

In the online social world, once a user has been activated in one of the social sites, the user account owner will receive the information and diffuse it to other users in the other networks. The network coupling model to be introduced in this part proposes to combine multiple social networks together, and treat the information diffusion in each of the networks independently. If one of a user's account has been activated in a network, the user will be treated as activated. To simplify the activation checking criterion, the *network coupling based information diffusion model* [49] also introduces a relaxed criterion.

9.4.1.1 Single Network Diffusion Model

Formally, let $G^{(1)}, G^{(2)}, \ldots, G^{(k)}$ denote the k online social networks that we are focusing on in the information diffusion, whose network structures are all homogeneous involving users and friendship links only. For each of the network, e.g., $G^{(i)}$, we can represent its structure as $G^{(i)} = (\mathcal{V}^{(i)}, \mathcal{E}^{(i)})$, where $\mathcal{V}^{(i)}$ denotes the set of users in the network. Information diffusion process in network $G^{(i)}$ can be modeled with some existing models. In this part, we will use the LT model as the base diffusion model for each of the networks.

Based on network $G^{(i)}$, each user u in the network is associated with a threshold $\theta_u^{(i)}$ indicating the minimal amount of required information to activate the users. Meanwhile, the amount of information sent between the users (e.g., u and v) can be denoted as weight $w_{u,v}^{(i)}$, whose value can be determined in the same way as the LT model introduced before. For an inactive user u, he/she can be activated iff the amount of information propagated from their friends to him/her is greater than u's threshold, i.e.,

$$\sum_{v \in \Gamma(u;G^{(i)})} \mathbb{I}(v,t) \cdot w_{v,u}^{(i)} \geq \theta_u^{(i)}, \tag{9.25}$$

where $\Gamma(u; G^{(i)})$ represents the neighbors of user u and $\mathbb{I}(v,t)$ indicates whether v has been activated or not at time t.

9.4.1.2 Network Coupling Based Information Diffusion Model

Meanwhile, among these k different online social sites $G^{(1)}, G^{(2)}, \ldots, G^{(k)}$, if there exists one network $G^{(i)}$, in which the above equation holds, user u will become activated by the information. In other words, to determine whether user u has been activated or not, we need to check his/her status in all these k networks one by one as follows:

$$\text{Network } G^{(1)} : \sum_{v \in \Gamma(u;G^{(1)})} \mathbb{I}(v,t) \cdot w_{v,u}^{(1)} \geq \theta_u^{(1)},$$

$$\text{Network } G^{(2)} : \sum_{v \in \Gamma(u;G^{(2)})} \mathbb{I}(v,t) \cdot w_{v,u}^{(2)} \geq \theta_u^{(2)},$$

$$\cdots,$$

$$\text{Network } G^{(k)} : \sum_{v \in \Gamma(u;G^{(k)})} \mathbb{I}(v,t) \cdot w_{v,u}^{(k)} \geq \theta_u^{(k)}. \tag{9.26}$$

Such a status checking process can be very time-consuming. To reduce the activation checking works, in the lossy network coupling scheme, the users' activation checking criterion is relaxed to

$$\sum_{i=1}^{k} \alpha^{(i)} \cdot \sum_{v \in \Gamma(u)} \mathbb{I}(v,t) \cdot w_{v,u}^{(i)} \geq \sum_{i=1}^{k} \alpha^{(i)} \cdot \theta_u^{(i)}, \tag{9.27}$$

where $\alpha^{(1)}, \alpha^{(2)}, \ldots, \alpha^{(k)} > 0$ denote the weight parameters representing the importance of different networks in user activation.

Theorem 9.1 *Given the k networks, $G^{(1)}, G^{(2)}, \ldots, G^{(k)}$, if equation*

$$\sum_{i=1}^{k} \alpha^{(i)} \cdot \sum_{v \in \Gamma(u)} \mathbb{I}(v,t) \cdot w_{v,u}^{(i)} \geq \sum_{i=1}^{k} \alpha^{(i)} \cdot \theta_u^{(i)}, \tag{9.28}$$

holds, user u will be activated.

Proof The theorem can be proven with by contradiction. Let's assume the equation holds but u has not been activated in networks $G^{(1)}, G^{(2)}, \ldots, G^{(k)}$, then we have

$$\sum_{v \in \Gamma(u)} \mathbb{I}(v, t) \cdot w_{v,u}^{(i)} < \theta_u^{(i)}, \tag{9.29}$$

hold for all these networks.

By multiplying both sides of the inequality with a positive weight $\alpha^{(i)}$, and sum the equations across all these k networks, we have

$$\sum_{i=1}^{k} \alpha^{(i)} \cdot \sum_{v \in \Gamma(u)} \mathbb{I}(v, t) \cdot w_{v,u}^{(i)} < \sum_{i=1}^{k} \alpha^{(i)} \cdot \theta_u^{(i)}, \tag{9.30}$$

which contradicts the equation in the theorem.

Therefore, if the new activation criterion holds, user u will be activated (in at least one of these online social networks).

The relaxed activation criterion is actually a sufficient but not necessary condition when determining whether u is activated or not. In some cases, u has already been activated in some of the networks, but the criterion cannot meet, which will lead to some latency in status checking. One way to solve the problem is to assign an appropriate weight $\alpha^{(i)}$ by increasing the value proportion of these networks in the summation equation $\sum_{v \in \Gamma(u)} \mathbb{I}(v, t) \cdot w_{v,u}^{(i)}$. In the special case that user u can be activated in network $G^{(i)}$ already, we can assign the weight $\alpha^{(i)}$ with a very large value, where $\alpha^{(i)} \gg \alpha^{(j)}, j \in \{1, 2, \ldots, k\}, j \neq i$ is way larger compared with the remaining networks. So far, there don't exist any methods to adjust the parameters automatically in the *network coupling based information diffusion model*, and heuristics are applied in most of the cases.

9.4.2 Random Walk Based Cross-Network Information Diffusion

Different online social networks usually have their own characteristics, and users tend to have different status regarding the same information in different platforms. For instance, information about personal entertainments (like movies, pop stars) can be widely spread among users in Facebook, and users who are interested in them will be activated very easily and also share the information to their friends. However, such a kind of information is relatively rare in the professional social network LinkedIn, where people seldom share personal entertainment to their colleagues, even though they may have been activated already in Facebook. What's more, the structures of these online social networks are usually heterogeneous, containing various types of connections. Besides the direct follow relationships among the users, these diverse connections available among the users may create different types of communication channels for information diffusion. To model such an observation in information diffusion across multiple heterogeneous online social sites, in this part, we will introduce a new information diffusion model, IPATH [41], based on random walk. Since there exist multiple networks here and users will also have different status in different networks, we can denote the *objective target network* that we study as G^t and the external aligned social platform as G^s. Between them, there exists a set of anchor links connecting the shared anchor users, which can be denoted as $\mathcal{A}^{(t,s)}$. For the scenarios with more than one external source network, a simple extension to the following model will be applicable to depict the diffusion process across these networks.

9.4.2.1 Intra-Network Propagation

The traditional research works on homogeneous networks assume that information can only be spread by the social links among users. If user v follows user u, i.e., $(v, u) \in \mathcal{E}$, the message can spread from u to v, i.e. $u \rightarrow v$. However in a heterogeneous social network, the multi-typed and interconnected entities can create various information propagation channels among the users. For instance, if user u recommends a good restaurant to his friend v by checking in at this place, information will flow from u to v through the location entity l, which can be expressed by $u \xrightarrow[l]{check-in} v$. Similarly, we can represent the information diffusion routes among users via other information entities, which can be formally represented as the diffusion route set $\mathcal{R} = \{r_1, r_2, \ldots, r_m\}$, where m is the total route number.

According to each diffusion route, we can represent the connections among users as an adjacency matrix actually. We can take the source network $G^s = (\mathcal{V}^s, \mathcal{E}^s)$ as an example. For any relation $r_i \in \mathcal{R}^s$, the adjacency matrix defined based on r_i among the set of users (i.e., $\mathcal{U}^s$) can be represented as $\mathbf{A}_i^s \in \mathbb{R}^{|\mathcal{U}^s| \times |\mathcal{U}^s|}$, where $A_i^s(u, v)$ is a binary-value variable and $A_i^s(u, v) = 1$ iff u and v are connected with each other via relation r_i. The weighted diffusion matrix can be represented as the normalized matrix of $\mathbf{A}_i^s$, i.e., $\mathbf{W}_i^s = \mathbf{A}_i^s \mathbf{D}^{-1}$, where $\mathbf{D}$ is a diagonal matrix with $D(u, u) = \sum_{v \in \mathcal{U}^s} A_i^s(v, u)$ denoting the in-degree of u on its diagonal. The entry $W_i^s(u, v)$ denotes the probability of going from v to u in one step. In a similar way, we can represent the weighted diffusion matrices for other relations, which altogether can be represented as $\{\mathbf{W}_1^s, \mathbf{W}_2^s, \ldots, \mathbf{W}_m^s\}$. To fuse the information diffused from different relations, IPATH uses linearly combination to integrate these weighted matrices as follows:

$$\mathbf{W}^s = \lambda_1 \times \mathbf{W}_1^s + \lambda_2 \times \mathbf{W}_2^s + \cdots + \lambda_m \times \mathbf{W}_m^s, \quad (9.31)$$

where λ_i denotes the aggregation weight of matrix corresponding to relation r_i. In real scenarios, different relations play different roles in the information propagation for different users. However, to simplify the settings, in this part, we treat all these relations to be equally important, and the aggregated matrix $\mathbf{W}^s$ takes the average of all these weighted diffusion matrices. In a similar way, we can define the weight matrix $\mathbf{W}^t$ for the target network G^t.

9.4.2.2 Inter-Network Propagation

Across the aligned networks, information can propagate not only within networks but also across networks. Based on the known anchor links between networks G^t and G^s, we can define the binary adjacency matrix $\mathbf{A}^{s \to t} \in \mathbb{R}^{|\mathcal{U}^s| \times |\mathcal{U}^t|}$, where $A^{s \to t}(u, v) = 1$ if $(u^s, v^t) \in \mathcal{A}^{(t,s)}$. In IPATH, we assume that each anchor user in G^s only has one corresponding account in G^t. Therefore $A^{s \to t}$ is already normalized (into binary) and the weight matrix $\mathbf{W}^{s \to t} = \mathbf{A}^{s \to t}$, denoting the chance of information propagating from G^s to G^t. Furthermore, we can represent the weighted diffusion matrix from networks G^s to G^t as $\mathbf{W}^{t \to s} = (\mathbf{W}^{s \to t})^\top$, considering that the anchor links are undirected.

Both the intra-network propagation relations, represented by weight matrices $\mathbf{W}^s$ and $\mathbf{W}^t$ in networks G^s and G^t, respectively, and the inter-network propagation relations, represented by weight matrix $\mathbf{W}^{s \to t}$ and $\mathbf{W}^{t \to s}$, have been constructed already in the previous subsection. As shown in Fig. 9.5, to model the cross-network information diffusion process involving both the intra- and inter-network relations simultaneously, IPATH proposes to combine these weighted diffusion matrices to build an integrated matrix $\mathbf{W} \in \mathbb{R}^{(|\mathcal{U}^s| + |\mathcal{U}^t|)^2}$. In the integrated matrix $\mathbf{W}$, the parameter $\alpha \in [0, 1]$ denotes the probability that the message stay in the original network, thus $1 - \alpha$ represents the chance of being transmitted across networks (i.e., the probability of activated anchor user passing the influence to the target network). In real scenarios, the probabilities for different users to repost

Fig. 9.5 The weight
matrix and the information
distribution vector

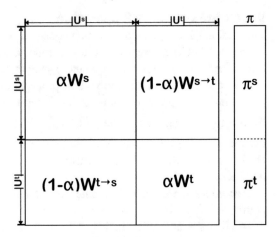

information across aligned networks can be quite diverse. However, to simplify the problem setting, in IPATH, we will unify these probabilities with parameter α.

Let vector $\pi^{(k)} \in \mathbb{R}^{(|\mathcal{U}^s|+|\mathcal{U}^t|)}$ represent the information that users in G^s and G^t can receive after k steps. As shown in Fig. 9.5, vector $\pi^{(k)}$ consists of two parts $\pi^{(k)} = [\pi^{s,(k)}, \pi^{t,(k)}]$, where $\pi^{s,(k)} \in \mathbb{R}^{|\mathcal{U}^s|}$ and $\pi^{t,(k)} \in \mathbb{R}^{|\mathcal{U}^t|}$. The initial state of the vector can be denoted as $\pi^{(0)}$, which is defined based on the seed user set $\mathcal{Z}$ with function $g(\cdot)$ as follows:

$$\pi^{(0)} = g(\mathcal{Z}), \text{ where } \pi^{(0)}(u) = \begin{cases} 1 & \text{if } u \in \mathcal{Z}, \\ 0 & \text{otherwise.} \end{cases} \tag{9.32}$$

Seed set $\mathcal{Z}$ can also be represented as $\mathcal{Z} = g^{-1}(\pi^{(0)})$. Users from G^s and G^t both have the chance of being selected as seeds, but when the structure information of G^t is hard to obtain, the seed users will be only chosen from G^s. In IPATH, the information diffusion process is modeled by *random walk*, because it is widely used in which the total probability of the diffusing through different relations remains constant 1 [8, 32]. Therefore in the information propagation process, vector π will be updated stepwise with the following equation:

$$\pi^{(k+1)} = (1 - a) \times \mathbf{W}\pi^{(k)} + a \times \pi^{(0)}, \tag{9.33}$$

where constant a denotes the probability of returning to the initial state. By keeping updating π according to (9.33) until convergence, we can present the stationary state of vector π to be π^*,

$$\pi^* = a[\mathbf{I} - (1 - a)\mathbf{W}]^{-1}\pi^{(0)}, \tag{9.34}$$

where matrix $\mathbf{I} \in \{0, 1\}^{(|\mathcal{U}^s|+|\mathcal{U}^t|) \times (|\mathcal{U}^s|+|\mathcal{U}^t|)}$ is an identity matrix. The value of entry $\pi^*[u]$ denotes the activation probability of u, and user u will be activated if $\pi^*[u] \geq \theta$, where θ denotes the threshold of accepting the message. In IPATH, parameter θ is randomly sampled from range $[0, \theta_{bound}]$. The threshold bound θ_{bound} is a small constant value, as the amount of information each user can get at the stationary state in IPATH can be very small. In addition, we can further represent the activation status of user u as vector π', where

$$\pi'[u] = \begin{cases} 1 & \text{if } \pi^*[u] \geq \theta, \\ 0 & \text{otherwise.} \end{cases} \tag{9.35}$$

In Eq. (9.35), $\pi'[u] = 1$ denotes that user u is activated. In practice, the value of $\pi^*[u]$ is usually in [0, 1] when the networks are sparse and the size of the seed set is small, and it can be represented approximately as following:

$$\pi'[u] \approx \lfloor \pi^*[u] - \theta + 1 \rfloor. \tag{9.36}$$

Based on this, we define the mapping function h between two vectors, where the floor function is applied to each element in the vector.

$$\pi' = h(\pi^*) = \lfloor \pi^* + \mathbf{c} \rfloor. \tag{9.37}$$

Here, $\mathbf{c}$ is a constant vector where each entry equals to $1 - \theta$.

9.5 Information Diffusion Across Online and Offline World

Besides the online world, information can also propagate within the offline word as well as between the online and offline world. In this section, we will use the workplace as one example to illustrate the information diffusion across both the online and offline world simultaneously.

9.5.1 Background Knowledge

On average, people nowadays need to spend more than 30% of their time at work everyday. According to the statistical data in [16], the total amount of time people spent at workplace in their life is tremendously large. For instance, a young man who is 20 years old now will spend 19.1% of his future time working [16]. Therefore, workplace is actually an easily neglected yet important social occasion for effective communication and information exchange among people in our social life.

Besides the traditional offline contacts, like face-to-face communication, telephone calls and messaging, to facilitate the cooperation and communications among employees, a new type of online social networks named enterprise social networks (ESNs) has been launched inside the firewalls of many companies [44, 45]. A representative example is Yammer, which is used by over 500,000 leading businesses around the world, including 85% of the Fortune 500.[1] Yammer provides various online communication services for employees at workplace, which include instant online messaging, write/reply/like posts, file upload/download/share, etc. In summary, the communication means existing among employees at workplaces are so diverse, which can generally be divided into two categories [33]: (1) offline communication means, and (2) online virtual communication means.

In this section, we will study how information diffuses via both online and offline communication means among employees at workplace, which is formally defined as the "Information Diffusion in Enterprise" (IDE) problem [47].

Example 9.4 To help illustrate the IDE problem more clearly, we also give an example in Fig. 9.6. The left plot of Fig. 9.6 is about an online ESN, employees in which can perform various social activities. For instances, employees can follow each other, can write/reply/like posts online, and posts written by them can also @certain employees to send notifications, which create various online information diffusion channels (i.e., the green lines) among employees. Meanwhile, the relative management

---

[1]https://about.yammer.com/why-yammer/.

Fig. 9.6 An example of information diffusion at workplace

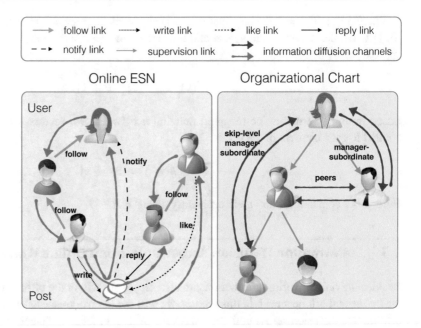

relationships among the employees in the company can be represented with the organizational chart (i.e., the right plot), which is a tree-structure diagram connecting employees via supervision links (from managers to subordinates). Colleagues who are physically close in the organizational chart (e.g., peers, manager-subordinates) may have more chance to meet in the offline workplace. For example, subordinates need to report to their managers regularly, peers may co-operate to finish projects together, which can form various offline information diffusion channels (i.e., the red lines) among employees at workplace.

Definition 9.8 (Enterprise Social Networks (ESNs)) Online *enterprise social networks* are a new type of online social networks used in enterprises to facilitate employees' communications and daily work, which can be represented as *heterogeneous information networks* $G = (\mathcal{V}, \mathcal{E})$, where $\mathcal{V} = \bigcup_i \mathcal{V}_i$ is the set of different kinds of nodes and $\mathcal{E} = \bigcup_j \mathcal{E}_j$ is the union of complex links in the network.

In this section, we will use Yammer as an example of online ESNs. Yammer can be represented as $G = (\mathcal{V}, \mathcal{E})$, where node set $\mathcal{V} = \mathcal{U} \cup \mathcal{O} \cup \mathcal{P}$ and $\mathcal{U}$, $\mathcal{O}$, and $\mathcal{P}$ are the sets of users, groups, and posts, respectively; link set $\mathcal{E} = \mathcal{E}_s \cup \mathcal{E}_j \cup \mathcal{E}_w \cup \mathcal{E}_r \cup \mathcal{E}_l$ denoting the union of social, group membership, write, reply, and like links in Yammer, respectively. At the workplace, information of various topics can propagate among the employees simultaneously.

Definition 9.9 (Organizational Chart) *Organizational chart* is a diagram outlining the structure of an organization as well as the relative ranks of employees' positions and jobs, which can be represented as a rooted tree $C = (\mathcal{N}, \mathcal{L}, root)$, where $\mathcal{N}$ denotes the set of employees and $\mathcal{L}$ is the set of directed *supervision links* from managers to subordinates in the company, *root* usually represents the CEO by default.

Each employee in the company can create exactly one account in Yammer with valid employment ID, i.e., there is *one-to-one* correspondence between the users in Yammer and employees in the organization chart. For simplicity, in this section, we assume the user set in online ESN to be identical to the employee set in the organizational chart (i.e., $\mathcal{U} = \mathcal{N}$) and we will use "Employee" to denote individuals in both online ESN and offline organizational chart by default.

To address all the above challenges, a novel information diffusion model MUSE (Multi-source Multi-channel Multi-topic diffUsion SElection) proposed in [47] will be introduced in this section. MUSE extracts and infers sets of online, offline, and hybrid (of online and offline) diffusion channels among employees across online ESN and offline organizational structure. Information propagated via different channels can be aggregated effectively in MUSE. Different diffusion channels will be weighted according to their importance learned from the social activity log data with optimization techniques and top-K effective diffusion channels will be selected in MUSE finally.

9.5.2 Preliminary

Before we talk about the detailed components involved in MUSE, we will introduce the general framework of the MUSE model in this part. Formally, we denote the set of topics diffusing in the workplace as set $\mathcal{T}$. Three different diffusion sources will be our main focus in this section: online source, offline source, and the hybrid source (across online and offline sources). The diffusion channel set of all these three sources can be represented as $\mathcal{C}^{(on)}$, $\mathcal{C}^{(off)}$ and $\mathcal{C}^{(hyb)}$, respectively, whose sizes are $\left|\mathcal{C}^{(on)}\right| = k^{(on)}$, $\left|\mathcal{C}^{(off)}\right| = k^{(off)}$, $\left|\mathcal{C}^{(hyb)}\right| = k^{(hyb)}$.

In MUSE, a set of users are activated initially, whose information will propagate in discrete steps within the network to other users. Let v be an employee at workplace who has been activated by topic $t \in \mathcal{T}$. For instance, at step τ, v will send an amount of $w^{(on),i}(v, u, t)$ information on topic t to u via the i_{th} channel in the online source (i.e., channel $c^{(on),i} \in \mathcal{C}^{(on)}$), where u is an employee following v in channel $c^{(on),i}$. The amount of information that u receives from v via all the channels in the online source at step τ can be represented as vector $\mathbf{w}^{(on)}(v, u, t) = [w^{(on),1}(v, u, t), w^{(on),2}(v, u, t), \ldots, w^{(on),k^{(on)}}(v, u, t)]$. Similarly, we can also represent the vectors of information u receives from v through channels in offline source and hybrid source as vectors $\mathbf{w}^{(off)}(v, u, t)$ and $\mathbf{w}^{(hyb)}(v, u, t)$, respectively.

Meanwhile, users in MUSE are associated thresholds to different topics, which are selected at random from the uniform distribution in range [0, 1]. Employee u can get activated by topic t if the information received from his active neighbors via diffusion channels of all these three sources can exceed his *activation threshold* $\theta(u, t)$ to topic t:

$$f\left(\mathbf{w}^{(on)}(\cdot, u, t), \mathbf{w}^{(off)}(\cdot, u, t), \mathbf{w}^{(hyb)}(\cdot, u, t)\right) \geq \theta(u, t), \tag{9.38}$$

where aggregation function $f(\cdot)$ maps the information u receives from all the channels to u's *activation probability* in range [0, 1]. Here, the vector $\mathbf{w}^{(on)}(\cdot, u, t) = [w^{(on),1}(\cdot, u, t), w^{(on),2}(\cdot, u, t), \ldots, w^{(on),k^{(on)}}(\cdot, u, t)]$, where $w^{(on),i}(\cdot, u, t)$ denotes the information received from all the employees u follows in channel $c^{(on),i}$, i.e.,

$$w^{(on),i}(\cdot, u, t) = \sum_{v \in \Gamma_{out}^{(on),i}(u)} w^{(on),i}(v, u, t). \tag{9.39}$$

Vectors $\mathbf{w}^{(off)}(\cdot, u, t)$ and $\mathbf{w}^{(hyb)}(\cdot, u, t)$ can be represented in a similar way. Once being activated, a user will stay active in the remaining rounds and each user can be activated at most once. Such a process will end if no new activations are possible.

Considering that individuals' *activation thresholds* $\theta(u, t)$ to topic t is pre-determined by the uniform distribution, next we will focus on studying the information received via channels of the *online*, *offline*, and *hybrid* sources and the *aggregation function* $f(\cdot)$ in detail.

9.5.3   Online Diffusion Channel

Online ESNs provide various communication tools for employees to contact each other, where individuals who have no social connections can still pass information via many other connections. Each connection among employees can form an information diffusion channel in online ESN. In this section, we will introduce the various diffusion channels among employees extracted based on a set of *online social meta paths* [30] from the heterogeneous information in the online ESN. Before that, we first introduce the schema of enterprise social network as follows.

Based on an online ESN $G = (\mathcal{V}, \mathcal{E})$, we can represent its network schema as $S_G = (\mathcal{T}_G, \mathcal{R}_G)$, where $\mathcal{T}_G$ and $\mathcal{R}_G$ represent the sets of node types and link types in network G, respectively. For example, for the Yammer network, we can define its schema as S_G, where the node type set $\mathcal{T}_G = \{Employee, Post\}$, link type set $\mathcal{R}_G = \{Social^{1/-1}, Write^{1/-1}, Reply^{1/-1}, Like^{1/-1}, Notify^{1/-1}\}$, and the superscript -1 denotes the reverse of the corresponding link type in online ESN.

In enterprise social networks, individuals can (1) get information from employees they follow (i.e., their followees) and (2) people that their "followees" follow (i.e., 2nd level followees), and obtain information from employees by (3) viewing and replying their posts, (4) viewing and liking their posts, as well as (5) getting notified by their posts (i.e., explicitly @ certain users in posts). MUSE extracts 5 different *online social meta paths* from the online ESN, whose physical meanings, representations, and abbreviated notations are listed as follows:

- Followee: $Employee \xleftarrow{Social^{-1}} Employee$, whose notation is Φ_1.
- Followee-Followee: $Employee \xleftarrow{Social^{-1}} Employee \xleftarrow{Social^{-1}} Employee$, whose notation is Φ_2.
- Reply Post: $Employee \xleftarrow{Reply^{-1}} Post \xleftarrow{Write} Employee$, whose notation is Φ_3.
- Like Post: $Employee \xleftarrow{Like^{-1}} Post \xleftarrow{Write} Employee$, whose notation is Φ_4.
- Post Notification: $Employee \xleftarrow{Notify} Post \xleftarrow{Write} Employee$, whose notation is Φ_5.

The direction of the links denotes the information diffusion direction and end of the diffusion links (i.e., the first employee of the above paths) represents the target employee to receive the information. For example, Φ_1 denotes the target user receives information from his followees, while Φ_5 means that the target employee receives information from employees who have ever written posts @ the target employee.

Each of the above *online social meta path* defines an information diffusion channel among individuals in online ESN. As a result, in this section, $\mathcal{C}^{(on)} = \{\Phi_1, \Phi_2, \Phi_3, \Phi_4, \Phi_5\}$ and $k^{(on)} = 5$ and Φ_i is identical to $c^{(on),i}$ mentioned before (denoting the i_{th} online diffusion channel). Based on each of these online social meta paths, we can extract the corresponding path instances connecting employees u and v (i.e., the concrete information diffusion traces from v to u), which can be represented as set $\mathcal{P}_{\Phi_i}^{(on)}(v \rightsquigarrow u)$, for $\forall \Phi_i \in \mathcal{C}^{(on)}$. Furthermore, let $\mathcal{P}_{\Phi_i}^{(on)}(v \rightsquigarrow \cdot)$ and $\mathcal{P}_{\Phi_i}^{(on)}(\cdot \rightsquigarrow u)$ be the sets of path instances of Φ_i going out from v and going into u, respectively, with which we can define the amount of information propagating from v to u via diffusion channel $c^{(on),i} = \Phi_i$ to be

$$w^{(on),i}(v, u, t) = \frac{2 \left| \mathcal{P}_{\Phi_i}^{(on)}(v \rightsquigarrow u) \right| \cdot I(v, t)}{\left| \mathcal{P}_{\Phi_i}^{(on)}(v \rightsquigarrow \cdot) \right| + \left| \mathcal{P}_{\Phi_i}^{(on)}(\cdot \rightsquigarrow u) \right|}, \tag{9.40}$$

where binary function $I(v, t) = 1$ if v has been activated by topic t and 0 otherwise.

In the above definition, the proportion of information propagated from v to u via the communication channels (i.e., $w^{(on),i}(v, u, t)$) can denote how close these two users are, which depends on (1) the number of concrete diffusion path instances between them (i.e., $\mathcal{P}_{\Phi_i}^{(on)}(v \rightsquigarrow u)$); (2) the out-degree in the channel from v (i.e., $\mathcal{P}_{\Phi_i}^{(on)}(v \rightsquigarrow \cdot)$); and (3) the in-degree in the channel to u (i.e., $\mathcal{P}_{\Phi_i}^{(on)}(\cdot \rightsquigarrow u)$).

9.5.4 Offline Diffusion Channel

Employees' offline interactions are actually confidential to both companies and the public, which is very hard to know exactly. To infer the potential offline information diffusion channels at workplace, a set of potential information diffusion channels among individuals are extracted based on the organizational chart of the company. Similar to online enterprise social networks, we can define the schema of the organization chart as $S_C = (\mathcal{T}_C, \mathcal{R}_C)$, where $\mathcal{T}_C = \{Employee\}$ and $\mathcal{R}_C = \{Supervision^{1/-1}\}$. In offline workplace, the most common social interaction should happen between close colleagues, e.g., peers, manager-subordinate, and skip-level manager-subordinates, etc. The physical meaning and notations of offline social meta paths extracted in this section are listed as follows:

- Manager: $Employee \xleftarrow{Supervision} Employee$, whose notation is Ω_1.
- Subordinate: $Employee \xleftarrow{Supervision^{-1}} Employee$, whose notation is Ω_2.
- Peer: $Employee \xleftarrow{Supervision} Employee \xleftarrow{Supervision^{-1}} Employee$, whose notation is Ω_3.
- 2nd-Level Manager: $Employee \xleftarrow{Supervision} Employee \xleftarrow{Supervision} Employee$, whose notation is Ω_4.
- 2nd-Level Subordinate: $Employee \xleftarrow{Supervision^{-1}} Employee \xleftarrow{Supervision^{-1}} Employee$, whose notation is Ω_5.

Similarly, the direction of links represents the information flow direction and the ending employees of the paths denotes the target employee, who receives information. For instance, meta path Ω_1 means that the target employee receives information from his manager, while Ω_3 denotes that the target employee receives information from his peers.

Each employee at the workplace can be influenced by both his manager and his subordinates (if exist) and to clarify the difference between these two different diffusion channels, we define both Ω_1 and Ω_2 (as well as Ω_4 and Ω_5). Based on the above introduced offline social meta paths, the offline diffusion channel can be represented as set $\mathcal{G}^{(off)} = \{\Omega_1, \Omega_2, \Omega_3, \Omega_4, \Omega_5\}$ and $k^{(off)} = 5$, where Ω_i denotes the i_{th} offline diffusion channel among employees.

Based on offline social meta path, e.g., Ω_i, the amount of information on topic t propagating from employee v to u can be represented as

$$w^{(off),i}(v,u,t) = \frac{2\left|\mathcal{P}_{\Omega_i}^{(off)}(v \rightsquigarrow u)\right| \cdot I(v,t)}{\left|\mathcal{P}_{\Omega_i}^{(off)}(v \rightsquigarrow \cdot)\right| + \left|\mathcal{P}_{\Omega_i}^{(off)}(\cdot \rightsquigarrow u)\right|}, \tag{9.41}$$

where $\mathcal{P}_{\Omega_i}^{(off)}(v \rightsquigarrow u)$ denotes the offline social meta path instance set of Ω_i connecting v to u in the chart.

9.5.5   Hybrid Diffusion Channel

Besides the pure online/offline diffusion channels, information can also propagate across both online and offline worlds simultaneously. Consider, for example, two employees v and u who are not connected by any diffusion channels in online ESN or offline workplace, v can still influence u by activating u's manager via online contacts and the manager will further propagate the influence to v via offline interactions. To represent such a kind of diffusion channels, a set of *hybrid social meta paths* are also extracted in MUSE. As proposed in [16], every pair of people in the worlds can get connected via 6 hops (i.e., six degrees of separation theory). To avoid connecting all the employees by hybrid diffusion channels, we limit its length (i.e., the number of relations in the meta path) to 3 only. The set of hybrid social meta path used in this section, together with their physical meanings, notations are listed as follows:

- Followee-Manager: $Employee \xleftarrow{Social^{-1}} Employee \xleftarrow{Supervision} Employee$, whose notation is Ψ_1,
- Followee-Subordinate: $Employee \xleftarrow{Social^{-1}} Employee \xleftarrow{Supervision^{-1}} Employee$, whose notation is Ψ_2,
- Manager-Followee: $Employee \xleftarrow{Supervision} Employee \xleftarrow{Social^{-1}} Employee$, whose notation is Ψ_3,
- Subordinate-Followee: $Employee \xleftarrow{Supervision^{-1}} Employee \xleftarrow{Social^{-1}} Employee$, whose notation is Ψ_4,
- Followee-Peer: $Employee \xleftarrow{Social^{-1}} Employee \xleftarrow{Supervision} Employee \xleftarrow{Supervision^{-1}} Employee$, whose notation is Ψ_5,
- Peer-Followee: $Employee \xleftarrow{Supervision} Employee \xleftarrow{Supervision^{-1}} Employee \xleftarrow{Social^{-1}} Employee$, whose notation is Ψ_6.

where meta path, e.g., Ψ_1, denotes that the target employee receives information from his followee in online ESN, who gets information from his manager in the offline workplace. We can get $C^{(hyb)} = \{\Psi_1, \Psi_2, \Psi_3, \Psi_4, \Psi_5, \Psi_6\}$ and $k^{(hyb)} = 6$. Based on each hybrid diffusion channel, e.g., Ψ_i, the amount of information on topic t that v sends to u can be represented as

$$w^{(hyb),i}(v,u,t) = \frac{2\left|\mathcal{P}_{\Psi_i}^{(hyb)}(v \rightsquigarrow u)\right| \cdot I(v,t)}{\left|\mathcal{P}_{\Psi_i}^{(hyb)}(v \rightsquigarrow \cdot)\right| + \left|\mathcal{P}_{\Psi_i}^{(hyb)}(\cdot \rightsquigarrow u)\right|}. \tag{9.42}$$

9.5.6 Channel Aggregation

Different diffusion channels deliver various amounts of information among employees via the online communications in ESN and offline contacts. In this subsection, we will focus on aggregating information propagated via different channels with the information aggregation function $f(\cdot)$: $\mathbb{R}^{n \times 1} \rightarrow [0, 1]$, which can map the amount of information received by employees to their activation probabilities. Generally, any function that can map real number to probabilities in range $[0, 1]$ can be applied and without loss of generality, we will use the logistic function $f(x) = \frac{e^x}{1+e^x}$ [6] in this section.

Based on the information on topic t received by u via the online, offline, and hybrid diffusion channels, we can represent u's activation probability to be:

$$
f\left(\mathbf{w}^{(on)}(\cdot, u, t), \mathbf{w}^{(off)}(\cdot, u, t), \mathbf{w}^{(hyb)}(\cdot, u, t)\right)
$$

$$
= \frac{e^{\left(g(\mathbf{w}^{(on)}(\cdot,u,t))+g(\mathbf{w}^{(off)}(\cdot,u,t))+g(\mathbf{w}^{(hyb)}(\cdot,u,t))+\theta_0\right)}}{1 + e^{\left(g(\mathbf{w}^{(on)}(\cdot,u,t))+g(\mathbf{w}^{(off)}(\cdot,u,t))+g(\mathbf{w}^{(hyb)}(\cdot,u,t))+\theta_0\right)}}, \tag{9.43}
$$

where function $g(\cdot)$ linearly combines the information in different channels belonging to certain sources and θ_0 denotes the weight of the constant factor. Terms $g(\mathbf{w}^{(on)}(\cdot, u, t))$, $g(\mathbf{w}^{(off)}(\cdot, u, t))$, and $g(\mathbf{w}^{(hyb)}(\cdot, u, t))$ can be represented as follows:

$$
g(\mathbf{w}^{(on)}(\cdot, u, t)) = \sum_{i=1}^{k^{(on)}} \alpha_i \cdot \sum_{v \in \Gamma_{out}^{(on),i}(u)} w^{(on),i}(v, u, t), \tag{9.44}
$$

$$
g(\mathbf{w}^{(off)}(\cdot, u, t)) = \sum_{i=1}^{k^{(off)}} \beta_i \cdot \sum_{v \in \Gamma_{out}^{(off),i}(u)} w^{(off),i}(v, u, t), \tag{9.45}
$$

$$
g(\mathbf{w}^{(hyb)}(\cdot, u, t)) = \sum_{i=1}^{k^{(hyb)}} \gamma_i \cdot \sum_{v \in \Gamma_{out}^{(hyb),i}(u)} w^{(hyb),i}(v, u, t), \tag{9.46}
$$

where $\alpha_i, \beta_i, \gamma_i$ are the weights of different *online*, *offline*, and *hybrid* diffusion channels, respectively and $\sum_{i=1}^{k^{(on)}} \alpha_i + \sum_{i=1}^{k^{(off)}} \beta_i + \sum_{i=1}^{k^{(hyb)}} \gamma_i + \theta_0 = 1$. Depending on the roles of different diffusion channels, the weights can be

- > 0, if positive information in the channel will increase employees' activation probability;
- $= 0$, if positive information in the channel will not change employees' activation probability;
- < 0, if positive information in the channel will decrease employees' activation probability.

In MUSE, weights of certain diffusion channels can be negative. As a result, the likelihood for a node to become active will no longer grow monotonically in the MUSE diffusion model. The optimal weights of different diffusion channels can be learned from the users' infection record data. Different diffusion channels will be ranked according to their importance and top-k diffusion channels which can increase individuals' activation probabilities will be selected in the next subsection.

9.5.7   Channel Weighting and Selection

The historical users' infection records data can be represented as a set of tuples $\{(u, t)\}_{u,t}$, where tuple (u, t) represents that user u gets activated by topic t. Such a tuple set can be split into three parts according to ratio 3:1:1 in the order of the timestamps, where threefolds are used as the training set, onefold is used as the validation set and onefold as the test set. We will use the training set data to calculate the activation probabilities of individuals getting activated by topics in both the validation set and test set, while validation set is used to learn the weights of different diffusion channels and test set is used to evaluate the learned model.

Let $\mathcal{V} = \{(u, t)\}_{u,t}$ be the validation set. Based on the amount of information propagating among employees in the workplace calculated with the training set, we can infer the probability of user u's (who has not been activated yet) get activated by topic t, for $\forall(u, t) \in \mathcal{V}$, which can be represented with matrix $\mathbf{F} \in \mathbb{R}^{|\mathcal{U}| \times |\mathcal{T}|}$, where $F(i, j)$ denotes the inferred activation probability of tuple (u_i, t_j) in the validation set. Meanwhile, based on the validation set itself, we can get the ground-truth of users' infection records, which can be represented as a binary matrix $\mathbf{H} \in \{0, 1\}^{|\mathcal{U}| \times |\mathcal{T}|}$. In matrix $\mathbf{H}$, only entries corresponding tuples in the validation set are filled with value 1 and the remaining entries are all filled with 0. The optimal weights of information delivered in different diffusion channels (i.e., $\boldsymbol{\alpha}^*$, $\boldsymbol{\beta}^*, \boldsymbol{\gamma}^*, \theta_0^*$) can be obtained by solving the following objective function:

$$\boldsymbol{\alpha}^*, \boldsymbol{\beta}^*, \boldsymbol{\gamma}^*, \theta_0^* = \arg \min_{\boldsymbol{\alpha}, \boldsymbol{\beta}, \boldsymbol{\gamma}, \theta_0} \|\mathbf{F} - \mathbf{H}\|_F^2$$

$$s.t. \sum_{i=1}^{k^{(on)}} \alpha_i + \sum_{i=1}^{k^{(off)}} \beta_i + \sum_{i=1}^{k^{(hyb)}} \gamma_i + \theta_0 = 1. \tag{9.47}$$

The final objective function is not convex and can have multiple local optima, as the aggregation function (i.e., the logistic function) is not convex actually. MUSE solves the objective function and handle the non-convex issue by using a two-stage process to ensure the robust of the learning process as much as possible.

(1) Firstly, the above objective function can be solved by using the method of Lagrange multipliers [1], where the corresponding Lagrangian function of the objective function can be represented as

$$\mathcal{L}(\boldsymbol{\alpha}, \boldsymbol{\beta}, \boldsymbol{\gamma}, \theta_0, \eta)$$

$$= \|\mathbf{F} - \mathbf{H}\|_F^2 + \eta \left(\sum_{i=1}^{k^{(on)}} \alpha_i + \sum_{i=1}^{k^{(off)}} \beta_i + \sum_{i=1}^{k^{(hyb)}} \gamma_i + \theta_0 - 1 \right),$$

$$= \text{Tr}(\mathbf{F}\mathbf{F}^{\top} - \mathbf{F}\mathbf{H}^{\top} - \mathbf{H}\mathbf{F}^{\top} + \mathbf{H}\mathbf{H}^{\top}) + \eta \left(\sum_{i=1}^{k^{(on)}} \alpha_i + \sum_{i=1}^{k^{(off)}} \beta_i + \sum_{i=1}^{k^{(hyb)}} \gamma_i + \theta_0 - 1 \right). \tag{9.48}$$

By taking the partial derivatives of the Lagrange function with regard to variable $\alpha_i, i \in \{1, 2, \ldots, k^{(on)}\}$, we can get

$$\frac{\partial \mathcal{L}(\boldsymbol{\alpha}, \boldsymbol{\beta}, \boldsymbol{\gamma}, \theta_0, \eta)}{\partial \alpha_i} = \frac{\partial \text{Tr}(\mathbf{F}\mathbf{F}^{\top})}{\partial \alpha_i} - \frac{\partial \text{Tr}(\mathbf{F}\mathbf{H}^{\top})}{\partial \alpha_i} - \frac{\partial \text{Tr}(\mathbf{H}\mathbf{F}^{\top})}{\partial \alpha_i} + \frac{\partial \text{Tr}(\mathbf{H}\mathbf{H}^{\top})}{\partial \alpha_i}$$

$$+ \frac{\partial \eta \left(\sum_{i=1}^{k^{(on)}} \alpha_i + \sum_{i=1}^{k^{(off)}} \beta_i + \sum_{i=1}^{k^{(hyb)}} \gamma_i + \theta_0 - 1 \right)}{\partial \alpha_i}. \tag{9.49}$$

Term

$$\frac{\partial \eta \left(\sum_{i=1}^{k^{(on)}} \alpha_i + \sum_{i=1}^{k^{(off)}} \beta_i + \sum_{i=1}^{k^{(hyb)}} \gamma_i + \theta_0 - 1 \right)}{\partial \alpha_i} = \eta \tag{9.50}$$

$$\frac{\partial \mathrm{Tr}(\mathbf{FF}^\top)}{\partial \alpha_i} = \sum_{j=1}^{|\mathcal{U}|} \sum_{l=1}^{|\mathcal{T}|} \frac{\partial \mathbf{F}^2(j,l)}{\partial \alpha_i}$$

$$= \sum_{j=1}^{|\mathcal{U}|} \sum_{l=1}^{|\mathcal{T}|} 2 f(\mathbf{w}^{(on)}(\cdot, u_j, t_l), \mathbf{w}^{(off)}(\cdot, u_j, t_l), \mathbf{w}^{(hyb)}(\cdot, u_j, t_l)) \cdot \frac{e^y}{(1+e^y)^2} \cdot \frac{\partial y}{\partial \alpha_i}, \tag{9.51}$$

where the introduced term y denotes

$$y = g(\mathbf{w}^{(on)}(\cdot, u_j, t_l)) + g(\mathbf{w}^{(off)}(\cdot, u_j, t_l)) + g(\mathbf{w}^{(hyb)}(\cdot, u_j, t_l)) + \theta_0 \tag{9.52}$$

and its derivative can be represented as

$$\frac{\partial y}{\partial \alpha_i} = \frac{\partial g(\mathbf{w}^{(on)}(\cdot, u_j, t_l))}{\partial \alpha_i} = \sum_{v \in \Gamma_{out}^{(on),i}(u)} w^{(on),i}(v, u_j, t_k). \tag{9.53}$$

Similarly, we can obtain terms $\frac{\partial \mathrm{Tr}(\mathbf{FH}^\top)}{\partial \alpha_i}$, $\frac{\partial \mathrm{Tr}(\mathbf{HF}^\top)}{\partial \alpha_i}$, and $\frac{\partial \mathrm{Tr}(\mathbf{HH}^\top)}{\partial \alpha_i}$. By making $\frac{\partial \mathcal{L}(\boldsymbol{\alpha}, \boldsymbol{\beta}, \boldsymbol{\gamma}, \theta_0, \eta)}{\partial \alpha_i} = 0$, we can obtain an equation involving variables α_i, β_i, γ_i, θ_0 and η. Furthermore, we can calculate the partial derivatives of the Lagrange function with regards to variable β_i, γ_i, θ_0, and η, respectively, and make the equation equal to 0, which will lead to an equation group about variables α_i, β_i, γ_i, θ_0, and η. The equation group can be solved with open source toolkits, e.g., SciPy Nonlinear Solver,[2] effectively. By giving the variables with different initial values, multiple solutions (i.e., multiple local optimal points) can be obtained by resolving the objective function.

(2) Secondly, the local optimal points obtained are further applied to the objective function and the one achieving the lowest objective function value is selected as the final results (i.e., the weights of different channels).

According to the learned weights, different diffusion channels can be ranked according to their importance in delivering information to activate employees in the workplace. Considering that, some diffusion channels may not perform very well in information propagation (e.g., those with negative or zero learned weights), top-k channels that can increase employees' activation probabilities are selected as the effective channels used in MUSE model finally. In other words, k equals to the number of diffusion channels with positive weights learnt from the above objective function. Such a process is formally called diffusion channel weighting and selection in this section.

9.6   Summary

In this section, we introduced various information diffusion models, which can depict the information propagation process in online social networks. Via the interactions among users, information of

---

[2]http://docs.scipy.org/doc/scipy-0.14.0/reference/optimize.nonlin.html.

various topics can diffuse among users in online social networks. In the information diffusion process, the information topics, the propagation channels, the target users, and the whole network properties can all affect the information diffusion and lead to different observations.

For the traditional single-homogeneous networks, we introduced several classic information diffusion models, including the threshold based diffusion model, cascade based diffusion model, epidemic diffusion model, and heat diffusion model. For these diffusion models, they assume there exist one single type of diffusion channels among users inside the network, and they can work well for the single-network scenarios with one single topic.

To model the information diffusion of multiple topics with intertwined relationships, we introduced an intertwined information diffusion model TLT by extending the classic LT model. The TLT diffusion model can update users' thresholds towards different topics with considerations about the relationships among the topics. To describe the information diffusion process in signed networks, we introduced the MFC diffusion model, which incorporates the link polarities into the model.

To model the information diffusion across multiple social networks, we introduced two different information diffusion models based on network coupling and cross-network random walk, respectively. In the network coupling based diffusion model, we introduced an efficient method to check the status of users in the diffusion process. In the random walk based diffusion model, users may propagate information either within networks or across networks with a certain chance.

Finally, at the last section, we introduced the information diffusion model MUSE at the workplace across the online and offline world. The introduced MUSE model is built based on the meta path concept. By aggregating the information propagated from the diverse meta paths, MUSE is able to compute the optimal weights assigned for these diverse channels and sources, respectively.

9.7   Bibliography Notes

Information diffusion has been a hot research topic in the last decade and dozens of papers have been published on this topic so far. Domingos and Richardson [7, 28] are the first to propose to study the influence propagation based on knowledge-sharing sites. Kempe et al. [13] are the first to study the influence propagation problem through social networks and propose two famous diffusion models: Independent cascade (IC) model and linear threshold (LT) model, which have been the basis of many diffusion models proposed later.

In recent years, signed networks have gained increasing attention. Li et al. [19] studied the influence diffusion dynamics in social networks with friend and foe relationships. Polarity related influence maximization problem in signed social networks is studied in [20], where a new diffusion model, corresponding to the polarity independent cascade (P-IC) model, is proposed. If the readers are interested in these models proposed for the signed networks, you may take a look at these articles to get more information.

Meanwhile, some works have also been done on studying information diffusion problems by considering multiple networks/sources. Zhan et al. [38] propose to study the information diffusion problem across two partially aligned social networks based on the extracted multi-relations among users. Zhan et al. [41] also introduce a novel information diffusion model based on random walk, which can be used to depict the cross-network information diffusion process. Nguyen et al. [25] propose a coupling-based diffusion models to study the information diffusion problem in multiplex social networks. Myers et al. [23] and Lin et al. [22] present two different information diffusion models incorporating both external influence sources and the internal influence among users in online social networks.

In addition, a comprehensive survey about the existing information diffusion models and research works has been provided in [11, 42] and the readers may also refer to these articles when reading this chapter.

9.8 Exercises

1. (Easy) In both LT and IC diffusion models, let $\mathcal{V}^a(t)$ denote the set of activated users at step t in the diffusion process. Please prove that the following statement holds for both LT and IC models:

$$\mathcal{V}^a(t-1) \subseteq \mathcal{V}^a(t), \forall t \in \{1, 2, \ldots\} \tag{9.54}$$

2. (Easy) Please think about the *heat diffusion model*, and explain what is the potential problem of this model when applied to depict the information diffusion process in online social networks (Hint: What will be the equilibrium status when the diffusion ends? Can the heat diffusion equilibrium status model the information diffusion ending status?)
3. (Easy) Please explain how the MFC handles the asymmetrical effects of positive and negative links in the information diffusion process.
4. (Easy) Please explain why the relaxed activation criterion adopted in the *network coupling based information diffusion* model will create some latency in users' status checking.
5. (Easy) Please briefly explain why the meta path selection is necessary in the MUSE model, especially when there exist a large number of meta paths defined in the model.
6. (Medium) Please implement the IC and LT diffusion models, and output the expected number of infected users based on a synthetic social network dataset.
7. (Medium) Please try to implement the TLT diffusion model with a preferred programming language, and test the algorithm with a synthetic social network dataset.
8. (Medium) Please implement the *random walk* based information diffusion model IPATH with a preferred programming language.
9. (Hard) Please revise the SIRS model by incorporating the population birth and death into considerations, and try to derive the partial derivative of the population at different stages regarding the time.
10. (Hard) Please try to implement the MUSE model together with its meta path weighting and selection algorithm.

References

1. D. Bertsekas, *Constrained Optimization and Lagrange Multiplier Methods (Optimization and Neural Computation Series)* (Athena Scientific, Nashua, 1996)
2. S. Bharathi, D. Kempe, M. Salek, Competitive influence maximization in social networks, in *Internet and Network Economics. WINE 2007* (Springer, Berlin, 2007)
3. T. Carnes, R. Nagarajan, S. Wild, A. Zuylen, Maximizing influence in a competitive social network: a follower's perspective, in *Proceedings of the Ninth International Conference on Electronic Commerce (ICEC '07)* (ACM, New York, 2007)
4. W. Chen, A. Collins, R. Cummings, T. Ke, Z. Liu, D. Rincon, X. Sun, Y. Wang, W. Wei, Y. Yuan, Influence maximization in social networks when negative opinions may emerge and propagate – Microsoft research, in *Proceedings of the Eleventh SIAM International Conference on Data Mining (SDM 2011)*, 2011.
5. S. Datta, A. Majumder, N. Shrivastava, Viral marketing for multiple products, in *2010 IEEE International Conference on Data Mining* (IEEE, Piscataway, 2010)
6. J.S. deCani, R.A. Stine, A note on deriving the information matrix for a logistic distribution. Am. Stat. **40**(3), 220–222 (1986)

7. P. Domingos, M. Richardson, Mining the network value of customers, in *Proceedings of the Seventh ACM SIGKDD International Conference on Knowledge Discovery and Data Mining (KDD '01)* (ACM, New York, 2001)

8. R. Ghosh, K. Lerman, Non-conservative diffusion and its application to social network analysis. arXiv preprint arXiv:1102.4639 (2011)

9. J. Goldenberg, B. Libai, E. Muller, Talk of the network: a complex systems look at the underlying process of word-of-mouth. Mark. Lett. **12**(3), 211–223 (2001)

10. M. Granovetter, Threshold models of collective behavior. Am. J. Sociol. **83**(6), 1420–1443 (1978)

11. A. Guille, H. Hacid, C. Favre, D. Zighed, Information diffusion in online social networks: a survey. SIGMOD Rec. **42**(2), 17–28 (2013)

12. H. Hethcote, *Three Basic Epidemiological Models* (Springer, Berlin, 1989)

13. D. Kempe, J. Kleinberg, É. Tardos, Maximizing the spread of influence through a social network, in *Proceedings of the Ninth ACM SIGKDD International Conference on Knowledge Discovery and Data Mining (KDD '03)* (ACM, New York, 2003)

14. D. Kempe, J. Kleinberg, É. Tardos, Influential nodes in a diffusion model for social networks, in *Proceedings of the 32nd International Conference on Automata, Languages and Programming (ICALP'05)* (Springer, Berlin, 2005)

15. W. Kermack, A. McKendrick, A contribution to the mathematical theory of epidemics. Proc. Roy. Soc. Lond. A Math. Phys. Eng. Sci. **115**(772), 700–721 (1927)

16. J. Kleinberg, The small-world phenomenon: an algorithmic perspective, in *Proceedings of the Thirty-Second Annual ACM Symposium on Theory of Computing (STOC '00)* (ACM, New York, 2000)

17. J. Leskovec, D. Huttenlocher, J. Kleinberg, Predicting positive and negative links in online social networks, in *Proceedings of the 19th International Conference on World Wide Web (WWW '10)* (ACM, New York, 2010)

18. J. Leskovec, D. Huttenlocher, J. Kleinberg, Signed networks in social media, in *Proceedings of the SIGCHI Conference on Human Factors in Computing Systems (CHI '10)* (ACM, New York, 2010)

19. Y. Li, W. Chen, Y. Wang, Z. Zhang, Influence diffusion dynamics and influence maximization in social networks with friend and foe relationships, in *Proceedings of the Sixth ACM International Conference on Web Search and Data Mining (WSDM '13)* (ACM, New York, 2013)

20. D. Li, Z. Xu, N. Chakraborty, A. Gupta, K. Sycara, S. Li, Polarity related influence maximization in signed social networks. PLoS One **9**(7), e102199 (2014)

21. J. Lienhard, *A Heat Transfer Textbook: Fourth Edition* (Dover Civil and Mechanical Engineering. Dover Publications, New York, 2013)

22. S. Lin, F. Wang, Q. Hu, P. Yu, Extracting social events for learning better information diffusion models, in *Proceedings of the 19th ACM SIGKDD International Conference on Knowledge Discovery and Data Mining (KDD '13)* (ACM, New York, 2013)

23. S. Myers, C. Zhu, J. Leskovec, Information diffusion and external influence in networks, in *Proceedings of the 18th ACM SIGKDD International Conference on Knowledge Discovery and Data Mining (KDD '12)* (ACM, New York, 2012)

24. R. Narayanam, A. Nanavati, Viral marketing for product cross-sell through social networks, in *Machine Learning and Knowledge Discovery in Databases (ECML PKDD 2012)* (Springer, Berlin, 2012)

25. D. Nguyen, H. Zhang, S. Das, M. Thai, T. Dinh, Least cost influence in multiplex social networks: model representation and analysis, in *2013 IEEE 13th International Conference on Data Mining* (IEEE, Piscataway, 2013)

26. R. Pastor-Satorras, C. Castellano, P. Mieghem, A. Vespignani, Epidemic processes in complex networks. CoRR, abs/1408.2701 (2014)

27. D. Peleg, Local majority voting, small coalitions and controlling monopolies in graphs: a review, in *Proceedings 3rd Colloquium on Structural Information and Communication Complexity* (Carleton University Press, Siena, 1996)

28. M. Richardson, P. Domingos, Mining knowledge-sharing sites for viral marketing, in *Proceedings of the Eighth ACM SIGKDD International Conference on Knowledge Discovery and Data Mining (KDD '02)* (ACM, New York, 2002)

29. D. Shah, T. Zaman, Rumors in a network: who's the culprit? IEEE Trans. Inf. Theory **57**(8), 5163–5181 (2011)

30. Y. Sun, J. Han, X. Yan, P. Yu, T. Wu, PathSim: meta path-based top-k similarity search in heterogeneous information networks, in *Proceedings of the VLDB Endowment* (2011)

31. J. Tang, Y. Chang, C. Aggarwal, H. Liu, A survey of signed network mining in social media. *ACM Computing Surveys, to appear,* CoRR abs/1511.07569 (2015)

32. H. Tong, C. Faloutsos, J.-Y. Pan, Fast random walk with restart and its applications (2006)

33. T. Turner, P. Qvarfordt, J. Biehl, G. Golovchinsky, M. Back, Exploring the workplace communication ecology, in *Proceedings of the SIGCHI Conference on Human Factors in Computing Systems (CHI '10)* (ACM, New York, 2010)

34. R. West, H. Paskov, J. Leskovec, C. Potts, Exploiting social network structure for person-to-person sentiment analysis. Trans. Assoc. Comput. Linguist. **2**, 297–310 (2014)

35. K. Wilcox, A. T. Stephen, Are close friends the enemy? Online social networks, self-esteem, and self-control. J. Consum. Res. **40**(1), 90–103 (2012)

36. Y. Yao, H. Tong, X. Yan, F. Xu, J. Lu, Matri: a multi-aspect and transitive trust inference model. in *Proceedings of the 22nd International Conference on World Wide Web* (ACM, New York, 2013)

37. J. Ye, H. Cheng, Z. Zhu, M. Chen, Predicting positive and negative links in signed social networks by transfer learning, in *Proceedings of the 22nd International Conference on World Wide Web (WWW '13)* (ACM, New York, 2013)

38. Q. Zhan, J. Zhang, S. Wang, P. Yu, J. Xie, Influence maximization across partially aligned heterogeneous social networks, in *Pacific-Asia Conference on Knowledge Discovery and Data Mining* (Springer, Cham, 2015)

39. Q. Zhan, J. Zhang, X. Pan, P. Yu, Discover tipping users for cross network influencing, in *2016 IEEE 17th International Conference on Information Reuse and Integration (IRI)* (IEEE, Piscataway, 2016)

40. Q. Zhan, J. Zhang, P. Yu, S. Emery, J. Xie, Inferring social influence of anti-tobacco mass media campaigns, in *2016 IEEE International Conference on Bioinformatics and Biomedicine (BIBM)* (IEEE, Piscataway, 2016)

41. Q. Zhan, J. Zhang, P. Yu, J. Xie, Viral marketing through aligned networks (2018)

42. J. Zhang, Social network fusion and mining: a survey (2018)

43. J. Zhang, P. Yu, Community detection for emerging networks, in *2015 SIAM International Conference on Data Mining* (2015)

44. J. Zhang, Y. Lv, P. Yu, Enterprise social link recommendation, in *Proceedings of the 24th ACM International on Conference on Information and Knowledge Management (CIKM '15)* (ACM, New York, 2015)

45. J. Zhang, P. Yu, Y. Lv, Organizational chart inference, in *Proceedings of the 21th ACM SIGKDD International Conference on Knowledge Discovery and Data Mining (KDD '15)* (ACM, New York, 2015)

46. J. Zhang, S. Wang, Q. Zhan, P. Yu, Intertwined viral marketing in social networks, in *2016 IEEE/ACM International Conference on Advances in Social Networks Analysis and Mining (ASONAM)* (IEEE, Piscataway, 2016)

47. J. Zhang, P. Yu, Y. Lv, Q. Zhan, Information diffusion at workplace, in *Proceedings of the 25th ACM International on Conference on Information and Knowledge Management (CIKM '16)* (ACM, New York, 2016)

48. J. Zhang, Q. Zhan, L. He, C. Aggarwal, P. Yu, Trust hole identification in signed networks, in *Machine Learning and Knowledge Discovery in Databases (ECML PKDD 2016)* (Springer, Cham, 2016)

49. H. Zhang, D. Nguyen, H. Zhang, M. Thai, Least cost influence maximization across multiple social networks. IEEE ACM Trans. Netw. **24**(2), 929–939 (2016)

50. J. Zhang, C. Aggarwal, P. Yu, Rumor initiator detection in infected signed networks, in *2017 IEEE 37th International Conference on Distributed Computing Systems (ICDCS)* (IEEE, Piscataway, 2017)

Viral Marketing

10

10.1 Overview

Via the social interactions among users, information of various topics, e.g., personal interests, products, commercial services, etc. can extensively propagate throughout the networks, where lots of users can get infected and become activated. Meanwhile, the social information diffusion can bring about great commercial values, and create lots of *viral marketing* [29] opportunities. Lots of commercial companies are utilizing the information diffusion phenomenon in online social networks to promote their products or services. For instance, Apple and Huawei have been promoting their latest cell phones via Facebook and Twitter. They can provide some free cell phone samples, coupons, or even cash to certain users (with lots of followers) in Facebook, and ask them to post some good review comments or advertising photos about the cell phone. Such information will propagate to their friends and followers, who may get activated to purchase the cell phone. Commercial promotions via the online social networks have become more and more important in recent years, which even surpass the traditional print media (like newspaper, magazine, TV, and radio). At the same time, viral marketing has also become one of the most important and secure revenue sources for many online social platforms, like Facebook and Twitter.

To achieve the maximum influence in the online social networks, the commercial companies may need to carry out serious investigations to select the initial user set for information spread. Formally, these information diffusion initiators are called the *seed users* in the existing research works [29]. The problem of selecting the optimal set of *seed users* is called the *influence maximization* problem [29,36] (or the *viral marketing* problem). Furthermore, in commercial promotion campaigns, besides releasing their own advertisements, the competitors may release lots of fake news [52] (i.e., rumors [28,46]) about the other competing products to cheat the consumers. Identification of these rumor initiators in the online social networks timely can avoid the negative impacts on the marketing activities greatly.

In this chapter, we will study the *seed user* and *rumor initiators* identification problems in *viral marketing*, which are all the crucial problems for designing the optimal marketing strategies for companies in carrying out their promotion campaigns. The problems to be studied in this chapter are mostly based on the information diffusion models introduced in the previous chapter.

In Sect. 10.2, we will first introduce the formulation of the influence maximization problem [29], and introduce several existing seed user selection strategies based on either approximation or heuristics [11, 22, 23, 29]. In Sect. 10.3, we will introduce the *intertwined influence maximization* problem [50] for the seed user selection in promoting multiple products with intertwined relationships.

© Springer Nature Switzerland AG 2019
J. Zhang, P. S. Yu, *Broad Learning Through Fusions*,
https://doi.org/10.1007/978-3-030-12528-8_10

By considering the information diffusion across networks, the *cross-network influence maximization* and seed user selection strategies [47,48] will be introduced in Sect. 10.4. To effectively and efficiently detect the *rumor initiators* [51], a new *rumor initiator* detection algorithm [51] is to be introduced in Sect. 10.5, which is introduced based on the signed network setting but can be applied to other networks as well.

10.2   Traditional Influence Maximization

The *influence maximization* problem first proposed in [29] has been studied for several years, and dozens of algorithms have been introduced to select the optimal marketing strategies for the problem. In the *influence maximization* problem, the *marketing strategy* usually refers to the set of seed users selected by the companies involved in the promotion campaign. In this section, we will introduce the traditional *influence maximization* problem, and provide a brief review of the existing seed user selection algorithms proposed for the problem.

10.2.1 Influence Maximization Problem

The *influence maximization* problem is one of the most fundamental research problems, which studies the word-of-mouth effects on promoting new products and making profitable services. Assuming the information diffusion model has been provided, the *influence maximization* problem aims at identifying the optimal marketing strategy (i.e., the seed users), who can lead to the maximal influence in the social networks.

Definition 10.1 (Influence Maximization) Given a network structure $G = (\mathcal{V}, \mathcal{E})$, where $\mathcal{V}$ denotes the users and $\mathcal{E}$ denotes the relationships among them. The *influence maximization* problem aims at selecting a set of seed users $\mathcal{S} \subset \mathcal{V}$, who can lead to the maximal influence inside the network. Generally, the size of the seed user set is limited by some budget, e.g., $|\mathcal{S}| \leq k$.

In the influence maximization problem, the diffusion model is not the focus, which will be provided as black-box taking the initial seed users as the input and producing the influence number as the output. To quantify the impact achieved by a diffusion model, we can introduce the *influence function* [29] here, which projects the initial seed users to the influence (i.e., the number of infected users).

Definition 10.2 (Influence Function) Let $\mathcal{S} \subset \mathcal{V}$ denote the set of seed users who will initiate the influence propagation. Given a diffusion model M, the *influence function* can be represented as $\sigma_M : \mathcal{S} \to \mathbb{R}$, which projects the seed user set $\mathcal{S}$ to the expected number of infected users after the diffusion process stops. With an input seed user set $\mathcal{S}$, based on the provided diffusion model M, the number of users who can be infected by these seed users can be represented as $\sigma_M(\mathcal{S})$.

With the *influence function* defined above, the *influence maximization* problem can be formally defined as the following optimization problem:

$$\max_{\mathcal{S} \subset \mathcal{V}} \sigma_M(\mathcal{S})$$

$$s.t. |\mathcal{S}| \leq k. \tag{10.1}$$

Depending on the specific representation of the influence function $\sigma_M(\cdot)$, the above optimization function can have different varying degrees of difficulty. In most cases, based on the diffusion models, like LT and IC [29], introduced in the previous chapter, solving the objective function may need to enumerate all the potential combination of the seed user set, which renders the problem to be NP-hard.

Generally, by selecting more users to initiate the promotion campaign, the diffusion model will achieve a larger influence and can infect more people. For some diffusion models, the *influence function* usually has the *monotonicity* property. Meanwhile, the total number of users in the social network is limited, and the influence cannot keep increasing as the seed user set size increases. As more users are selected as the seed user, the influence gain obtained by involving these extra seed users will degrade steadily. For some diffusion models, the *influence function* normally follows the *marginal decline*, and has the *submodularity* property. These two properties about the *influence function* are very important for the *influence maximization* problem, which serve as the foundations for many approximation solutions to the problem.

Definition 10.3 (Monotonicity) Given an influence function $\sigma_M(\mathcal{S})$ based on the diffusion model M, the function has the *monotonicity* property iff $\sigma_M(\mathcal{S}) < \sigma_M(\mathcal{S}')$ holds for any $\mathcal{S} \subset \mathcal{S}' \subset \mathcal{V}$.

Definition 10.4 (Submodularity) Given an influence function $\sigma_M(\mathcal{S})$ based on the diffusion model M, the function has the *submodularity* property iff $\sigma_M(\mathcal{S} \cup \{u\}) - \sigma_M(\mathcal{S}) \geq \sigma_M(\mathcal{S}' \cup \{u\}) - \sigma_M(\mathcal{S}')$, for all user $u \in \mathcal{V}$, $u \notin \mathcal{S}$, $u \notin \mathcal{S}'$ and $\mathcal{S} \subset \mathcal{S}' \subset \mathcal{V}$.

Currently, most of the existing algorithms proposed for the influence maximization problem are based on either approximation algorithms or heuristics. In the following subsections, we will introduce some representative algorithms belonging to these two categories.

10.2.2 Approximated Seed User Selection

Based on many of the diffusion models, like LT or IC, the *influence maximization* problem is NP-hard to solve. In the case when the *influence function* is both *monotone* and *submodular*, according to the existing works [29], algorithms applying *greedy strategy* can achieve $(1 - \frac{1}{e})$-approximation of the optimal result. In this section, we will introduce several approximation-based seed user selection algorithms, which include *greedy* [29] and *CELF* [11] algorithms.

10.2.2.1 Greedy Algorithm
Based on the classic diffusion models M, like LT and IC models introduced in Sects. 9.2.1.1 and 9.2.2.1, respectively, the *influence maximization* problem is shown to be NP-hard [29]. Given the objective seed user set size k and the diffusion model, the *influence maximization* problem aims at identifying the optimal seed users of size k who can lead to the maximal influence inside the social network. Let $\mathcal{S}$ denote the set of selected seed users, the influence obtained by these seed users can be represented by the influence function $\sigma_M(\mathcal{S})$. Generally, the value of $\sigma_M(\mathcal{S})$ can be obtained by running the diffusion model M on the selected seed user set $\mathcal{S}$. Based on the LT and IC diffusion models, the influence function has both the *monotonicity* and *submodularity* properties, and the *greedy* algorithm can achieve a constant approximation ratio.

Algorithm 1 Greedy algorithm

Require: Input network G with user node set $\mathcal{V}$
          Influence function $\sigma_M(\cdot)$
          Seed use set size k
Ensure: Seed user set $\mathcal{S}$
 1: initialize $\mathcal{S} = \emptyset$
 2: **while** $|\mathcal{S}| \leq k$ **do**
 3:    $u^* = \arg\max_{u \in \mathcal{V} \backslash \mathcal{S}} \sigma_M(\mathcal{S} \cup \{u\}) - \sigma_M(\mathcal{S})$
 4:    $\mathcal{S} = \mathcal{S} \cup \{u\}$
 5: **end while**
 6: Return $\mathcal{S}$

In the *greedy* algorithm, the k seed users are selected in k rounds, and the user who can lead to the maximum marginal influence gain (of the influence function $\sigma_M(\mathcal{S})$) will be selected in each of the rounds as the seed user. For instance,

- *Round 1*: In round 1, the original seed user set $\mathcal{S}^{(0)}$ is initialized as an empty set, i.e., $\mathcal{S}^{(0)} = \emptyset$, which can achieve 0 influence, $\sigma_M(\mathcal{S}^{(0)}) = 0$. The *greedy* algorithm will enumerate all the users in the social network greedily. User u will be selected if $\sigma_M(\mathcal{S}^{(0)} \cup \{u\}) \geq \sigma_M(\mathcal{S}^{(0)} \cup \{v\}), \forall v \in \mathcal{V}$, $u \neq v$. The time cost of round 1 will be $O(|\mathcal{V}|(|\mathcal{V}| + |\mathcal{E}|))$, where $O(|\mathcal{V}| + |\mathcal{E}|)$ denotes the time cost in diffusing the information throughout the network.
- *Round i*: In round i ($i > 1$), the seed user set obtained from the last round can be represented as set $\mathcal{S}^{(i-1)}$, and the *greedy* algorithm will enumerate all the remaining users in the social network and add them to the seed user set. The optimal seed user to be selected in this round can be represented as $u = \arg\max_{u \in \mathcal{V} \backslash \mathcal{S}^{(i-1)}} \sigma_M(\mathcal{S}^{(i-1)} \cup \{u\}) - \sigma_M(\mathcal{S}^{(i-1)})$. The time cost of round i will be $O((|\mathcal{V}| - (i - 1))(|\mathcal{V}| + |\mathcal{E}|))$.

Such an iteratively process continues until the required k seed users have been selected, and these selected *seed user* sets can be formally represented as $\mathcal{S}$. The pseudo-code of the *greedy* algorithm is available in Algorithm 1. Formally, let $\mathcal{S}^*$ denote the optimal seed user solution to the *influence maximization* problem, and $\mathcal{S}^g$ represent the seed user set selected by the greedy algorithm. According to the existing works [29], the approximation ratio of the performance achieved by the *greedy algorithm* is shown to be

$$\frac{\sigma_M(\mathcal{S}^g)}{\sigma_M(\mathcal{S}^*)} \geq 1 - \frac{1}{e}. \tag{10.2}$$

Actually, the exact computation complexity of $\sigma_M(\mathcal{S})$ is left as an open problem [11], in the context of influence maximization. Later on, Chen et al. [11] demonstrate that the exact computation of $\sigma_M(\mathcal{S})$ is actually #-hard. According to the step-wise analysis of the *greedy* algorithm, the running time of the algorithm at the worst case will be $O\left(|\mathcal{V}|^2(|\mathcal{V}| + |\mathcal{E}|)\right)$, which renders the *greedy* algorithm hardly applicable to large-scale social network data sets.

10.2.2.2 CELF

Due to the *submodularity* property of the influence function $\sigma_M(\cdot)$, given two seed user sets $\mathcal{S}^{(i)}$ and $\mathcal{S}^{(i+1)}$ in rounds i and $i + 1$, the influence gain introduced by adding user u (where $u \in \mathcal{V}$, $u \notin \mathcal{S}^{(i)}$, $u \notin \mathcal{S}^{(i+1)}$) to $\mathcal{S}^{(i+1)}$ will not surpass the introduced influence gain by adding user u to $\mathcal{S}^{(i)}$. Such a property can be utilized in the seed user selection. For instance, assuming there are two seed user candidates u and v, if the influence gain introduced by u in the current round is greater than the influence gain obtained by v in the previous round, then v will not be selected definitely in the current round. Therefore, based on such an intuition, when choosing the seed users in each round, we don't need to enumerate all the remaining users to identify the one achieving the maximum influence gain. It is also the basic idea of the "cost-effective lazy forward" (*CELF*) algorithm [11] to be introduced here.

In the *CELF* algorithm, a heap data structure is maintained, where the node achieving the maximum influence gain is placed at the root. Formally, in the heap, the tree node is represented as a triple $(u, \Delta(\mathcal{S}, u), r)$, where $u \in \mathcal{V} \setminus \mathcal{S}$ represents the id of the remaining nodes, $\Delta(\mathcal{S}, u) = \sigma_M(\mathcal{S} \cup \{u\}) - \sigma_M(\mathcal{S})$ denotes the influence gain by adding u to the current seed user set $\mathcal{S}$, and r denotes the most recent round updating the triple. In each round, the *CELF* algorithm will pick some node triples form the heap to update, and select the optimal one which can introduce the maximum influence gain.

- *Round 1*: In round 1, *CELF* algorithm constructs the heap data structure involving all the nodes in the network based on the calculated influence introduced by them. For all the use node, we can represent the triples as set $\{(u, \sigma_M(\{u\}), 1)\}_{u \in \mathcal{V}}$, which will be used to construct the heap H.
- *Round i*: In round i ($i > 1$), *CELF* algorithm keeps picking the node triple from the root of the heap, updating the influence gain and round number of the node triple, reinserting the node back to the heap. Such a process continues until the node at the root is the current round number i (i.e., we have just updated it, and it still achieves the maximum influence gain among the remaining nodes), which will be deleted from the heap and added to the seed user set.

The pseudo-code of the *CELF* algorithm is available in Algorithm 2, which is shown to be over 700 times faster than the *greedy* algorithm in identifying the same seed user set [11]. Besides the *greedy* and *CELF* algorithms, many other approximation-based seed user identification algorithms have been proposed, which further optimize the *greedy* algorithm to lower down the time complexity, like *CELF++* [22], *SIMPATH* [23]. If the readers are interested in these algorithms, please refer to these reference papers for more detailed information.

Algorithm 2 CELF algorithm

Require: Input network G with user node set $\mathcal{V}$
          Influence function $\sigma_M(\cdot)$
          Seed use set size k
Ensure: Seed user set $\mathcal{S}$
 1: initialize $\mathcal{S} = \emptyset$, heap $H = \emptyset$
 2: **for** each $u \in \mathcal{V}$ **do**
 3:     calculate the influence gain measure gain=$\sigma_M(\{u\})$
 4:     add tuple $(u, \text{gain}, 1)$ to the heap in decreasing order of influence gain measure
 5: **end for**
 6: **while** $|\mathcal{S}| < k$ **do**
 7:     pick node tuple (u, gain, r) from the heap root
 8:     **if** $r == |\mathcal{S}| + 1$ **then**
 9:        $\mathcal{S} = \mathcal{S} \cup \{u\}$
10:       delete node tuple (u, gain, r) from the heap H
11:     **else**
12:       delete node tuple (u, gain, r) from the heap H
13:       update gain=$\sigma_M(\mathcal{S} \cup \{u\}) - \sigma_M(\mathcal{S})$
14:       update $r = |\mathcal{S}| + 1$
15:       insert node tuple (u, gain, r) back to the heap H
16:     **end if**
17: **end while**
18: Return $\mathcal{S}$

10.2.3 Heuristics-Based Seed User Selection

The algorithms introduced in the previous parts are mostly based on the greedy seed user selection strategy, and are not scalable to large-scale networks. Even though some speed-up techniques have been proposed, e.g., CELF, the time complexity of these algorithms can still be very high. In this part, we will introduce a number of seed user selection algorithms based on heuristics, which can select the promising seed users with a much faster speed.

10.2.3.1 Centrality Heuristics

In our daily life, the important users (e.g., the famous *celebrities*) can usually have much more influence in disseminating information. In the real-world online social networks, the posts from famous people (e.g., celebrities, politician, and movie stars) can always influence more people, and people tend to follow them. Viewed in such a perspective, when selecting the seed users, selecting the nodes with large *centrality* [5] measures will be a good choice. As introduced in Sect. 3.3.2, the node centrality score can be defined based on different kinds of metrics, like *node degree* [1] and *PageRank score* [8].

For the user nodes with larger degrees, there will exist more neighbors that these nodes can spread their influence to, as introduced in the LT and IC model. Lots of commercial brands tend to invite them to help share some advertising posts and photos to promote products, as they can infect more people in the network. Viewed in such a perspective, choosing the nodes with large degrees is a good way for selecting the seed users.

However, when calculating the weight among users in LT and IC models, if we apply the Jaccard's coefficient [26] as the weight measure of social links among users, the weight of links incident to the large-degree nodes will be small since their degrees will penalize the diffusion weight greatly. In other words, selecting the nodes merely based on the node degree may have some problems. Therefore, some other works propose to apply PageRank score to select the seed users, where users with larger PageRank scores [8] tend to be selected in advance. Another method proposed for the networks with small diffusion weights is the *degree discount* [11] heuristics to be introduced in the following part.

Besides the diffusion models have introduced in Chap. 9, there also exist many other diffusion models, like the *path*-based models. In the *shortest path* (SP) model proposed in [30], the nodes are activated through the shortest path form the initial seed user set. Based on these diffusion models, some other types of heuristics have been applied, like *distance-based centrality*. Nodes can be sorted according to the average distance from them to all the other nodes in the network, where those with smaller average distances will be picked as the *seed user nodes*.

10.2.3.2 Degree Discount Heuristics

Both the *node degree* and *PageRank score*-based heuristics work very well in the experimental simulations, and they can achieve much larger influence than the other heuristics. However, the influence obtained by them is still much smaller than the *greedy* algorithm. Furthermore, for the nodes with large degrees, the influence they can send out to their neighbors will be relatively small and can hardly activate their neighbors. To resolve such a problem, some works propose to further improve the pure degree-based heuristics, and introduce the *degree discount* method [11].

Let v be a neighbor of user node u in the network. If u has been selected as a seed user, when considering adding node v as a new seed user based on his/her degree, we should not count seed user u as his neighbor towards the degree. Since u has been added to the seed user set already, node v' degree should be discounted by 1 for u, and similarly for the other neighbors who have been selected as the seed nodes. Such a heuristic is applicable to all the diffusion model introduced before.

Algorithm 3 Degree discount heuristics

Require: Input network G with user node set $\mathcal{V}$
              Seed use set size k
              Diffusion weight w
Ensure: Seed user set $\mathcal{S}$
1: initialize $\mathcal{S} = \emptyset$
2: **for** each node $u \in \mathcal{V}$ **do**
3:      compute degree $D(u) = |\Gamma(u)|$
4:      initialize discounted degree $DD(u) = D(u)$
5:      initialize $T(u) = 0$
6: **end for**
7: **while** $|\mathcal{S}| < k$ **do**
8:      select $u^* = \arg\max_{u \in \mathcal{V} \setminus \mathcal{S}} DD(u)$
9:      $\mathcal{S} = \mathcal{S} \cup \{u\}$
10:     **for** neighbor $v \in \Gamma(u) \setminus \mathcal{S}$ **do**
11:        $T(v) = T(v) + 1$
12:        $DD(v) = 1 + (D(v) - 2T(v) - (D(v) - T(v))T(v)p)p.$
13:     **end for**
14: **end while**
15: Return $\mathcal{S}$

Specifically, for the IC model with a relatively small diffusion weight $w \to 0$, a more accurate *degree discount* heuristic has been proposed in [11]. In the IC mode, user u will activate his/her neighbor v with a probability w. If user u has been selected into the seed user set and u can activate v, then we don't need to further add v into the seed user set. In the case that w is small, the two-hop diffusion can be ignored, and the *degree discount* is applied to a local subgraph.

Let v be a user who has not been selected as the seed user yet, we can represent the his/her neighbor as set $\Gamma(v)$. The number of v's neighbors who have been selected as the seed user can be represented as $T(v)$, while the original degree of node v is $D(v) = |\Gamma(v)|$. The expected number of additional nodes in $\Gamma(v)$ to be infected by adding v into the seed user set can be approximately represented as v's *discounted degree*

$$DD(v) = 1 + (D(v) - 2T(v) - (D(v) - T(v))T(v)p) \cdot p. \tag{10.3}$$

where p denotes the activation probability between users. For all the nodes in the network, the *degree discount* method will pick the seed users with larger *discounted degree* iteratively. The pseudo-code of the *degree discount* method is available in Algorithm 3.

10.3 Intertwined Influence Maximization

Besides the traditional *influence maximization* problems about one single product studied based on the online social networks, in the real scenarios, the promotions of multiple products can co-exist in the social networks at the same time. The relationships among the products to be promoted in the network can be very complicated. In this section, we want to maximize the influence of one specific product that we target on in online social networks, where many other products are being promoted simultaneously. The relationships among these product can be obtained in advance via effective market research, which can be *independent*, *competitive*, or *complementary* as introduced in Sect. 9.3. Formally, we define this problem as the inter<u>T</u>wined <u>I</u>nfluence <u>M</u>aximization (Tim) problem [50].

More specifically, depending on the promotional order of other products and the target product, the TIM problem can have two different variants (we don't care about the case that other products are promoted after the target product):

- C-TIM *problem*: In some cases, the other products have been promoted ahead of the target products, where their selected seed users are known and product information has already been propagated within the network. In such a case, the variant of TIM is defined as the Conditional interTwined Influence Maximization (C-TIM) problem.
- J-TIM *problem*: However, in some other cases, the promotion activities of multiple products occur simultaneously, where the *marketing strategies* of all these products are confidential to each other. Such a variant of TIM is defined as the Joint interTwined Influence Maximization (J-TIM) problem.

To solve the above two sub-problems, in this section, we will introduce a unified *greedy* framework interTwined Influence EstimatoR (TIER) proposed in [50]. The TIER method also has two variants: (1) C-TIER (Conditional TIER) for the C-TIM problem, and (2) J-TIER (Joint TIER) for the J-TIM problem. TIER is based on the diffusion model TLT [50] introduced in Sect. 9.3, which quantifies the *impacts* among products with the *intertwined threshold updating strategy* and can handle the intertwined diffusion of these products at the same time. To solve the C-TIM problem, C-TIER will select seed users greedily and is proved to achieve a $(1 - \frac{1}{e})$-approximation to the optimal result. For the J-TIM problem, we show that the theoretical influence of upper and lower bounds calculation is *NP-hard*. Alternatively, we formulate the J-TIM problem as a game among different products and propose to infer the potential *marketing strategies* of other products. The *step-wise greedy* method J-TIER can achieve promising results by selecting seed users wisely according to the inferred marketing strategies of other products.

10.3.1 Conditional TIM

Formally, we can represent the online *social network* as $G = (\mathcal{V}, \mathcal{E})$, where $\mathcal{V}$ is the set of users and $\mathcal{E}$ contains the interactions among users in $\mathcal{V}$. The set of n different products to be promoted in network G can be represented as $\mathcal{P} = \{p^1, p^2, \ldots, p^n\}$. For a given product $p^j \in \mathcal{P}$, users who are influenced to buy p^j are defined to be "*active*" to p^j, while the remaining users who have not bought p^j are defined to be "*inactive*" to p^j. User u_i's status towards all the products in $\mathcal{P}$ can be represented as "*user status vector*" $\mathbf{s}_i = (s_i^1, s_i^2, \ldots, s_i^n)$, where s_i^j is u_i's status to product p^j. Users can be activated by multiple products at the same time (even competing products), i.e., multiple entries in *status vector* $\mathbf{s}_i$ can be "*active*" concurrently.

In traditional single-product viral marketing problems, the selected *seed users* will propagate the influence of the target product in the network and the number of users getting activated can be obtained with the *influence function* $\sigma : \mathcal{S} \to \mathbb{R}$, which maps the selected seed users to the number of influenced users. Traditional one single-product viral marketing problem aims at selecting the optimal seed users $\bar{\mathcal{S}}$ for the target product, who can achieve the maximum influence:

$$\bar{\mathcal{S}} = \arg_{\mathcal{S}} \max \sigma(\mathcal{S}). \tag{10.4}$$

However, in the TIM problem, promotions of multiple products in $\mathcal{P}$ co-exist simultaneously. The influence function of the target product $p^j \in \mathcal{P}$ depends on not only the seed user set $\mathcal{S}^j$ selected for itself but also the seed users of other products in $\mathcal{P} \setminus \{p^j\}$. In the case that the other products are promoted ahead of the target product, we formally define the TIM problem as the *conditional intertwined influence function* C-TIER problem and the corresponding influence function is called the *conditional intertwined influence maximization function*.

Definition 10.5 (Conditional Intertwined Influence Function) Formally, let the notation $\mathcal{S}^{-j} = (\mathcal{S}^1, \ldots, \mathcal{S}^{j-1}, \mathcal{S}^{j+1}, \ldots, \mathcal{S}^n)$ be the known seed user sets selected for all products in $\mathcal{P} \setminus \{p^j\}$, the *influence function* of the target product p^j given the known *seed user sets* $\mathcal{S}^{-j}$ is defined as the *conditional intertwined influence function*: $\sigma(\mathcal{S}^j | \mathcal{S}^{-j})$.

C-TIM Problem: C-TIM problem aims at selecting the optimal *marketing strategy* $\bar{\mathcal{S}}^j$ to maximize the *conditional intertwined influence function* of p^j in the network, i.e.,

$$\bar{\mathcal{S}}^j = \arg_{\mathcal{S}^j} \max \sigma(\mathcal{S}^j | \mathcal{S}^{-j}). \tag{10.5}$$

10.3.1.1 Conditional TIM Problem Analysis

In the C-TIM problem, the promotion activities of other products have been done before we start to promote our target product. Subject to the TLT diffusion model, users' thresholds to the target product can be updated with the *threshold updating strategy* after the promotions of other products. Based on the updated network, the C-TIM can be mapped to the *tradition single-product viral marketing*, which has been proved to be *NP-hard* already.

Theorem 10.1 *The C-TIM problem is NP-hard based on the TLT diffusion model.*

The proof of Theorem 10.1 is omitted and will be left as an exercise for the readers. Meanwhile, based on the TLT diffusion model, the *conditional influence function* of the target product $\sigma(\mathcal{S}^j | \mathcal{S}^{-j})$ is observed to be both *monotone* and *submodular*.

Theorem 10.2 *For the TLT diffusion model, the conditional influence function is monotone and submodular.*

Proof We will prove the theorem from two perspectives:

(1) *monotone*: Given the existing seed user sets $\mathcal{S}^{-j}$ for existing products $\mathcal{P} - \{p^j\}$ in the market, let $\mathcal{T}$ be a seed user set of product p^j. Users in the network who are not involved in $\mathcal{T}$ can be represented as $\mathcal{V} - \mathcal{T}$. For the given seed user set $\mathcal{T}$ and the fixed seed users set $\mathcal{S}^{-j}$ of other products, adding a new seed user, e.g., $u \in \mathcal{V} - \mathcal{T}$, to the seed user set $\mathcal{T}$ will not decrease the number of influenced users, i.e., $\sigma(\mathcal{T} \cup \{u\} | \mathcal{S}^{-j}) \geq \sigma(\mathcal{T} | \mathcal{S}^{-j})$.
(2) *submodular*: After the diffusion process of the existing products in $\mathcal{P} - \{p^j\}$ users the thresholds towards product p^j will be updated. Based on the updated network, for two given seed user sets $\mathcal{R}$ and $\mathcal{T}$, where $\mathcal{R} \subseteq \mathcal{T} \subseteq \mathcal{V}$, it is easy to show that $\sigma(\mathcal{R} \cup \{v\} | \mathcal{S}^{-j}) - \sigma(\mathcal{R} | \mathcal{S}^{-j}) \geq \sigma(\mathcal{T} \cup \{v\} | \mathcal{S}^{-j}) - \sigma(\mathcal{T} | \mathcal{S}^{-j})$ with the *"live-edge path"* [29].

10.3.1.2 The C-TIER Algorithm

According to the above analysis, a greedy algorithm C-TIER is proposed to solve the problem C-TIM in this section, whose pseudo-code is available in Algorithm 4. In C-TIER, we select the user u who can lead to the maximum increase of the conditional influence function $\sigma(\mathcal{S}^j \cup \{u\} | \mathcal{S}^{-j})$ at each step as the new seed user. This process repeats until either no potential seed user is available or all the k^j required seed users have been selected. The time complexity of C-TIER is $O(k^j |\mathcal{V}|(|\mathcal{V}| + |\mathcal{E}|))$. Since

---

Algorithm 4 The C-TIER algorithm

Require: input social network $G = (\mathcal{V}, \mathcal{P}, \mathcal{E})$
 target product: p^j
 known seed user sets of $\mathcal{P} - \{p^j\}$: $\mathcal{S}^{-j}$
 conditional influence function of p^j: $I(\mathcal{S}^j | \mathcal{S}^{-j})$
 seed user set size of p^j: k^j
Ensure: selected seed user set $\mathcal{S}^j$ of size k^j
1: initialize seed user set $\mathcal{S}^j = \emptyset$
2: propagate influence of products $\mathcal{P} - \{p^j\}$ with $\mathcal{S}^{-j}$ and update users' thresholds with intertwined threshold updating strategy
3: **while** $\mathcal{V} \setminus \mathcal{S}^j \neq \emptyset \wedge |\mathcal{S}^j| \neq k^j$ **do**
4:   pick a user $u \in \mathcal{V} - \mathcal{S}^j$ according to equation $\arg\max_{u \in \mathcal{V}} I(\mathcal{S}^j \cup \{u\} | \mathcal{S}^{-j}) - I(\mathcal{S}^j | \mathcal{S}^{-j})$
5:   $\mathcal{S}^j = \mathcal{S}^j \cup \{u\}$
6: **end while**
7: return $\mathcal{S}^j$.

---

the *conditional influence function* is *monotone* and *submodular* based on the TLT diffusion model, then the *step-wise greedy* algorithms C-TIER, which select the users who can lead to the maximum increase of influence, can achieve a $(1 - \frac{1}{e})$-approximation of the optimal result for the target product.

10.3.2 Joint TIM

C-TIM studies a common case in real-world viral marketing, where different companies have different schedules to promote their products and some can be conducted ahead of the target product. Meanwhile, in this part, we will study a more challenging case: J-TIM, where other products are being promoted at the same time as our target product and the marketing strategies of different products are totally confidential.

Definition 10.6 (Joint Intertwined Influence Function) When the seed user sets of products $\mathcal{P} \setminus \{p^j\}$ are unknown, i.e., $\mathcal{S}^{-j}$ is not given, the *influence function* of product p^j together with other products in $\mathcal{P} \setminus \{p^j\}$ is defined as the *joint intertwined influence function*: $\sigma(\mathcal{S}^j; \mathcal{S}^{-j})$.

J-TIM Problem: J-TIM problem aims at choosing the optimal *marketing strategy* $\bar{\mathcal{S}}^j$ to maximize the *joint intertwined influence function* of p^j in the network, i.e.,

$$\bar{\mathcal{S}}^j = \arg_{\mathcal{S}^j} \max \sigma(\mathcal{S}^j; \mathcal{S}^{-j}), \tag{10.6}$$

where set $\mathcal{S}^{-j}$ can take any possible value.

10.3.2.1 Joint TIM Problem Analysis

When the *marketing strategies* of other products are unknown, the *influence function* of the target product and other products co-exist in the network is defined as the *joint influence function*: $\sigma(\mathcal{S}^j; \mathcal{S}^{-j})$. Meanwhile, by setting $\mathcal{S}^1 = \cdots = \mathcal{S}^{j-1} = \mathcal{S}^{j+1} = \cdots = \mathcal{S}^n = \emptyset$, the J-TIM problem can be mapped to the traditional *single-product influence maximization* problem in polynomial time, which is an NP-hard problem.

Theorem 10.3 *The* J-TIM *problem is NP-hard based on the* TLT *diffusion model.*

Proof We construct an instance of the J-TIM problem by setting $\mathcal{S}^1 = \cdots = \mathcal{S}^{j-1} = \mathcal{S}^{j+1} = \cdots = \mathcal{S}^n = \emptyset$, which will map the J-TIM problem to the traditional *single-product influence maximization* problem in polynomial time. Meanwhile, as proved in [29], the traditional *single-product influence maximization* problem is *NP-hard*. As a result, the J-TIM is also a *NP-hard* problem.

Meanwhile, if all the products in $\mathcal{P} \setminus \{p^j\}$ are *independent* to p^j, the *joint influence function* $\sigma(\mathcal{S}^j; \mathcal{S}^{-j})$ will be both *monotone* and *submodular*.

Theorem 10.4 *Based on the* TLT *diffusion model, the joint influence function is monotone and submodular if all the other products are independent to p^j.*

Proof If all the other products are *independent* to product p^j, then the promotion process of all the other products has no effects on the promotion of p^j. According to the *threshold updating strategy* in the TLT diffusion model, users' thresholds to the target product p^j will not be affected by all the remaining products, i.e., the TIM problem identical to the *traditional single-product viral marketing* problem. Furthermore, the *joint influence function* of p^j and all the other products will be reduced to the *influence function* of product p^j in the traditional *single-product influence maximization* setting:

$$\sigma(\mathcal{S}^j; \mathcal{S}^{-j}) \rightarrow \sigma(\mathcal{S}^j). \tag{10.7}$$

Meanwhile, as proved in [29], the *influence function* of p^j in the *single-product influence maximization* setting is both *monotone* and *submodular*. As a result, based on the TLT diffusion model, the *joint influence function* is *monotone* and *submodular* if all the other products are *independent* to p^j.

However, when there exist products in $\mathcal{P} \setminus \{p^j\}$ to be either *competing* or *complementary* to p^j, the *joint influence function* $\sigma(\mathcal{S}^j; \mathcal{S}^{-j})$ will be neither *monotone* nor *submodular*.

Theorem 10.5 *Based on the* TLT *diffusion model, the joint influence function is not monotone if there exist products which are either competing or complementary to the target product p^j.*

Proof We propose to prove the above Theorem 10.5 with counterexamples shown in Fig. 10.1a, where we can find one product p^i to be either *competing* or *complementary* to p^j.

Case (1): when competing products exist: as shown in the upper two plots in Fig. 10.1a, we have four users in the network $\{A, B, C, D\}$ and we want to select seed users for products p^i and p^j. The influence from A to B and C are 0.3 and 0.5, whose original thresholds to the target product p^j are 0.25 and 0.45, respectively. In the example, the *seed users* selected for two *competing* products p^j and p^i are (1) $\{A\}$ and $\{D\}$, respectively, in competing case 1 at the upper left corner; and (2) $\{A, B\}$ and $\{C\}$ in competing case 2 at the upper right corner. In competing case 1, p^j can influence three users $\{A, B, C\}$ as the influence from A to B and C can both exceed their thresholds, i.e., $\sigma(\mathcal{S}^j = \{A\}; \mathcal{S}^i = \{D\}) = 3$. However, in competing case 2, p^j can only influence two users even though the seed use set has been expanded by adding B as a *seed user*, i.e., $\sigma(\mathcal{S}^j = \{A, B\}; \mathcal{S}^i = \{C\}) = 2$. The reason is that the competing product p^i selects C as the seed user which increase C's threshold towards p^j from 0.45 to 0.55.

So, we can find a counterexample where $\{A\} \subset \{A, B\}$ but $\sigma(\mathcal{S}^j = \{A\}; \mathcal{S}^i = \{D\}) > \sigma(\mathcal{S}^j = \{A, B\}; \mathcal{S}^i = \{C\})$, when there exists *competing* product p^i in the network.

Case (2): when complementary products exist: similar counterexamples are shown in the lower two plots of Fig. 10.1a, which are identical to the upper two plots except that the influence from A to C for product p^j is changed to 0.4 and p^i is *complementary* to p^j instead. In complementary case 1,

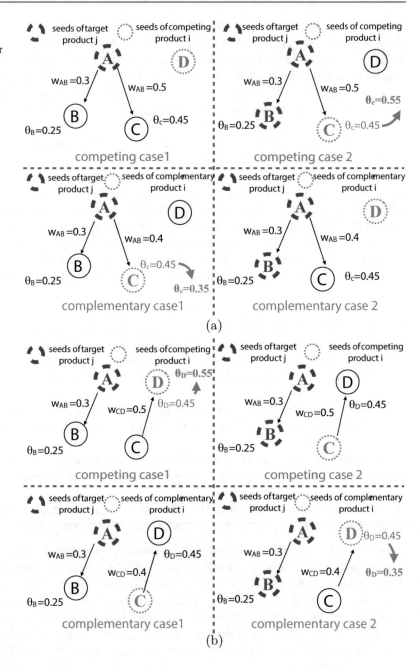

Fig. 10.1
Counterexamples of
monotone and submodular
properties.
(**a**) Counterexamples of
monotone property.
(**b**) Counterexamples of
submodular property

p^i selects C as the seed user, which can decrease C's threshold towards p^j and p^j can achieve a
influence of three by choosing A as the seed user. However, in complementary case 2, p^i selects
D as the seed user and p^j can only influence two users even though the seed user set has been
expanded by adding B to the set.

So, we can find a counterexample where $\{A\} \subset \{A, B\}$ but $\sigma(\mathcal{S}^j = \{A\}; \mathcal{S}^i = \{C\}) > \sigma(\mathcal{S}^j = \{A, B\}; \mathcal{S}^i = \{D\})$ when there exists *complementary* product p^i in the network.

As a result, based on the TLT diffusion model, the *joint influence function* is not *monotone* if these
exist products which are either *competing* or *complementary* to p^j.

Theorem 10.6 *For the* TLT *diffusion model, the joint influence function is not submodular if these exist products which are either competing or complementary to the target product* p^j.

Proof We propose to prove Theorem 10.6 with counterexamples shown in Fig. 10.1b, where we can find one product p^i to be either *competing* or *complementary* to p^j.

Case (1): when competing products exist: Let $\mathcal{T} = \{A\} \subset \mathcal{S} = \{A, B\}$ and $u = C$. In the competing case 1, $\mathcal{T}$ is the seed user set selected by product p^j and $\{D\}$ is selected as the seed user by product p^i, which increase D's threshold to p^j from 0.45 to 0.55. As a result, p^j can only influence two users ($\{A, B\}$) when using $\mathcal{T}$ as the seed user set and influence three users ($\{A, B, C\}$) when using $\mathcal{T} \cup \{u\}$ as the seed user set. However, in the competing case 2, where p^i selects C as the seed user, p^j can activate two users ($\{A, B\}$) when using $\mathcal{S}$ as the seed user set but can activate four users ($\{A, B, C, D\}$) when using $\mathcal{S} \cup \{u\}$ as the seed user set.

So, we can find a counterexample where $\mathcal{T} = \{A\} \subset \mathcal{S} = \{A, B\}$ and $u = C$, but $\sigma(\mathcal{S}^j = \mathcal{T} \cup \{u\}; \mathcal{S}^i = \{D\}) - \sigma(\mathcal{S}^j = \mathcal{T}; \mathcal{S}^i = \{D\}) < \sigma(\mathcal{S}^j = \mathcal{S} \cup \{u\}; \mathcal{S}^i = \{C\}) - \sigma(\mathcal{S}^j = \mathcal{S}; \mathcal{S}^i = \{C\})$.

Case (2): when complementary products exist: similar counterexample is shown in the lower two plots of Fig. 10.1b, where p^i is *complementary* to p^i. We can also find a counterexample where $\mathcal{T} = \{A\} \subset \mathcal{S} = \{A, B\}$ and $u = C$, and $\sigma(\mathcal{S}^j = \mathcal{T} \cup \{u\}, \mathcal{S}^i = \{C\}) - \sigma(\mathcal{S}^j = \mathcal{T}, \mathcal{S}^i = \{D\}) < \sigma(\mathcal{S}^j = \mathcal{S} \cup \{u\}, \mathcal{S}^i = \{C\}) - \sigma(\mathcal{S}^j = \mathcal{S}, \mathcal{S}^i = \{D\})$.

As a result, for the TLT diffusion model, the *joint influence function* is not *submodular* if these exist products which are either *competing* or *complementary* to p^j. ∎

When all the other products are *independent* to p^j, the *joint influence function* of p^j will be *monotone* and *submodular*, which is solvable with the *traditional greedy algorithm* proposed [29] and can achieve a $(1 - \frac{1}{e})$-approximation of the optimal results. However, when there exists at least one product which is either *competing* or *complementary* to p^j, the *joint influence function* will be no longer *monotone* or *submodular*. In such a case, the J-TIM will be very hard to solve and no promising optimality bounds of the results are available.

By borrowing ideas from the game theory studies [6, 40], for product p^j, the lower bound and upper bound of influence the J-TIM problem can be achieved by selecting seed users of size k can be represented as

$$\max_{\mathcal{S}^j} \min_{\mathcal{S}^{-j}} \sigma(\mathcal{S}^j; \mathcal{S}^{-j}), \quad \max_{\mathcal{S}^j} \max_{\mathcal{S}^{-j}} \sigma(\mathcal{S}^j; \mathcal{S}^{-j}) \tag{10.8}$$

respectively, which denotes the maximum influence p^j can achieve in the worst (and the best) cases where all the remaining products work together to make p^j's influence as low (and high) as possible. The *seed user set* selected by p^j when achieving the lower bound and upper bound of influence can be represented as

$$\hat{\mathcal{S}}^j_{low} = \arg\max_{\mathcal{S}^j} \min_{\mathcal{S}^{-j}} \sigma(\mathcal{S}^j; \mathcal{S}^{-j}), \quad \hat{\mathcal{S}}^j_{up} = \arg\max_{\mathcal{S}^j} \max_{\mathcal{S}^{-j}} \sigma(\mathcal{S}^j; \mathcal{S}^{-j}). \tag{10.9}$$

However, the lower and upper bounds of the optimal results of the J-TIM problem are hard to calculate mathematically.

Theorem 10.7 *Computing the Max-Min for three or more player games is NP-hard.*

Proof As proposed in [6], the problem of finding any (approximate) Nash equilibrium for a three-player game is computationally intractable and it is NP-hard to approximate the min-max payoff value for each of the player [6, 9, 10, 16].

10.3.2.2 The J-Tier Algorithm

In addition, in the real world, the other products will not co-operate together in designing their marketing strategies to create the worst or the best situations for the target product p^j, i.e., choosing the *marketing strategies* S^{-j} such that the *joint influence function* $\sigma(S^j; S^{-j})$ is minimized or maximized. To address the J-Tim problem, in this part, we propose the J-Tier algorithm to simulate the intertwined round-wise greedy seed user selection process of all the products.

In J-Tier, all products are assumed to be *selfish* and want to maximize their own influence when selecting seed users based on the *"current"* situation created by all the products. J-Tier will infer the next potential *marketing strategies* of other products round by round and select the *optimal* seed users for each product based on the inference.

In algorithm J-Tier, we let all products in $\mathcal{P}$ choose their optimal *seed users* randomly at each round. For example, let $(S)^{\tau-1}$ be the seed users selected by products in $\mathcal{P}$ at round $\tau - 1$. At round τ, a random product p^i can select one seed user. To achieve the largest influence, product p^i will infer the next potential seed users to be selected by other products based on the assumption that they are all selfish. For example, based on p^i's inference, the next seed user to be selected by p^j can be represented as $\bar{u}^j$, i.e.,

$$\arg \max_{u \in \mathcal{V} - (S^j)^{\tau-1}} [I((S^j)^{\tau-1} \cup \{u\}; (S^{-j})^{\tau-1}) - I((S^j)^{\tau-1}; (S^{-j})^{\tau-1})]. \tag{10.10}$$

Similarly, p^i can further infer the potential seed users to be selected next by products in $\mathcal{P} \setminus \{p^i, p^j\}$, and these selected seed users can be represented as a set$\{\bar{u}_1, \bar{u}_2, \ldots, \bar{u}_{i-1}, \bar{u}_{i+1}, \ldots, \bar{u}_{j-1}, \bar{u}_{j+1}, \ldots, \bar{u}_n\}$, respectively. Based on such inference, p^i knows who are the next seed users to be selected by other products and will make use of the "prior knowledge" to select its own seed user $\hat{u}^i$ in round τ:

$$\hat{u}^i = \arg \max_{u \in \mathcal{V} - (S^i)^{\tau-1}} [I((S^i)^{\tau-1} \cup \{u\}; \bar{S}^{-i}) - I((S^i)^{\tau-1}; \bar{S}^{-i})]. \tag{10.11}$$

where $\bar{S}^{-i}$ is the "inferred" seed user sets of other products inferred by p^i based on current situation by "adding" these inferred potential seed users to their seed user sets.

The selected $(\hat{u}^i)^\tau$ will be added to the seed user set of product p^i, i.e.,

$$(S^i)^\tau = (S^i)^{\tau-1} \cup \{(\hat{u}^i)^\tau\}. \tag{10.12}$$

And the *"current"* seed user sets of all the products, i.e., S, are updated as follows:

$$S = ((S^1)^\tau, (S^2)^{\tau-1}, \ldots, (S^n)^{\tau-1}). \tag{10.13}$$

The selected $(\hat{u}^i)^\tau$ will propagate his influence in the network and all the users just activated to product p^i will update their thresholds to other products in $\mathcal{P} \setminus \{p^i\}$.

Next, we let another random product (which has not selected seed users yet) to infer the next seed users to be selected by other products and choose its seed user based on the inferred situation. In each round, each product will have a chance to select one seed user and the user selection order of

---

Algorithm 5 The J-Tier algorithm

Require: input social network $G = (\mathcal{V}, \mathcal{P}, \mathcal{E})$
         target product: p^j
         set of other products: $\mathcal{P} - \{p^j\}$
         joint influence function of p^j: $I(\mathcal{S}^j; \mathcal{S}^{-j})$
         seed user set size of products in $\mathcal{P}$:$k^1, k^2, \ldots, k^j, \ldots, k^n$
Ensure: selected seed user sets $\{\mathcal{S}^1, \mathcal{S}^2, \ldots, \mathcal{S}^n\}$ of products in $\mathcal{P}$ respectively
 1: initialize seed user set $\mathcal{S}^1, \mathcal{S}^2, \ldots, \mathcal{S}^n = \emptyset$
 2: **while** $(\mathcal{V} \setminus \mathcal{S}^1 \neq \emptyset \vee \cdots \vee \mathcal{V} \setminus \mathcal{S}^n \neq \emptyset) \wedge (|\mathcal{S}^1| \neq k^1 \vee \cdots \vee |\mathcal{S}^n| \neq k^n)$ **do**
 3:      **for** random $i \in \{1, 2, \ldots, n\}$ (p^i has not selected seeds in the round yet) **do**
 4:          **if** $\mathcal{V} \setminus \mathcal{S}^i \neq \emptyset \wedge |\mathcal{S}^i| \neq k^i$ **then**
 5:              p^i infers the seed user sets $\bar{\mathcal{S}}^{-i}$ of other products
 6:              p^i selects its seed user $u^i \in \mathcal{V} - \mathcal{S}^i$, who can maximize $I(\mathcal{S}^i \cup \{u^i\}; \bar{\mathcal{S}}^{-i}) - I(\mathcal{S}^i; \bar{\mathcal{S}}^{-i})$
 7:              $\mathcal{S}^i = \mathcal{S}^i \cup \{u^i\}$
 8:              propagate influence of u in G and update influenced users' thresholds to products in $\mathcal{P}$ with the *intertwined threshold updating strategy*.
 9:          **end if**
10:      **end for**
11: **end while**
12: return $\mathcal{S}^1, \mathcal{S}^2, \ldots, \mathcal{S}^n$.

---

different products in each round is totally random. Such a process will stop when all the products either have selected the required number of *seed users* or no users are available to be chosen. With the J-Tier model, we simulate an alternative seed user selection procedure of multiple products in viral marketing and the pseudo-code J-Tier method is given in Algorithm 5. The time complexity of the J-Tier algorithm is $O((\sum_i k_i \cdot n)|\mathcal{V}|(|\mathcal{V}| + |\mathcal{E}|))$, where $k_i = |\mathcal{S}^i|$ is the number of seed users to be selected for product p^i.

10.4   Cross-Network Influence Maximization

Traditional viral marketing problem aims at selecting the set of seed users to maximize the awareness of ideas or products merely based on the *social connections* among users in *one single social network* [12, 22, 27]. However, in the real world, social networks usually contain heterogeneous information [45,49], e.g., various types of nodes and complex links, via which users are extensively connected and have multiple channels to influence each other [24]. Meanwhile, as studied in [31,49], users nowadays are also involved in multiple social networks simultaneously to enjoy more social network services. Via these shared users, information can propagate not only within but also across social networks [39].

Example 10.1 To support such a claim, we investigate a partially aligned network data set (i.e., Twitter and Foursquare) and the results are given in Fig. 10.2. In Fig. 10.2a, we randomly sample a subset of anchor users from Foursquare and observe that 409 out of 500 (i.e., 81.8%) sampled users have reposted their activities (e.g., tips, location check-ins, etc.) to Twitter. Meanwhile, the activities reposted by these 409 anchor users only account for a small proportion of their total activities in Foursquare, as shown in Fig. 10.2b. In other words, these anchor users will repose the information to other networks selectively.

In this section, we study the influence maximization problem across multiple partially aligned heterogeneous social networks simultaneously. This is formally defined as the *aligned heterogeneous network influence maximization (ANIM)* problem [48]. The *ANIM* problem studied in this section

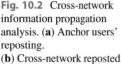

Fig. 10.2 Cross-network information propagation analysis. (**a**) Anchor users' reposting. (**b**) Cross-network reposted activities

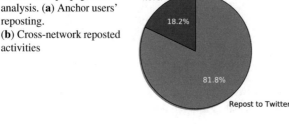

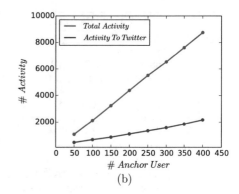

(a)　　　　　　　　　　　　　　　　　　　　　　(b)

is very important and has extensive concrete applications in real-world social networks, e.g., *cross-community* [3], even *cross-platform* [39], *product promotion* [42], and *opinion diffusion* [13]. Based on different inter-network information diffusion models introduced in Sect. 9.4, two different seed user selection algorithms will be introduced in this section for the inter-network information diffusion scenario, including the *greedy seed user selection* algorithm [47] and the *dynamic programming-based seed user selection* algorithm [48].

10.4.1 Greedy Seed User Selection Across Networks

In this part, a new information diffusion model named $\underline{M}$ulti-aligned $\underline{M}$ulti-relational network (M&M) [47] will be introduced to address the cross-network *seed user selection* challenges. M&M first extracts multi-aligned multi-relational networks with the heterogeneous information across the input *online social networks* based on a set of inter- and intra-network social meta paths [45, 49]. M&M extends the traditional linear threshold (LT) model to depict the information propagation within and across these multi-aligned multi-relational networks. Based on the extended diffusion model, the influence function which maps seed user set to the number of activated users is proved to be both *monotone* and *submodular* [47]. Thus the greedy algorithm used in M&M, which selects seed users greedily at each step, is proved to achieve a $(1 - \frac{1}{e})$-approximation of the optimal result. The M&M diffusion model to be introduced in this part is very similar to the MUSE diffusion model introduced in Sect. 9.5, which is also defined based on the meta path concept.

Formally, given two partially aligned networks $G^{(1)}$ and $G^{(2)}$ together with the undirected anchor link set $\mathcal{A}$ between $G^{(1)}$ and $G^{(2)}$, the user sets of $G^{(1)}$ and $G^{(2)}$ can be represented as $\mathcal{U}^{(1)}$ and $\mathcal{U}^{(2)}$, respectively. Let $\sigma(\cdot) : \mathcal{S} \to \mathbb{R}, \mathcal{S} \subset \mathcal{U}^{(1)} \cup \mathcal{U}^{(2)}$ be the *influence function*, which maps the seed user set $\mathcal{S}$ to the number of users influenced by users in $\mathcal{S}$. The *ANIM* problem aims at selecting the optimal set $\mathcal{S}^*$ with size k to maximize the propagation of information across the networks, i.e.,

$$\mathcal{S}^* = \arg \max_{\mathcal{S} \subseteq \mathcal{U}^{(1)} \cup \mathcal{U}^{(2)}} \sigma(\mathcal{S}). \tag{10.14}$$

10.4.1.1 Multi-Aligned Multi-Relational Networks Extraction

We utilize the meta paths [45, 49] defined based on the *network schema* to extract multi-aligned multi-relational networks with the heterogeneous information in aligned networks. In both Foursquare and Twitter, users can follow other users and check-in at locations, forming two intra-network influence channels among users. Meanwhile, (1) in Foursquare, users can create/like lists containing a set of locations; (2) while in Twitter, users can retweet other users' tweets, both of which will form an intra-network influence channel among users in Foursquare and Twitter, respectively. The set of intra-network social meta paths considered here as well as their physical meanings are listed as follows:

Intra-Network Social Meta Paths in Foursquare

(1) *follow*: User $\xrightarrow{follow^{-1}}$ User,

(2) *co-location check-ins*: User $\xrightarrow{check-in}$ Location $\xrightarrow{check-in^{-1}}$ User,

(3) *co-location via shared lists*: User $\xrightarrow{create/like}$ List $\xrightarrow{contain}$ Location $\xrightarrow{contain^{-1}}$ List $\xrightarrow{create/like^{-1}}$ User.

Intra-Network Social Meta Paths in Twitter

(1) *follow*: User $\xrightarrow{follow^{-1}}$ User,

(2) *co-location check-ins*: User $\xrightarrow{check-in}$ Location $\xrightarrow{check-in^{-1}}$ User,

(3) *contact via tweet*: User $\xrightarrow{write}$ Tweet $\xrightarrow{retweet}$ Tweet $\xrightarrow{write^{-1}}$ User.

Users can diffuse information across networks via the anchor links formed by anchor users. This can be abstracted as

$$\text{\textbf{inter-network social meta path} (1) User} \xleftrightarrow{Anchor} \text{User.}$$

By taking the inter-network meta paths into account, the studied problem becomes even more complex due to the fact that anchor users in both networks can also be connected via intra- and inter-network meta paths. As a result, the number of social meta path instances grows mightily. Each meta path defines an influence propagation channel among linked users. If linked users u, v are connected by only intra-network meta path, we say u has *intra-network relation* to v, otherwise there is an *inter-network relation* between them. Based on these relations, we can construct multi-aligned multi-relational networks for the aligned heterogeneous networks. The formal definition of multi-aligned multi-relational networks is given as follows.

Definition 10.7 (Multi-Aligned Multi-Relational Networks (MMNs)) For two given heterogeneous networks $G^{(1)}$ and $G^{(2)}$, we can define the multi-aligned multi-relational network constructed based on the above intra- and inter-network social meta paths as $\mathcal{G} = (\mathcal{U}, \mathcal{E}, \mathcal{R})$, where $\mathcal{U} = \mathcal{U}^{(1)} \cup \mathcal{U}^{(2)}$ denotes the set of user nodes in the MMNs $\mathcal{G}$. Set $\mathcal{E}$ contains the links among nodes in $\mathcal{U}$ and element $e \in \mathcal{E}$ can be represented as $e = (u, v, r)$ denoting that there exists at least one link (u, v) of link type $r \in \mathcal{R} = \mathcal{R}^{(1)} \cup \mathcal{R}^{(2)} \cup \{Anchor\}$, where $\mathcal{R}^{(1)}$, $\mathcal{R}^{(2)}$, and *Anchor* denote the *intra-network* and *inter-network social meta paths* defined above.

10.4.1.2 M&M Diffusion Model

In this subsection, we will extend the traditional *linear threshold* (LT) model to handle the information diffusion across the multi-aligned multi-relational networks. In the traditional *linear threshold* (LT) model, for one single homogeneous network $G = (\mathcal{V}, \mathcal{E})$, user $u_i \in \mathcal{V}$ can influence his neighbor $u_k \in \Gamma_{in}(u_i) \subseteq \mathcal{V}$ according to weight $w_{i,k} \geq 0$ ($w_{i,k} = 0$ if u_i is *inactive*), where $\Gamma_{in}(u_i)$ represents the users following u_i (i.e., set of users that u_i can influence) and $\sum_{u_k \in \Gamma_{in}(u_i)} w_{i,k} \leq 1$. Each user, e.g., u_i, is associated with a *static threshold* θ_i, which represents the minimal required influence for u_i to become *active*.

Meanwhile, based on the MMNs $\mathcal{G} = (\mathcal{U}, \mathcal{E}, \mathcal{R})$, the weight of each pair of users with different diffusion relations is estimated by PathSim [45]. Formally, the intra-network (inter-network) diffusion weight between user u and v with relation $i(j)$ is defined as

$$\phi^i_{(u,v)} = \frac{2|\mathcal{P}^i_{(u,v)}|}{|\mathcal{P}^i_{(u,)}| + |\mathcal{P}^i_{(,v)}|}, \quad \psi^j_{(u,v)} = \frac{2|\mathcal{Q}^j_{(u,v)}|}{|\mathcal{Q}^j_{(u,)}| + |\mathcal{Q}^j_{(,v)}|}, \tag{10.15}$$

where $\mathcal{P}^i_{(u,v)}(\mathcal{Q}^j_{(u,v)})$ denotes the set of intra-network (inter-network) diffusion meta path instances starting from u and ending at v with relation $i(j)$. $|\cdot|$ denotes the size of the set. Thus, $\mathcal{P}^i_{(u,)}(\mathcal{Q}^j_{(u,)})$ and $\mathcal{P}^i_{(,v)}(\mathcal{Q}^j_{(,v)})$ means the number of meta path instances with users u, v as the starting and ending users, respectively.

Based on the traditional LT model, influence propagates in discrete steps in the network. In step t, all *active* users remain *active* and *inactive* user can be *activated* if the received influence exceeds his threshold. Only activated users at step t can influence their neighbors at step $t + 1$ and the activation probability for user v in one network (e.g., $G^{(1)}$) with intra-network relation i and inter-network relation j can be represented as $g^{(1),i}_v(t + 1)$ and $h^{(1),j}_v(t + 1)$, respectively:

$$g^{(1),i}_v(t + 1) = \frac{\sum_{u \in \Gamma_{in}(v,i)} \phi^i_{(u,v)} \mathbb{I}(u, t)}{\sum_{u \in \Gamma_{in}(v,i)} \phi^i_{(u,v)}}, \tag{10.16}$$

$$h^{(1),j}_v(t + 1) = \frac{\sum_{u \in \Gamma_{in}(v,j)} \phi^j_{(u,v)} \mathbb{I}(u, t)}{\sum_{u \in \Gamma_{in}(v,j)} \phi^j_{(u,v)}} \tag{10.17}$$

where $\Gamma_{in}(v, i)$, $\Gamma_{in}(v, j)$ are the neighbor sets of user v in relations i and j, respectively and $\mathbb{I}(u, t)$ denotes if user u is activated at timestamp t. Note that anchor user $v^{(1)}$ is activated does not mean that his/her corresponding account in network $G^{(2)}$, i.e., $v^{(2)}$, will be activated at the same time, but $v^{(2)}$ will get influence from $v^{(1)}$ via the anchor link.

By aggregating all kinds of intra-network and inter-network relations, we can obtain the integrated activation probability of $v^{(1)}$ [24]. Here logistic function is used as the aggregation function.

$$p^{(1)}_v(t + 1) = \frac{e^{\sum_{(i)} \rho^{(1)}_i g^{(1),i}_v(t+1) + \sum_{(j)} \omega^{(1)}_j h^{(1),j}_v(t+1)}}{1 + e^{\sum_{(i)} \rho^{(1)}_i g^{(1),i}_v(t+1) + \sum_{(j)} \omega^{(1)}_j h^{(1),j}_v(t+1)}}, \tag{10.18}$$

where $\rho^{(1)}_i$ and $\omega^{(1)}_j$ denote the weights of each relation in the diffusion process, whose values satisfy $\sum_{(i)} \rho^{(1)}_i + \sum_{(j)} \omega^{(1)}_j = 1$, $\rho^{(1)}_i \geq 0$, $\omega^{(1)}_j \geq 0$. Similarly, we can get the activation probability of a user $v^{(2)}$ in $G^{(2)}$.

10.4.1.3 Problem Solution and Algorithm Analysis

Kempe et al. [29] proved that traditional influence maximization problem is an NP-hard for LT model, where the objective function of influence $\sigma(S)$ is *monotone* and *submodular*. Based on these properties, the greedy approximation algorithms can achieve an approximation ratio of $1 - 1/e$. With the above background knowledge, we will show that the influence maximization problem under the M&M model is also NP-hard and prove the influence spread function $\sigma(S)$ is both *monotone* and *submodular*.

Theorem 10.8 *Influence Maximization Problem across Partially Aligned Heterogeneous Social Networks is NP-hard.*

Theorem 10.9 *For the M&M model, the influence function $\sigma(S)$ is monotone and submodular.*

The proofs of Theorems 10.8 and 10.9 will be left as an exercise for the readers. Since the influence function is both *monotone* and *submodular*, as well as *non-negative*, based on the M&M model, step-wise greedy algorithm introduced in Sect. 10.2.2.1 can be applied to select the seed users who can lead to the maximum marginal influence increase in each step from both networks $G^{(1)}$ and $G^{(2)}$. According to the analysis provided before, such a *step-wise greedy seed user selection* approach can achieve a $(1 - \frac{1}{e})$-approximation of the optimal result.

10.4.2 Dynamic Programming-Based Seed User Selection

In the real world, selecting users as the seed user may introduce certain costs but the cost can be different for users in different networks. Normally, the mature online social networks with a large number of active users may cost more than other smaller-sized online social networks in commercial promotion. In this part, we will still focus on the *influence maximization* problem across multiple aligned social networks. Here, we will introduce another *influence maximization* method to activate users in a specific target network only. We propose to select seed users from both the target network and other aligned source networks subject to certain budget constraint, and these selected users will propagate information to activate users in the target network via both intra- and inter-network information diffusion routes.

Formally, let $G^{(t)}$ and $G^{(s)}$ denote the target and source network, respectively, whose involved user sets can be represented as $\mathcal{U}^{(t)}$ and $\mathcal{U}^{(s)}$, respectively. We can represent the influence function defined based on a certain information diffusion model as $\sigma(\cdot)$, which projects the selected seed user set to the number of infected users in the target network. According to the random walk based information diffusion model introduced in Sect. 9.4.2, to calculate the final number of activated users in $G^{(t)}$, we define a $(|\mathcal{U}^{(s)}| + |\mathcal{U}^{(t)}|)$-dimensional constant vector $\mathbf{b} = [0, 0, \ldots, 0, 1, 1, \ldots, 1]$, where the number of 0 is $|\mathcal{U}^{(s)}|$ and the number of 1 is $|\mathcal{U}^{(t)}|$. Thus the influence function of the IPATH model can be denoted as

$$\sigma(\mathcal{Z}) = \mathbf{b} \cdot h(\pi^*) = \mathbf{b} \cdot h\left(a[\mathbf{I} - (1-a)\mathbf{W}]^{-1} \cdot g(\mathcal{Z})\right). \tag{10.19}$$

From the above function, we can achieve the number of users who can be activated by the seed user set, while the specific user status can be obtained from the status vector π.

The objective function of the *influence maximization* problem studied in this part can be represented as

$$\begin{aligned} \max_{S} \quad & \sigma(S) \\ s.t. \quad & S \subset \mathcal{U}^{(t)} \cup \mathcal{U}^{(s)} \\ & \sum_{u \in S} c_u \leq b, \end{aligned} \tag{10.20}$$

where c_u denotes the introduced cost in adding u in the seed node set S and b represents the pre-specified budget.

To solve the above problem, we will provide a theoretic analysis about it first and then introduce a new viral marketing method "**I**nfluence **M**aximization algorithm based on **D**ynamic **P**rogramming" (IMDP) proposed in [48]. In IMDP, the information diffusion process is described by the random walk-based diffusion model IPATH introduced in Sect. 9.4.2. Furthermore, IMDP employs dynamic programming to address the problem, and can identify a fully polynomial approximation of the optimal seed user set.

10.4.2.1 Problem Analysis

This section will analyze the problem based on the IPATH model. We first prove that the studied problem is NP-hard. To simplify the notations, let $\mathbf{D} = a[\mathbf{I} - (1 - a)\mathbf{W}]^{-1} \in \mathbb{R}^{(|\mathcal{U}^{(s)}|+|\mathcal{U}^{(t)}|)^2}$ (where $a[\mathbf{I} - (1 - a)\mathbf{W}]^{-1} \in \mathbb{R}^{(|\mathcal{U}^{(s)}|+|\mathcal{U}^{(t)}|)^2}$ is a term used in the IPATH model as described in Sect. 9.4.2) and $\pi^{(0)} \in \{0, 1\}^{(|\mathcal{U}^{(s)}|+|\mathcal{U}^{(t)}|)^2}$ denote the vector indicating the initially selected seed users from both of these networks G^t and $G^{(s)}$. For the entries in vector $\pi^{(0)}$ filled with value 0, the corresponding users are selected as the seed users.

Theorem 10.10 *The problem denoted by Eq. (10.20) is NP-hard.*

Proof 0–1 Knapsack Problem, which is NP-hard, can be reduced to the problem in polynomial time. *0–1 Knapsack Problem* is a combination optimization problem: Given a set of items, each with mass w_i and value v_i, the aim of the problem is to determine the number of copies x_i of each kind of item to include in a collection, where x_i is restricted to zero or one, so that the total weight is less than a given limit and their total value is as large as possible, i.e.,

$$
\begin{aligned}
\max_{x} \quad & \sum_{i=1}^{n} v_i x_i \\
\text{s.t.} \quad & \sum_{i=1}^{n} w_i x_i \leq W \text{ and } x_i \in \{0, 1\}
\end{aligned}
\tag{10.21}
$$

The above objective function is actually equivalent to Eq. (10.20). The constraint of budget in (10.20) is equivalent to the weight limit. In the IPATH model, entry $D(j, u)$ is the information that user j can get when u is the only seed. Thus $\mathbf{b} \cdot h(D(:, u))$ is the number of users activated by u, which is mapped to the value v_i of item i. In the notation, function $h(\cdot)$ denotes the floor function introduced in the IPATH diffusion model. Hence the *0–1 Knapsack Problem* can be reduced to the studied problem in polynomial time, and the problem is also *NP-hard*.

According to Theorem 10.10, there is no polynomial algorithm which can give an optimal solution to the problem, if P$\neq$NP. While the greedy seed user selection algorithm for traditional influence maximization problem is proved to achieve an approximated optimal solution with a factor $(1 - 1/e)$ [29]. This approximation is achieved when the influence function $\sigma(S)$ is monotone and submodular.

A function $f(\cdot)$ is *monotone* iff $f(\mathcal{A}) \leq f(\mathcal{B})$ when $\mathcal{A} \subseteq \mathcal{B}$. We observe that the influence function $\sigma(S)$ is monotone, which means that adding a new seed user, the number of activated users will not decrease. It is also rational intuitively because all values in the weight matrix $\mathbf{W}$ are non-negative, involving more seed users will increase the activation probability of other users. Formally, we can prove such a claim as follows.

Theorem 10.11 *Based on the IPATH diffusion model, the influence function $\sigma(S)$ is monotone.*

Proof Given the current seed user set $\mathcal{S} \subset (\mathcal{U}^{(t)} \cap \mathcal{U}^{(s)})$, by incorporating user $u \in \mathcal{U}^{(t)} \cap \mathcal{U}^{(s)} \setminus \mathcal{S}$ to $\mathcal{S}$, we can represent the influence gain as

$$
\sigma(\mathcal{S} \cup \{u\}) - \sigma(\mathcal{S})
$$
$$
= \mathbf{b} \cdot h \left(\mathbf{D} \cdot \pi^{(0)} + \mathbf{D} \cdot \mathbf{u} \right) - \mathbf{b} \cdot h \left(\mathbf{D} \cdot \pi^{(0)} \right)
$$
$$
= \mathbf{b} \cdot h \left(\pi^* + \mathbf{u}^* \right) - \mathbf{b} \cdot h \left(\pi^* \right)
$$
$$
= \mathbf{b} \cdot \left(\lfloor \pi^* + \mathbf{u}^* + \mathbf{c} \rfloor - \lfloor \pi^* + \mathbf{c} \rfloor \right) \tag{10.22}
$$

The binary vector $\mathbf{u}$ denotes the new seed user u, where only $\mathbf{u}[u] = 1$, and other values are 0. According to Eq. (9.33), $\pi^* = \mathbf{D} \cdot \pi^{(0)}$ represents the information amount of each user can get at convergence with initial state $\pi^{(0)}$. Similarly, $\mathbf{u}^* = \mathbf{D} \cdot \mathbf{u}$ denotes the information amount of each user can get at the convergence state merely with the new seed user u.

Since $\lfloor x + y \rfloor \geq \lfloor x \rfloor + \lfloor y \rfloor$, we can have

$$
\sigma(\mathcal{S} \cup \{u\}) - \sigma(\mathcal{S})
$$
$$
\geq \mathbf{b} \cdot \left(\lfloor \pi^* + \mathbf{c} \rfloor + \lfloor \mathbf{u}^* \rfloor - \lfloor \pi^* + \mathbf{c} \rfloor \right) = \mathbf{b} \cdot \lfloor \mathbf{u}^* \rfloor \tag{10.23}
$$

As all elements in both $\mathbf{b}$ and $\mathbf{u}^*$ are non-negative, so $\sigma(\mathcal{S} + \{u\}) - \sigma(\mathcal{S}) \geq 0$, i.e., the influence function $\sigma(\mathcal{S})$ is monotone.

Meanwhile, a function $f(\cdot)$ is submodular, iff $f(\mathcal{A} \cup \{a\}) - f(\mathcal{A}) \geq f(\mathcal{B} \cup \{a\}) - f(\mathcal{B})$ for $\forall \mathcal{A} \subseteq \mathcal{B}$. It implies that for a specific seed user, his marginal contribution will be larger when being added into a smaller seed user set. Meanwhile, we observe that the influence function $\sigma(\mathcal{S})$ does not have such a property, and will prove it with a counterexample as follows.

Theorem 10.12 *Based on the IPATH diffusion model, the influence function $\sigma(\mathcal{S})$ is not submodular.*

Proof To prove the influence function is not submodular, we need to find a pair of seed sets $\mathcal{S}_1, \mathcal{S}_2$ and user u $(u \notin \mathcal{S}_1, u \notin \mathcal{S}_2)$, where $\mathcal{S}_1 \subseteq \mathcal{S}_2$ and $\sigma(\mathcal{S}_1 \cup \{u\}) - \sigma(\mathcal{S}_1) < \sigma(\mathcal{S}_2 \cup \{u\}) - \sigma(\mathcal{S}_2)$.

$$
\sigma(\mathcal{S}_1 \cup \{u\}) - \sigma(\mathcal{S}_1) - \sigma(\mathcal{S}_2 \cup \{u\}) + \sigma(\mathcal{S}_2)
$$
$$
= \mathbf{b} \cdot h \left(\mathbf{D} \cdot \pi_1^{(0)} + \mathbf{D} \cdot \mathbf{u} \right) - \mathbf{b} \cdot h \left(\mathbf{D} \cdot \pi_1^{(0)} \right)
$$
$$
- \mathbf{b} \cdot h \left(\mathbf{D} \cdot \pi_2^{(0)} + \mathbf{D} \cdot \mathbf{u} \right) + \mathbf{b} \cdot h \left(\mathbf{D} \cdot \pi_2^{(0)} \right)
$$
$$
= \mathbf{b} \cdot h \left(\pi_1^* + \mathbf{u}^* \right) - \mathbf{b} \cdot h \left(\pi_1^* \right) - \mathbf{b} \cdot h \left(\pi_2^* + \mathbf{u}^* \right) + \mathbf{b} \cdot h \left(\pi_2^* \right)
$$
$$
= \mathbf{b} \cdot \left(\lfloor \pi_1^* + \mathbf{u}^* + \mathbf{c} \rfloor - \lfloor \pi_1^* + \mathbf{c} \rfloor - \lfloor \pi_2^* + \mathbf{u}^* + \mathbf{c} \rfloor + \lfloor \pi_2^* + \mathbf{c} \rfloor \right) \tag{10.24}
$$

We suggest the following counterexample, involving four users $\{u_1, u_2, u_3, u_4\}$. Let $\theta = \frac{3}{4}$, $\mathbf{b} = [0, 0, 1, 1]$, $\mathcal{S}_1 = \emptyset$, $\mathcal{S}_2 = \{u_1\}$ and $u = u_2$, i.e., $\pi_1^{(0)} = [0, 0, 0, 0]^\top$, $\pi_2^{(0)} = [1, 0, 0, 0]^\top$ and $\mathbf{u} = [0, 1, 0, 0]^\top$. In addition, let the weighted diffusion matrix among these four users be

$$
\mathbf{D} = \begin{bmatrix} 0 & 0 & 0 & 1 \\ 0 & 0 & 1 & 0 \\ \frac{1}{2} & \frac{1}{2} & 0 & 0 \\ \frac{1}{2} & \frac{1}{2} & 0 & 0 \end{bmatrix} \tag{10.25}
$$

Thus the convergence state $\pi_1^* = [0, 0, 0, 0]^\top$, $\pi_2^* = [0, 0, \frac{1}{2}, \frac{1}{2}]^\top$ and $\mathbf{u} = [0, 0, \frac{1}{2}, \frac{1}{2}]^\top$. When we substitute them into (10.24), we can have

$$\sigma(\mathcal{S}_1 \cup \{u\}) - \sigma(\mathcal{S}_1) - \sigma(\mathcal{S}_2 \cup \{u\}) + \sigma(\mathcal{S}_2)$$

$$= \mathbf{b} \cdot \left([0, 0, 0, 0]^\top - [0, 0, 0, 0]^\top - [0, 0, 1, 1]^\top + [0, 0, 0, 0]^\top \right)$$

$$= [0, 0, 1, 1] \cdot [0, 0, -1, -1]^\top = -2 \tag{10.26}$$

Therefore, in such a counterexample, we have $\sigma(\mathcal{S}_1 \cup \{u\}) - \sigma(\mathcal{S}_1) - [\sigma(\mathcal{S}_2 \cup \{u\}) - \sigma(\mathcal{S}_2)] < 0$. It shows the influence function $\sigma(\mathcal{S})$ of the IPATH is not submodular.

Since the influence function is monotone but not submodular, no theoretic performance guarantee exists for the traditional step-wise greedy seed user selection algorithm [29] any more.

10.4.2.2 The IMDP Optimization Algorithm

In the proof of *NP-hardness*, we mentioned user u's contribution is the number of users activated by u, i.e., $p_u = \mathbf{b} \cdot h(\mathbf{D}(:, u))$. We propose to adjust the contribution with a factor $\delta = \frac{\epsilon \cdot p_{\max}}{n}$, where $n = |V_s| + |V_t|$ is the number of all users, $p_{\max} = \max_{u \in (V_s \cup V_t)} p_u$ denotes the largest contribution from all the users. We define $\bar{p}_u = \lfloor \frac{p_u}{\delta} \rfloor$, for $u = 1, 2, \ldots, n$. As $p_{\max}$ is the largest contribution, we cannot get the profit larger than $np_{\max}$, which denotes the contribution upper bound in the dynamic programming. Let $f(i, \rho)$, $(1 \le i \le n, 1 \le \rho \le np_{\max})$ be the smallest cost sum, so that a solution with scaled contribution sum equal to ρ can be obtained by users $j = 1, 2, \ldots, i$. Thus, $f(i, p)$ can be represented as

$$f(i, \rho) = \min \left\{ \sum_{j=1}^{i} c_j : \sum_{j=1}^{i} \bar{p}_j x_j = \rho, x_j \in \{0, 1\}, j = 1, 2, \ldots, i \right\} \tag{10.27}$$

All values of $f(i, \rho)$ can be calculated through the following recurrence:

$$f(i, \rho) = \begin{cases} \min\{f(i-1, \rho), f(i-1, \rho - \bar{p}_i) + c_i\} & \text{if} \quad \bar{p}_i < \rho \\ f(i-1, \rho) & \text{otherwise.} \end{cases} \tag{10.28}$$

When there are several choices (i.e., users) introducing the same amount of contribution, the method will pick one of them randomly as the seed user. We initialize the base case $f(1, \rho)$ as follows:

$$f(1, \rho) = \begin{cases} c_i & \text{if } \rho = \bar{p}_i, \\ \infty & \text{otherwise.} \end{cases} \tag{10.29}$$

Thus the IMDP algorithm first gets the intra-network weight matrix $\mathbf{W}^s$ and $\mathbf{W}^t$, and constructs the inter-network weight matrix $\mathbf{W}^{s \to t}$ and $\mathbf{W}^{s \to t}$. Then the final weight matrix is built with four components as Fig. 9.5 shows in Sect. 9.4.2. At last IMDP uses dynamic programming to identify the optimal seed users across the networks.

10.4.2.3 Analysis of the IMDP Algorithm

In this part, we will analyze the performance of IMDP from the theoretical perspective. We will prove that by scaling with respect to the desired ϵ, and we will be able to get a solution that is at least $(1 - \epsilon) \cdot OPT$ (the optimal solution) in polynomial time with respect to both n and $1/\epsilon$.

Lemma 10.1 *The set $\mathcal{S}$ output by the* IMDP *satisfies*

$$\sigma(\mathcal{S}) \geq (1 - \epsilon) \cdot OPT. \tag{10.30}$$

Proof Let O be the optimal seed set activating the maximum users. For any user $\bar{p}_u = \lfloor \frac{p_u}{\delta} \rfloor$, thus $0 \leq p_u - \delta \bar{p}_u \leq \delta$. Therefore the profit of the optimal set O can decrease is at most $n\delta$:

$$\sigma(O) - \delta \cdot \bar{\sigma}(O) \leq n\delta \tag{10.31}$$

where $\bar{\sigma}(\cdot)$ denotes the influence function scaled by the factor δ.

$$\sigma(\mathcal{S}) \geq \delta \cdot \bar{\sigma}(O) \tag{10.32}$$

$$\geq \sigma(O) - n\delta = OPT - \epsilon \cdot p_{\max} \tag{10.33}$$

$$\geq (1 - \epsilon) \cdot OPT \tag{10.34}$$

Inspired by [34], the seed set $\mathcal{S}$ selected from the dynamic programming is optimal for the scaled instance and therefore must be at least as good as choosing the set O with the smaller profits.

The approximation algorithm IMDP is said to be a *polynomial time approximation scheme*, if for each fixed $\epsilon > 0$, its running time is bounded by a polynomial in the size n. And the *fully polynomial time approximation scheme* is an approximation scheme for which the algorithm is bounded polynomially in both the size n and $1/\epsilon$. We prove that the IMDP method is a fully polynomial approximation scheme for the ANIM problem, with the following Theorem 10.13.

Theorem 10.13 *The* IMDP *method is a fully polynomial approximation scheme for the ANIM problem.*

Proof Since $\delta = (\epsilon \times p_{\max})/n$, the running time of IMDP is $O(n^2 \lfloor \frac{p_{\max}}{\delta} \rfloor) = O(n^2 \lfloor \frac{n}{\epsilon} \rfloor)$, which is polynomial in both n and $1/\epsilon$. As shown in Lemma 10.1, the IMDP framework can achieve a $(1 - \epsilon)$-approximation of the optimal result.

10.5 Rumor Initiator Detection

This section is a follow-up problem based on the MFC diffusion model introduced in Sect. 9.3.2 based on signed networks. Rumor initiation and incorrect information dissemination are both common in social networks [44]. Incorrect rumors sometimes can bring about devastating effects, and an important goal in improving the credibility of the social channel is to identify rumor initiators [35, 41, 43, 44] in signed social networks. This section studies the detection of rumor initiators in infected signed social networks, given the state of the network at a specific moment in time.

To identify the rumor initiators, we study the problem based on the MFC diffusion model introduced in Sect. 9.3.2. Although the exact identification of the rumor initiators is NP-hard for general graphs, but it can be resolved in polynomial time for binary tree structured networks, and it provides the insights for high quality solutions in the general case. We leverage these insights to introduce the RID framework [51] to identify the optimal *rumor initiators*, including their number, identities, and initial states. The readers are suggested to read this section together with Sect. 9.3.2 introduced in the previous chapter. Here, we will not introduce the definitions of the *weighted signed social network* and *weighted signed diffusion network* concepts again, the readers may refer to Sect. 9.3.2 for more information.

The social psychology literature defines a *rumor* as a story or a statement in general circulation without confirmation or certainty of facts [2]. The originators of rumors are formally defined as *rumor initiators*, which can be individuals, groups, or institutes. In this section, we refer to *rumor initiators* as the users who initially spread the rumor to other users in online social networks. Within the *diffusion networks*, *rumors* can spread from the *initiators* to other users via diffusion links, which will lead to *infected signed diffusion networks*. Since all networks studied in this section are all weighted and signed by default, we will refer to them as *diffusion networks* for simplicity.

Definition 10.8 (Infected Diffusion Network) The *infected diffusion network* $G_I = (\mathcal{V}_I, \mathcal{E}_I, s_I, w_I)$ is a subgraph of the complete diffusion network G_D, where $\mathcal{V}_I \subseteq \mathcal{V}_D$ is the set of infected users, $\mathcal{E}_I \subseteq \mathcal{E}_D$ is the set of potential diffusion links among these infected users. s_I, w_I are the *sign* and *weight* mappings, whose domains are all those diffusion links in $\mathcal{E}_I$.

Definition 10.9 (Activation Link) Among all the links $\mathcal{E}_I$ in the infected diffusion networks, link (u, v) is called an activation link iff u activates v in the screenshot of the infected diffusion network.

Based on the MFC model introduced before, each node in the infected diffusion network screenshot can be activated by exactly one node via the *activation link* and the *rumor initiators* have no incoming activation links. As a result, all the nodes in $\mathcal{V}_I$ together with the activation links among them can actually form a set of cascade trees, where nodes at higher levels are activated by nodes in the lower levels and *rumor initiators* are the roots (at level 1).

In this section, our main goal is to work backwards from the available state of the network given at any moment in time, and we will use the developed diffusion model to track down the rumor initiators. Let $\mathcal{I} \subseteq \mathcal{V}_I \subseteq \mathcal{V}$ be the potential set of *rumor initiators*, whose initial states towards the rumor can be represented as $\mathcal{S} = \{+1, -1\}^{|\mathcal{I}|}$, where $+1$ indicates a belief in the fact at hand, and -1 denotes belief in the opposite fact. We use binary modes of information propagation because of its relative simplicity and intuitive appeal in modeling a variety of situations. The ISOMIT problem aims at inferring the optimal *rumor initiator* set $\mathcal{I}^*$ as well as their initial states $\mathcal{S}^*$, which can maximize the likelihood that it will lead to the current state of the *infected signed network* G_I:

$$\mathcal{I}^*, \mathcal{S}^* = \arg \max_{\mathcal{I}, \mathcal{S}} \mathbf{P}(G_I | \mathcal{I}, \mathcal{S}), \tag{10.35}$$

Here, $\mathbf{P}(G_I | \mathcal{I}, \mathcal{S})$ represents the likelihood of obtaining the infected network G_I based on the influence propagated from $\mathcal{I}$ with states $\mathcal{S}$.

Formally, we will call the above problem as the "Infected Signed netwOrk ruMor Initiator deTection" (ISOMIT) problem. In summary, the input of the ISOMIT problem is the infected signed network G_I, while the objective output is the inferred rumor initiators $\mathcal{I}$ together with their initial states $\mathcal{S}$ which can maximize the likelihood $\mathbf{P}(G_I | \mathcal{I}, \mathcal{S})$.

10.5.1 The ISOMIT Problem

Given the *rumor initiators* $\mathcal{I}$ together with their initial states $\mathcal{S}$, influence can propagate from them to other users in the network via different paths. For any user u in the infected network, the influence propagation paths from *initiators* to u can be represented as the set $\{\mathcal{P}(u_i, u)\}_{u_i \in \mathcal{I}}$, where $\mathcal{P}(u_i, u)$ represents the set of paths from initiator u_i to user u specifically. Each path (e.g., $p \in \mathcal{P}(u_i, u)$) is a sequence of directed diffusion links from u_i to u. We use the notation $(x, y) \in p$ to denote the fact that the diffusion link (x, y) lies on path p. Depending on the sign of link (u, v) as well as the states of u and v, link (u, v) can be either *sign consistent* or *sign inconsistent*.

Definition 10.10 (Sign Inconsistent Diffusion Link) Diffusion link (u, v) is defined to be *sign inconsistent* if $s(u) \cdot s(u, v) \neq s(v)$.

The probability that $u \in \mathcal{V}$ is infected with state $s(u)$ because of influence from the initiators $\mathcal{I}$ with state $\mathcal{S}$ can be computed as

$$\mathbf{P}(u, s(u)|\mathcal{I}, \mathcal{S})$$

$$= 1 - \prod_{i \in \mathcal{I}} \prod_{p \in \mathcal{P}(i,u)} \left(1 - \prod_{(x,y) \in p} g\left(s(x), s_I(x, y), s(y), w_I(x, y)\right) \right), \tag{10.36}$$

where the function

$$g\left(s(x), s_I(x, y), s(y), w_I(x, y)\right)$$

$$= \begin{cases} \min\{1, \alpha \cdot w_I(x, y)\}, & \text{if } s(x) \cdot s_I(x, y) = s(y), s_I(x, y) = +1, \\ w_I(x, y), & \text{if } s(x) \cdot s_I(x, y) = s(y), s_I(x, y) = -1, \\ 0, & \text{if } s(x) \cdot s_I(x, y) \neq s(y). \end{cases} \tag{10.37}$$

Consider a link (x, y) lying on the path from rumor initiators in $\mathcal{I}$ to u, such that states of x and y are consistent (i.e., $s(x) \cdot s_I(x, y) = s(y)$). In such a case, the probability of link (x, y) being an activation link would be $\min\{1, \alpha \cdot w_I(x, y)\}$ if (x, y) is a positive link (due to the boosting of positive links in MFC model), and it would be $w_I(x, y)$, otherwise. However, in case of inconsistency (i.e., $s(x) \cdot s_I(x, y) \neq s(y)$), link (x, y) will be either not an activation link or was an activation link originally but y's state is flipped by some other nodes. In other words, y would not be activated by x in the screenshot of the infected diffusion network, and the $g(\cdot)$ is assigned with value one in the *sign inconsistent* case.

One can model the probability of the current state of the *infected signed network* G_I, conditional on the *rumor initiators* $\mathcal{I}$ with initial states $\mathcal{S}$ as follows:

$$\mathbf{P}(G_I|\mathcal{I}, \mathcal{S}) = \prod_{u \in \mathcal{V}_I} \mathbf{P}(u, s(u)|\mathcal{I}, \mathcal{S}). \tag{10.38}$$

10.5.2 NP-Hardness of Exact ISOMIT Problem

Based on the aforementioned remarks, we will show that obtaining the whole *infected networks* exactly based on $\mathcal{I}$ and $\mathcal{S}$ achieving 100% inference probability with minimum number of rumor initiators is an NP-hard problem.

Lemma 10.2 *Based on the* MFC *diffusion model, the* ISOMIT *problem of achieving probability* $\mathbf{P}(G_I|\mathcal{I}, \mathcal{S}) = 1$ *with the minimum number of initiators is NP-hard.*

Proof We will prove the lemma by showing that the set-cover problem (which is known to be NP-hard) can be reduced to the ISOMIT problem in polynomial time. Formally, given a set of elements $\mathcal{E} = \{e_1, e_2, \ldots, e_n\}$ and a set of m subsets of $\mathcal{E}$, $\mathcal{L} = \{\mathcal{L}_1, \mathcal{L}_2, \ldots, \mathcal{L}_m\}$, where $\mathcal{L}_i \subseteq \mathcal{E}, i \in \{1, 2, \ldots, m\}$. The set-cover problem aims at finding as few subsets as possible from $\mathcal{L}$, so that the union of the selected subsets is equal to $\mathcal{E}$, i.e., $\bigcup \mathcal{L}_i = \mathcal{E}$ [21].

For an arbitrary instance of the set-cover problem, we define an instance of the infected signed graph to be a directed graph, denoted by G_I. The graph G_I contains $n + m + 1$ nodes: (1) for each element $e_i \in \mathcal{E}$, we construct a corresponding node n_i; (2) for each set $\mathcal{L}_j \in \mathcal{L}$, we construct node n_{j+n}; and (3) a dummy node d (i.e., the $(n + m + 1)_{th}$ node) is added to the infected network. The links in G_I include: (1) for all the elements in each set, e.g., $e_i \in \mathcal{L}_j$), we add a directed link connecting their corresponding nodes in the graph from n_i to n_{j+n}; (2) all the corresponding nodes of elements in $\mathcal{E}$ are connected to d via a directed link; and (3) d connects to the corresponding nodes of sets in $\mathcal{L}$ by directed links as well. The signs of all these links are all assigned $+1$, whose weights are: (1) $w(n_i, n_{j+n}) = 1$, for $\forall e_i \in \mathcal{E}, \forall e_i \in \mathcal{L}_j, \mathcal{L}_j \in \mathcal{L}$; (2) $w(n_i, d) = \frac{1}{n}$, for $\forall e_i \in \mathcal{E}$; (3) $w(n_{j+n}, d) = 1$, for $\forall \mathcal{L}_j \in \mathcal{L}$.

Now, we want to activate all the nodes in G_I with state $+\mathbf{1}$ (i.e., all trust the rumor) with as few rumor initiators as possible. Based on G_I, the solution to the ISOMIT problem will be equivalent to the set-cover problem based on elements $\mathcal{E}$ and subsets $\mathcal{L}$.

10.5.3 A Special Case: k-ISOMIT-BT Problem

In the previous section, the ISOMIT problem of achieving probability 100% with the minimum number of initiators is proven to be NP-hard. In this part, we will study a special case of the ISOMIT problem, where the number of *rumor initiators* is known to be k and the network is a binary tree, i.e., the k-ISOMIT-BT (k ISOMIT on Binary Tree) problem. We will show that the k-ISOMIT-BT problem can be addressed efficiently in polynomial time. This will also provide the insight needed to solve the general case to be introduced in the next section.

Let $T_I = (\mathcal{V}_I, \mathcal{E}_I, s_I, w_I)$ be an infected signed binary tree. If the user node $u \in \mathcal{V}_I$ is regarded as the root in the tree, its left and right children can be represented as $left(u)$ and $right(u)$, respectively. At the beginning, the *rumor initiator set* and the *state set* is empty, i.e., $\mathcal{I} = \emptyset$ and $\mathcal{S} = \emptyset$. The cost of the optimal solution (i.e., the inferred initiators $\mathcal{I}$ and states $\mathcal{S}$) can be recursively computed with the following dynamic programming equation:

$$\mathbf{OPT}(u, \mathcal{I}, \mathcal{S}, k) = \max \Bigg\{$$

$$\min_{m=0}^{k} \Big\{ \mathbf{OPT}\big(left(u), \mathcal{I}, \mathcal{S}, m\big) + \mathbf{OPT}\big(right(u), \mathcal{I}, \mathcal{S}, k - m\big) + \mathbf{P}\big(u, s(u)|\mathcal{I}, \mathcal{S}\big) \Big\}; \quad (10.39)$$

$$\mathbf{P}\big(u, s(u) = +1|\mathcal{I} \cup \{u\}, \mathcal{S} \cup \{s(u) = +1\}\big) + \min_{m=0}^{k-1} \Big\{ \mathbf{OPT}\big(left(u), \mathcal{I} \cup \{u\},$$

$$\mathcal{S} \cup \{s(u) = +1\}, m\big) + \mathbf{OPT}\big(right(u), \mathcal{I} \cup \{u\}, \mathcal{S} \cup \{s(u) = +1\}, k - 1 - m\big) \Big\}; \quad (10.40)$$

$$\mathbf{P}\big(u, s(u) = -1 | \mathcal{I} \cup \{u\}, \mathcal{S} \cup \{s(u) = -1\}\big) + \min_{m=0}^{k-1} \Big\{ \mathbf{OPT}\big(left(u), \mathcal{I} \cup \{u\},$$

$$\mathcal{S} \cup \{s(u) = -1\}, m\big) + \mathbf{OPT}\big(right(u), \mathcal{I} \cup \{u\}, \mathcal{S} \cup \{s(u) = -1\}, k - 1 - m\big)\Big\}\Big\}. \quad (10.41)$$

From root u, the optimal *rumor initiator* detection can generally follow one of three cases:

- u is not the *initiator*: The root u is not added to the *rumor initiator* set, and we make recursive calls with its left and right children nodes to identify the k *rumor initiators*.
- u is the *initiator* with state $s(u) = +1$: The root u and its state are added into the *rumor initiator* set and the *state* set, respectively (i.e., $\mathcal{I} \cup \{u\}$, and $\mathcal{S} \cup \{s(u) = +1\}$). Furthermore, we make recursive calls with its left and right children nodes to identify the remaining $k - 1$ *rumor initiators* based on the updated *rumor initiator* and their *state*.
- u is the *initiator* with state $s(u) = -1$: The root u and its state are added into the *rumor initiator* set and *state* set, respectively (i.e., $\mathcal{I} \cup \{u\}$, and $\mathcal{S} \cup \{s(u) = -1\}$). Furthermore, we make recursive calls with its left and right children nodes to identify the remaining $k - 1$ *rumor initiators* based on the updated *rumor initiator* and their *state*.

The formal definition of $\mathbf{P}(u, s(u) | \mathcal{I}, \mathcal{S})\}$ is available in Sect. 10.5.1. Meanwhile, the special case $\mathbf{P}(u, s(u) | \{u\}, \{s(u)\})$, for a *single node* u, is computed as follows:

$$\mathbf{P}(u, s(u) | \{u\}, \{s(u)\}) = \begin{cases} 1, & \text{if } s_I(u) = s(u); \\ 0, & \text{if } s_I(u) \neq s(u), \end{cases} \quad (10.42)$$

where $s_I(u)$ is the real state of u in the infected network.

The aforementioned dynamic programming objective function can be addressed in polynomial time, and we will not introduce the details involved in solving it here due to the limited space.

10.5.4 RID Method for General Networks

For the ISOMIT problems in social networks of general structure and an unknown number of rumor initiators, the method introduced in the previous section cannot be directly applied. In this section, we will introduce the RID framework to address the ISOMIT problem. We propose to first detect the infected connected components from the whole network. For each detected connected component, we propose to further prune the non-existing activation links among users to extract the "*infected cascade trees*" in the signed networks. From each infected cascade tree, we introduce the objective function to detect the optimal rumor initiators (the number, identities as well as their states).

10.5.4.1 Infected Connected Components Detection

The infected diffusion network can contain multiple infected connected components, where users in each component can be connected to each other via potential diffusion links among them. In this part, we will introduce the method to detect the infected connected components from the network.

Definition 10.11 (Infected Connected Components) An *infected connected component* is a subgraph of the *infected network* and, by ignoring the directions of diffusion links, any two vertices in the component are connected to each other.

The *signed connected components* in the pruned networks can be detected with algorithms, like breadth-first search (BFS) [15] and depth-first search (DFS) [15], in linear time. For instance, based on the BFS algorithm, we will loop through all the infected vertices in the pruned infected signed network and once we reach an unvisited vertex, e.g., u, we will call BFS function to find the entire connected component containing u. The time cost of BFS-based connected component detection algorithm will be $O(n+m)$, where n and m are the numbers of user nodes and diffusion links in the infected diffusion network.

10.5.4.2 Signed Infected Cascade Forest Extraction

Let $\mathcal{C} = \{C_1, C_2, \ldots, C_l\}$ be the set of l connected components detected in the pruned infected signed network. As introduced earlier, the real information diffusion process in the infected connected component based on MFC can form a set of infected cascade trees. We show how to extract such trees later in this section.

Definition 10.12 (Infected Cascade Tree) The *signed infected cascade tree* summarizes the state of the information propagation and user activation process in the network. Let $T = (\mathcal{V}_T, \mathcal{E}_T, s, w)$ be a *signed infected cascade tree*. The node set $\mathcal{V}_T \subseteq \mathcal{V}_D$ consists of all the infected users in the tree and the directed activation link $(u, v) \in \mathcal{E}_T \subseteq \mathcal{E}_D$ if and only if u succeeds in activating v.

The signed infected cascade trees can be inferred from the infected network, and we propose to extract the trees capturing the most information (i.e., the most likely trees) for each connected component. Let $C_i = (\mathcal{V}_{C_i}, \mathcal{E}_{C_i}, s, w)$ be a detected connected component consisting of multiple infected cascade trees, and let $T = (\mathcal{V}_T, \mathcal{E}_T, s, w)$ be one of the trees extracted from C_i, where $\mathcal{V}_T \subseteq \mathcal{V}_{C_i}$ and $\mathcal{V}_T \subseteq \mathcal{V}_{C_i}$. The likelihood of tree T is $\mathcal{L}(T) = \prod_{(u,v) \in \mathcal{E}_T} w(u, v)$. Furthermore, the *optimal* infected cascade tree T^* in C_i can be defined as:

$$T^* = \arg \max_{T \in \mathcal{T}} \mathcal{L}(T), \qquad (10.43)$$

where $\mathcal{T}$ denotes the set of all potential trees that can be detected from component C_i. The maximum likelihood *infected cascade trees* can be extracted using the Chu-Liu/Edmonds' algorithm [14, 20] from the directed connected components. The pseudo-code of the *infected cascade trees* extraction method is available in Algorithm 8, which will call the functions in Algorithms 6 and 7 to get the maximum weight spanning graphs and resolve the circles in the graph.

10.5.4.3 Rumor Initiator Inference

Based on the methods introduced in the previous sections, we are able to detect a set of diffusion trees from the network, the roots of which without incoming edges represent the rumor initiators. Meanwhile, besides the roots, multiple *rumor initiators* can co-exist in one *infected cascade tree*. In other words, the number of extracted diffusion trees is a lower bound on the number of rumor

---

Algorithm 6 Maximum weight spanning graph (MWSG)

Require: Graph $G = (\mathcal{V}, \mathcal{E}, s, w)$
Ensure: Maximum weight spanning graph $G' = (\mathcal{N}, \mathcal{L}, w)$
1: initialize node set $\mathcal{N} = \emptyset$, link set $\mathcal{L} = \emptyset$
2: **for** $u \in \mathcal{V} \setminus \mathcal{N}$ **do**
3:      $\mathcal{N} = \mathcal{N} \cup \{u\}$
4:      find edge $e = \arg \max_{e \in \mathcal{E}} w(e)$
5:      $\mathcal{L} = \mathcal{L} \cup \{e\}$
6: **end for**

Algorithm 7 Contract circles (CC)

Require: Graph containing circles $G = (\mathcal{N}, \mathcal{L}, w)$
Ensure: Contracted graph without circles $G' = (\mathcal{N}', \mathcal{L}', w')$
1: $\mathcal{L}' = \emptyset$ and new link weight mapping w'
2: **for** each circle $O = (\mathcal{N}_O, \mathcal{L}_O)$ in $(\mathcal{N}, \mathcal{L})$ **do**
3:      contract all nodes in O into a pseudo-node u_o
4:      **for** each link $(u_x, u_y) \in \mathcal{L}$ **do**
5:          **if** $u_x \notin \mathcal{N}_O$ and $u_y \in \mathcal{N}_O$ **then**
6:              $\mathcal{L}' = \mathcal{L}' \cup \{(u_x, u_o)\}$
7:              $w'(u_x, u_o) = w(u_x, u_y) - w(\pi(u_y), u_y)$, where $(\pi(u_y), u_y) \in \mathcal{L}$ is the link with the maximum weight linked to u_y
8:          **else**
9:              **if** $u_x \in \mathcal{N}_O$ and $u_y \notin \mathcal{N}_O$ **then**
10:                  $\mathcal{L}' = \mathcal{L}' \cup \{(u_o, u_y)\}$
11:                  $w'(u_o, u_y) = w(u_x, u_y)$
12:              **else**
13:                  $\mathcal{L}' = \mathcal{L}' \cup \{(u_x, u_y)\}$
14:                  $w'(u_x, u_y) = w(u_x, u_y)$
15:              **end if**
16:          **end if**
17:      **end for**
18: **end for**

Algorithm 8 Infected cascade trees extraction

Require: infected connected component set $\mathcal{C}$
Ensure: infected cascade tree set $\mathcal{T}$
1: initialize tree set $\mathcal{T} = \emptyset$
2: **for** component $C_i = (\mathcal{V}_{C_i}, \mathcal{E}_{C_i}, s_{C_i}, w_{C_i}) \in \mathcal{C}$ **do**
3:      $(\mathcal{N}, \mathcal{L}, w) = \text{MWSG}(C_i)$
4:      **if** $(\mathcal{N}, \mathcal{L}, w)$ contains circles $\mathcal{O}$ **then**
5:          $(\mathcal{N}', \mathcal{L}', w') = \text{CC}(\mathcal{N}, \mathcal{L}, w_{C_i})$
6:          $(\mathcal{N}', \mathcal{L}', w') = \text{MWSG}((\mathcal{N}', \mathcal{L}', w'))$
7:      **end if**
8:      **for** circle $O \in \mathcal{O}$ **do**
9:          **for** link $(u_x, u_o) \in \mathcal{L}'$ **do**
10:              get the corresponding link (u_x, u_y), where u_y is in the circle
11:              remove link $(\pi(u_y), u_y)$ from $\mathcal{L}$ to break the circle O
12:          **end for**
13:      **end for**
14:      $\mathcal{T} = \mathcal{T} \cup \{(\mathcal{N}, \mathcal{L})\}$
15: **end for**

---

initiators. The detected cascade tree can actually be partitioned into several isolated sub-trees instead. The roots of these sub-trees provide additional candidates for being rumor initiators. Such a partitioning process can be achieved with the algorithm introduced in Sect. 10.5.3 effectively. However, the extracted *infected cascade trees* from the infected signed network may not necessarily be binary trees, and this can be very complex to deal with [35]. Next, we propose to transform each cascade tree into a binary tree first and then identify the optimal rumor initiators.

To transform a general tree into a binary tree without distorting information about the relative influence relationships, we propose to add extra dummy nodes to the trees, which have no effect on information diffusion, and they cannot be selected as *rumor initiators*.

Example 10.2 For example, in Fig. 10.3, the tree in the left figure is not a binary tree, where the root node has three children nodes. To transform it into a binary tree, between the root and its children, $\lceil \log_2 3 \rceil$ extra nodes are added to the tree as the root's new children and the root's children nodes are

Fig. 10.3 Example of the
binary tree transformation

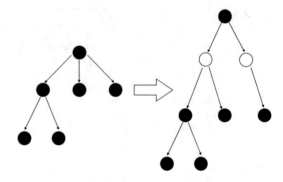

assigned as the new nodes' children. These newly added nodes will not participate in the information diffusion and they cannot be selected as *rumor initiators*.

Meanwhile, to avoid the case of having too many *rumor initiators* (e.g., every user in the component is a *rumor initiators*), we will add a penalty term to constrain the number of detected rumor initiators. For each tree $T \in \mathcal{T}$ rooted at u, we can represent the optimal rumor initiators $\mathcal{I}^*$ of size k^* with initial state $\mathcal{S}^*$ as follows:

$$k^*, \mathcal{I}^*, \mathcal{S}^* = \arg \min_{k, \mathcal{I}, \mathcal{S}} -\mathbf{OPT}(u, \mathcal{I}, \mathcal{S}, k) + (k - 1) \cdot \beta, \tag{10.44}$$

Here, parameter β denotes the penalty of each introduced rumor initiator and term $(k - 1)$ represents the extra initiators detected besides the original root of tree T. Function $\mathbf{OPT}(u, \mathcal{I}, \mathcal{S}, k)$ can be computed with the dynamic programming based method introduced in the previous section. By enumerating k from 1 to the number of nodes in T (i.e., $|\mathcal{V}_T|$), we are able to obtain the optimal solution of the above objective function. However, such a process can be very time consuming. To balance between the time cost and quality of the result, we propose to increase k from 1 to $|\mathcal{V}_T|$ and stop once the increase in k cannot lead to increase in the objective function.

10.6   Summary

In this chapter, we talked about the viral marketing problem and provided an introduction to several viral marketing algorithms for seed user selection based on various learning settings. The viral marketing problem is formulated as an optimization problem, which aims at selecting the optimal seed user set, who can maximize the impact in the information diffusion process.

Based on the classic information diffusion models, like LT and IC, we introduced two viral marketing algorithms, greedy and CELF, to pick the seed users. The greedy algorithms will select the seed user with multiple rounds, where the users who can introduce the maximum influence gain will be added into the seed user set in each round. To resolve the high time cost problem, CELF uses a heap data structure to keep record of the users' influence, and the heap structure will be updated dynamically in the seed user selection process. To further lower down the time complexity, we also introduced two algorithms based on heuristics, where one is based on the node centrality and the other one is based on degree discount, respectively.

In the multi-product information diffusion setting, the products may have intertwined relationships with each other, including independent, competing, and complimentary, respectively. Depending on the promotion orders of the other products and the target product, we categorized the viral marketing problem in such a learning setting into the C-TIM and J-TIM, respectively. To resolve the problem, the C-TIER and J-TIER algorithms were introduced in this chapter.

To study the viral marketing problem across multiple aligned social networks, we introduced the M&M diffusion model. Based on a set of meta paths, M&M defines the multi-relational network with the meta path concept, based on which the viral marketing problem is addressed with a greedy algorithm. Meanwhile, based on the IPATH diffusion model, we introduced a dynamic programming-based seed user selection algorithm named IMDP.

At the end of this chapter, we talked about the rumor initiator detection problem. Given an infected network, the rumor initiator detection problem aims at identifying the rumor initiators IDs, numbers, and initial status concurrently. By assuming that the rumor diffusion process is based on the MFC model, to address the problem, we introduced a multi-phase rumor initiator detection algorithm, which partitions the infected network into several components and further identifies the rumor initiators from them with a dynamic program algorithm.

10.7 Bibliography Notes

Viral marketing (i.e., influence maximization) problem in customer networks first proposed by Domingos et al. [19] has been a hot research topic. Richardson et al. [42] study the viral marketing based on knowledge-sharing sites and propose a new model which needs less computational cost than the model proposed in [19]. Kempe et al. propose to study the influence maximization problem through a social network [29] and propose two different diffusion models: independent cascade (IC) model and linear threshold (LT) model, which have been widely used in later influence maximization papers. Zhan et al. propose to extend the traditional single-network viral marketing problem to multiple aligned networks in [47].

Meanwhile, the promotions of multiple products can exist in social networks simultaneously, which can be independent, competing, or complementary. Datta et al. [17] study the viral marketing for multiple independent products at the same time and aim at selecting seed users for each products to maximize the overall influence. Bharathi et al. [4] propose to study the competitive influence maximization in social networks, where multiple competing products are to be promoted. He et al. [25] propose to study the influence blocking maximization problem in social networks with the competitive linear threshold model. Chen et al. [13] study the influence maximization in social networks when negative opinions can emerge and propagate. Multiple threshold models for competitive influence in social networks are proposed in [7], whose submodularity and monotonicity are studied in details. Meanwhile, Narayanam et al. [38] study the viral marketing for product cross-sell through social networks to maximize the revenue, where products can have promotion cost, benefits, and promotion budgets. Lu et al. [37] study the influence propagation and maximization problem in the setting from competition to complementarity.

Among these works on information diffusion and viral marketing problems, rumor propagation in online social networks is of practical importance. Kwon et al. identify characteristics of rumors by examining temporal, structural, and linguistic aspects of rumors [33]. Rumors can spread very fast in online social networks, and Doerr et al. propose to study the structural and algorithmic properties of networks which accelerate such a propagation in [18]. To maximize the influence or rumors, the diffusion of competing rumors in social networks is studied in [32].

Influence source identification in unsigned networks has been studied in the existing works. Lappas et al. [35] propose the problem of finding effectors in social networks. In [35], the k-effectors problem is formally defined and the time complexity of the problem for different types of graphs is analyzed in detail. Shah et al. study similar problems in [44] to infer the sources of a rumor in a network, where a SIR-based rumor diffusion model is introduced. They propose to detect the rumor sources by identifying users with high "*rumor centrality*," which is also used in their computer virus sources discovery [43]. Prakash et al. propose to study the culprits in epidemics in [41]. The underlying structure of cascades in social networks is studied in [53].

10.8   Exercises

1. (Easy) Please try to complete the proof of the Theorem 10.8 we introduced in Sect. 10.4.1.2.
2. (Easy) Please try to complete the proof of the Theorem 10.9 we introduced in Sect. 10.4.1.2.
3. (Medium) Please try to prove Theorem 10.1 we introduced in Sect. 10.3.1.1.
4. (Medium) Please try to implement the *greedy* algorithm we introduced in Sect. 10.2.2.1 based on the LT model with a preferred programming language, and test its performance with some simulations on a toy network data set.
5. (Medium) Please try to implement the *greedy* algorithm we introduced in Sect. 10.2.2.1 based on the IC model with a preferred programming language, and test its performance with some simulations on a toy network data set.
6. (Medium) Please refer to [29], and try to prove that for the LT and IC diffusion models, their influence function is *monotone*.
7. (Medium) Please refer to [29], and try to prove that for the LT and IC diffusion models, their influence function is *submodular*.
8. (Hard) Based on the LT model, if adding each user into the seed user set will bring about a certain cost, please try to consider to extend and improve the *step-wise greedy influence maximization* algorithm for such a scenario.
9. (Hard) Please try to implement the *CELF* algorithm with a preferred programming language, and compare its efficiency with the *greedy* algorithm.
10. (Hard) Please try to implement the dynamic programing algorithm introduced in Sect. 10.5.3.

References

1. L. Adamic, R. Lukose, A. Puniyani, B. Huberman, Search in power-law networks. CoRR, cs.NI/0103016 (2001)
2. G. Allport, L. Postman, *The Psychology of Rumor* (Henry Holt, New York, 1947)
3. V. Belak, S. Lam, C. Hayes, Towards maximising cross-community information diffusion, in *2012 IEEE/ACM International Conference on Advances in Social Networks Analysis and Mining* (2012)
4. S. Bharathi, D. Kempe, M. Salek, Competitive influence maximization in social networks, in *International Workshop on Web and Internet Economics* (2007)
5. S. Borgatti, M. Everett, A graph-theoretic perspective on centrality. Soc. Netw. **28**(4), 466–484 (2006)
6. C. Borgs, J. Chayes, N. Immorlica, A. Kalai, V. Mirrokni, C. Papadimitriou, The myth of the folk theorem. Games Econ. Behav. **70**, 34–43 (2010)
7. A. Borodin, Y. Filmus, J. Oren, Threshold models for competitive influence in social networks, in *International Workshop on Internet and Network Economics* (2010)
8. S. Brin, L. Page, The anatomy of a large-scale hypertextual web search engine, in *Proceedings of the Seventh International Conference on World Wide Web 7 (WWW7)* (1998)
9. X. Chen, X. Deng, S. Teng, Computing Nash equilibria: Approximation and smoothed complexity, in *2006 47th Annual IEEE Symposium on Foundations of Computer Science (FOCS'06)* (2006)
10. X. Chen, S. Teng, P. Valiant, The approximation complexity of win-lose games, in *Proceedings of the Eighteenth Annual ACM-SIAM Symposium on Discrete Algorithms (SODA '07)* (2007)
11. W. Chen, Y. Wang, S. Yang, Efficient influence maximization in social networks, in *Proceedings of the 15th ACM SIGKDD International Conference on Knowledge Discovery and Data Mining (KDD '09)* (2009)
12. W. Chen, C. Wang, Y. Wang, Scalable influence maximization for prevalent viral marketing in large-scale social networks, in *Proceedings of the 16th ACM SIGKDD International Conference on Knowledge Discovery and Data Mining (KDD '10)* (2010)

13. W. Chen, A. Collins, R. Cummings et al., Influence maximization in social networks when negative opinions may emerge and propagate – Microsoft research, in *Conference: Proceedings of the Eleventh SIAM International Conference on Data Mining, SDM 2011* (2011)

14. Y. Chu, T. Liu, On the shortest arborescence of a directed graph. Sci. Sin. **14**, 1396–1400 (1965)

15. T. Cormen, C. Stein, R. Rivest, C. Leiserson, *Introduction to Algorithms*, 2nd edn. (McGraw-Hill Higher Education, New York, 2001)

16. C. Daskalakis, P. Goldberg, C. Papadimitriou, The complexity of computing a Nash equilibrium, in *Proceedings of the Thirty-Eighth Annual ACM Symposium on Theory of Computing (STOC '06)* (2006)

17. S. Datta, A. Majumder, N. Shrivastava, Viral marketing for multiple products, in *2010 IEEE International Conference on Data Mining* (2010)

18. B. Doerr, M. Fouz, T. Friedrich, Why rumors spread so quickly in social networks. Commun. ACM **55**, 70–75 (2012)

19. P. Domingos, M. Richardson, Mining the network value of customers, in *Proceedings of the Seventh ACM SIGKDD International Conference on Knowledge Discovery and Data Mining (KDD '01)* (2001)

20. J. Edmonds, Optimum branchings. J. Res. Natl. Bur. Stand. **71**, 233–240 (1967)

21. U. Feige. A threshold of ln n for approximating set cover. J. ACM **45**, 634–652 (1998)

22. A. Goyal, W. Lu, L. Lakshmanan, Celf++: optimizing the greedy algorithm for influence maximization in social networks, in *Proceedings of the 20th International Conference Companion on World Wide Web (WWW '11)* (2011)

23. A. Goyal, W. Lu, L. Lakshmanan, Simpath: an efficient algorithm for influence maximization under the linear threshold model, in *2011 IEEE 11th International Conference on Data Mining* (2011)

24. H. Gui, Y. Sun, J. Han, G. Brova, Modeling topic diffusion in multi-relational bibliographic information networks, in *Proceedings of the 23rd ACM International Conference on Information and Knowledge Management (CIKM '14)* (2014)

25. X. He, G. Song, W. Chen, Q. Jiang, Influence blocking maximization in social networks under the competitive linear threshold model, in *Conference: Proceedings of SIAM International Conference on Data Mining, SDM 2012* (2012)

26. P. Jaccard, Étude comparative de la distribution florale dans une portion des alpes et des jura. Bull. Soc. Vaud. Sci. Nat. **37**, 547–579 (1901)

27. Q. Jiang, G. Song, G. Cong, Y. Wang, W. Si, K. Xie, Simulated annealing based influence maximization in social networks, in *Twenty-Fifth AAAI Conference on Artificial Intelligence* (2011)

28. F. Jin, E. Dougherty, P. Saraf, Y. Cao, N. Ramakrishnan, Epidemiological modeling of news and rumors on twitter, in *Proceedings of the 7th Workshop on Social Network Mining and Analysis (SNAKDD '13)* (2013)

29. D. Kempe, J. Kleinberg, É. Tardos, Maximizing the spread of influence through a social network, in *Proceedings of the ACM SIGKDD International Conference on Knowledge Discovery and Data Mining* (2003)

30. M. Kimura, K. Saito, Approximate solutions for the influence maximization problem in a social network, in *Knowledge-Based Intelligent Information and Engineering Systems*, ed. by B. Gabrys, R. Howlett, L. Jain (Springer, Berlin, 2006)

31. X. Kong, J. Zhang, P. Yu, Inferring anchor links across multiple heterogeneous social networks, in *Proceedings of the 22nd ACM International Conference on Information & Knowledge Management (CIKM '13)* (2013)

32. J. Kostka, Y. Oswald, R. Wattenhofer, Word of mouth: rumor dissemination in social networks, in *International Colloquium on Structural Information and Communication Complexity* (2008)

33. S. Kwon, M. Cha, K. Jung, W. Chen, Y. Wang, Prominent features of rumor propagation in online social media, in *IEEE 13th International Conference on Data Mining* (2013)

34. K. Lai, The knapsack problem and fully polynomial time approximation schemes (fptas). Technical report (2006)

35. T. Lappas, E. Terzi, D. Gunopulos, H. Mannila, Finding effectors in social networks, in *Proceedings of the 16th ACM SIGKDD International Conference on Knowledge Discovery and Data Mining (KDD '10)* (2010)

36. J. Leskovec, L. Adamic, B. Huberman, The dynamics of viral marketing. ACM Trans. Web **1**(1), 5 (2007)

37. W. Lu, W. Chen, L. Lakshmanan, From competition to complementarity: comparative influence diffusion and maximization, in *Proceedings of VLDB Endowment* (2015)

38. R. Narayanam, A. Nanavati, Viral marketing for product cross-sell through social networks, in *Joint European Conference on Machine Learning and Knowledge Discovery in Databases (ECML PKDD 2012)* (2012)

39. D. Nguyen, H. Zhang, S. Das, M. Thai, T. Dinh, Least cost influence in multiplex social networks: model representation and analysis, in *2013 IEEE 13th International Conference on Data Mining* (2013)

40. N. Nisan, T. Roughgarden, E. Tardos, V. Vazirani, *Algorithmic Game Theory* (Cambridge University Press, New York, 2007)

41. B. Prakash, J. Vreeken, C. Faloutsos, Spotting culprits in epidemics: how many and which ones? in *2012 IEEE 12th International Conference on Data Mining* (2012)

42. M. Richardson, P. Domingos, Mining knowledge-sharing sites for viral marketing, in *Proceedings of the Eighth ACM SIGKDD International Conference on Knowledge Discovery and Data Mining* (2002)

43. D. Shah, T. Zaman, Detecting sources of computer viruses in networks: theory and experiment, in *Proceedings of the ACM SIGMETRICS International Conference on Measurement and Modeling of Computer Systems, SIGMETRICS '10*, New York (2010), pp. 203–214. http://doi.acm.org/10.1145/1811039.1811063

44. D. Shah, T. Zaman, Rumors in a network: who's the culprit? IEEE Trans. Inf. Theory **57**(8), 5163–5181 (2011)

45. Y. Sun, J. Han, X. Yan, P. Yu, T. Wu, Pathsim: meta path-based top-k similarity search in heterogeneous information networks, in *Proceedings of VLDB Endowment* (2011)

46. F. Yang, Y. Liu, X. Yu, M. Yang, Automatic detection of rumor on Sina Weibo, in *Proceedings of the ACM SIGKDD Workshop on Mining Data Semantics (MDS '12)* (2012)

47. Q. Zhan, J. Zhang, S. Wang, P. Yu, J. Xie, Influence maximization across partially aligned heterogeneous social networks, in *Pacific-Asia Conference on Knowledge Discovery and Data Mining* (2015)

48. Q. Zhan, J. Zhang, P. Yu, J. Xie, Viral marketing through aligned networks. Technical report (2018)

49. J. Zhang, P. Yu, Z. Zhou, Meta-path based multi-network collective link prediction, in *Proceedings of the 20th ACM SIGKDD International Conference on Knowledge Discovery and Data Mining (KDD '14)* (2014)

50. J. Zhang, S. Wang, Q. Zhan, P. Yu, Intertwined viral marketing in social networks, in *2016 IEEE/ACM International Conference on Advances in Social Networks Analysis and Mining (ASONAM)* (2016)

51. J. Zhang, C. Aggarwal, P. Yu, Rumor initiator detection in infected signed networks, in *2017 IEEE 37th International Conference on Distributed Computing Systems (ICDCS)* (2017)

52. J. Zhang, L. Cui, Y. Fu, F. Gouza, Fake news detection with deep diffusive network model. CoRR, abs/1805.08751 (2018)

53. B. Zong, Y. Wu, A. Singh, X. Yan, Inferring the underlying structure of information cascades, in *12th IEEE International Conference on Data Mining (ICDM 2012)* (2012)

Network Embedding

11

11.1 Overview

In the era of big data, information from diverse disciplines is generated at an extremely fast pace, lots of which are highly structured and can be represented as massive and complex networks. The representative examples include online social networks, like Facebook and Twitter, academic retrieval sites, like DBLP and Google Scholar, as well as bio-medical data, e.g., human brain networks. These networks/graphs are usually very challenging to handle due to their extremely large-scale (involving millions even billions of nodes), complex structures (containing heterogeneous links) as well as the diverse attributes (attached to the nodes or links). For instance, the Facebook social network involves more than 1 billion active users; DBLP contains about 2.8 billions of papers; and human brain has more than 16 billion neurons.

Great challenges exist when handling these network-representation data with traditional machine learning algorithms, which usually take feature vector representation data as the input. A general representation of heterogeneous networks as feature vectors is desired for knowledge discovery from such complex network structured data. In recent years, many research works propose to embed the online social network data into a lower-dimensional feature space [4, 17, 22], in which the user node is represented as a unique feature vector, and the network structure can be reconstructed from these feature vectors. With the embedded feature vectors, classic machine learning models can be applied to deal with the social network data directly, and the storage space can be saved greatly.

In this chapter, we will talk about the *network embedding* problem [17], aiming at projecting the nodes and links in the network data in low-dimensional feature spaces. Depending on the application settings, existing graph embedding works can be categorized into the embedding of *multi-relational networks* [1, 8, 16], *homogeneous networks* [6, 11, 14], *heterogeneous networks* [2, 3, 7], and *multiple aligned heterogeneous networks* [21]. Meanwhile, depending on the models being applied, the current embedding works can be divided into the *translation based embedding* [1, 8, 16], *random walk based embedding* [6, 11], *proximity based embedding* [14], and *deep learning based embedding* [21].

In the following parts in this chapter, we will first introduce the *translation based graph embedding* models in Sect. 11.2, which are mainly proposed for the multi-relational knowledge graphs, including *TransE* [1], *TransH* [16], and *TransR* [8]. After that, in Sect. 11.3, we will introduce three homogeneous network embedding models, including *DeepWalk* [11], *LINE* [14], and *node2vec* [6]. Three embedding models, HNE [2], PANE [3], and HEBE [7], for the heterogeneous networks

© Springer Nature Switzerland AG 2019
J. Zhang, P. S. Yu, *Broad Learning Through Fusions*,
https://doi.org/10.1007/978-3-030-12528-8_11

will be introduced in Sect. 11.4, which project the nodes to feature vectors based on the heterogeneous information inside the networks. Finally, we will talk about the model proposed for embedding the multiple aligned heterogeneous network in Sect. 11.5, where the anchor links are utilized to transfer information across different sites for mutual refinement of the embedding results synergistically.

11.2 Relation Translation Based Graph Entity Embedding

Multi-relational data refers to the graph structured data whose nodes correspond to entities and links denote the relationships. The multi-relational data can be represented as a graph $G = (\mathcal{V}, \mathcal{E})$, where $\mathcal{V}$ denotes the node set and $\mathcal{E}$ represents the link set. For the link in the graph, e.g., $r = (h, t) \in \mathcal{E}$, we can represent the corresponding entity-relation as a triple (h, r, t), where h denotes the link initiator entity, t denotes the link recipient entity, and r represents the link. The embedding problem studied in this section is to learn a feature representation of both entities and relations in the triples, i.e., h, r, and t.

Model *TransE* [1] is the initial translation based embedding work, which projects the entity and relation into a common feature space. *TransH* [16] improves *TransE* by considering the link cardinality constraint in the embedding process, and can achieve a comparable time complexity. In the real-world multi-relational networks, the entities can have multiple aspects, and the different relations can express different aspects of the entity. Model *TransR* [8] proposes to build the entity and relation embeddings in separate entity and relation spaces instead. Next, we will introduce the embedding models *TransE*, *TransH*, and *TransR* one by one as follows, where the relation is more like a translation of entities in the embedding space. It is the reason why these models are called the *translation based embedding models*.

11.2.1 TransE

The *TransE* [1] model is an energy-based model for learning low-dimensional embeddings of entities and relations, where the relations are represented as the *translations* of entities in the embedding space. Given an entity-relation triple (h, r, t), as shown in Fig. 11.1, we can represent the embedding feature representations of the entities and relations as vectors $\mathbf{h} \in \mathbb{R}^k, \mathbf{r} \in \mathbb{R}^k$, and $\mathbf{t} \in \mathbb{R}^k$, respectively ($k$ denotes the objective vector dimension). If the triple (h, r, t) holds, i.e., there exists a link r starting from h to t in the network, the corresponding embedding vector $\mathbf{h} + \mathbf{r}$ should be as close to vector $\mathbf{t}$ as possible.

Fig. 11.1 An example of TransE

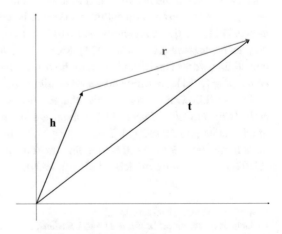

Let $\mathcal{S}^+ = \{(h, r, t)\}_{r=(h,t)\in\mathcal{E}}$ represent the set of positive training data, which contains the triples existing in the networks. The *TransE* model aims at learning the embedding feature vectors of the entities h, t and the relation r, i.e., $\mathbf{h}$, $\mathbf{r}$, and $\mathbf{t}$. For the triples in the positive training set, we want to ensure that the learnt embedding vectors $\mathbf{h} + \mathbf{r}$ are very close to $\mathbf{t}$. Let $d(\mathbf{h} + \mathbf{r}, \mathbf{t})$ denote the distance between vectors $\mathbf{h} + \mathbf{r}$ and $\mathbf{t}$. The loss introduced for the triples in the positive training set can be represented as

$$\mathcal{L}(\mathcal{S}^+) = \sum_{(h,r,t)\in\mathcal{S}^+} d(\mathbf{h} + \mathbf{r}, \mathbf{t}). \tag{11.1}$$

Here, the distance function can be defined in different ways, like the L_2 norm of the difference between vectors $\mathbf{h} + \mathbf{r}$ and $\mathbf{t}$, i.e.,

$$d(\mathbf{h} + \mathbf{r}, \mathbf{t}) = \|\mathbf{h} + \mathbf{r} - \mathbf{t}\|_2. \tag{11.2}$$

By minimizing the above loss function, the optimal feature representations of the entities and relations can be learnt. To avoid trivial solutions, like the zero vector $\mathbf{0}$ for $\mathbf{h}$, $\mathbf{r}$, and $\mathbf{t}$, additional constraints, e.g., the L_2-norm of the embedding vectors of the entities should be 1, can be added in the function. Furthermore, a negative training set is also sampled to differentiate the learnt embedding vectors. For a triple $(h, r, t) \in \mathcal{S}^+$, we can denote the corresponding sampled negative training set as $\mathcal{S}^-_{(h,r,t)}$, which contains the triples formed by replacing the initiator entity h or the recipient entity t with the random entities. In other words, we can represent the negative training set $\mathcal{S}^-_{(h,r,t)}$ as

$$\mathcal{S}^-_{(h,r,t)} = \{(h', r, t)|h' \in \mathcal{V}\} \cup \{(h, r, t')|t' \in \mathcal{V}\}. \tag{11.3}$$

The loss function involving both the positive and negative training set can be represented as

$$\mathcal{L}(\mathcal{S}^+, \mathcal{S}^-) = \sum_{(h,r,t)\in\mathcal{S}^+} \sum_{(h',r,t')\in\mathcal{S}^-_{(h,r,t)}} \max\left(\gamma + d(\mathbf{h} + \mathbf{r}, \mathbf{t}) - d(\mathbf{h}' + \mathbf{r}, \mathbf{t}'), 0\right), \tag{11.4}$$

where γ is a margin hyperparameter and $\max(\cdot, 0)$ will count the positive loss values only.

The optimization is carried out by stochastic gradient descent (in minibatch mode). The embedding vectors of entities and relationships are initialized with a random procedure. At each iteration of the algorithm, the embedding vectors of the entities are normalized and a small set of triplets is sampled from the training set, which will serve as the training triplets of the minibatch. The parameters are then updated by taking a gradient step with a constant learning rate.

11.2.2 TransH

TransE is a promising method proposed recently, which is very efficient while achieving state-of-the-art predictive performance. However, in the embedding process, *TransE* fails to consider the *cardinality constraint* on the relations, like *one-to-one*, *one-to-many*, and *many-to-many*. The *TransH* [16] model to be introduced in this part considers such properties on relations in the embedding process. Furthermore, different from the other complex models, which can handle these properties but sacrifice efficiency, *TransH* achieves comparable time complexity as *TransE*. *TransH* models the relation as a hyperplane together with a translation operation on it, where the correlation among the entities can be effectively preserved.

Fig. 11.2 An example of
TransH

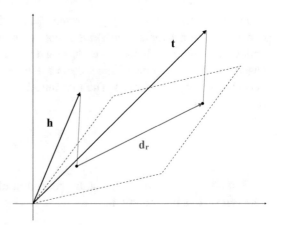

In *TransH*, different from the embedding space of entities, the relations, e.g., r, is denoted as
a transition vector $\mathbf{d}_r$ in the hyperplane $\mathbf{w}_r$ (a normal vector). For each of the triple (h, r, t), as
illustrated in Fig. 11.2, the embedding vectors $\mathbf{h}$, $\mathbf{t}$ are first projected to the hyperplane $\mathbf{w}_r$, whose
corresponding projected vectors can be represented as $\mathbf{h}_\perp$ and $\mathbf{t}_\perp$, respectively. The vectors $\mathbf{h}_\perp$ and
$\mathbf{t}_\perp$ can be connected by the translation vector $\mathbf{d}_r$ on the hyperplane. Depending on whether the triple
appears in the positive or negative training set, the distance $d(\mathbf{h}_\perp + \mathbf{d}_r, \mathbf{t}_\perp)$ should be either minimized
or maximized.

Formally, given the hyperplane $\mathbf{w}_r$, we can represent the projection vectors $\mathbf{h}_\perp$ and $\mathbf{t}_\perp$ as

$$\mathbf{h}_\perp = \mathbf{h} - \mathbf{w}_r^\top \mathbf{h} \mathbf{w}_r, \tag{11.5}$$

$$\mathbf{t}_\perp = \mathbf{t} - \mathbf{w}_r^\top \mathbf{t} \mathbf{w}_r. \tag{11.6}$$

Furthermore, the L_2 norm based distance function can be represented as

$$d(\mathbf{h}_\perp + \mathbf{d}_r, \mathbf{t}_\perp) = \|\mathbf{h}_\perp + \mathbf{d}_r - \mathbf{t}_\perp\|_2^2$$
$$= \|(\mathbf{h} - \mathbf{w}_r \mathbf{h} \mathbf{w}_r) + \mathbf{d}_r - (\mathbf{t} - \mathbf{w}_r \mathbf{t} \mathbf{w}_r)\|_2^2. \tag{11.7}$$

The variables to be learnt in the *TransH* model include the embedding vectors of all the entities, the
hyperplane, and translation vectors for each of the relations. To learn these variables simultaneously,
the objective function of *TransH* can be represented as

$$\mathcal{L}(\mathcal{S}^+, \mathcal{S}^-)$$
$$= \sum_{(h,r,t)\in\mathcal{S}^+} \sum_{(h',r',t')\in\mathcal{S}^-_{(h,r,t)}} \max\left(\gamma + d(\mathbf{h}_\perp + \mathbf{d}_r, \mathbf{t}_\perp) - d(\mathbf{h}'_\perp + \mathbf{d}'_r, \mathbf{t}'_\perp), 0\right), \tag{11.8}$$

where $\mathcal{S}^-_{(h,r,t)}$ denotes the negative set constructed for triple (h, r, t). Different from *TransE*, *TransH*
applies a different approach to sample the negative training triples with considerations of the relation
cardinality constraint. For the relations with *one-to-many*, *TransH* will give more chance to replace
the initiator node; and for the *many-to-one* relations, *TransH* will give more chance to replace the
recipient node instead.

Besides the loss function, the variables to be learnt are subject to some constraints, like the embedding vectors for entities are normal vectors, $\mathbf{w}_r$ and $\mathbf{d}_r$ should be orthogonal, and $\mathbf{w}_r$ is also a normal vector. We summarize the constraints of the *TransH* model as follows:

$$\|\mathbf{h}\|_2 \leq 1, \|\mathbf{t}\|_2 \leq 1, \forall h, t \in \mathcal{V}, \tag{11.9}$$

$$\frac{|\mathbf{w}_r^\top \mathbf{d}_r|}{\|\mathbf{d}_r\|_2} \leq \epsilon, \forall r \in \mathcal{E}, \tag{11.10}$$

$$\|\mathbf{w}_r\|_2 = 1, \forall r \in \mathcal{E}, \tag{11.11}$$

where the second constraint guarantees that the translation vector $\mathbf{d}_r$ is in the hyperplane. The constraints can be relaxed as some penalty terms, which can be added to the objective function with a relatively large penalty weight. The final objective function can be learnt with the stochastic gradient descent, and by minimizing the loss function, we can learn the variables and get the final embedding results.

11.2.3 TransR

Both *TransE* and *TransH* introduced in the previous subsections assume embeddings of entities and relations to be within the same space $\mathbb{R}^k$. However, entities and relations are actually totally different objects, and they may be not capable to be represented in a common semantic space. To address such a problem, *TransR* [8] is proposed, which models the entities and relations in distinct spaces, i.e., the entity space and relation space, and performs the translation between the relation spaces.

As shown in Fig. 11.3, in *TransR*, given a triple (h, r, t), the entities h and t are embedded as vectors $\mathbf{h}, \mathbf{t} \in \mathbb{R}^{k_e}$, and the relation r is embedded as vector $\mathbf{r} \in \mathbb{R}^{k_r}$, where the dimension of the entity space and relation space is not the same, i.e., $k_e \neq k_r$. To project the entities from the entity space to the relation space, a projection matrix $\mathbf{M}_r \in \mathbb{R}^{k_e \times k_r}$ is defined in *TransR*. With the projection matrix, we can define the projected entity embedding vectors as

$$\mathbf{h}_r = \mathbf{M}_r \mathbf{h}, \tag{11.12}$$

$$\mathbf{t}_r = \mathbf{M}_r \mathbf{t}. \tag{11.13}$$

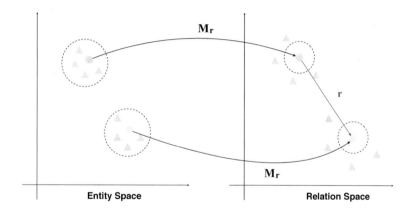

Fig. 11.3 An example of TransR

Entity Space   Relation Space

The loss function is defined as

$$d(\mathbf{h}_r + \mathbf{r}, \mathbf{t}_r) = \|\mathbf{h}_r + \mathbf{r} - \mathbf{t}_r\|_2^2$$
$$= \|\mathbf{M}_r\mathbf{h} + \mathbf{r} - \mathbf{M}_r\mathbf{t}\|_2^2 . \tag{11.14}$$

The constraints involved in *TransR* include

$$\|\mathbf{h}\|_2 = 1, \|\mathbf{t}\|_2 = 1, \forall h, t \in \mathcal{V}, \tag{11.15}$$

$$\|\mathbf{M}_r\mathbf{h}\|_2 = 1, \|\mathbf{M}_r\mathbf{t}\|_2 = 1, \forall h, t \in \mathcal{V}, \tag{11.16}$$

$$\|\mathbf{w}_r\|_2 \leq 1, \forall r \in \mathcal{E}. \tag{11.17}$$

The negative training set $\mathcal{S}^-$ in *TransR* can be obtained in a similar way as *TransH*, where the variables can be learnt with the stochastic gradient descent. We will not introduce the information here to avoid content duplication.

11.3 Homogeneous Network Embedding

Besides these introduced translation based network embedding models, in this section, we will introduce three other recent embedding models proposed for homogeneous network data, including *DeepWalk* [11], *LINE* [14], and *node2vec* [6]. Formally, the networks studied in this part are all homogeneous networks, which is represented as $G = (\mathcal{V}, \mathcal{E})$. $\mathcal{V}$ denotes the set of nodes in the network, and $\mathcal{E}$ represents the set of links among the nodes inside the network.

11.3.1 DeepWalk

The *DeepWalk* [11] algorithm consists of two main components: (1) a random walk generator, and (2) random walk based node representation learning. In the first step, the DeepWalk model randomly selects a node, e.g., $u \in \mathcal{V}$, as the starting node of a random walk W_u in the network. Random walk W_u will sample the neighbors of the node last visited uniformly until the maximum length l is met. In the second step, the sampled neighbors are used to update the representations of the nodes inside the graph, where *Skip-Gram* is applied here.

The pseudo-code of the *DeepWalk* algorithm is available in Algorithm 1, which illustrates the general procedure of the algorithm. In the algorithm, line 1 initializes the representation matrix $\mathbf{X}$ for all the nodes, and line 2 builds a binary tree involving all the nodes in the network as the leaves, which will be introduced in more detail in Sect. 11.3.1.3. Lines 3–9 denote the main part of the *DeepWalk* algorithm, where the random walk starting randomly at each node is generated for γ times by calling the function *WalkGenerator*. For each node u, a random walk W_u is generated whose length is bounded by parameter l. The random walk will be applied to update the node representation with the *Skip-Gram* function to be introduced in Sect. 11.3.1.2.

11.3.1.1 Random Walk Generator
The random walk model has been introduced in Sect. 3.3.3.3. Formally, we can represent the random walk starting at node $u \in \mathcal{V}$ as W_u, which actually denotes a stochastic process with random status

---

Algorithm 1 DeepWalk

Require: Input homogeneous network $G = (\mathcal{V}, \mathcal{E})$
      Window size s; Embedding size d
      Walk length l; Walks per node γ
Ensure: Matrix of node representations $\mathbf{X} \in \mathbb{R}^{|\mathcal{V}| \times d}$
 1: Initialize $\mathbf{X}$ with random values following the uniform distribution
 2: Build a binary tree T from node set $\mathcal{V}$
 3: **for** Round $i = 1$ to γ **do**
 4:    $\mathcal{O} = \text{shuffle}(\mathcal{V})$
 5:    **for** Node $u \in \mathcal{O}$ **do**
 6:       $W_u = \text{WalkGenerator}(G, u, l)$
 7:       $\text{SkipGram}(\mathbf{X}, W_u, w)$
 8:    **end for**
 9: **end for**
10: Return $\mathbf{X}$

$W_u^0, W_u^1, \ldots, W_u^k$. Formally, at the very beginning, i.e., in step 0, the random walk is at the initial node, i.e., $W_u^0 = u$. The status variable W_u^k denotes the node where the walker is at the step k.

Random walk can capture the local network structures effectively, where the neighborhood and social connection closeness can affect the next nodes that the random walk will move to in the following steps. Therefore, in the *DeepWalk* algorithm, random walk is applied to sample a stream of short random walks as the tool for extracting information from a network. Random walk can provide two very desirable properties, besides the ability to capture the local network structures. Firstly, the random walk based local network exploration is easy to parallelize. Several random walks can simultaneously explore different parts of the same network in different threads, processes, and machines. Secondly, with the information obtained from short random walks, it is possible to accommodate small changes in the network structure without the need for global recomputation.

11.3.1.2 Skip-Gram Technique

The node representation learning step involved in the DeepWalk algorithm is very similar to the word appearance prediction in language modeling. In this part, we will first provide some basic knowledge about the language modeling problem first, and then introduce the *Skip-Gram* technique.

Formally, the objective of language modeling is to estimate the likelihood of a specific sequence of words appearing in a corpus. More specifically, given a sequence of words $(w_1, w_2, \ldots, w_{n-1})$ where word $w_i \in \mathcal{V}$ ($\mathcal{V}$ denotes the vocabulary set), the word appearance prediction problem aims at inferring the word w_n that will appear next. An intuitive idea to model the problem is to maximize the estimation likelihood for the next word w_n given $w_1, w_2, \ldots, w_{n-1}$, and the problem can be formally represented as

$$w_n^* = \arg_{w_n \in \mathcal{V}} P(w_n | w_1, w_2, \ldots, w_{n-1}), \tag{11.18}$$

where term $P(w_n | w_1, w_2, \ldots, w_{n-1})$ denotes the conditional probability of having w_n attached to the observed word sequence $w_1, w_2, \ldots, w_{n-1}$.

Meanwhile, in neural networks, the words will have a latent representation denoted as a vector, like $\mathbf{x}_{w_i} \in \mathbb{R}^d$ for word $w_i \in \mathcal{V}$. Furthermore, computation of the above conditional probability is very challenging, especially as the observed word sequence gets longer, i.e., n is large. Therefore, a window is proposed to limit the length of word sequence in probability computation. We can denote s as the size of the window. Therefore, we can rewrite the above objective function as

$$w_n^* = \arg_{w_n \in \mathcal{V}} P(w_n | \mathbf{x}_{w_{n-s}}, \mathbf{x}_{w_{n-s+1}}, \ldots, \mathbf{x}_{w_{n-1}}). \tag{11.19}$$

A recent relaxation to the above problem in language modeling turns the prediction problem on its head. Three big changes are applied to the model: (1) instead of predicting the objective word with the context, the relaxation predicts the context based on the objective word; (2) the context denotes the words appearing before and after the objective word limited by the window size s; and (3) the order of words is removed and the context denotes a set of words instead. Formally, we can rewrite the objective function as

$$w_n^* = \arg_{w_n \in \mathcal{V}} P(\{w_{n-s}, w_{n-s+1}, \dots, w_{n+s}\} \setminus \{w_n\} | \mathbf{x}_{w_n}). \tag{11.20}$$

Skip-Gram is a language model that maximizes the co-occurrence probability of words appearing in the time window s in a sentence. Here, when applying the *Skip-Gram* technique in the *DeepWalk* model, we can treat the nodes $u \in \mathcal{V}$ in the network as the words w denoted in the aforementioned equations. Meanwhile, for the nodes sampled by the random walk model within the window size s before and after node v, they will be treated as the context nodes appearing ahead of and after node v. Furthermore, Skip-Gram assumes the appearance of the words (or nodes for networks) to be independent, and we can rewrite the above probability equations as follows:

$$P(\{u_{n-s}, u_{n-s+1}, \dots, u_{n+s}\} \setminus \{u_n\} | \mathbf{x}_{u_n}) = \prod_{i=n-s, i \neq n}^{n+s} P(u_i | \mathbf{x}_{u_n}), \tag{11.21}$$

where $u_{n-s}, u_{n-s+1}, \dots, u_{n+s}$ denotes the sequence of nodes sampled by the random walk model.

The learning process of the Skip-Gram algorithm is provided in Algorithm 2, where we enumerate all the co-locations of nodes in the sampled node sequence $u_{n-s}, u_{n-s+1}, \dots, u_{n+s}$ by a random walk W_u (starting from node u in the network). With the gradient descent based algorithm, we can update the representations of nodes according to their neighbor representations with the stochastic gradient descent learning algorithm. The derivatives are estimated with the back-propagation algorithm. However, in the equation, we need to have the conditional probabilities of the nodes and their representations. A concrete representation of the probability can be a great challenging problem. As proposed in [11], such a distribution can be learnt with some existing models, like logistic regression. However, since the labels used here involve all the nodes in the network, it will lead to a very large label space with $|\mathcal{V}|$ different labels, which renders the learning process extremely time-consuming and ineffective. To solve such a problem, some techniques, like hierarchical softmax [10], have been proposed, which represents the nodes in the network as a binary tree and can lower down the probability computation time complexity effectively from $O(|\mathcal{V}|)$ to $O(\log |\mathcal{V}|)$.

Algorithm 2 Skip-Gram

Require: Representations of nodes: $\mathbf{X}$
               Random walk starting from node u: W_u
               Window size s
Ensure: Updated matrix of node representations $\mathbf{X}$
1: **for** Each node $u_i \in W_u$ **do**
2:      W_u will generate a sampled sequence before and after u_j bounded by window size s: $(u_{i-s}, \dots, u_{i+s})$
3:      **for** Each node $u_j \in (u_{i-s}, \dots, u_{i+s})$ **do**
4:          $J(\mathbf{X}) = -\log P(u_j | \mathbf{x}_{u_i})$
5:          $\mathbf{X} = \mathbf{X} - \alpha \frac{J(\mathbf{X})}{\partial \mathbf{X}}$
6:      **end for**
7: **end for**
8: Return $\mathbf{X}$

11.3.1.3 Hierarchical Softmax

In the Skip-Gram algorithm, calculating probability $P(u_i|\mathbf{x}_{u_n})$ is infeasible. Therefore, in the DeepWalk model, *hierarchical softmax* is used to factorize the conditional probability. In *hierarchical softmax*, a binary tree is constructed, where the number of leaves equals to the network node set size, and each network node is assigned to a leaf node. The prediction problem is turned into a path probability maximization problem. If a path $(b_0, b_1, \ldots, b_{\lceil \log |\mathcal{V}| \rceil})$ is identified from the tree root to the node u_k, i.e., $b_0 = \text{root}$ and $b_{\lceil \log |\mathcal{V}| \rceil} = u_k$, then the probability can be rewritten as

$$P(u_i|\mathbf{x}_{u_n}) = \prod_{l=1}^{\lceil \log |\mathcal{V}| \rceil} P(b_l|\mathbf{x}_{u_n}), \tag{11.22}$$

where $P(b_l|\mathbf{x}_{u_n})$ can be modeled by a binary classifier denoted as

$$P(b_l|\mathbf{x}_{u_n}) = \frac{1}{1 + e^{-\mathbf{x}_{b_l}^\top \cdot \mathbf{x}_{u_n}}}. \tag{11.23}$$

Here, the parameters involved in the learning process include the representations for both the nodes in the network and the nodes in the constructed binary trees.

11.3.2 LINE

To handle the real-world information networks, the embedding models need to have several requirements: (1) preserve the *first-order* and *second-order* proximity between the nodes, (2) scalable to large sized networks, and (3) should be able to handle networks with different links: *directed* and *undirected*, *weighted* and *unweighted*. In this part, we will introduce another homogeneous network embedding model, named *LINE* [14], which can meet those requirements.

11.3.2.1 First-Order Proximity

In the network embedding process, the network structure should be effectively preserved, where the node closeness is defined as the node *proximity* concept in *LINE*. The *first-order proximity* in a network denotes the *local* pairwise proximity between nodes. For a link $(u, v) \in \mathcal{E}$ in the network, the *first-order proximity* denotes the weight of link (u, v) in the network (or 1 if the network is unweighted). Meanwhile, if link (u, v) doesn't exist in the network, the *first-order proximity* between them will be 0 instead. To model the *first-order proximity*, for a given link $(u, v) \in \mathcal{E}$ in the network G, *LINE* defines the joint probability between nodes u and v as

$$p_1(u, v) = \frac{1}{1 + e^{-\mathbf{x}_u^\top \cdot \mathbf{x}_v}}, \tag{11.24}$$

where $\mathbf{x}_u, \mathbf{x}_v \in \mathbb{R}^d$ denote the vector representations of nodes u and v, respectively.

Function $p_1(\cdot, \cdot)$ defines the proximity distribution in the space of $\mathcal{V} \times \mathcal{V}$. Meanwhile, given a network G, the *empirical proximity* between nodes u and v can be denoted as

$$\hat{p}_1(u, v) = \frac{w_{(u,v)}}{\sum_{(u,v) \in \mathcal{E}} w_{(u,v)}}. \tag{11.25}$$

To preserve the *first-order proximity*, *LINE* defines the objective function for the network embedding as

$$J_1 = d(p_1(\cdot, \cdot), \hat{p}_1(\cdot, \cdot)), \qquad (11.26)$$

where function $d(\cdot, \cdot)$ denotes the distance between the introduced proximity distribution and the empirical proximity distribution, respectively. By replacing the distance function $d(\cdot, \cdot)$ with the KL-divergence and omitting some constants, the objective function can be rewritten as

$$J_1 = - \sum_{(u,v)\in\mathcal{E}} w_{(u,v)} \log p_1(u, v). \qquad (11.27)$$

By minimizing the objective function, *LINE* can learn the feature representation $\mathbf{x}_u$ for each node $u \in \mathcal{V}$ in the network.

11.3.2.2 Second-Order Proximity

In the real-world social networks, the links among the nodes can be very sparse, where the *first-order proximity* can hardly preserve the complete structure information of the network. LINE introduces the concept of *second-order proximity*, which denotes the similarity between the neighborhood structure of nodes. Given a user pair (u, v) in the network, the more the common neighbors shared by them, the closer the users u and v will be in the network. Besides the original representation $\mathbf{x}_u$ for node $u \in \mathcal{V}$, the nodes are also associated with a feature vector representing its context in the network (i.e., the node neighborhood), which is denoted as $\mathbf{y}_u \in \mathbb{R}^d$.

Formally, for a given link $(u, v) \in \mathcal{E}$, we can represent the probability of context $\mathbf{y}_v$ generated by node u as

$$p_2(v|u) = \frac{e^{\mathbf{x}_u^\top \cdot \mathbf{y}_v}}{\sum_{v'\in\mathcal{V}} e^{\mathbf{x}_u^\top \cdot \mathbf{y}_{v'}}}. \qquad (11.28)$$

Slightly different from the *first-order proximity*, the *second-order* empirical proximity is denoted as

$$\hat{p}_2(v|u) = \frac{w_{(u,v)}}{D(u)}. \qquad (11.29)$$

By minimizing the difference between the introduced proximity distribution and the empirical proximity distribution, the objective function for the *second-order* proximity can be represented as

$$J_2 = \sum_{u\in\mathcal{V}} \lambda_u d(p_2(\cdot|u), \hat{p}_2(\cdot|u)), \qquad (11.30)$$

where λ_u denotes the prestige of node u in the network. Here, by replacing the distance function $d(\cdot|\cdot)$ with the KL-divergence and setting $\lambda_u = D(u)$, the *second-order proximity* based objective function can be represented as

$$J_2 = - \sum_{(u,v)\in\mathcal{E}} w_{(u,v)} \log p_2(v|u). \qquad (11.31)$$

11.3.2.3 Model Optimization

Instead of combining the *first-order proximity* and *second-order proximity* into a joint optimization function, *LINE* learns the embedding vectors based on Eqs. (11.27) and (11.31), respectively, which will be further concatenated together to obtain the final embedding vectors.

In optimizing objective function in Eq. (11.31), *LINE* needs to calculate the conditional probability $P(\cdot|u)$ for all nodes $u \in \mathcal{V}$ in the network, which is computationally infeasible. To solve the problem, *LINE* uses the negative sampling approach instead. For each link $(u, v) \in \mathcal{E}$, *LINE* samples a set of negative links according to some noisy distributions.

Formally, for link $(u, v) \in \mathcal{E}$, we can represent the set of negative links sampled for it as $\mathcal{L}^-_{(u,v)} \subset \mathcal{V} \times \mathcal{V}$. The objective function defined for link (u, v) can be represented as

$$\log \sigma(\mathbf{y}_v^\top \cdot \mathbf{x}_u) + \sum_{(u,v') \in \mathcal{L}^-_{(u,v)}} \log \sigma(-\mathbf{y}_{v'}^\top \cdot \mathbf{x}_u), \tag{11.32}$$

where $\sigma(\cdot)$ is the sigmoid function. The first term in the above equation denotes the observed links, and the second term represents the negative links drawn from the noisy distribution. Similar approach can also be applied to solve the objective function in Eq. (11.27) as well. The new objective function can be solved with the asynchronous stochastic gradient algorithm (ASGD), which samples a minibatch of links and then updates the parameters.

11.3.3 node2vec

In *LINE*, the closeness among nodes in the networks is preserved based on either the *first-order proximity* or the *second-order proximity*. In a recent work, *node2vec* [6], the researchers propose to preserve the proximity between nodes with a sampled set of nodes in the network.

11.3.3.1 node2vec Framework

Model *node2vec* is based on the *Skip-Gram* [10] in language modeling, and the objective function of *node2vec* can be formally represented as

$$\max \sum_{u \in \mathcal{V}} \log P(\Gamma(u)|\mathbf{x}_u), \tag{11.33}$$

where $\mathbf{x}_u$ denotes the latent feature vector learnt for node u and $\Gamma(u)$ represents the neighbor set of node u in the network.

To simplify the problem and make the problem solvable, some assumptions are made to approximate the objective function into a simpler form.

- *Conditional Independence Assumption*: Given the latent feature vector $\mathbf{x}_u$ of node u, by assuming the observation of node in set $\Gamma(u)$ to be independent, the probability equation can be rewritten as

$$P(\Gamma(u)|\mathbf{x}_u) = \prod_{v \in \Gamma(u)} P(v|\mathbf{x}_u). \tag{11.34}$$

- *Symmetric Node Effect*: Furthermore, by assuming that the source and neighbor nodes have a symmetric effect on each other in the feature space, the conditional probability $P(v|\mathbf{x}_u)$ can be rewritten as

$$P(v|\mathbf{x}_u) = \frac{e^{\mathbf{x}_v^\top \cdot \mathbf{x}_u}}{\sum_{v' \in \mathcal{V}} e^{\mathbf{x}_{v'}^\top \cdot \mathbf{x}_u}}. \tag{11.35}$$

Therefore, the objective log likelihood function can be simplified as

$$\max_{\mathbf{X}} \sum_{u \in \mathcal{V}} \left[-\log Z_u + \sum_{v' \in \Gamma(u)} \mathbf{x}_{v'}^\top \cdot \mathbf{x}_u \right], \tag{11.36}$$

where $Z_u = \sum_{v' \in \mathcal{V}} e^{\mathbf{x}_{v'}^\top \cdot \mathbf{x}_u}$. The term Z_u will be different for different nodes $u \in \mathcal{V}$, which is expensive to compute for large networks, and *node2vec* proposes to apply the negative sampling technique instead. The main issue discussed in *node2vec* is about sampling the neighborhood set $\Gamma(u)$ from the network, which can be obtained with either BFS or DFS based sampling strategies to be introduced as follows.

11.3.3.2 BFS and DFS Based Neighborhood Sampling

In *Skip-Gram*, neighborhood set $\Gamma(u)$ denotes the direct neighbors of u in the network, i.e., the *first-order proximity* of network local structures. Besides the local structure, *node2vec* can also capture other network structures with set $\Gamma(u)$ depending on the sampling strategy being applied. To fairly compare different sampling strategies, the neighborhood set $\Gamma(u)$ is usually limited with size k, i.e., $|\Gamma(u)| = k$. Two extreme sampling strategies for the neighborhood set $\Gamma(u)$ are

- *BFS*: BFS samples the nodes directly connected to node u and involves them in the neighborhood set $\Gamma(u)$ first, and then go to the second layer, where the nodes are two hopes away from u in the network, until the size k is met. Generally, the $\Gamma(u)$ sampled via BFS can sufficiently characterize the local neighborhood structure of the network. The *node2vec* model learnt based on the DFS sampling strategy provides a micro-view of the network structure.
- *DFS*: DFS samples the nodes which are sequentially reachable from u at an increasing distance and involves them into the neighborhood set $\Gamma(u)$ first. In DFS, the sampled nodes reflect a more global neighborhood of users in the network. The *node2vec* model learnt based on BFS sampling strategy provides a macro-view of the network neighborhood structure of the network, which can be essential for inferring the communities based on homophily.

However, the BFS and DFS sampling strategies may also suffer from some shortcomings. For BFS, only a small proportion of the network is explored surrounding node u in the sampling. Meanwhile, for DFS, the sampled nodes far away from the source node u tend to involve complex dependencies relationships.

11.3.3.3 Random Walk Based Search

To overcome the shortcomings of BFS and DFS, *node2vec* proposes to apply random walk to sample the neighborhood set $\Gamma(u)$ instead. Given a random walk W, we can represent the node where the random walk W resides at in step i as variable $s_i \in \mathcal{V}$. The complete sequence of nodes that W has resides at can be represented as $s_0, s_1, \ldots, s_k$, where s_0 denotes the initial node starting the random walk. The transitional probability from node u to v in random walk W at the ith step can be represented as

$$P(s_i = v | s_{i-1} = u) = \begin{cases} w_{(u,v)} & \text{if } (u, v) \in \mathcal{E}, \\ 0, & \text{otherwise}, \end{cases} \tag{11.37}$$

where $w_{(u,v)}$ denotes the normalized weight of link (u, v) in the network.

Traditional random walk model doesn't take account for the network structure and can hardly explore different network neighborhoods. *node2vec* adapts the random walk model and introduces the 2nd order random walk model with parameters p and q, which will help to guide the random walk. In *node2vec*, let's assume that the random walk just traversed link (t, u) and can go to node v in the next step. Formally, the transitional probability of link (u, v) is adjusted with parameter $\alpha_{p,q}(t, v)$ (i.e., $w_{(u,v)} = \alpha_{p,q}(t, v) \cdot w_{(u,v)}$), where

$$\alpha_{p,q}(t, v) = \begin{cases} \frac{1}{p}, & \text{if } d_{t,v} = 0, \\ 1, & \text{if } d_{t,v} = 1, \\ \frac{1}{q}, & \text{if } d_{t,v} = 2, \end{cases} \tag{11.38}$$

where $d_{t,v}$ denotes the shortest distance between nodes t and v in the network. Since the walk can go from t to u, and then from u to v, the distance from t to v will be at most 2.

Parameters p and q control the random walk transition sequence effectively, where parameter p is also called the *return parameter* and q is called the *in-out parameter* in *node2vec*.

- *Return Parameter p*: In the case that $d_{t,v} = 0$, i.e., $t = v$, the probability adjusting parameter $\frac{1}{p}$ controls the chance for the random walk to return to the node t. By assigning p with a large value, the random walk model will have a lower chance to go back to node t that the model has just visited. Meanwhile, by assigning p with a small value, the random walk model will backtrack a step and keep exploring the local nodes that it has visited already.
- *In-out Parameter q*: In the case that $d_{t,v} = 2$, nodes t and v are not directly connected but are reachable via the intermediate node u. Therefore, parameter q controls the chance of exploring the structure that is far away from the visited nodes. If $q > 1$, the random walk model is biased to explore nodes that are closer to t, since $\frac{1}{q}$ is smaller than the probability of visiting nodes in case that $d_{t,v} = 1$. Meanwhile, if $q < 1$, the random walk will be inclined to visit nodes that are far away from t in the network instead.

Based on a well-selected parameters p and q, the *node2vec* model will be able to utilize both local and global network structures in the node representation learning.

11.4 Heterogeneous Network Embedding

The embedding models introduced in the previous sections are mainly proposed for homogeneous networks, which will encounter great challenges when applied to embed the heterogeneous networks. In this section, we will talk about the recent development of embedding problems for heterogeneous networks, and introduce three latest *heterogeneous network embedding models*, including HNE (heterogeneous information network embedding) [2], path-augmented heterogeneous network embedding [3], and HEBE (HyperEdge based embedding) [7].

11.4.1 HNE: Heterogeneous Information Network Embedding

Generally, the data available in the online social networks doesn't exist in isolation, and different types of data may co-exist simultaneously. For instances, in the posts and articles written by users online, there may exist both text and image. The co-existence interactions of text and image in the same articles can be formed either explicitly or implicitly with the linkages between text and images. Meanwhile, there also exist correlations between the text data and image data due to the hyperlinks among the text and common tags/categories shared by different images. The *HNE* [2] model is initially proposed for a heterogeneous information network involving text and image.

11.4.1.1 Terminology Definition and Problem Formulation

The network studied in HNE involves both text and images, which can be represented as the *text–image heterogeneous information network* as follows:

Definition 11.1 (Text–Image Heterogeneous Information Network) Let $G = (\mathcal{V}, \mathcal{E})$ denote the heterogeneous information network involving text and image as the nodes, as well as diverse categories of links among them. Formally, the node set $\mathcal{V}$ can be decomposed into two disjoint subsets $\mathcal{V} = \mathcal{T} \cup \mathcal{I}$, where $\mathcal{T}$ denotes the text node set and $\mathcal{I}$ represents the image node set. Meanwhile, among the text, image as well as between text and images, there may exist different kinds of connections, which can be denoted as sets $\mathcal{E}_{T,T}$, $\mathcal{E}_{I,I}$, and $\mathcal{E}_{T,I}$, respectively, in the link set $\mathcal{E}$. In other words, we have $\mathcal{E} = \mathcal{E}_{T,T} \cup \mathcal{E}_{T,I} \cup \mathcal{E}_{I,I}$.

Furthermore, the text and image nodes are also summarized by unique content information. For instance, for each image $i_k \in \mathcal{I}$, it can be represented as a 3-way tensor $\mathbf{x}_k \in \mathbb{R}^{d_I \times d_I \times 3}$, where d_I denotes the dimension of the image and values in $\mathbf{x}_k$ correspond to the pixels of the image in the RGB color space. Meanwhile, for each text $t_k \in \mathcal{T}$, it can be represented as a raw feature vector $\mathbf{z}_k \in \mathbb{R}^{d_T}$ as well, where d_T denotes the dimension of the text represented with the bag-of-words vectors normalized by TF-IDF [12]. For the images involved in set $\mathcal{I}$, the connections among them can be represented as matrix $\mathbf{A}_{I,I} \in \{+1, -1\}^{|\mathcal{I}| \times |\mathcal{I}|}$, where entry $A_{I,I}(j, k) = +1$ if there exist a link connecting nodes i_j and i_k in the network; otherwise, we will have $A_{I,I}(j, k) = -1$. In a similar way, we can also define the adjacency matrices $A_{T,T}$ and $A_{I,T}$ to represent the connections among texts as well as those between images and texts.

For all the connections among nodes in set $\mathcal{V}$, they can be represented with matrix $\mathbf{A} \in \{+1, -1\}^{|\mathcal{V}| \times |\mathcal{V}|}$, which groups matrices $\mathbf{A}_{I,I}$, $\mathbf{A}_{T,I}$, $\mathbf{A}_{I,T}$, and $\mathbf{A}_{T,T}$ together. In the matrix, entry $A(i, j) = +1$ if the corresponding nodes are connected by a link in the network; and $A(i, j) = -1$, otherwise.

To handle the diverse information in the *text–image heterogeneous information network*, a good way is to learn the feature vector representations of nodes inside the network. Formally, the network embedding problem studied here includes the learning of mappings $\mathbf{U} : \mathbf{x} \rightarrow \mathbb{R}^r$ and $\mathbf{V} : \mathbf{z} \rightarrow \mathbb{R}^r$ which will project the images and texts into a shared feature space of dimension r. Furthermore, the network structure can be preserved in the embedding results, where the connected nodes should be projected to a close region and unconnected nodes will be separated apart instead.

11.4.1.2 HNE Model

For each image $i_k \in \mathcal{I}$, HNE proposes to transform its representation from 3-way tensor $\mathbf{x}_k$ into a column vector $\mathbf{x}_k \in \mathbb{R}^{d'_I}$, where d'_I denotes the dimension of the feature vector space. Different methods can be applied in the transformation. For instance, a simple way to do the transformation is to stack the column vectors of the image and append them together, in which case d'_I will be equal to $d_I \times d_I \times 3$. Some other advanced techniques have also been proposed, like feature extraction of the images as well as pre-embedding of images, which will not be introduced here since they are not part of the network embedding problem studied in this section.

Formally, we can denote the linear mapping functions for the image and text data as matrices $\mathbf{U} : \mathbf{x} \rightarrow \mathbb{R}^r$ and $\mathbf{V} : \mathbf{z} \rightarrow \mathbb{R}^r$, which project the data into a feature space of dimension r. The embedding process of image $i_j \in \mathcal{I}$ and text $t_k \in \mathcal{T}$ can be denoted as

$$\tilde{\mathbf{x}}_j = \mathbf{U}^\top \mathbf{x}_j, \tag{11.39}$$

$$\tilde{\mathbf{z}}_k = \mathbf{V}^\top \mathbf{z}_k, \tag{11.40}$$

where vectors $\tilde{\mathbf{x}}_j$ and $\tilde{\mathbf{z}}_k$ denote the embedded feature representations of image i_j and text t_k, respectively.

The similarity between the embedded feature representation of images and texts can be defined as

$$s(\mathbf{x}_j, \mathbf{x}_k) = \tilde{\mathbf{x}}_j^\top \tilde{\mathbf{x}}_k = \mathbf{x}_j^\top (\mathbf{U}\mathbf{U}^\top)\mathbf{x}_k = \mathbf{x}_j^\top \mathbf{M}_{I,I}\mathbf{x}_k, \tag{11.41}$$

$$s(\mathbf{z}_j, \mathbf{z}_k) = \tilde{\mathbf{z}}_j^\top \tilde{\mathbf{z}}_k = \mathbf{z}_j^\top (\mathbf{V}\mathbf{V}^\top)\mathbf{z}_k = \mathbf{z}_j^\top \mathbf{M}_{T,T}\mathbf{z}_k, \tag{11.42}$$

respectively. Furthermore, since the images and texts are embedded into a common feature space, the similarity between the nodes of different categories can be represented as

$$s(\mathbf{x}_j, \mathbf{z}_k) = \tilde{\mathbf{x}}_j^\top \tilde{\mathbf{z}}_k = \mathbf{x}_j^\top (\mathbf{U}\mathbf{V}^\top)\mathbf{z}_k = \mathbf{x}_j^\top \mathbf{M}_{I,T}\mathbf{z}_k. \tag{11.43}$$

In the above equations, via the positive semi-definite matrices $\mathbf{M}_{I,I}$, $\mathbf{M}_{T,T}$, $\mathbf{M}_{I,T}$ the similarity of the texts and images can be effectively captured.

Meanwhile, based on the network structure, the empirical similarities of the nodes in the networks can be denoted by their structures. For instance, the empirical similarity between images $i_j, i_k \in \mathcal{I}$ can be denoted as

$$\hat{s}(\mathbf{x}_j, \mathbf{x}_k) = A_{I,I}(j, k). \tag{11.44}$$

To ensure similar images will be projected into a close region, the loss function introduced by the image pair i_j, i_k is defined as

$$L(\mathbf{x}_j, \mathbf{x}_k) = \log\left(1 + e^{(-A_{I,I}(j,k)s(\mathbf{x}_j, \mathbf{x}_k))}\right). \tag{11.45}$$

In a similar way, the loss functions for the text pairs, and image–text pairs can also be defined. By combining the loss functions together, the objective function of HNE can be represented as

$$\min_{\mathbf{U},\mathbf{V}} \frac{1}{N_{I,I}} \sum_{i_j,i_k \in \mathcal{I}} L(\mathbf{x}_j, \mathbf{x}_k) + \frac{\lambda_1}{N_{T,T}} \sum_{t_j,t_k \in \mathcal{T}} L(\mathbf{z}_j, \mathbf{z}_k) + \frac{\lambda_2}{N_{I,T}} \sum_{i_j \in \mathcal{I}, t_k \in \mathcal{T}} L(\mathbf{x}_j, \mathbf{z}_k)$$

$$+ \lambda_3(\|\mathbf{U}\|_F^2 + \|\mathbf{V}\|_F^2), \tag{11.46}$$

where $N_{I,I} = |\mathcal{I} \times \mathcal{I} \setminus \{(i_j, i_j)\}_{i_j \in \mathcal{I}}|$ denotes the number of image pairs, and $\lambda_1, \lambda_2, \lambda_3$ denote the weights of the loss terms introduced by texts, image–text, and the regularization terms, respectively.

The function can be solved alternatively with coordinate descent by fixing one variable and updating the other variable. More detailed information about the solution is available in [2].

11.4.2 Path-Augmented Heterogeneous Network Embedding

For most of the embedding models, they are based on the assumptions that the node feature representations can be learnt with the neighborhood. Here, the neighborhood denotes either the set of nodes directly connected to the target node or the nodes accessible to the target node via a random walk. In [3], a new heterogeneous network embedding model has been introduced, which uses the meta path to exploit the rich information in heterogeneous networks.

In the path-augmented network embedding model (PANE), a set of meta paths are defined based on the heterogeneous network schema. For the node pairs in the network which are connected based on each of the meta paths, their correlation is represented with a meta path-augmented adjacency matrix. For instance, based on the rth type of meta path, the corresponding adjacency matrix can be denoted as $\mathbf{M}^r$. In heterogeneous networks, some of the meta paths will have lots of concrete meta path instances connecting nodes. For instance, in the online social networks, the meta path "User $\xrightarrow{\text{write}}$ Post $\xrightarrow{\text{contain}}$ Word $\xleftarrow{\text{contain}}$ Post $\xleftarrow{\text{write}}$ User" will have lots of instances, since users write lots of posts and each post will contain many words. The raw adjacency matrix will contain very large numbers in its representations. Therefore, matrix $\mathbf{M}^r$ is usually normalized to ensure $\sum_{i,j} M^r(i,j) = 1$.

The learning framework used here is very similar to those introduced *LINE* and *node2vec* in Sects. 11.3.2 and 11.3.3. The proximity between nodes $n_i, n_j \in \mathcal{V}$ based on the rth meta path can be denoted as

$$P(n_j|n_i; r) = \frac{e^{\mathbf{x}_i^\top \mathbf{x}_j}}{\sum_{j' \in DST(r)} e^{\mathbf{x}_i^\top \mathbf{x}_j}}, \tag{11.47}$$

where $\mathbf{x}_i$ and $\mathbf{x}_j$ denote the embedding vectors of nodes n_i and n_j, respectively, and $DST(r)$ denotes the set of all possible nodes that are in the destination side of path r.

In the real world, set $DST(r)$ is usually very large, which renders the above conditional probability very expensive to compute. In [3], the authors propose to follow the techniques proposed in the existing works and apply negative sampling to reduce the computational costs. Formally, the approximated objective function can be represented as

$$\log \tilde{P}(n_j|n_i; r) \approx \log \sigma(\mathbf{x}_i^\top \mathbf{x}_j) + \sum_{l=1}^{k} \mathbb{E}_{n_{j'} \sim P_n^r(n_{j'})}[\log \sigma(-\mathbf{x}_i^\top \mathbf{x}_{j'} - b_r)], \tag{11.48}$$

where j' denotes the negative node sampled from the pre-defined noise distribution, k denotes the number of sampled nodes, and b_r is the bias term added for the rth meta path. The embedding vectors $\mathbf{x}_{n_i}$ for node n_i in the network as well as the bias terms b_r for the rth meta path can be learnt with the stochastic gradient descent method.

11.4.3 HEBE: HyperEdge Based Embedding

The embedding models proposed so far mostly only consider the *single typed* objective interactions, while the *strongly typed* objects involving multiple kinds of interactions among different objectives have achieved an increasing interest in recent years. In this part, we will introduce a new embedding framework HEBE (HyperEdge based embedding) [7], which captures strongly typed objective interactions as a whole in the embedding process.

11.4.3.1 Terminology Definition and Problem Formulation

In HEBE, the subgraph centered with one certain type of target object in the whole network is defined as an *event* [7]. Depending on the number of node types involved in the *event*, they can be further categorized into *homogeneous event* and *heterogeneous event*.

Definition 11.2 (Event) Formally, the objects involved in the network can be represented as set $\mathcal{X} = \{\mathcal{X}_t\}_{t=1}^T$, where $\mathcal{X}_t$ denotes the set of objects belonging to the tth type. An event Q_i is denoted as a subset of nodes involved in it and can be represented as $(\mathcal{V}_i, w_i)$, where $\mathcal{V}_i$ denotes the set of involved objects and w_i is the occurrence number of event Q_i in the network. The object set $\mathcal{V}_i$ can be further divided into several subsets $\mathcal{V}_i = \bigcup_{t=1}^T \mathcal{V}_i^t$ depending on the object categories.

In the above event definition, links connecting the nodes in the network are involved in by default, which are not mentioned here for simplicity reasons. For event $Q_i = (\mathcal{V}_i, w_i)$, if more than one type of nodes are covered, it will be called a homogeneous event; otherwise, it is a heterogeneous event.

Formally, the set of events involved in the network can be represented as *event data* $\mathcal{D} = \{Q_i\}_i^N$. In the embedding problem, the objective is to learn a function $f : \mathcal{X} \to \mathbb{R}^d$ to project the different types of objects involved in the *event data* $\mathcal{D}$ into a shared feature space of dimension d. Meanwhile, the *proximity* of each event should be preserved. Here, the *proximity* of an event is defined as the likelihood of observing a target object given all other participating objects in the same event.

Example 11.1 For instance, as introduced in [7], in Fig. 11.4, we illustrate two examples about the events involved in a location-based social network. As shown in the plot, there are two types of events. The first event type (left) is business profile, the participating object types of which include terms in name and business; the second event type (right) is the review event, including user, business, and term types. The business objects type participates in both event types.

Fig. 11.4 Event schema of location-based social networks with two event types: business profile (left) and review (right)

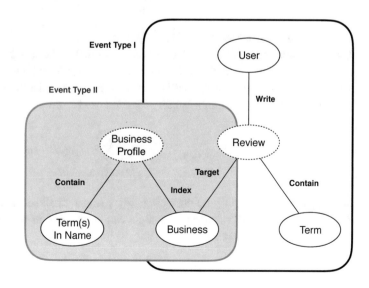

11.4.3.2 Objective Function

Given an event $Q_i = (\mathcal{V}_i, w_i)$, let $u \in \mathcal{V}_i$ denote an object involved in the event. The remaining nodes in the event can be denoted as the context of u, i.e., $\mathcal{C} = \mathcal{V}_i \setminus \{u\}$. Let's assume object u belongs to category $\mathcal{X}_1$ (i.e., $u \in \mathcal{X}_1$), the probability of predicting the target object u given its context $\mathcal{C}$ is defined as

$$P(u|\mathcal{C}) = \frac{e^{S(u,\mathcal{C})}}{\sum_{v \in \mathcal{X}_1} e^{S(v,\mathcal{C})}}, \tag{11.49}$$

where $S(u, \mathcal{C})$ denotes the similarity between u and context $\mathcal{C}$ and can be calculated by summing the inner products of object pairs in $\{u\} \times \mathcal{C}$.

The loss function defined in HEBE is based on the Kullback–Leibler (KL) divergence between the conditional probability $P(\cdot|\mathcal{C})$ and the empirical probability $\hat{P}(\cdot|\mathcal{C})$, which can be defined as

$$\mathcal{L} = -\sum_{t=1}^{T} \sum_{\mathcal{C}_t \in \mathcal{P}_t} \lambda_{\mathcal{C}_t} KL(P(\cdot|\mathcal{C}), \hat{P}(\cdot|\mathcal{C})), \tag{11.50}$$

where $\lambda_{\mathcal{C}_t}$ denotes the weight of context $\mathcal{C}_t$ and is defined as the occurrence of it in the event data $\mathcal{D}$

$$\lambda_{\mathcal{C}_t} = \sum_{i=1}^{N} \frac{w_i \mathbb{I}(\mathcal{C}_t \in \mathcal{V}_i)}{|\mathcal{P}_{i,t}|}. \tag{11.51}$$

In the above equation, $\mathcal{P}_t$ denotes the sample space of context $\mathcal{C}_t$ and $\mathcal{P}_{i,t}$ is the constrained sample space by object set $\mathcal{V}_i$. Function $\mathbb{I}(\cdot)$ is a binary function which takes value 1 if the condition holds. By replacing $\lambda_{\mathcal{C}_t}$, the loss function can be rewritten as follows:

$$\mathcal{L} = -\sum_{i=1}^{N} w_i \sum_{t=1}^{T} \frac{1}{|\mathcal{P}_{i,t}|} \sum_{\mathcal{C}_t \in \mathcal{P}_t} P(\cdot|\mathcal{C}). \tag{11.52}$$

11.4.3.3 Learning Algorithm Description

The conditional probability involved in the loss function is very hard to calculate especially in the case that the object set $\mathcal{X}_1$ that u belongs to is very big. To address the problem, HEBE proposes to use the *noise pairwise ranking* (NPR) to approximate the probability calculation instead.

Formally, the conditional probability function can be rewritten as

$$P(u|\mathcal{C}) = \left(1 + \sum_{v \in \mathcal{X}_1 \setminus \{u\}} e^{S(v,\mathcal{C})-S(u,\mathcal{C})}\right)^{-1}. \tag{11.53}$$

Instead of enumerating all the nodes $v \in \mathcal{X}_1 \setminus \{u\}$, a small set of noise samples are selected from $\mathcal{X}_1 \setminus \{u\}$, where an individual noise sample can be denoted as v_n. HEBE proposes to maximize the following probability instead:

$$P(u > u_n|\mathcal{C}) = \sigma(-S(v_n, \mathcal{C}) + S(u, \mathcal{C})). \tag{11.54}$$

It is shown that

$$P(u|\mathcal{C}) > \prod_{v_n \neq u} P(u > v_n|\mathcal{C}). \tag{11.55}$$

And the conditional probability can be approximated as follows:

$$P(u|\mathcal{C}) \propto \mathbb{E}_{v_n \sim P_n} \log P(u > v_n|\mathcal{C}), \tag{11.56}$$

where P_n denotes the noise distribution and it is set as $P_n \propto D(u)^{\frac{3}{4}}$ with regarding to the degree of u. By replacing the probability into the loss function, the loss function will be

$$\tilde{\mathcal{L}} = -\sum_{i=1}^{N} w_i \sum_{t=1}^{T} \frac{1}{|\mathcal{P}_{i,t}|} \sum_{\mathcal{C}_t \in \mathcal{P}_t} \mathbb{E}_{v_n \sim P_n} \log P(u > v_n|\mathcal{C}). \tag{11.57}$$

The objective function can be solved with the asynchronous stochastic gradient descent (ASGD) algorithm in HEBE, and we will not talk about the learning detail here, since it is out of the scope of this chapter.

11.5 Emerging Network Embedding Across Networks

We have introduced several network embedding models in the previous sections already. However, when applied to handle real-world social network data, these existing embedding models can hardly work well. The main reason is that the network internal social links are usually very sparse in online social networks [14]. For a pair of users who are not directly connected, these models will not be able to determine the closeness of these users' feature vectors in the embedding space. Such a problem will be more severe when it comes to the *emerging social networks* [18, 21], which denote the newly created online social networks containing very few social connections.

In this section, we will study the emerging network embedding problem across multiple aligned heterogeneous social networks simultaneously, namely the MNE problem [21]. In the concurrent embedding process, MNE aims at distilling relevant information from both the emerging and other aligned mature networks to derive compliment knowledge and learn a good vector representation for user nodes in the emerging network.

The MNE problem studied in this section is significantly different from existing network embedding problems [1–3, 6, 8, 11, 14, 16] in several perspectives. First of all, the target network studied in MNE are emerging networks suffering from the information sparsity problem a lot, which is different from the embedding problems for regular networks [1–3, 8, 16]. Secondly, the networks studied in MNE are all heterogeneous networks containing complex links and diverse attributes, which renders MNE different from existing homogeneous network embedding problems [6, 11, 14]. Furthermore, MNE is based on the multiple aligned networks setting, where information from aligned networks will be exchanged to refine the embedding results mutually, and it is different from the existing single-network based embedding problems [1–3, 6, 8, 11, 14, 16].

To solve the problem, in this section, we introduce a novel multiple aligned heterogeneous social network embedding framework, named DIME [21]. To handle the heterogeneous link and attribute information in the networks in a unified analytic, DIME introduces the *aligned attribute augmented heterogeneous network* concept. From these networks, a set of meta paths are introduced to represent the diverse connections among users in online social networks (via social links, other heterogeneous connections, and diverse attributes). A set of *meta proximity* measures are defined for each of the meta paths denoting the closeness among users. These meta proximity information will be fed into a deep learning framework, which takes the input information from multiple aligned heterogeneous social networks simultaneously, to achieve the embedding feature vectors for all the users in these aligned networks. Based on the connection among users, framework DIME aims at embedding close user nodes to a close area in the lower-dimensional feature space for each of the social network, respectively. Meanwhile, framework DIME also poses constraints on the feature vectors corresponding to the shared users across networks to map them to a relatively close region as well. In this way, information can be transferred from the mature networks to the emerging network and solve the *information sparsity* problem.

11.5.1 Concept Definition and Problem Formulation

The social networks studied here contain different categories of nodes and links, as well as very diverse attributes attached to the nodes. To handle the diverse links and attributes in a unified analytic, we can represent these network structured data as the *attributed heterogeneous social networks* formally.

Definition 11.3 (Attributed Heterogeneous Social Networks) The *attributed heterogeneous social network* can be represented as a graph $G = (\mathcal{V}, \mathcal{E}, \mathcal{T})$, where $\mathcal{V} = \bigcup_i \mathcal{V}_i$ denotes the set of nodes belonging to various categories and $\mathcal{E} = \bigcup_i \mathcal{E}_i$ represents the set of diverse links among the nodes. What's more, $\mathcal{T} = \bigcup_i \mathcal{T}_i$ denotes the set of attributes attached to the nodes in $\mathcal{V}$. For user u in the network, we can represent the ith type of attribute associated with u as $T_i(u)$, and all the attributes u has can be represented as $T(u) = \bigcup_i T_i(u)$.

The social network data sets used in this section include Foursquare and Twitter. Formally, the Foursquare and Twitter can both be represented as the *attributed heterogeneous social networks* $G = (\mathcal{V}, \mathcal{E}, \mathcal{T})$, where $\mathcal{V} = \mathcal{U} \cup \mathcal{P}$ involves the user and post nodes, and $\mathcal{E} = \mathcal{E}_{u,u} \cup \mathcal{E}_{u,p}$ contains the links among users and those between users and posts. In addition, the nodes in $\mathcal{V}$ are also attached with a set of attributes, i.e., $\mathcal{T}$. For instance, for the posts written by users, we can obtain the contained textual contents, timestamps, and check-ins, which can all be represented as the attributes of the post nodes.

Between Foursquare and Twitter, there may exist a large number of shared common users, who can align the networks together. The user account correspondence relationships can be denoted as the *anchor links* across networks. Meanwhile, the networks connected by the *anchor links* are called the multiple *aligned attributed heterogeneous social networks* (or *aligned social networks* for short).

For the Foursquare and Twitter social networks used here, we can represent them as two aligned social networks $\mathcal{G} = ((G^{(1)}, G^{(2)}), (\mathcal{A}^{(1,2)}))$, which will be used as an example to illustrate the models. A simple extension of the proposed framework can be applied to k *aligned networks* very easily.

Formally, given two aligned networks $\mathcal{G} = ((G^{(1)}, G^{(2)}), (\mathcal{A}^{(1,2)}))$, where $G^{(1)}$ is an emerging network and $G^{(2)}$ is a mature network. In the MNE problem, we aim at learning a mapping function $f^{(i)} : \mathcal{U}^{(i)} \rightarrow \mathbb{R}^{d^{(i)}}$ to project the user node in $G^{(i)}$ to a feature space of dimension $d^{(i)}$ ($d^{(i)} \ll |\mathcal{U}|^{(i)}$). The objective of mapping functions $f^{(i)}$ is to ensure that the embedding results can preserve the network structural information, where similar user nodes will be projected to close regions. Furthermore, in the embedding process, MNE also wants to transfer information between $G^{(2)}$ and $G^{(1)}$ to overcome the information sparsity problem in $G^{(1)}$.

11.5.2 Deep DIME for Emerging Network Embedding

For each attributed heterogeneous social network, the closeness among users can be denoted by the friendship links among them, where friends tend to be closer compared with user pairs without connections. Meanwhile, for the users who are not directly connected by the friendship links, few existing embedding methods can figure out their closeness, as these methods are mostly built based on the direct friendship link only. In this section, we will introduce how to infer the potential closeness scores among the users with the heterogeneous information in the networks based on meta path concept [13], which are formally called the *meta proximity*.

11.5.2.1 Friendship Based Meta Proximity

In online social networks, the friendship links are the most obvious indicator of the social closeness among users. Online friends tend to be closer with each other compared with the user pairs who are not friends. Users' friendship links also carry important information about the local network structure information, which should be preserved in the embedding results. Based on such an intuition, the *friendship based meta proximity* concept can be defined as follows.

Definition 11.4 (Friendship Based Meta Proximity) For any two user nodes $u_i^{(1)}, u_j^{(1)}$ in an online social network (e.g., $G^{(1)}$), if $u_i^{(1)}$ and $u_j^{(1)}$ are friends in $G^{(1)}$, the *friendship based meta proximity* between $u_i^{(1)}$ and $u_j^{(1)}$ in the network is 1; otherwise, the *friendship based meta proximity* score between them will be 0 instead. To be more specific, we can represent the *friendship based meta proximity* score between users $u_i^{(1)}, u_j^{(1)}$ as $p^{(1)}(u_i^{(1)}, u_j^{(1)}) \in \{0, 1\}$, where $p^{(1)}(u_i^{(1)}, u_j^{(1)}) = 1$ iff $(u_i^{(1)}, u_j^{(1)}) \in \mathcal{E}_{u,u}^{(1)}$.

Based on the above definition, the *friendship based meta proximity* scores among all the users in network $G^{(1)}$ can be represented as matrix $\mathbf{P}_{\Phi_0}^{(1)} \in \mathbb{R}^{|\mathcal{U}^{(1)}| \times |\mathcal{U}^{(1)}|}$, where entry $P_{\Phi_0}^{(1)}(i, j)$ equals to $p^{(1)}(u_i^{(1)}, u_j^{(1)})$. Here Φ_0 denotes the simplest meta path of length 1 in the form U $\xrightarrow{\text{follow}}$ U, and its formal definition will be introduced in the following subsection.

When network $G^{(1)}$ is an emerging online social network which has just started to provide services for a very short time, the friendship links among users in $G^{(1)}$ tend to be very limited (majority of the users are isolated in the network with few social connections). In other words, the *friendship based meta proximity* matrix $\mathbf{P}_{\Phi_0}^{(1)}$ will be extremely sparse, where very few entries will have value 1 and most of the entries are 0s. With such a sparse matrix, most existing embedding models will fail to work. The reason is that the sparse friendship information available in the network can hardly categorize the relative closeness relationships among the users (especially for those who are even not connected by friendship links), which renders that these existing embedding models may project all the nodes to random regions.

To overcome such a problem, besides the social links, DIME also proposes to calculate the potential proximity scores for the users with the diverse link and attribute information available in the heterogeneous networks. Based on a new concept named *attribute augmented meta path*, a set of *meta proximity* measures will be defined with each of the meta paths, which will be introduced in the following sections.

11.5.2.2 Attribute Augmented Meta Path

To handle the diverse links and attributes simultaneously in a unified analytic, DIME treats the attributes as nodes and introduces the *attribute augmented network*. If a node has certain attributes, a new type of link "*have*" will be added to connect the node and the newly added attribute node. The structure of the *attribute augmented network* can be described with the *attribute augmented network schema* as follows.

Definition 11.5 (Attribute Augmented Network Schema) Formally, the network schema of a given online social network $G^{(1)} = (\mathcal{V}, \mathcal{E})$ can be represented as $S_{G^{(1)}} = (\mathcal{N}_{\mathcal{V}} \cup \mathcal{N}_{\mathcal{T}}, \mathcal{R}_{\mathcal{E}} \cup \{have\})$, where $\mathcal{N}_{\mathcal{V}}$ and $\mathcal{N}_{\mathcal{T}}$ denote the set of node and attribute categories in the network, while $\mathcal{R}_{\mathcal{E}}$ represents the set of link types in the network, and $\{have\}$ represents the relationship between node and attribute node types.

For instance, about the *attributed heterogeneous social network* introduced studied in this section, we can represent its network schema as $S_{G^{(1)}} = (\mathcal{N}_{\mathcal{V}} \cup \mathcal{N}_{\mathcal{T}}, \mathcal{R}_{\mathcal{E}} \cup \{have\})$. The node type set $\mathcal{N}_{\mathcal{V}}$ involves node types $\{\text{User}, \text{Post}\}$ (or $\{U, P\}$ for simplicity), while the node attribute type set $\mathcal{N}_{\mathcal{T}}$ includes $\{\text{Word}, \text{Time}, \text{Location}\}$ (or $\{W, T, L\}$ for short). As to the link types involved in the network, the link type set $\mathcal{R}_{\mathcal{E}}$ contains $\{\text{follow}, \text{write}\}$, which represents the friendship link type and the write link type, respectively.

Based on the *attribute augmented network schema*, a set of different *social meta path* $\{\Phi_0, \Phi_1, \Phi_2, \ldots, \Phi_7\}$ can be extracted from the network, whose notations, concrete representations, and the physical meanings are illustrated in Table 11.1. Here, meta paths $\Phi_0 - \Phi_4$ are all based on the user node type and follow link type; meta paths $\Phi_5 - \Phi_7$ involve the user, post node type, attribute node type, as well as the *write* and *have* link type. Based on each of the meta paths, there will exist a set of concrete meta path instances connecting users in the networks. For instance, given a user pair u and v, they may have been checked-in at 5 different common locations, which will introduce 5 concrete meta path instances of meta path Φ_7 connecting u and v indicating their strong closeness (in location check-ins). In the next subsection, we will introduce how to calculate the proximity scores for the users based on these extracted meta paths.

11.5.2.3 Heterogeneous Network Meta Proximity

The set of *attribute augmented social meta paths* $\{\Phi_0, \Phi_1, \Phi_2, \ldots, \Phi_7\}$ extracted in the previous subsection create different kinds of correlations among users (especially for those who are not directly

Table 11.1 Summary of social meta paths (for both Foursquare and Twitter)

ID	Notation	Heterogeneous network meta path	Semantics
Φ_0	$U \rightarrow U$	User $\xrightarrow{\text{follow}}$ User	Follow
Φ_1	$U \rightarrow U \rightarrow U$	User $\xrightarrow{\text{follow}}$ User $\xrightarrow{\text{follow}}$ User	Follower of follower
Φ_2	$U \rightarrow U \leftarrow U$	User $\xrightarrow{\text{follow}}$ User $\xrightarrow{\text{follow}^{-1}}$ User	Common out neighbor
Φ_3	$U \leftarrow U \rightarrow U$	User $\xrightarrow{\text{follow}^{-1}}$ User $\xrightarrow{\text{follow}}$ User	Common in neighbor
Φ_4	$U \leftarrow U \leftarrow U$	User $\xrightarrow{\text{follow}^{-1}}$ User $\xrightarrow{\text{follow}^{-1}}$ User	Common in neighbor
Φ_5	$U \rightarrow P \rightarrow W \leftarrow P \leftarrow U$	User $\xrightarrow{\text{write}}$ Post $\xrightarrow{\text{have}}$ Word $\xrightarrow{\text{have}^{-1}}$ Post $\xrightarrow{\text{write}^{-1}}$ User	Posts containing Common words
Φ_6	$U \rightarrow P \rightarrow T \leftarrow P \leftarrow U$	User $\xrightarrow{\text{write}}$ Post $\xrightarrow{\text{have}}$ Time $\xrightarrow{\text{have}^{-1}}$ Post $\xrightarrow{\text{write}^{-1}}$ User	Posts containing Common timestamps
Φ_7	$U \rightarrow P \rightarrow L \leftarrow P \leftarrow U$	User $\xrightarrow{\text{write}}$ Post $\xrightarrow{\text{have}}$ Location $\xrightarrow{\text{have}^{-1}}$ Post $\xrightarrow{\text{write}^{-1}}$ User	Posts attaching Common check-ins

connected by friendship links). With these *social meta paths*, different types of proximity scores among the users can be captured. For instance, for the users who are not friends but share lots of common friends, they may also know each other and can be close to each other; for the users who frequently checked-in at the same places, they tend to be more close to each other compared with those isolated ones with nothing in common. Therefore, these meta paths can help capture much broader network structures compared with the local structure captured by the *friendship based meta proximity* talked about in Sect. 11.5.2.1. In this part, we will introduce the method to calculate the proximity scores among users based on these *social meta paths*.

As shown in Table 11.1, all the social meta paths extracted from the networks can be represented as set $\{\Phi_1, \Phi_2, \ldots, \Phi_7\}$. Given a pair of users, e.g., $u_i^{(1)}$ and $u_j^{(1)}$, based on meta path $\Phi_k \in \{\Phi_1, \Phi_2, \ldots, \Phi_7\}$, we can represent the set of meta path instances connecting $u_i^{(1)}$ and $u_j^{(1)}$ as $\mathcal{P}_{\Phi_k}^{(1)}\left(u_i^{(1)}, u_j^{(1)}\right)$. Users $u_i^{(1)}$ and $u_j^{(1)}$ can have multiple meta path instances going into/out from them. Formally, we can represent all the meta path instances going out from user $u_i^{(1)}$ (or going into $u_j^{(1)}$), based on meta path Φ_k, as set $\mathcal{P}_{\Phi_k}^{(1)}\left(u_i^{(1)}, \cdot\right)$ (or $\mathcal{P}_{\Phi_k}^{(1)}\left(\cdot, u_j^{(1)}\right)$). The proximity score between $u_i^{(1)}$ and $u_j^{(1)}$ based on meta path Φ_k can be represented as the following *meta proximity* concept formally.

Definition 11.6 (Meta Proximity) Based on social meta path Φ_k, the meta proximity between users $u_i^{(1)}$ and $u_j^{(1)}$ in network $G^{(1)}$ can be represented as

$$p_{\Phi_k}^{(1)}\left(u_i^{(1)}, u_j^{(1)}\right) = \frac{2\left|\mathcal{P}_{\Phi_k}^{(1)}\left(u_i^{(1)}, u_j^{(1)}\right)\right|}{\left|\mathcal{P}_{\Phi_k}^{(1)}\left(u_i^{(1)}, \cdot\right)\right| + \left|\mathcal{P}_{\Phi_k}^{(1)}\left(\cdot, u_j^{(1)}\right)\right|}. \tag{11.58}$$

Meta proximity considers not only the meta path instances between users but also penalizes the number of meta path instances going out from/into $u_i^{(1)}$ and $u_j^{(1)}$ at the same time. It is also reasonable. For instance, sharing some common location check-ins with some extremely active users (who have tens of thousands check-ins) may not necessarily indicate closeness with them, since they may have common check-ins with so many other users due to his very large check-in record volume.

With the above meta proximity definition, we can represent the meta proximity scores among all users in the network $G^{(1)}$ based on meta path Φ_k as matrix $\mathbf{P}_{\Phi_k}^{(1)} \in \mathbb{R}^{|\mathcal{U}^{(1)}| \times |\mathcal{U}^{(1)}|}$, where entry $P_{\Phi_k}^{(1)}(i, j) = p_{\Phi_k}^{(1)}\left(u_i^{(1)}, u_j^{(1)}\right)$. All the meta proximity matrices defined for network $G^{(1)}$ can be represented as $\left\{\mathbf{P}_{\Phi_k}^{(1)}\right\}_{\Phi_k}$. Based on the meta paths extracted for network $G^{(2)}$, similar matrices can be defined as well, which can be denoted as $\left\{\mathbf{P}_{\Phi_k}^{(2)}\right\}_{\Phi_k}$.

11.5.2.4 Deep DIME-SH Model

With these calculated *meta proximity* introduced in the previous section, we will introduce the embedding framework DIME next. DIME is based on the *aligned autoencoder model*, which extends the traditional *deep autoencoder model* to the *multiple aligned heterogeneous networks* scenario. In this part, we will talk about the embedding model component for one heterogeneous information network in Sect. 11.5.2.4, which takes the various meta proximity matrices as the input. DIME effectively couples the embedding process of the emerging network with other aligned mature networks, where cross-network information exchange and result refinement is achieved via the loss term defined based on the anchor links, which will be introduced in the next part.

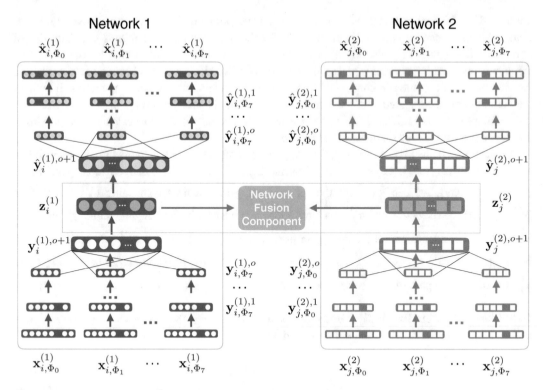

Fig. 11.5 The DIME framework

When applying the autoencoder model for one single homogeneous network node embedding, e.g., for $G^{(1)}$, we can fit the model with the node meta proximity feature vectors, i.e., rows corresponding to users in matrix $\mathbf{P}_{\Phi_0}^{(1)}$ (introduced in Sect. 11.5.2.1). In the case that $G^{(1)}$ is heterogeneous, multiple node *meta proximity* matrices have been defined before (i.e., $\left\{ \mathbf{P}_{\Phi_0}^{(1)}, \mathbf{P}_{\Phi_1}^{(1)}, \ldots, \mathbf{P}_{\Phi_7}^{(1)} \right\}$), how to fit these matrices simultaneously to the autoencoder models is an open problem. In this part, we will introduce the single-heterogeneous-network version of framework DIME, namely DIME-SH, which will be used as an important component of framework DIME as well. For each user node in the network, DIME-SH computes the embedding vector based on each of the proximity matrix independently first, which will be further fused to compute the final latent feature vector in the output hidden layer.

As shown in the architecture in Fig. 11.5 (either the left component for network 1 or the right component for network 2), about the same instance, DIME-SH takes different feature vectors extracted from the meta paths $\{\Phi_0, \Phi_1, \ldots, \Phi_7\}$ as the input. For each meta path, a series of separated encoder and decoder steps are carried out simultaneously, whose latent vectors are fused together to calculate the final embedding vector $\mathbf{z}_i^{(1)} \in \mathbb{R}^{d^{(1)}}$ for user $u_i^{(1)} \in \mathcal{V}^{(1)}$. In the DIME-SH model, the input feature vectors (based on meta path $\Phi_k \in \{\Phi_0, \Phi_1, \ldots, \Phi_7\}$) of user u_i can be represented as $\mathbf{x}_{i,\Phi_k}^{(1)}$, which denotes the row corresponding to users $u_i^{(1)}$ in matrix $\mathbf{P}_{\Phi_k}^{(1)}$ defined before. Meanwhile, the latent representation of the instance based on the feature vector extracted via meta path Φ_k at different hidden layers can be represented as $\left\{ \mathbf{y}_{i,\Phi_k}^{(1),1}, \mathbf{y}_{i,\Phi_k}^{(1),2}, \ldots, \mathbf{y}_{i,\Phi_k}^{(1),o} \right\}$.

One of the significant difference of model DIME-SH from traditional autoencoder model lies in the (1) combination of various hidden vectors $\left\{ \mathbf{y}_{i,\Phi_0}^{(1),o}, \mathbf{y}_{i,\Phi_1}^{(1),o}, \ldots, \mathbf{y}_{i,\Phi_7}^{(1),o} \right\}$ to obtain the final embedding vector $\mathbf{z}_i^{(1)}$ in the encoder step, and (2) the dispatch of the embedding vector $\mathbf{z}_i^{(1)}$ back to the hidden vectors in the decoder step. As shown in the architecture, formally, these extra steps can be represented as

$$
\begin{cases}
\text{\# extra encoder steps} \\
\mathbf{y}_i^{(1),o+1} = \sigma\left(\sum_{\Phi_k \in \{\Phi_0,\ldots,\Phi_7\}} \mathbf{W}_{\Phi_k}^{(1),o+1} \mathbf{y}_{i,\Phi_k}^{(1),o} + \mathbf{b}_{\Phi_k}^{(1),o+1} \right), \\
\mathbf{z}_i^{(1)} = \sigma\left(\mathbf{W}^{(1),o+2} \mathbf{y}_i^{(1),o+1} + \mathbf{b}^{(1),o+2} \right). \\
\text{\# extra decoder steps} \\
\hat{\mathbf{y}}_i^{(1),o+1} = \sigma\left(\hat{\mathbf{W}}^{(1),o+2} \mathbf{z}_i^{(1)} + \hat{\mathbf{b}}^{(1),o+2} \right), \\
\hat{\mathbf{y}}_{i,\Phi_k}^{(1),o} = \sigma\left(\hat{\mathbf{W}}_{\Phi_k}^{(1),o+1} \hat{\mathbf{y}}_i^{(1),o+1} + \hat{\mathbf{b}}_{\Phi_k}^{(1),o+1} \right).
\end{cases}
\tag{11.59}
$$

What's more, since the input feature vectors are extremely sparse (lots of the entries have value 0s), simply feeding them to the model may lead to some trivial solutions, like $\mathbf{0}$ vectors for both $\mathbf{z}_i^{(1)}$ and the decoded vectors $\hat{\mathbf{x}}_{i,\Phi_k}^{(1)}$. To overcome such a problem, another significant difference of model DIME-SH from traditional autoencoder model lies in the loss function definition, where the loss introduced by the non-zero features will be assigned with a larger weight. In addition, by adding the loss function for each of the meta paths, the final loss function in DIME-SH can be formally represented as

$$
\mathcal{L}^{(1)} = \sum_{\Phi_k \in \{\Phi_0,\ldots,\Phi_7\}} \sum_{u_i \in \mathcal{V}} \left\| \left(\mathbf{x}_{i,\Phi_k}^{(1)} - \hat{\mathbf{x}}_{i,\Phi_k}^{(1)} \right) \odot \mathbf{b}_{i,\Phi_k}^{(1)} \right\|_2^2,
\tag{11.60}
$$

where vector $\mathbf{b}_{i,\Phi_k}^{(1)}$ is the weight vector corresponding to feature vector $\mathbf{x}_{i,\Phi_k}^{(1)}$. Entries in vector $\mathbf{b}_{i,\Phi_k}^{(1)}$ are filled with value 1 except the entries corresponding to non-zero element in $\mathbf{x}_{i,\Phi_k}^{(1)}$, which will be assigned with value γ ($\gamma > 1$ denoting a larger weight to fit these features). In a similar way, we can define the loss function for the embedding result in network $G^{(2)}$, which can be formally represented as $\mathcal{L}^{(2)}$.

11.5.2.5 Deep DIME Framework

DIME-SH has incorporate all these heterogeneous information in the model building, the meta proximity calculated based on which can help differentiate the closeness among different users. However, for the emerging networks which just start to provide services, the information sparsity problem may affect the performance of DIME-SH significantly. In this part, we will introduce DIME, which couples the embedding process of the emerging network with another mature aligned network. By accommodating the embedding between the aligned networks, information can be transferred from the aligned mature network to refine the embedding result in the emerging network effectively. The complete architecture of DIME is shown in Fig. 11.5, which involves the DIME-SH components for each of the aligned networks, where the information transfer component aligns these separated DIME-SH models together.

To be more specific, given a pair of aligned heterogeneous networks $\mathcal{G} = ((G^{(1)}, G^{(2)}), \mathcal{A}^{(1,2)})$ ($G^{(1)}$ is an emerging network and $G^{(2)}$ is a mature network), we can represent the embedding results as matrices $\mathbf{Z}^{(1)} \in \mathbb{R}^{|\mathcal{U}^{(1)}| \times d^{(1)}}$ and $\mathbf{Z}^{(2)} \in \mathbb{R}^{|\mathcal{U}^{(2)}| \times d^{(2)}}$ for all the user nodes in $G^{(1)}$ and $G^{(2)}$, respectively. The ith row of matrix $\mathbf{Z}^{(1)}$ (or the jth row of matrix $\mathbf{Z}^{(2)}$) denotes the encoded feature vector of user $u_i^{(1)}$ in $G^{(1)}$ (or $u_j^{(2)}$ in $G^{(2)}$). If $u_i^{(1)}$ and $u_j^{(2)}$ are the same user, i.e., $(u_i^{(1)}, u_j^{(2)}) \in \mathcal{A}^{(1,2)}$, by placing vectors $\mathbf{Z}^{(1)}(i, :)$ and $\mathbf{Z}^{(2)}(j, :)$ in a close region in the embedding space, we can use the information from $G^{(2)}$ to refine the embedding result in $G^{(1)}$.

Information transfer is achieved based on the anchor links, and we only care about the anchor users. To adjust the rows of matrices $\mathbf{Z}^{(1)}$ and $\mathbf{Z}^{(2)}$ to remove non-anchor users and make the same rows correspond to the same user, we introduce the binary inter-network transitional matrix $\mathbf{T}^{(1,2)} \in \mathbb{R}^{|\mathcal{U}^{(1)}| \times |\mathcal{U}^{(2)}|}$. Entry $T^{(1,2)}(i, j) = 1$ if the corresponding users are connected by anchor links, i.e., $\left(u_i^{(1)}, u_j^{(2)} \right) \in \mathcal{A}^{(1,2)}$. Furthermore, the encoded feature vectors for users in these two networks can be of different dimensions, i.e., $d^{(1)} \neq d^{(2)}$, which can be accommodated via the projection $\mathbf{W}^{(1,2)} \in \mathbb{R}^{d^{(1)} \times d^{(2)}}$.

Formally, the introduced *information fusion loss* between networks $G^{(1)}$ and $G^{(2)}$ can be represented as

$$\mathcal{L}^{(1,2)} = \left\| (\mathbf{T}^{(1,2)})^\top \mathbf{Z}^{(1)} \mathbf{W}^{(1,2)} - \mathbf{Z}^{(2)} \right\|_F^2. \tag{11.61}$$

By minimizing the *information fusion loss* function $\mathcal{L}^{(1,2)}$, we can use the anchor users' embedding vectors from the mature network $G^{(2)}$ to adjust his embedding vectors in the emerging network $G^{(1)}$. Even though in such a process the embedding vector in $G^{(2)}$ can be undermined by $G^{(1)}$, it will not be a problem since $G^{(1)}$ is our target network and we only care about the embedding result of the emerging network $G^{(1)}$.

The complete objective function of framework includes the loss terms introduced by the component DIME-SH for networks $G^{(1)}$, $G^{(2)}$, and the *information fusion loss*, which can be denoted as

$$\mathcal{L}(G^{(1)}, G^{(2)}) = \mathcal{L}^{(1)} + \mathcal{L}^{(2)} + \alpha \cdot \mathcal{L}^{(1,2)} + \beta \cdot \mathcal{L}_{reg}. \tag{11.62}$$

Parameters α and β denote the weights of the *information fusion loss* term and the regularization term. In the objective function, term $\mathcal{L}_{reg}$ is added to the above objective function to avoid overfitting, which can be formally represented as

$$\begin{cases} \mathcal{L}_{reg} = \mathcal{L}_{reg}^{(1)} + \mathcal{L}_{reg}^{(2)} + \mathcal{L}_{reg}^{(1,2)}, \\ \mathcal{L}_{reg}^{(1)} = \sum_i^{o^{(1)}+2} \sum_{\Phi_k \in \{\Phi_0, \ldots, \Phi_7\}} \left(\left\| \mathbf{W}_{\Phi_k}^{(1),i} \right\|_F^2 + \left\| \hat{\mathbf{W}}_{\Phi_k}^{(1),i} \right\|_F^2 \right), \\ \mathcal{L}_{reg}^{(2)} = \sum_i^{o^{(2)}+2} \sum_{\Phi_k \in \{\Phi_0, \ldots, \Phi_7\}} \left(\left\| \mathbf{W}_{\Phi_k}^{(2),i} \right\|_F^2 + \left\| \hat{\mathbf{W}}_{\Phi_k}^{(2),i} \right\|_F^2 \right), \\ \mathcal{L}_{reg}^{(1,2)} = \left\| \mathbf{W}^{(1,2)} \right\|_2^2. \end{cases} \tag{11.63}$$

To optimize the above objective function, we utilize stochastic gradient descent (SGD). To be more specific, the training process involves multiple epochs. In each epoch, the training data is shuffled and a minibatch of the instances is sampled to update the parameters with SGD. Such a process continues until either convergence or the training epochs have been finished.

11.6 Summary

In this chapter, we introduced the network embedding problems, whose objective is to learn a low-dimensional feature representation of nodes in the network structured data. Meanwhile, in the embedding process, the network structure can be preserved at the same time, and the complete network structure can be recovered from the embedding results. With the embedding feature representations, traditional machine learning algorithms can be applied to deal with the network data directly.

We introduced 3 different translation based network embedding algorithms, which treat the relation as a translation among the entities. Assuming that entities and relations can be embedded into the same feature space, TransE learns the embeddings of entities and relations as an optimization problem. TransH models relations as a hyperplane together with a translation operation on it, where the correlation among the entities can be effectively preserved. TransR models the entities and relations in distinct spaces, projection between which can be achieved with a linear mapping function.

We also introduced 3 network embedding models for the homogeneous networks specifically. DeepWalk proposes to sample node sequences from the network structured data, from which the Skip-Gram model can be applied to learn the embedding feature representations. LINE computes both the first-order proximity and second-order proximity among nodes in the networks. By projecting similar nodes into closer regions, LINE is able to learn the embedding representations of the nodes in homogeneous networks. node2vec also adopts the Skip-Gram model to learn the node representations based node sequences sampled from the network data based on a random walk, where the sampled sequences can be adjusted by controlling two parameters in the random walk model.

To learn the node representations in heterogeneous networks, 3 different embedding models were introduced in this chapter. HNE learns the representations of images and texts based on a heterogeneous network involving images and texts, as well as the diverse connections among them. The path-augmented heterogeneous network embedding model learns the node representations based on node sequences generated by the meta paths. HEBE learns the node representations by considering their roles in different events.

Finally, at the end of this chapter, we introduced a network embedding model DIME across multiple aligned heterogeneous networks, where one of the network is an emerging network lacking enough information for effective representation learning. DIME is built based on the autoencoder model, which projects and fuses the node adjacency vectors achieved based on diverse meta paths into a shared low-dimensional space. DIME accommodates the representations of the shared anchor nodes with a network fusion component.

11.7 Bibliography Notes

A comprehensive survey about the network embedding problems and algorithms is provided in [4, 17, 22]. For the readers who are interested in the translation based embedding models for multi-relational networks, please refer to articles *TransE* [1], *TransH* [16], and *TransR* [8] for more information.

The homogeneous network embedding models introduced in this chapter are all based on the recent research papers, including *DeepWalk* [11], *LINE* [14], and *node2vec* [6]. Meanwhile, the heterogeneous network embedding works are based on the articles HNE [2], PANE [3], and HEBE [7], respectively. The community has actually maintained a GitHub page for the recent network embedding articles, and the readers can access the page via link.[1] Many of these proposed embedding models are actually based on the Skip-Gram model [10], which was initially proposed for the representation learning in text data. Besides the Skip-Gram model, continuous bag-of-words is also frequently used for text representation learning, and the readers may refer to [9] for more information.

The emerging network embedding model is based on the latest embedding paper [21]. For the readers who are interested in the autoencoder model used in the algorithm, please refer to [15] for more information. A comprehensive review of deep learning models is provided in [5]. The emerging

---

[1]https://github.com/chihming/awesome-network-embedding.

network concept is initially proposed in [18], which actually describes the scenario where the networks are suffering from the new network problem [19, 20].

11.8 Exercises

1. (Easy) To train the TransE model, besides the positive triple set, a set of negative triples are sampled from the multi-relational network. Please briefly introduce how TransE samples these negative triples.
2. (Easy) Please summarize the similarity and differences of the TransE, TransH, and TransR models.
3. (Easy) Please briefly introduce what are the advantages of adopting the *hierarchical softmax* to compute the probabilities in DeepWalk.
4. (Easy) What are the first-order proximity and second-order proximity measures used in the LINE model? Please provide a brief introduction about these two proximity measures, and their applications in the LINE model learning.
5. (Medium) Please summarize the differences between DeepWalk and node2vec in both the embedding model and the node sequence sampling approaches.
6. (Medium) Please briefly introduce the DIME model architecture, and introduce how DIME transfers information from the mature networks to the emerging networks in the representation learning process.
7. (Hard) Please implement the TransE, TransH, and TransR models with your preferred programming language, and compare their performance with a toy multi-relational network data set.
8. (Hard) Please implement the DeepWalk, LINE, and node2vec algorithms with your preferred programming language, and compare their performance with a toy homogeneous network data set.
9. (Hard) Please implement the HNE, path-augmented network embedding, and HEBE algorithms with your preferred programming language, and compare their performance with a toy heterogeneous network data set.
10. (Hard) Please implement the DIME algorithm, and test its performance on a multiple aligned heterogeneous network data set.

References

1. A. Bordes, N. Usunier, A. Garcia-Durán, J. Weston, O. Yakhnenko, Translating embeddings for modeling multi-relational data, in *Advances in Neural Information Processing Systems (NIPS'13)* (2013), pp. 2787–2795
2. S. Chang, W. Han, J. Tang, G. Qi, C. Aggarwal, T. Huang, Heterogeneous network embedding via deep architectures, in *Proceedings of the 21th ACM SIGKDD International Conference on Knowledge Discovery and Data Mining* (ACM, New York, 2015), pp. 119–128
3. T. Chen, Y. Sun, Task-guided and path-augmented heterogeneous network embedding for author identification. CoRR, abs/1612.02814 (2016)
4. P. Cui, X. Wang, J. Pei, W. Zhu, A survey on network embedding. CoRR, abs/1711.08752 (2017)
5. I. Goodfellow, Y. Bengio, A. Courville, *Deep Learning* (MIT Press, Cambridge, 2016). http://www.deeplearningbook.org
6. A. Grover, J. Leskovec, Node2vec: scalable feature learning for networks, in *Proceedings of the 22nd ACM SIGKDD International Conference on Knowledge Discovery and Data Mining* (ACM, New York, 2016), pp. 855–864
7. H. Gui, J. Liu, F. Tao, M. Jiang, B. Norick, J. Han, Large-scale embedding learning in heterogeneous event data, in *2016 IEEE 16th International Conference on Data Mining (ICDM)* (IEEE, Piscataway, 2016), pp. 907–912

8. Y. Lin, Z. Liu, M. Sun, Y. Liu, X. Zhu, Learning entity and relation embeddings for knowledge graph completion, in *Twenty-Ninth AAAI Conference on Artificial Intelligence* (2015)

9. T. Mikolov, K. Chen, G. Corrado, J. Dean, Efficient estimation of word representations in vector space. CoRR, abs/1301.3781 (2013)

10. T. Mikolov, I. Sutskever, K. Chen, G. Corrado, J. Dean, Distributed representations of words and phrases and their compositionality, in *Advances in Neural Information Processing Systems* (Curran Associates, Inc., Red Hook, 2013), pp. 3111–3119

11. B. Perozzi, R. Al-Rfou, S. Skiena, DeepWalk: Online learning of social representations, in *Proceedings of the 20th ACM SIGKDD International Conference on Knowledge Discovery and Data Mining* (ACM, New York, 2014), pp. 701–710

12. J. Ramos, Using TF-IDF to determine word relevance in document queries, in *Proceedings of the First Instructional Conference on Machine Learning* (1999), pp. 133–142

13. Y. Sun, C. Aggarwal, J. Han, Relation strength-aware clustering of heterogeneous information networks with incomplete attributes. Proc. VLDB Endow. **5**(5), 394–405 (2012)

14. J. Tang, M. Qu, M. Wang, M. Zhang, J. Yan, Q. Mei, Line: large-scale information network embedding, in *Proceedings of the 24th International Conference on World Wide Web* (International World Wide Web Conferences Steering Committee, Geneva, 2015), pp. 1067–1077

15. P. Vincent, H. Larochelle, I. Lajoie, Y. Bengio, P. Manzagol, Stacked denoising autoencoders: learning useful representations in a deep network with a local denoising criterion. J. Mach. Learn. Res. **11**(Dec), 3371–3408 (2010)

16. Z. Wang, J. Zhang, J. Feng, Z. Chen, Knowledge graph embedding by translating on hyperplanes, in *Twenty-Eighth AAAI Conference on Artificial Intelligence* (2014)

17. J. Zhang, Social network fusion and mining: a survey. arXiv preprint, arXiv:1804.09874 (2018)

18. J. Zhang, P. Yu, Community detection for emerging networks, in *Proceedings of the 2015 SIAM International Conference on Data Mining* (Society for Industrial and Applied Mathematics, Philadelphia, 2015), pp. 127–135

19. J. Zhang, X. Kong, P. Yu, Predicting social links for new users across aligned heterogeneous social networks, in *2013 IEEE 13th International Conference on Data Mining* (IEEE, Piscataway, 2013), pp. 1289–1294

20. J. Zhang, X. Kong, P. Yu, Transferring heterogeneous links across location-based social networks, in *Proceedings of the 7th ACM International Conference on Web Search and Data Mining* (ACM, New York, 2014), pp. 303–312

21. J. Zhang, C. Xia, C. Zhang, L. Cui, Y. Fu, P. Yu, BL-MNE: emerging heterogeneous social network embedding through broad learning with aligned autoencoder, in *2017 IEEE International Conference on Data Mining (ICDM)* (IEEE, Piscataway, 2017), pp. 605–614

22. D. Zhang, J. Yin, X. Zhu, C. Zhang, Network representation learning: a survey. CoRR, abs/1801.05852 (2018)

Part IV

Future Directions

Frontier and Future Directions

12

12.1 Overview

In this book, we have introduced the current research works on broad learning and its applications on online social networks. This book has covered 4 main parts, where the first 3 parts include 6 main research directions about broad learning based social network mining problems, including (1) *network alignment*, (2) *link prediction*, (3) *community detection*, (4) *information diffusion*, (5) *viral marketing*, and (6) *network embedding*. These problems introduced in this book are all very important for many concrete real-world social network applications and services. A number of state-of-the-art algorithms have been proposed to solve these problems, which are introduced in great detail in this book. *Broad learning* is a very promising research area, and several potential future development directions about broad learning will be illustrated in the following sections.

12.2 Large-Scale Broad Learning

Data generated nowadays is usually of very large scale, and fusion of such big data from multiple sources together will render the scalability problem much more challenging. For instance, the online social networks (like Facebook) usually involve millions even billions of active users, and the social data generated by these users in each day will consume more than 600 TB storage space (in Facebook). One of the major future developments about *broad learning* is to develop scalable fusion and mining algorithms that can handle such a large volume challenge (of big data). One tentative approach is to develop information fusion and mining algorithms based on distributed platforms, like Spark and Hadoop [1], and handle the data with a large distributed computing cluster. Another method to resolve the scalability challenge is from the model optimization perspective. Optimizing the existing learning algorithms and proposing new approximated learning algorithms with a much lower time complexity are desirable in the future research projects. In addition, applications of the latest deep learning models to learn the low-dimensional latent representations of such large-scale datasets can be another alternative approach to achieve the scalable *broad learning* objective.

© Springer Nature Switzerland AG 2019
J. Zhang, P. S. Yu, *Broad Learning Through Fusions*,
https://doi.org/10.1007/978-3-030-12528-8_12

12.3 Multi-Source Broad Learning

Current research works on multiple source data fusion and mining mainly focus on aligning entities in one single pair of data sources (i.e., two sources), where information exchange between the sources mainly relies on the anchor links between these aligned entities. Meanwhile, when it comes to fusion and mining of multiple (more than two) sources, the problem setting will be quite different and become more challenging. For example, in the alignment of more networks, the transitivity property of the inferred anchor links needs to be preserved [3]. Meanwhile, in the information transfer from multiple external aligned sources to the target source, the information sources should be weighted differently according to their importance. How to determine the different weight parameters of these sources is still an open problem by this context so far. Therefore, the diverse variety of the multiple sources will lead to more research challenges and opportunities. New information fusion and mining algorithms for the multi-source scenarios can be another great opportunity to explore *broad learning* in the future.

12.4 Broad Learning Applications

Besides the research works on online social networks, the third potential future development direction of broad learning lies its broader applications on various categories of datasets, like enterprise internal data [4, 5, 8, 9], geo-spatial data [2, 6, 7], knowledge graph base data, financial time series data, images/video data, as well as pure text data. Some prior research works on fusing enterprise context information sources, like enterprise social networks, organizational chart, and employee profile information have been done already [4, 5, 8, 9]. Several interesting problems, like organizational chart inference [5], enterprise link prediction [4], information diffusion at workplace [8], and enterprise employee training [9], have been studied based on the fused enterprise internal information. In addition, analysis of the correlation of different traveling modalities (like shared bicycles [2, 6, 7], bus and metro train) with the city zonings in smart city; and fusing multiple knowledge bases, like Douban and IMDB, for knowledge discovery and truth finding can all be good application scenarios for broad learning.

12.5 Summary

We would like to conclude this chapter and this book with the acknowledgement for the readers, who have finished reading the whole book. We really hope the readers can learn useful knowledge from the book, and use the knowledge in your courses, your research projects, and your work.

12.6 Exercises

1. (Easy) Please briefly describe the broad learning problems we have covered in this textbook.
2. (Hard) Please think about some potential problems we can study based on broad learning in the future.

References

1. S. Jin, J. Zhang, P. Yu, S. Yang, A. Li, Synergistic partitioning in multiple large scale social networks, in *2014 IEEE International Conference on Big Data (Big Data)* (2014)
2. Q. Zhan, J. Zhang, X. Pan, P. Yu, Discover tipping users for cross network influencing, in *2016 IEEE 17th International Conference on Information Reuse and Integration (IRI)* (2016)
3. J. Zhang, P. Yu, Multiple anonymized social networks alignment, in *2015 IEEE International Conference on Data Mining* (2015)
4. J. Zhang, Y. Lv, P. Yu, Enterprise social link prediction, in *Conference on Information and Knowledge Management* (2015)
5. J. Zhang, P. Yu, Y. Lv, Organizational chart inference, in *Proceedings of the 21th ACM SIGKDD International Conference on Knowledge Discovery and Data Mining (KDD '15)* (2015)
6. J. Zhang, X. Pan, M. Li, P. Yu, Bicycle-sharing system analysis and trip prediction, in *2016 17th IEEE International Conference on Mobile Data Management (MDM)* (2016)
7. J. Zhang, X. Pan, M. Li, P. Yu, Bicycle-sharing systems expansion: station re-deployment through crowd planning, in *Proceedings of the 24th ACM SIGSPATIAL International Conference on Advances in Geographic Information Systems (SIGSPACIAL '16)* (2016)
8. J. Zhang, P. Yu, Y. Lv, Q. Zhan, Information diffusion at workplace, in *Proceedings of the 25th ACM International on Conference on Information and Knowledge Management (CIKM '16)* (2016)
9. J. Zhang, P. Yu, Y. Lv, Enterprise employee training via project team formation, in *Proceedings of the Tenth ACM International Conference on Web Search and Data Mining (WSDM '17)* (2017)

Printed in the United States
By Bookmasters